Advance Praise for
State of the World 2013: Is Sustainability Still Possible?

"*State of the World 2013* cuts through the rhetoric surrounding sustainability, providing a broad and realistic look at how close we are to achieving it and outlining practices and policies that can steer us in the right direction. . . . A must-read for those seeking authentic sustainability."

—**Hunter Lovins,** President, Natural Capital Solutions and Author of *Climate Capitalism*

"This is a book of hope for a world in profound crisis. It gives honest assessments of the enormous challenges we face and points us toward institutional and cultural changes that are proportional to our dire situation. *State of the World 2013* reaffirms that we are not helpless but that we have real choices—and that transformation is both possible and desirable."

—**Reverend Peter S. Sawtell,** Executive Director, Eco-Justice Ministries

"*State of the World 2013* cuts through 'sustainababble' with crisp coverage that puts the news of the year in context and provides an expert survey of today's and tomorrow's big issues. It's a perennial resource for everyone concerned about our common future."

—**Karen Christensen,** publisher of the 10-volume *Berkshire Encyclopedia of Sustainability*

"Every elected official in the world needs to read this book. Mass denial is no longer an option. An 'all hands on deck' approach to transforming our culture and economy is the only path to a safe, resilient future. This book is the blueprint for that safe path forward."

—**Betsy Taylor,** President, Breakthrough Strategies & Solutions and Founder, Center for a New American Dream

Look for "State of the World 2014: Governing for Sustainability" coming in April 2014. Follow the development of the book here: http://islandpress.org/sotw2014

About Island Press

Since 1984, the nonprofit Island Press has been stimulating, shaping, and communicating the ideas that are essential for solving environmental problems worldwide. With more than 800 titles in print and some 40 new releases each year, we are the nation's leading publisher on environmental issues. We identify innovative thinkers and emerging trends in the environmental field. We work with world-renowned experts and authors to develop cross-disciplinary solutions to environmental challenges.

Island Press designs and implements coordinated book publication campaigns in order to communicate our critical messages in print, in person, and online using the latest technologies, programs, and the media. Our goal: to reach targeted audiences—scientists, policymakers, environmental advocates, the media, and concerned citizens—who can and will take action to protect the plants and animals that enrich our world, the ecosystems we need to survive, the water we drink, and the air we breathe.

Island Press gratefully acknowledges the support of its work by the Agua Fund, Inc., The Margaret A. Cargill Foundation, Betsy and Jesse Fink Foundation, The William and Flora Hewlett Foundation, The Kresge Foundation, The Forrest and Frances Lattner Foundation, The Andrew W. Mellon Foundation, The Curtis and Edith Munson Foundation, The Overbrook Foundation, The David and Lucile Packard Foundation, The Summit Foundation, Trust for Architectural Easements, The Winslow Foundation, and other generous donors.

The opinions expressed in this book are those of the author(s) and do not necessarily reflect the views of our donors.

State of the World 2013

Is Sustainability Still Possible?

Other Worldwatch Books

State of the World 1984 through *2012*
(an annual report on progress toward a sustainable society)

Vital Signs 1992 through *2003* and *2005* through *2012*
(a report on the trends that are shaping our future)

Saving the Planet
Lester R. Brown
Christopher Flavin
Sandra Postel

How Much Is Enough?
Alan Thein Durning

Last Oasis
Sandra Postel

Full House
Lester R. Brown
Hal Kane

Power Surge
Christopher Flavin
Nicholas Lenssen

Who Will Feed China?
Lester R. Brown

Tough Choices
Lester R. Brown

Fighting for Survival
Michael Renner

The Natural Wealth of Nations
David Malin Roodman

Life Out of Bounds
Chris Bright

Beyond Malthus
Lester R. Brown
Gary Gardner
Brian Halweil

Pillar of Sand
Sandra Postel

Vanishing Borders
Hilary French

Eat Here
Brian Halweil

Inspiring Progress
Gary Gardner

Is Sustainability Still Possible?

Erik Assadourian and Tom Prugh, *Project Directors*

Rebecca Adamson
Gar Alperovitz
Olivia Arnow
David Christian
Dwight E. Collins
Robert Costanza
Larry Crowder
Herman Daly
Robert Engelman
Joshua Farley
Carl Folke
Carol Franco
Gary Gardner
Russell M. Genet
Paula Green
Jeff Hohensee
Tim Jackson

Ida Kubiszewski
Melissa Leach
Annie Leonard
Shakuntala Makhijani
Michael Maniates
Jack P. Manno
Brian Martin
Pamela Martin
Laurie Mazur
Jennie Moore
Kathleen Dean Moore
Faith Morgan
Pat Murphy
T. W. Murphy, Jr.
Melissa Nelson
Michael P. Nelson
Simon Nicholson

Danielle Nierenberg
Alexander Ochs
David W. Orr
Sandra Postel
Thomas Princen
Kate Raworth
William E. Rees
Michael Renner
Kim Stanley Robinson
Phillip Saieg
Juliet Schor
Antonia Sohns
Pavan Sukhdev
Bron Taylor
Peter Victor
Eric Zencey

Linda Starke, *Editor*

ISLANDPRESS

Washington | Covelo | London

ISBN 13: 978-1-61091-449-9

ISBN 10: 1-61091-449-X

The text of this book is composed in Minion, with the display set in Myriad Pro. Book design and composition by Lyle Rosbotham.

✸ Printed on recycled, acid-free paper

Manufactured in the United States of America

Worldwatch Institute Staff

Andrew Alesbury
Customer Relations Assistant

Katie Auth
Research Associate, Climate and Energy Program

Adam Dolezal
Research Associate and Central America Project Manager, Climate and Energy Program

Courtney Dotson
Development Associate

Robert Engelman
President

Barbara Fallin
Director of Finance and Administration

Mark Konold
Research Associate and Caribbean Program Manager, Climate and Energy Program

Supriya Kumar
Communications Manager

Matt Lucky
Research Associate, Climate and Energy Program

Haibing Ma
China Program Manager

Shakuntala Makhijani
Research Associate and India Project Manager, Climate and Energy Program

Lisa Mastny
Senior Editor

Evan Musolino
Research Associate and Renewable Energy Indicators Project Manager, Climate and Energy Program

Alexander Ochs
Director, Climate and Energy Program

Ramon Palencia
Central America Fellow, Climate and Energy Program

Grant Potter
Development Associate and Assistant to the President

Tom Prugh
Codirector, State of the World

Laura Reynolds
Staff Researcher, Food and Agriculture Program

Mary C. Redfern
Director of Institutional Relations, Development

Michael Renner
Senior Researcher

Reese Rogers
MAP Sustainable Energy Fellow, Climate and Energy Program

Cameron Scherer
Marketing and Communications Associate

Michael Weber
Research Coordinator, Climate and Energy Program

Sophie Wenzlau
Staff Researcher, Food and Agriculture Program

Worldwatch Institute Fellows, Advisors, and Consultants

Erik Assadourian
Senior Fellow

Christopher Flavin
President Emeritus

Gary Gardner
Senior Fellow

Mia MacDonald
Senior Fellow

Bo Normander
Director, Worldwatch Institute Europe

Corey Perkins
Information Technology Manager

Sandra Postel
Senior Fellow

Lyle Rosbotham
Art and Design Consultant

Janet Sawin
Senior Fellow

Linda Starke
State of the World Editor

Acknowledgments

Each year *State of the World* comes together due to the efforts of scores of individuals and organizations that contribute directly or indirectly to the volume's theme, direction, support, content, shaping, or publication. Any book is a collaborative miracle of sorts, but *State of the World 2013* reflects the labor of more contributors than ever appeared in a previous edition, as well as that of a wide variety of donors, partners, and advisors from around the globe.

None of this would have happened without the support of the Town Creek Foundation, the V. Kann Rasmussen Foundation, the Victoria and Roger Sant Founders Fund of the Summit Fund of Washington, and Peter Seidel—all of whom gave generously to underwrite this edition of *State of the World* and the associated outreach work. A special note of thanks goes to Stuart Clarke and his team at Town Creek, as well as to numerous other sustainability organizations in Maryland, for their help in conducting outreach events around that state.

In addition, we gratefully recognize the continued support of the Ray C. Anderson Foundation. Ray, who passed away in 2011, was a sustainable-business visionary, an active member of the Worldwatch board of directors, and a steadfast believer in our work. His voice and ideas are sorely missed. We hope *State of the World 2013* will be taken as an expression of the honor we feel he is due.

We are also deeply appreciative to our many institutional and foundation supporters, including the Barilla Center for Food & Nutrition; Caribbean Community; Climate and Development Knowledge Network; Compton Foundation, Inc.; The David B. Gold Foundation; Del Mar Global Trust; Elion Group; Energy and Environment Partnership with Central America; Ford Foundation and the Institute of International Education, Inc.; Green Accord International Secretariat; Hitz Foundation; Inter-American Development Bank; International Climate Initiative of the German Federal Ministry for the Environment, Nature Conservation and Nuclear Safety; International Renewable Energy Association; MAP Sustainable Energy Fellowship Program; Ministry for Foreign Affairs of Finland; Renewable Energy Policy Network for the 21st Century; Richard and Rhoda Goldman Fund and the Goldman Environmental Prize; Shenandoah Foundation; Small Planet Fund of RSF Social Finance; Steven C. Leuthold Family Foundation; Transatlantic Climate Bridge of the German Federal Ministry for the Environment, Nature Conservation and Nuclear Safety; United Nations Population Fund; Wallace Global Fund; Weeden Foundation; The William and Flora Hewlett Foundation; and Women Deliver, Inc.

We are delighted to partner, for our second year, with Island Press to publish and distribute *State of the World*. Island Press is a preeminent publisher of sustainability content, and it is a pleasure to continue in the ranks of their many estimable titles. We also owe a huge debt of gratitude to our publishing partners outside of North America; without their indispensable input and help in spreading the word, a volume about the state of the world would be hollow indeed. Specifically, many thanks to Universidade Livre da Mata Atlântica/Worldwatch Brasil; China Social Science Press; Worldwatch Institute Europe; Gaudeamus Helsinki University Press; Good Planet Foundation (France); Germanwatch, Heinrich Böll Foundation, and OEKOM Verlag GmbH (Germany); Organization Earth and the University of Crete (Greece); Earth Day Foundation (Hungary); Centre for Environment Education (India); WWF-Italia and Edizioni Ambiente; Worldwatch Japan; Korea Green Foundation Doyosae (South Korea); FUHEM Ecosocial and Icaria Editorial (Spain); Taiwan Watch Institute; and Turkiye Erozyonla Mucadele, Agaclandima ve Dogal Varliklari Koruma Vakfi (TEMA), and Kultur Yayinlari Is-Turk Limited Sirketi (Turkey).

Although not the very first time a cartoon has appeared in *State of the World*, this year is something of a departure from tradition in that we used several of them prominently for illustration and to help introduce the three sections. Given the rather sober message of this year's volume, an occasion or two for a laugh, or at least a wry smile, did not seem out of place. Special thanks for the cartoons go to Leo Murray, the webcomic xkcd.com, the Jay N. "Ding" Darling Wildlife Society, and the Cartoon Movement.

We would be remiss if we failed to mention John Graham, Alison Singer, and all the interns who work so hard to strengthen the Institute's research. Finally, our deepest gratitude goes to the authors of the 34 chapters and 30 text boxes who contributed so much of their learning, wisdom, time, and patience to the long and sometimes laborious production of this book. Every one of them has much more of value to say than we could print in their individual contributions here, and we wholeheartedly urge readers to explore their work further.

Erik Assadourian and Tom Prugh, *Project Directors*
www.worldwatch.org
www.sustainabilitypossible.org

Contents

Economics on 11, 12, 13

BOXES

TABLES

FIGURES

Units of measure throughout this book are metric unless common usage dictates otherwise.

State of the World: A Year in Review

Compiled by Alison Singer

This timeline covers some significant announcements and reports from December 2011 through November 2012. It is a mix of progress, setbacks, and missed steps around the world that are affecting environmental quality and social welfare.

Timeline events were selected to increase awareness of the connections between people and the environmental systems on which they depend.

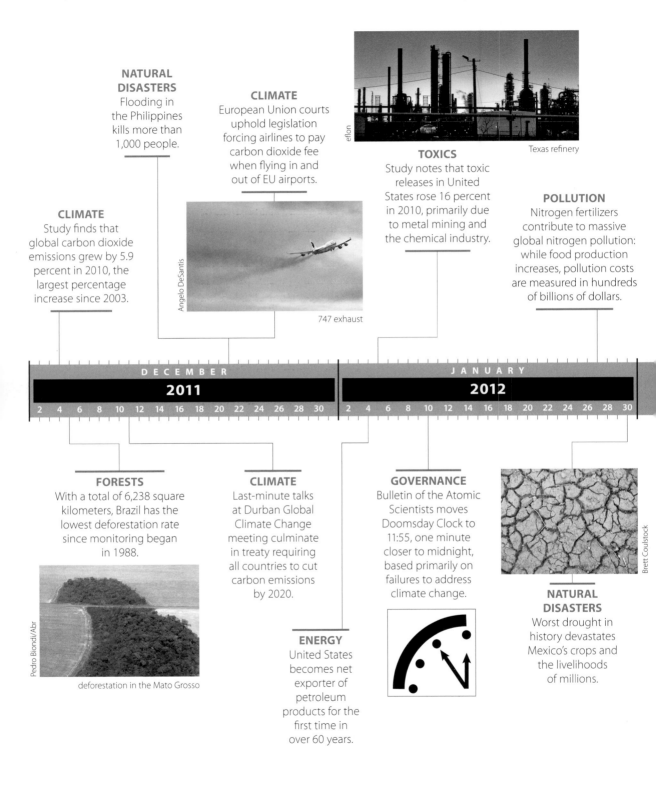

NATURAL DISASTERS
Flooding in the Philippines kills more than 1,000 people.

CLIMATE
European Union courts uphold legislation forcing airlines to pay carbon dioxide fee when flying in and out of EU airports.

eflon

Texas refinery

TOXICS
Study notes that toxic releases in United States rose 16 percent in 2010, primarily due to metal mining and the chemical industry.

POLLUTION
Nitrogen fertilizers contribute to massive global nitrogen pollution: while food production increases, pollution costs are measured in hundreds of billions of dollars.

CLIMATE
Study finds that global carbon dioxide emissions grew by 5.9 percent in 2010, the largest percentage increase since 2003.

Angelo DeSantis

747 exhaust

DECEMBER 2011

2 4 6 8 10 12 14 16 18 20 22 24 26 28 30

JANUARY 2012

2 4 6 8 10 12 14 16 18 20 22 24 26 28 30

FORESTS
With a total of 6,238 square kilometers, Brazil has the lowest deforestation rate since monitoring began in 1988.

Pedro Biondi/Abr

deforestation in the Mato Grosso

CLIMATE
Last-minute talks at Durban Global Climate Change meeting culminate in treaty requiring all countries to cut carbon emissions by 2020.

GOVERNANCE
Bulletin of the Atomic Scientists moves Doomsday Clock to 11:55, one minute closer to midnight, based primarily on failures to address climate change.

Brett Coulstock

NATURAL DISASTERS
Worst drought in history devastates Mexico's crops and the livelihoods of millions.

ENERGY
United States becomes net exporter of petroleum products for the first time in over 60 years.

HEALTH
Millennium Development
Goal to halve the
proportion of people with
no access to safe drinking
water is met ahead of time.

NATURAL DISASTERS
Hundreds die and
hundreds of thousands
are trapped in homes
in Europe's cold snap.

OCEANS
World Bank
announces global
partnership to
manage and
protect the
world's oceans.

Steve Drolet

Prague snowstorm

UK Department for International Development

AGRICULTURE
Australian team
develops strain of
salt-resistant wheat.

F E B R U A R Y   M A R C H

2012

2 4 6 8 10 12 14 16 18 20 22 24 26 28   2 4 6 8 10 12 14 16 18 20 22 24 26 28 30

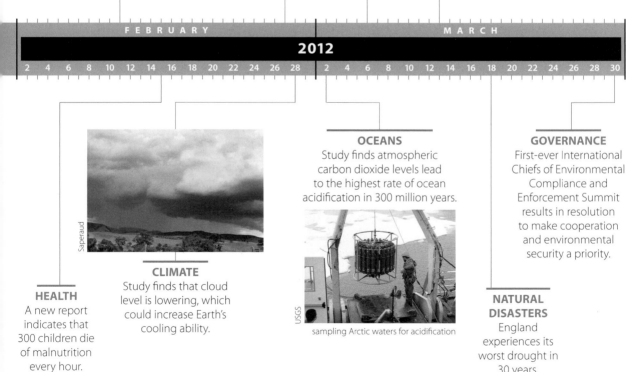

OCEANS
Study finds atmospheric
carbon dioxide levels lead
to the highest rate of ocean
acidification in 300 million years.

GOVERNANCE
First-ever International
Chiefs of Environmental
Compliance and
Enforcement Summit
results in resolution
to make cooperation
and environmental
security a priority.

Saperaud

USGS

sampling Arctic waters for acidification

CLIMATE
Study finds that cloud
level is lowering, which
could increase Earth's
cooling ability.

HEALTH
A new report
indicates that
300 children die
of malnutrition
every hour.

**NATURAL
DISASTERS**
England
experiences its
worst drought in
30 years.

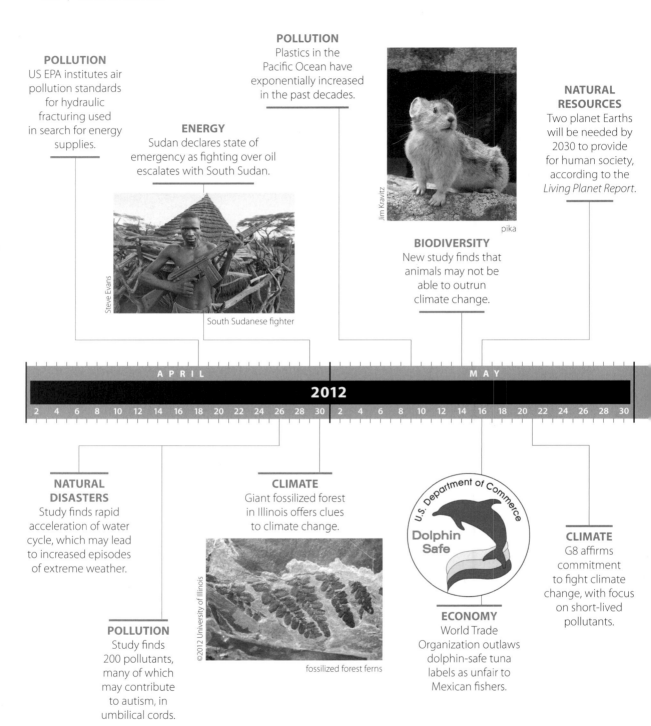

POLLUTION
US EPA institutes air pollution standards for hydraulic fracturing used in search for energy supplies.

ENERGY
Sudan declares state of emergency as fighting over oil escalates with South Sudan.

POLLUTION
Plastics in the Pacific Ocean have exponentially increased in the past decades.

Jim Kravitz

pika

NATURAL RESOURCES
Two planet Earths will be needed by 2030 to provide for human society, according to the *Living Planet Report*.

Steve Evans

South Sudanese fighter

BIODIVERSITY
New study finds that animals may not be able to outrun climate change.

APRIL

2012

MAY

2 4 6 8 10 12 14 16 18 20 22 24 26 28 30   2 4 6 8 10 12 14 16 18 20 22 24 26 28 30

NATURAL DISASTERS
Study finds rapid acceleration of water cycle, which may lead to increased episodes of extreme weather.

CLIMATE
Giant fossilized forest in Illinois offers clues to climate change.

POLLUTION
Study finds 200 pollutants, many of which may contribute to autism, in umbilical cords.

©2012 University of Illinois

fossilized forest ferns

U.S. Department of Commerce
Dolphin Safe

ECONOMY
World Trade Organization outlaws dolphin-safe tuna labels as unfair to Mexican fishers.

CLIMATE
G8 affirms commitment to fight climate change, with focus on short-lived pollutants.

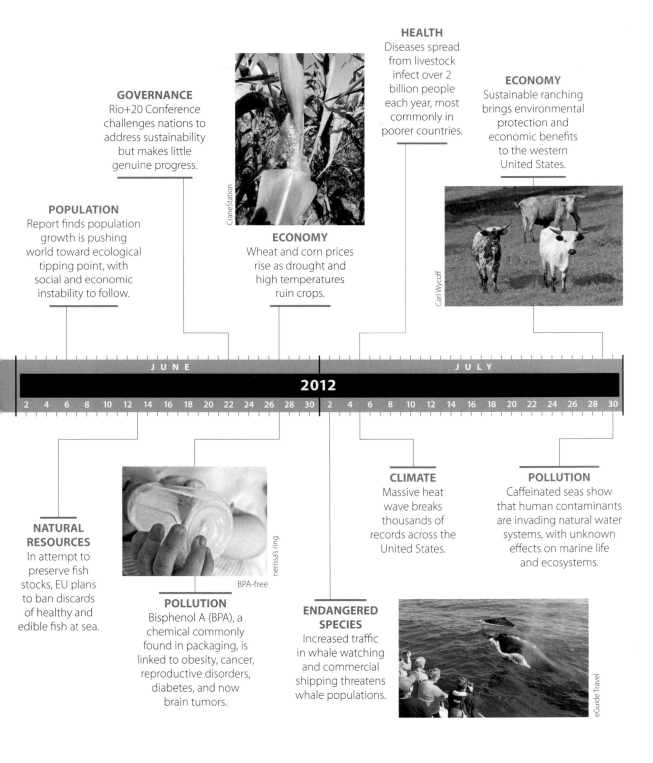

GOVERNANCE
Rio+20 Conference challenges nations to address sustainability but makes little genuine progress.

HEALTH
Diseases spread from livestock infect over 2 billion people each year, most commonly in poorer countries.

ECONOMY
Sustainable ranching brings environmental protection and economic benefits to the western United States.

CraneStation

POPULATION
Report finds population growth is pushing world toward ecological tipping point, with social and economic instability to follow.

ECONOMY
Wheat and corn prices rise as drought and high temperatures ruin crops.

Carl Wycoff

JUNE

2012

JULY

2 4 6 8 10 12 14 16 18 20 22 24 26 28 30 2 4 6 8 10 12 14 16 18 20 22 24 26 28 30

NATURAL RESOURCES
In attempt to preserve fish stocks, EU plans to ban discards of healthy and edible fish at sea.

nerissa's ring

BPA-free

POLLUTION
Bisphenol A (BPA), a chemical commonly found in packaging, is linked to obesity, cancer, reproductive disorders, diabetes, and now brain tumors.

ENDANGERED SPECIES
Increased traffic in whale watching and commercial shipping threatens whale populations.

CLIMATE
Massive heat wave breaks thousands of records across the United States.

POLLUTION
Caffeinated seas show that human contaminants are invading natural water systems, with unknown effects on marine life and ecosystems.

eGuide Travel

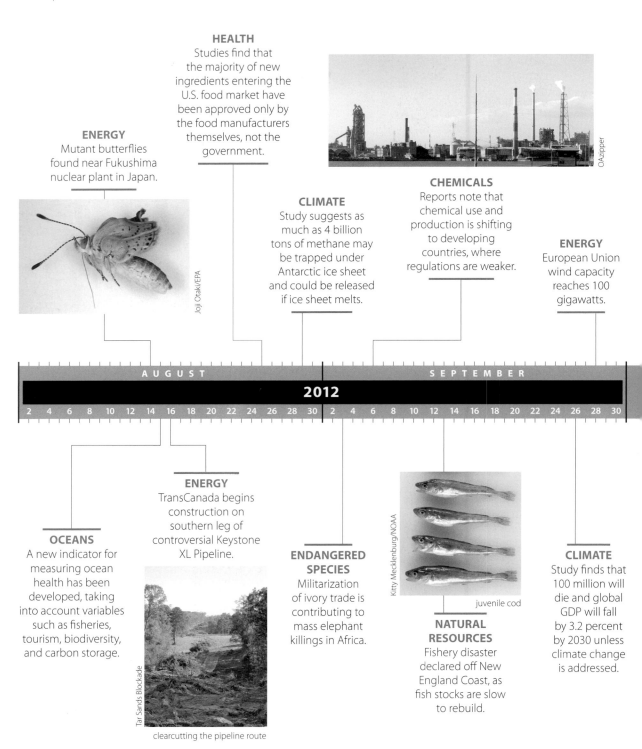

HEALTH
Studies find that the majority of new ingredients entering the U.S. food market have been approved only by the food manufacturers themselves, not the government.

ENERGY
Mutant butterflies found near Fukushima nuclear plant in Japan.

Joji Otaki/EPA

OAzipper

CHEMICALS
Reports note that chemical use and production is shifting to developing countries, where regulations are weaker.

CLIMATE
Study suggests as much as 4 billion tons of methane may be trapped under Antarctic ice sheet and could be released if ice sheet melts.

ENERGY
European Union wind capacity reaches 100 gigawatts.

AUGUST

2012

SEPTEMBER

2 4 6 8 10 12 14 16 18 20 22 24 26 28 30   2 4 6 8 10 12 14 16 18 20 22 24 26 28 30

ENERGY
TransCanada begins construction on southern leg of controversial Keystone XL Pipeline.

OCEANS
A new indicator for measuring ocean health has been developed, taking into account variables such as fisheries, tourism, biodiversity, and carbon storage.

Tar Sands Blockade

clearcutting the pipeline route

ENDANGERED SPECIES
Militarization of ivory trade is contributing to mass elephant killings in Africa.

Kitty Mecklenburg/NOAA

juvenile cod

NATURAL RESOURCES
Fishery disaster declared off New England Coast, as fish stocks are slow to rebuild.

CLIMATE
Study finds that 100 million will die and global GDP will fall by 3.2 percent by 2030 unless climate change is addressed.

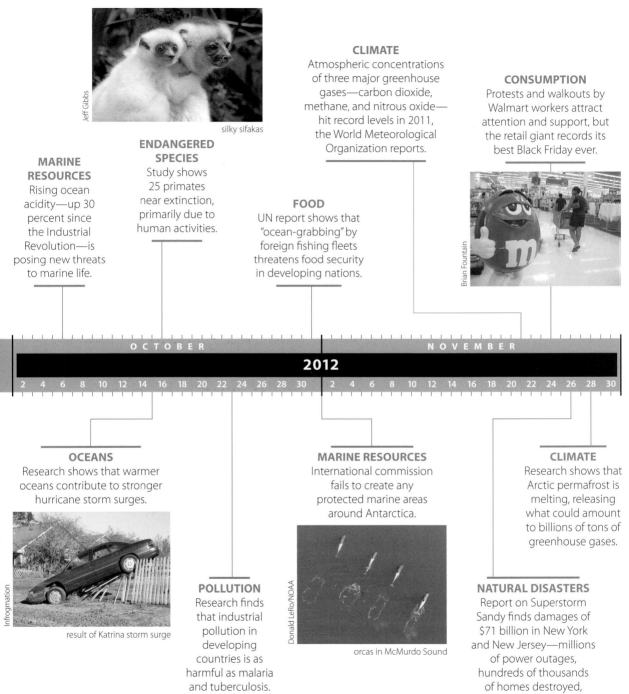

Jeff Gibbs

silky sifakas

CLIMATE
Atmospheric concentrations of three major greenhouse gases—carbon dioxide, methane, and nitrous oxide—hit record levels in 2011, the World Meteorological Organization reports.

CONSUMPTION
Protests and walkouts by Walmart workers attract attention and support, but the retail giant records its best Black Friday ever.

MARINE RESOURCES
Rising ocean acidity—up 30 percent since the Industrial Revolution—is posing new threats to marine life.

ENDANGERED SPECIES
Study shows 25 primates near extinction, primarily due to human activities.

FOOD
UN report shows that "ocean-grabbing" by foreign fishing fleets threatens food security in developing nations.

Brian Fountain

O C T O B E R   N O V E M B E R

2012

2 4 6 8 10 12 14 16 18 20 22 24 26 28 30   2 4 6 8 10 12 14 16 18 20 22 24 26 28 30

OCEANS
Research shows that warmer oceans contribute to stronger hurricane storm surges.

MARINE RESOURCES
International commission fails to create any protected marine areas around Antarctica.

CLIMATE
Research shows that Arctic permafrost is melting, releasing what could amount to billions of tons of greenhouse gases.

Infrogmation

result of Katrina storm surge

POLLUTION
Research finds that industrial pollution in developing countries is as harmful as malaria and tuberculosis.

Donald LeRo/NOAA

orcas in McMurdo Sound

NATURAL DISASTERS
Report on Superstorm Sandy finds damages of $71 billion in New York and New Jersey—millions of power outages, hundreds of thousands of homes destroyed, and transportation systems crippled.

Is Sustainability Still Possible?

Beyond Sustainababble

Robert Engelman

We live today in an age of *sustainababble*, a cacophonous profusion of uses of the word *sustainable* to mean anything from environmentally better to cool. The original adjective—meaning capable of being maintained in existence without interruption or diminution—goes back to the ancient Romans. Its use in the environmental field exploded with the 1987 release of *Our Common Future*, the report of the World Commission on Environment and Development. Sustainable development, Norwegian Prime Minister Gro Harlem Brundtland and the other commissioners declared, "meets the needs of the present without compromising the ability of future generations to meet their own needs."[1]

For many years after the release of the Brundtland Commission's report, environmental analysts debated the value of such complex terms as *sustainable*, *sustainability*, and *sustainable development*. By the turn of the millennium, however, the terms gained a life of their own—with no assurance that this was based on the Commission's definition. Through increasingly frequent vernacular use, it seemed, the word *sustainable* became a synonym for the equally vague and unquantifiable adjective *green*, suggesting some undefined environmental value, as in *green growth* or *green jobs*.

Today the term *sustainable* more typically lends itself to the corporate behavior often called *greenwashing*. Phrases like sustainable design, sustainable cars, even sustainable underwear litter the media. One airline assures passengers that "the cardboard we use is taken from a sustainable source," while another informs them that its new in-flight "sustainability effort" saved enough aluminum in 2011 "to build three new airplanes." Neither use sheds any light on whether the airlines' overall operations—or commercial aviation itself—can long be sustained on today's scale.[2]

The United Kingdom was said to be aiming for "the first sustainable Olympics" in 2012, perhaps implying an infinitely long future for the quadrennial event no matter what else happens to humanity and the planet.

Robert Engelman is president of the Worldwatch Institute.

www.sustainabilitypossible.org

(If environmental impact is indeed the operable standard, the Olympics games in classical Greece or even during the twentieth century were far more sustainable than today's.) The upward trend line of the use of this increasingly meaningless word led one cartoonist to suggest that in 100 years *sustainable* will be the only word uttered by anyone speaking American English. (See Figure 1–1.)[3]

By some metrics this might be considered success. To find *sustainable* in such common use indicates that a key environmental concept now enjoys general currency in popular culture. But sustainababble has a high cost. Through overuse, the words *sustainable* and *sustainability* lose meaning and impact. Worse, frequent and inappropriate use lulls us into dreamy belief that all of us—and everything we do, everything we buy, everything we use—are now able to go on forever, world without end, amen. This is hardly the case.

The question of whether civilization can continue on its current path without undermining prospects for future well-being is at the core of the world's current environmental predicament. In the wake of failed international environmental and climate summits, when national governments take no actions commensurate with the risk of catastrophic environmental change, are there ways humanity might still alter current behaviors to make

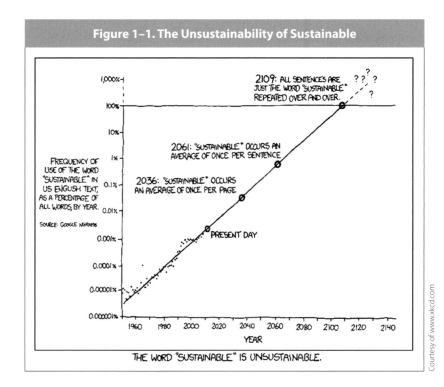

Figure 1–1. The Unsustainability of Sustainable

Courtesy of www.xkcd.com

them sustainable? Is sustainability still possible? If humanity fails to achieve sustainability, when—and how—will unsustainable trends end? And how will we live through and beyond such endings? Whatever words we use, we need to ask these tough questions. If we fail to do so, we risk self-destruction.

This year's *State of the World* aims to expand and deepen discussion of the overused and misunderstood adjective *sustainable*, which in recent years has morphed from its original meaning into something like "a little better for the environment than the alternative." Simply doing "better" environmentally will not stop the unraveling of ecological relationships we depend on for food and health. Improving our act will not stabilize the atmosphere. It will not slow the falling of aquifers or the rising of oceans. Nor will it return Arctic ice, among Earth's most visible natural features from space, to its pre-industrial extent.

In order to alter these trends, vastly larger changes are needed than we have seen so far. It is essential that we take stock, soberly and in scientifically measurable ways, of where we are headed. We desperately need—and are running out of time—to learn how to shift direction toward safety for ourselves, our descendants, and the other species that are our only known companions in the universe. And while we take on these hard tasks, we also need to prepare the social sphere for a future that may well offer hardships and challenges unlike any that human beings have previously experienced. While it is a subset of the biosphere, the social sphere is shaped as well by human capacities with few known limits. We can take at least some hope in that.

Birth of a Concept

Respect for sustainability may go back far in human cultures. North America's Iroquois expressed concern for the consequences of their decisionmaking down to the seventh generation from their own. A proverb often attributed to Native American indigenous cultures states, "We have not inherited the earth from our fathers, we are borrowing it from our children." In modern times, the idea of sustainability took root in the writings of naturalist and three-term U.S. Representative George Perkins Marsh in the 1860s and 1870s. Humans were increasingly competing with, and often outcompeting, natural forces in altering the earth itself, Marsh and later writers documented. This is dangerous in the long run, they argued, even if demographically and economically stimulating in the short run.[4]

"What we do will affect not only the present but future generations," President Theodore Roosevelt declared in 1901 in his first Message to Congress, which called for conservation of the nation's natural resources. The value of conserving natural resources for future use—and the dangers of failing to do so—even made it into political cartoons in the decades that followed. (See Figure 1–2.) The U.S. National Environmental Policy Act of 1969 echoed

Courtesy of the Jay N. "Ding" Darling Wildlife Society, originally published 15 September 1936

Figure 1–2. 1936 Cartoon by Jay N. "Ding" Darling

Roosevelt's words, affirming that "it is the continuing policy of the Federal Government . . . to create and maintain conditions under which man and nature can exist in productive harmony, and fulfill the social, economic, and other requirements of present and future generations of Americans."[5]

Two important points emerge from the definition of *sustainable development* found in *Our Common Future*, which is still the most commonly cited reference for sustainability and sustainable development. The first is that any environmental trend line can at least in theory be analyzed quantitatively through the lens of its likely impact on the ability of future generations to meet their needs. While we cannot predict the precise impacts of trends and the responses of future humans, this definition offers the basis for metrics of sustainability that can improve with time as knowledge and experience accumulate. The two key questions are, What's going on? And can it keep going on in this way, on this scale, at this pace, without reducing the likelihood that future generations will live as prosperously and comfortably as ours has? For sustainability to have any meaning, it must be tied to clear and rigorous definitions, metrics, and mileage markers.

The second point is the imperative of development itself. Environmental sustainability and economic development, however, are quite different objectives that need to be understood separately before they are linked. In the Chairman's Foreword to *Our Common Future*, Gro Harlem Brundtland defined development as "what we all do in attempting to improve our lot." It is no slight to either low- or high-income people to note that as 7.1 billion people "do what we all do . . . to improve our lot," we push more dangerously into environmentally unsustainable territory. We might imagine optimistically that through reforming the global economy we will find ways to "grow green" enough to meet everyone's needs without threatening the future. But we will be better served by thinking rigorously about biophysical boundaries, how to keep within them, and how—under these unforgiving realities—we can best ensure that all human beings have fair and equitable access to nourishing food, energy, and other prerequisites of a decent life. It will almost certainly take more cooperation and more sharing than we can imagine in a world currently driven by competition and individual accumulation of wealth.[6]

What right, we might then ask, do present generations have to improve their lot at the cost of making it harder or even impossible for all future generations to do the same? Philosophically, that's a fair question—especially from the viewpoint of the future generations—but it is not taken seriously. Perhaps if "improving our lot" could somehow be capped at modest levels of resource consumption, a fairer distribution of wealth for all would allow development that would take nothing away from future generations. That may mean doing without a personal car or living in homes that are unimaginably small by today's standards or being a bit colder inside during the winter and hotter during the summer. With a large enough human population, however, even modest per capita consumption may be environmentally unsustainable. (See Box 1–1.)[7]

Gro Brundtland, however, made the practical observation that societies are unlikely to enact policies and programs that favor the future (or nonhuman life) at the expense of people living in the present, especially the poorer among us. Ethically, too, it would be problematic for environmentalists, few of us poor ourselves, to argue that prosperity for those in poverty should take a back seat to protection of the development prospects of future generations. Unless, perhaps, we are willing to take vows of poverty.[8]

While sustainability advocates may work to enfranchise future generations and other species, we have little choice but to give priority to the needs of human beings alive today while trying to preserve conditions that allow future generations to meet their needs. It is worth recognizing, however, that there is no guarantee that this tension is resolvable and the goal achievable.

If Development Isn't Sustainable, Is It Development?

The world is large, yet human beings are many, and our use of the planet's atmosphere, crust, forests, fisheries, waters, and resources is now a force like that of nature. On the other hand, we are a smart and adaptive species, to say the least. Which perhaps helps explain why so many important economic and environmental trends seem headed in conflicting and even opposite directions. Are things looking up or down?

On the development side, the world has already met one of the Millennium Development Goals set for 2015 by the world's governments in 2000: by 2010 the proportion of people lacking access to safe water was cut in half from 1990 levels. And the last decade has witnessed so dramatic a reduction in global poverty, central to a second development goal, that the London-based Overseas Development Institute urged foreign assistance agencies to redirect their aid strategies over the next 13 years to a dwindling number of the lowest-income nations, mostly in sub-Saharan Africa. By some measures, it can be argued that economic prosperity is on the rise and basic needs in most parts of the world are increasingly being met.[9]

On the environment side, indicators of progress are numerous. They include rising public awareness of problems such as climate change, rainforest loss, and declining biological diversity. Dozens of governments on both sides of the development divide are taking steps to reduce their countries' greenhouse gas emissions—or at least the growth of those emissions. The use of renewable energy is growing more rapidly than that of fossil fuels (although from a much smaller base). Such trends do not themselves lead

Box 1–1. Toward a Sustainable Number of Us

To link environmental and social sustainability, think population. When we consider what levels of human activity are environmentally sustainable and then, for the sake of equity, calculate an equal allocation of such activity for all, we are forced to ask how many people are in the system.

Suppose for example, we conclude that 4.9 billion tons of carbon dioxide (CO_2) per year and its global-warming equivalent in other greenhouse gases—one tenth of the 49 billion tons emitted in 2010—would be the most that humanity could emit annually to avoid further increases in the atmospheric concentrations of these gases. We then need to divide this number by the 7.1 billion human beings currently alive to derive an "atmosphere-sustainable" per capita emission level. No one responsible for emissions greater than the resulting 690 kilograms annually could claim that his or her lifestyle is atmosphere-sustainable. To do so would be to claim a greater right than others to use the atmosphere as a dump.

One 1998 study used then-current population and emission levels and a somewhat different calculation of global emissions level that would lead to safe atmospheric stability. The conclusion: Botswana's 1995 per capita emission of 1.54 tons of CO_2 (based in this case on commercial energy and cement consumption only) was mathematically climate-sustainable at that time. Although population-based calculations are not always so informative with every resource or system (sustaining biodiversity, for example), similar calculations could work to propose sustainable per capita consumption of water, wood products, fish, and potentially even food.

Once we master such calculations, we begin to understand their implications: As population rises, so does the bar of per capita sustainable behavior. That is, the more of us there are, the less of a share of any fixed resource, such as

the atmosphere, is available for each of us to sustainably and equitably transform or consume in a closed system. All else being equal, the smaller the population in any such system, the more likely sustainability can be achieved and the more generous the sustainable consumption level can be for each person. With a large enough population there is no guarantee that even very low levels of equitable per capita greenhouse emissions or resource consumption are environmentally sustainable. If Ecological Footprint calculations are even roughly accurate, humanity is currently consuming the ecological capacity of 1.5 Earths. That suggests that no more than 4.7 billion people could live within the planet's ecological boundaries without substantially reducing average individual consumption.

Absent catastrophe, sustainable population anything like this size will take many decades to reach through declines in human fertility that reflect parents' intentions. There is good reason to believe, however, that a population peak below 9 billion might occur before mid-century if societies succeed in offering near-universal access to family planning services for all who want them along with near-universal secondary education for everyone. Also helpful would be greatly increased autonomy for women and girls and the elimination of fertility-boosting programs such as birth dividends and per child tax credits.

In the meantime, while population remains in the range of 7 billion, individual levels of greenhouse gas emissions and natural resource consumption will have to come way, way down to even begin to approach environmental sustainability. Consumption levels that would bring those of us in high-consuming countries into a sustainable relation with the planet and an equitable relation with all who live on it would undoubtedly be small fractions of what we take for granted today.

Source: See endnote 7.

directly in any measurable way to true sustainability (fossil fuel use is climbing fast as China and India industrialize, for example), but they may help create conditions for it. One important trend, however, is both measurable and sustainable by strict definition: thanks to a 1987 international treaty, the global use of ozone-depleting substances has declined to the point where the atmosphere's sun-screening ozone layer is considered likely to repair itself, after sizable human-caused damage, by the end of this century.[10]

It is not clear, however, that any of these development and environmental trends demonstrate that truly sustainable development is occurring. Safe water may be reaching more people, but potentially at the expense of maintaining stable supplies of renewable freshwater in rivers or underground aquifers for future generations. Reducing the proportion of people in poverty is especially encouraging, but what if the instruments of development—intense application of fossil fuels to industrial growth, for example—contribute significantly to increasing proportions of people in poverty in the future?

Moreover, economic development itself is running into constraints in many countries, as population and consumption growth inflate demand for food, energy, and natural resources beyond what supply—or at least the simple economics of price or the logistics of distribution—can provide. The price of resources has climbed for most of the last 10 years after sliding during the previous several decades. Results of rising prices for food, fossil fuels, minerals, and necessities that rely on nonrenewable resources for their production include food riots like those of 2008 and crippling power blackouts like the one in India that affected nearly a tenth of the world's population in 2012.[11]

Yet even as economic growth seems to be bumping into its own limiting constraints in much of the world, the most important environmental trends are discouraging and in many cases alarming. Human-caused climate change, in particular, shows no signs of slowing or beginning any soft landing toward sustainability, with global emissions of greenhouse gases continuing to climb in the upper range of past projections. The rise is slowed, on occasion and in some countries, mostly by recession or happenstance shifts in fossil-fuel economics (such as the recent ascendance of shale gas production in the United States) rather than any strategic intention or policy.

Despite all international efforts to rein in emissions of fossil-fuel-based carbon dioxide, for example, these emissions are today larger than ever and may be increasing at an accelerating pace. (See Figure 1–3.) A brief downward blip in 2009 was unrelated to coordinated government action but stemmed from global economic decline. The global increase in fossil-fuel-based CO_2 was estimated at 3 percent in 2011 compared with 2010—nearly three times the pace of population growth—despite a still sluggish global economy and absolute emissions reductions in the United States that year. This trend leads some scientists to suggest it may be too late to stop future warming in a safe temperature range for humanity.[12]

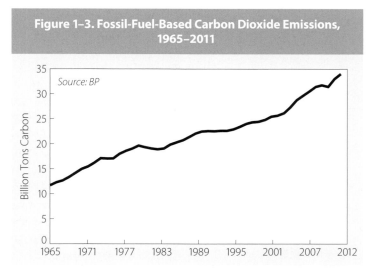

Figure 1–3. Fossil-Fuel-Based Carbon Dioxide Emissions, 1965–2011

Source: BP

Demographic and economic growth drives growth in greenhouse gas emissions and natural resource use. Aspirations over the past few decades that economic growth can be "decoupled" from energy and natural resource use, thus allowing the growth to continue indefinitely, have proved overly optimistic. An earlier trend toward energy decoupling reversed course during the global economic downturn that began in late 2007. This was partly because governments of industrial countries attempted to stimulate their sluggish economies through energy-intensive public works programs, but it was mostly due to massive industrialization in the emerging economies of China and India. Until the combined power of population and economic growth is reversed or a strong climate pact transforms the global economy, there seems to be little prospect for either true sustainability or truly sustainable development through ever-greater efficiency and decoupling.[13]

This logic is especially worrisome because we have already dug ourselves so deeply into unsustainability, based on the assessment of many scientists, that we are now passing critical environmental thresholds or "tipping points." We are starting to feel the weight of what was once balanced on Earth's seesaw now sliding down upon us. In 2009, a group of 30 scientists identified nine planetary boundaries where sustainability could be roughly measured and monitored. Human beings had already, by their calculation, crashed through two such boundaries and part of a third: in greenhouse gas loading of the atmosphere, in nitrogen pollution, and in the loss of biological diversity.[14]

Three years later, in the run-up to the U.N. Rio+20 Conference on sustainable development, another group of scientists, led by Anthony D. Barnofsky of the University of California, Berkeley, warned that based on land use and other indicators of human domination of natural systems, the planet may already be poised to undergo an imminent, human-induced state shift. That phrase refers to an abrupt and irreversible shift from an existing state to a new one. In this case, the shift would compare in magnitude (though not in comfort) to the rapid transition that ended the last Ice Age and ushered in the more temperate climate in which human civilization evolved.[15]

What the scientists found in physical and biological systems, U.N. Environment Programme analysts found in political ones. Rooting among the 90 most important international environmental commitments made by governments, the analysts could identify significant progress only in four, including halting further damage to the ozone layer and improving access to safe water.[16]

Other signs are positive, however, as noted earlier. The rapid growth of renewable energy, growing acceptance that human activities are warming the world, new efforts among many corporations to improve their environmental behavior and reputations (although sometimes this is more sustain-ababble than real), the seriousness with which Mexico and China are trying to rein in their greenhouse gas emissions, a recent slowdown in deforestation in Brazil—all these trends signal the possibility of shifts in unsustainable trends in the near future.[17]

But absent far more progress, the basic trends themselves remain clearly, measurably unsustainable: the shrinking of aquifers around the world as farmers are called on to produce more food while competing with other water users, the global declines of fisheries and of all biodiversity, the accelerating emergence of new infectious diseases over the last few decades, and—of course—the relentless march of warmer temperatures, higher oceans, and ever-more-intense downpours and droughts. People who survive in leadership roles at some point develop realistic strategies for likely eventualities. And it now seems pretty obvious that the time has arrived to prepare for the consequences of unsustainability, even while we refuse to give up the effort, however quixotic, to shift to true sustainability on some reasonable schedule.

Predicament and Possibility

Why has it proved so hard to conform human behavior to the needs of a life-supporting future? A major reason is simply the unprecedented scale that humanity has reached in the twenty-first century: We are 7.1 billion sizable individual organisms, each requiring thousands of kilocalories of food energy and several liters of water per day. The vast majority of us are unwilling to share our private living space with wild plants and animals. We like to live in a temperature range far narrower than that of the outdoors, and we like to be mobile. As we carve out land to grow our food, we fully convert it from wild nature to humanized territory.

In all these needs and wants, we are helped by the fact that much of the stored energy that living things gained from the sun over hundreds of millions of years has been unleashed for our enjoyment—to fuel our globe-spanning travel, to control the climates of our homes and workplaces, to allow many of us to enjoy pleasures and comforts unknown even to monarchs in the past. Our political and economic institutions evolved before anyone imagined the need to restrain human behavior out of concern for the future. An estimated 2.8 trillion tons of carbon dioxide emissions sleep in fossil fuel reserves—more than enough to guarantee climate catastrophe from a CO_2-saturated atmosphere—that companies and governments would gladly sell tomorrow for immediate combustion if they could bring the buried carbon to the surface and get the right price for it.[18]

With exceptions in a few countries, growing human populations are eating more meat, using more carbon-based energy, shouldering aside more natural landscapes, and tapping into more renewable and nonrenewable commodities than ever before in history. The momentum of a still-young global population all but guarantees demographic growth for decades to come. The momentum of the world's transportation networks, infrastructure, and built environment all but guarantees that shifts toward low-carbon energy will take decades. Individual aspirations for wealth and comfort all but guarantee increasing per capita global consumption, at least to the extent the world economy will support it. But ever-greater energy investments are needed to tap fossil fuels and other critical nonrenewable resources, raising the likelihood that these will become increasingly expensive with time.

Our predicament at least presents us with opportunity. In the words of poet W. H. Auden, "We must love one another or die." In order to survive, we may find ourselves dragged kicking and screaming into ways of relating to each other and the world around us that humanity has been aspiring to achieve since the emergence of the great ethical and spiritual traditions many centuries ago.[19]

Asking the Difficult Questions

In asking "Is Sustainability Still Possible?" we realized several other questions would also need to be grappled with in this report. The first section, The Sustainability Metric, explores what a rigorous definition of sustainability would entail, helping to make this critical concept measurable and hence meaningful. Though such measurement is often challenging to design and agree on, much less carry out, the objective would be to continually improve on it, for scientific measurement has always improved over time.

The first step toward survival is to define *environmentally sustainable* and to use this definition to measure and monitor whether current trends are heading toward or away from trajectories that could continue indefinitely without threatening future life. The second is to use these sustainability metrics to develop practical measures, whether politically feasible at the moment or not, that can bend the curve of current trends toward sustainability.

To help with measurement, we should look without blinking at what is unsustainable—at practices and patterns that, if we don't stop them, will stop us. The rarely voiced reality of environmental unsustainability is that we may have not just less prosperous and comfortable lives in the future but shorter and fewer lives altogether. If it proves too challenging to feed the projected 2050 world population of more than 9 billion people, for example, it is quite possible we will not have to—for the worst of reasons. The same can be said of "business-as-usual" greenhouse gas emission scenarios: by the time global thermometers register a hike of 4 degrees Celsius, business-as-usual will have ended a long time ago.[20]

Raising the specter of rising death rates and civilizational collapse underlines the need for rigor in assessing what true sustainability is and how to measure if we are heading there. In doing so we must accept that true sustainability may not arrive for decades or even centuries, yet we'll need to be vigilant about making progress toward it now and at each point along the way. The objective will then be to build popular support, make such measures feasible, and eventually transform them into effective policies and programs worldwide.

The second section of the book, Getting to True Sustainability, explores the implications of the gaps that remain between present realities and a truly sustainable future. What would it take—what actions, policies, institutional and behavioral changes, and reductions in the scale of human activity—to arrive at a truly sustainable society? In a world far more preoccupied with present economic and security conditions than with its own future capacity to support life, how can those who care about these issues help move societies in the right direction? How can we spur a sufficiently rapid transition toward a world in which humanity and the nature that supports it can thrive indefinitely?

Andrea O'Connell

Equipped with clearer definitions of true sustainability and clearer indicators of where we stand in relation to it, we can begin to "get real"—that is, more practical and ambitious—about making our actions and behaviors truly sustainable. Straight-forward objectives of where we need to be can help us separate marginal action, political showmanship, and feel-good aspirations from measurable progress. The danger of rigorous definition and measurement is, of course, the psychological impact of the awareness of how distant the goal of true sustainability is. The momentum and weight of that distance can be overwhelming and debilitating. But the fool's gold that sustainababble offers is poor medicine; far better to know where we stand—and to stop standing in a space in which we will not survive.

Are there really policy options for forging toward true sustainability? There are at least some good candidates, and attention to the sustainability metric will help us identify which ones are worth making a priority—whether relating to climate change, population growth, nitrogen runoff, or biodiversity loss. Detailed and productive policy proposals can emerge when we focus more on sustainability metrics and how to manage them to produce equitable outcomes. It will take time; as current environmental politics makes clear, not much is achievable with today's governments. Those who care about these issues need to think like eighteenth- and nineteenth-century abolitionists, who worked tirelessly on their cause for generations before legal slavery disappeared from the world. While time is in most ways the scarcest resource of all, achieving true sustainability will need a political movement that grows and gains power over time to make its influence decisive.

Centuries of human experience amid hardship do nonetheless suggest the possibility that we will muddle through whatever lies before us on the home planet. We have no way of knowing what inventions will arise to revolutionize our lives and maybe minimize our impacts. Perhaps ocean currents or cold fusion will offer supplies of energy that are safe, climate-neutral, and effectively inexhaustible. There is no basis for smug certainty that we face catastrophe. Yet based on what we have done and are doing ever-more intensively to the atmosphere, oceans, soils, forests, fisheries, and life itself, it takes an almost religious conviction to be confident that such sunny outcomes will unfold all over the environmental stage.

History also shows that even human resilience can have its downside. By adapting so well to past environmental losses (the extinction of large mammals in the Pleistocene, for example), we humans have been able to keep expanding our population, leading to ever-wider ripples and denser layers of long-term unsustainability. Unless scientists are way off track in their understanding of the biophysical world, we would be wise today to look to dramatic and rapid "demand contraction"—call it degrowth or simply an adaptive response to an overused planet—to shift toward a truly environmentally sustainable world that meets human needs. We need to understand the boundaries we face—and then craft ways to fairly share the burden of living within them so that the poor bear the least and the wealthy the most. That's only fair.

The stakes by their nature are higher the younger someone is—and highest still for those who are not yet but will be born. We are talking about the survival of human civilization as we know it, and possibly of the species itself. "There is . . . no certainty that adaptation to a 4°C world is possible," a recent World Bank report conceded, referring to a global average temperature increase of 7.2 degrees Fahrenheit from pre-industrial times that is considered likely by 2100 without policy change. And so the book's third section—Open in Case of Emergency—takes on a topic that most discussions of sustainability leave unsaid: whether and how to prepare for the possibility of a catastrophic global environmental disruption. We could define this as a sharp break with the past that reverses the long advance of human creature comforts, health, and life expectancy—and from which recovery might take centuries.[21]

In many parts of the world, the emergency has already arrived. There are places where violence is routine—and routinely unpunished—and where creature comforts are as distant as personal safety. Sustainability is a meaningless concept in such places, but scholars of sustainability could profitably study how people survive there. How do they adapt and stay resilient in the face of their struggles? How did cultures and societies survive during and after one of the worst civilizational reversals in history, the fourteenth-century Black Death, which may have cut European population by half?

It is through just such an exploration that the environmental movement enters fully into the social sphere, after a long history in which the objective was to protect nature from human influence. We are living in the Anthropocene now, the era in which humans are the main force shaping the future of life. And it is too late to wall off nature from human influence. Even if we could somehow cork all the world's tailpipes and smokestacks, quench all fires, and cap all other greenhouse gas emission sources, Earth will keep warming for decades and the oceans will rise for centuries to come. We need to focus on adapting to a dramatically changing climate and environment while simultaneously pressing ever harder to head off further change. If we fail to constrain the ways we are changing the planet, the planet will eventually overwhelm all our efforts to adapt.[22]

Such speculation may sound pessimistic, but neither fear of pessimism nor a dogged determination to remain optimistic are reasons for understating our predicament. Optimism and pessimism are equal distractions from what we need in our current circumstance: realism, a commitment to nature and to each other, and a determination not to waste more time. There seems little point in determining your gut feeling about the future when you can put your shoulder to the wheel to make sure the world will keep sustaining life. "Feeling that you have to maintain hope can wear you out," eco-philosopher Joanna Macy said in a recent interview with the wisdom of her 81 years. "Just be present. . . . When you're worrying about whether you're hopeful or hopeless or pessimistic or optimistic, who cares? The main thing is that you are showing up, that you're here, and that you're finding ever more capacity to love this world, because it will not be healed without that."[23]

Giacomo Cardelli/Cartoon Movement

The Sustainability Metric

"You cannot manage what you do not measure." So runs the business adage. Immeasurables, too, often need managing, but the point remains that metrics matter. Marketers and many of the rest of us blithely dub products, cities, activities, and almost anything else under the sun "sustainable" with no quantification that might allow independent verification. If we are to manage our way to a sound environment and a durable civilization, we'll need to weigh rigorously our progress in ways scientists can support and the rest of us agree on.

Some sustainability metrics are straightforward. The atmosphere will stabilize when the mass of greenhouse gases that humanity emits is no greater than the mass the earth reabsorbs. Global progress toward emissions sustainability can be tracked, leaving only the harder task of devising ways to mark individual and national sustainability. Since we emit more almost every year, we know we are less "emissions-sustainable" with each passing hour. How, though, do we track progress in sustaining biological diversity? With so much uncertainty about causes and rates of extinction, it is much harder to find the set point for "biodiversity-sustainable."

Developing sustainability metrics will be an evolutionary process, an objective to work toward and use for accountability in the long conversation ahead. The authors in this section ponder the task and its implications in a variety of environmental systems and natural resources. Carl Folke opens with an assessment of perhaps the broadest and most critical range of sustainability metrics: those defining literal boundary points on the planet that we pass only at peril to our future. Among these are the two systems just mentioned—climate and biodiversity—but also key mineral cycles and changes in land, oceans, and air. Marking these boundaries and our position relative to them sometimes requires subjective judgment, yet the process nonetheless contributes to better metrics. The concepts of planetary boundaries and of the Ecological Footprint, discussed here by Jennie Moore and William E. Rees, offer among the most influential sustainability metrics yet devised, and their implications are daunting.

Renewable freshwater especially lends itself to sustainability quantification. Hydrologists have carefully measured much of Earth's water cycle. We will never run out of water, but some societies drive themselves into scarcity by using so much water that precipitation fails to maintain levels in rivers, lakes, and aquifers. Sandra Postel explores these metrics—and finds hope for future sustainability in the fact that so much freshwater is wasted through inefficient use. Covering 71 percent of Earth's surface, salt water offers wide scope for sustainability metrics. As Antonia Sohns and Larry Crowder note, unsustainable human behaviors of many kinds ultimately leave their mark on the seas—in acidification, rising temperatures, declining oxygen content, the onset of red tides, and the ongoing decline of fisheries. More challenging is the task of connecting each of these trends and others with the metrics of the human activities that lead to them, but that too is part of our task.

On renewable energy, Shakuntala Makhijani and Alexander Ochs approach quantification from a different perspective, measuring the potential to expand access to "sustainable energy" to the point that this all-important sector no longer adds to the atmospheric burden of greenhouse gases. Eric Zencey develops metrics for energy-related principles such as Energy Return on Energy Invested (EROI), which like unforgiving physical laws may limit how much energy humanity can mobilize and for how long. Gary Gardner takes up EROI as well, in addressing quantification of natural resources that perhaps can only be used sustainably with perfect recycling—which of course excludes fossil fuels and other resources consumed entirely by use.

Kate Raworth tackles another kind of sustainability, that of the social sphere. She takes inspiration from the planetary boundaries work to explore metrics that might help us understand when our treatment of our fellow human beings exceeds the bounds of what is needed for long-term societal survival. Social sustainability may be the hardest type to submit to measurement, but without enduring societies, a supportive natural environment will matter to few human beings. The question of how we live together on a crowded planet that unravels even as we work to hold its strands in place may call forth the most important sustainability metric of all.

—*Robert Engelman*

CHAPTER 2

Respecting Planetary Boundaries and Reconnecting to the Biosphere

Carl Folke

The biosphere—the sphere of life—is the living part of the outermost shell of our rocky planet, the part of the Earth's crust, waters, and atmosphere where life dwells. It is the global ecological system integrating all living beings and their relationships. People and societies depend on its functioning and life support while also shaping it globally. Life on Earth interacts with the chemistry of the atmosphere, the circulation of the oceans, the water cycle (including the solid water in polar and permafrost regions), and geological processes to form favorable conditions on Earth.

The issue at stake for humanity with respect to the biosphere is broader than the climate change that is beginning to gain needed attention. It is about a whole spectrum of global environmental changes that interact with interdependent and rapidly globalizing human societies. A key challenge for humanity in this situation is to understand its new role as a dominant force in the operation of the biosphere, start accounting for and governing natural capital (the resources and services derived from and produced by ecosystems), and actively shape societal development in tune with the planet that we are part of. It is time to reconnect to the biosphere.[1]

During the last couple of generations there has been an amazing expansion of human activities into a converging globalized society, enhancing the material standard of living for most people and narrowing many gaps between rich and poor. The expansion, which predominantly benefited the industrialized world, has pushed humanity into a new geological era, the Anthropocene—the age in which human actions are a powerful planetary force shaping the biosphere—and has generated the bulk of the global environmental challenges confronting the future well-being of the human population on Earth.[2]

The Anthropocene is a manifestation of what could be called the Great Acceleration of human activity, in particular since the 1950s. It took humanity close to 200,000 years to reach a population of 1 billion in the early 1800s, and

Carl Folke is a professor at and director of the Beijer Institute of Ecological Economics, Royal Swedish Academy of Sciences, and the founder and science director of the Stockholm Resilience Centre, Stockholm University.

www.sustainabilitypossible.org

now that population is beyond 7 billion. A central factor behind the shift from a human-empty to a human-full world (see Chapter 11) was the discovery of fossil fuels, a major source of additional energy, which allowed humanity to take off into a truly globalized world. It is a remarkable achievement for a single species to become this dominant and, although there are conflicts, still exist in relative peace—with a stunning capacity for ingenuity, innovation, collaboration, and collective action. To a large extent this has been enabled by the human ability to draw on the functioning of the biosphere.[3]

Societies are now interconnected globally not only through political, economic, and technical systems but also through Earth's biophysical life-support systems. The increasingly urbanized global society—cities already accommodate more than 50 percent of the world's population—depends on the capacity of ecosystems of all kinds worldwide to support urban life with such essential ecosystem services as fertile soils, storm protection, and sinks for greenhouse gases and other wastes, even though people may not perceive this support or believe it valuable. For example, shrimp farmed in ponds in Thailand for export to cities in industrial countries are fed with fish meal derived from the harvests of fish in marine ecosystems worldwide. Or consider evolving changes in the variability of rainfall patterns that will likely trigger changes in the frequency, magnitude, and duration of droughts, fires, storms, floods, and other shocks and surprises, affecting food production, trade, migration, and possibly sociopolitical stability. And it has been suggested that the wildfires in Russia in 2010—fueled by record temperatures and a summer drought—burned away much of Russia's wheat harvest and halted exports, contributing to the rising food prices that are seen as one of the triggers of the Arab Spring.[4]

Such novel interactions play out in all corners of the world. Surprises, both positive and negative, are inevitable. And now, new forces are appearing on stage to accelerate the pace. Most of the world's population has started to move decisively out of poverty, leading to the rise of an affluent middle class aiming for material growth, new diets, and increased income. Simultaneously, information technology, nano-technology, and molecular science are accelerating with unknown potentials, while the speed of connectivity and the interactions of globalization create complex new dynamics across sectors, areas, and societies in yet unknown ways.[5]

Increases in connectivity, speed, and scale are by no means only bad news; they may enhance the capacity of societies to adapt and transform with changing circumstances. If globalization operates as if disconnected from the biosphere, however, it may undermine the capacity of the life-supporting ecosystems to sustain such adaptations and provide the essential ecosystem services that human well-being ultimately depends on. Shifting from managing natural resources one by one and treating the environment as an externality to stewardship of interdependent social-ecological systems is a prerequisite for long-term human well-being.[6]

The Human Expansion in a Planetary Context

At the global level there are so-called Earth System services operating on large temporal and spatial scales without the major direct influence of living organisms (unlike ecosystem services). These include the provision of fertile soils through glacial action, the upwelling of ocean circulation that brings nutrients from the deep ocean to support many of the marine ecosystems that provide protein-rich food, and glaciers that act as giant water storage facilities. Storage of carbon through the dissolving of atmospheric carbon dioxide into the ocean is also part of a larger Earth System regulatory service. Others include the chemical reactions in the atmosphere that continually form ozone (essential for filtering out ultraviolet radiation from the sun) and the role of large polar ice sheets in regulating temperature on Earth.[7]

During the last 10,000 years, these and other forces have allowed Earth to provide humanity with favorable environmental conditions and have—until recently—been resilient to human actions. This epoch, the Holocene (see Figure 2–1), has proved to be most accommodating for the development of human civilizations. It has allowed agriculture, villages, and cities to develop and thrive. Before the Holocene period, conditions on Earth were likely too unpredictable, with fluctuating temperatures, for humans to settle down and develop in one place. The much more stable environment of the Holo-

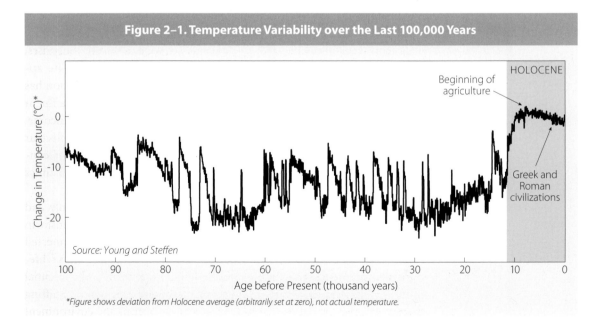

Figure 2–1. Temperature Variability over the Last 100,000 Years

HOLOCENE

Beginning of agriculture

Greek and Roman civilizations

Change in Temperature (°C)*

Source: Young and Steffen

Age before Present (thousand years)

*Figure shows deviation from Holocene average (arbitrarily set at zero), not actual temperature.

cene made it possible for people to invest in the capital of the biosphere and start to domesticate nature. Modern globalized society has developed within these unusually stable conditions, which are generally taken for granted in investment decisions, political actions, and international agreements.[8]

But it seems that humanity is prospering from an exception in the history of Earth and has become critically dependent on the support of the Holocene biosphere's natural capital. For the sake of future human development, it would be helpful if the planet remained in a Holocene-like state. As the Anthropocene unfolds, it is important to understand the envelope of variability that characterizes the Holocene as a baseline to interpret the global changes that are now under way.

The Envelope for Sustainability

The planetary boundaries framework is an approach that sheds light on the significance of the biosphere and how it operates in support of social and economic development. It is an attempt to make visible the biophysical preconditions of a Holocene-like state, the only state that we can be sure provides an accommodating environment for the further development of human societies.[9]

Nine planetary boundaries for critical biophysical processes in the Earth's system have been identified. (See Table 2–1.) Together, they describe an envelope for a safe operating space for humanity that, if respected, would likely ensure that Earth remains in a Holocene-like state. The safe operating space means avoiding moving into a zone of uncertainty where there may be large-scale and critical thresholds. The boundaries are set at the lower level of these zones and illuminate Earth's "rules of the game" for prosperous human development. (See also Chapter 3.) The proposed boundaries are rough first estimates only, marked by large uncertainties and knowledge gaps.[10]

Preliminary analyses have estimated quantitative planetary boundaries for seven of the nine processes or elements: climate change, stratospheric ozone, ocean acidification, the nitrogen and phosphorus cycles, biodiversity loss, land use change, and freshwater use. For some of these, this was the first attempt at quantifying boundaries of any kind. There was insufficient knowledge to propose quantitative boundaries for aerosol loading and chemical pollution. Three of the boundaries may already have been transgressed: those for climate change, changes of the global nitrogen cycle, and the rate of biodiversity loss.

The boundary estimates are based on an effort to synthesize current scientific understanding. They and the scientific analyses behind them were presented and discussed in two papers by Johan Rockström and colleagues in 2009. The following brief summary of the boundaries is derived from that work.[11]

	Table 2–1. The Nine Planetary Boundaries*			
Earth System Process	Parameters	Proposed Boundary	Current Status	Pre-industrial Value
Climate change	(i) Atmospheric carbon dioxide concentration (parts per million by volume)	350	387	280
	(ii) Change in radiative forcing (watts per meter squared)	1	1.5	0
Rate of biodiversity loss	Extinction rate (number of species per million species per year)	10	>100	0.1–1
Nitrogen cycle (part of a boundary with the phosphorus cycle)	Amount of N_2 removed from the atmosphere for human use (millions of tons per year)	35	121	0
Phosphorus cycle (part of a boundary with the nitrogen cycle)	Quantity of P flowing into the oceans (millions of tons per year)	11	8.5–9.5	–1
Stratospheric ozone depletion	Concentration of ozone (Dobson unit)	276	283	290
Ocean acidification	Global mean saturation state of aragonite in surface seawater	2.75	2.90	3.44
Global freshwater use	Consumption of freshwater by humans (km^3 per year)	4,000	2,600	415
Change in land use	Percentage of global land cover converted to cropland	15	11.7	low
Atmospheric aerosol loading	Overall particulate concentration in the atmosphere, on a regional basis	To be determined		
Chemical pollution	For example, amount emitted to, or concentration in, the global environment of persistent organic pollutants, plastics, endocrine disruptors, heavy metals, and nuclear waste, or their effects on the functioning of ecosystems and the Earth System	To be determined		

Boundaries of processes in gray have been crossed.
Source: See endnote 10.

Climate Change. The suggested climate change boundary of 350 parts per million of carbon dioxide in the atmosphere aims at minimizing the risk of getting into zones of uncertainty and crossing thresholds that could lead to major changes in regional climates, alter climate-dynamics patterns such as the oceanic thermohaline circulation, or cause rapid sea level rise. Current observations of a possible climate transition include the retreat of summer sea ice in the Arctic Ocean, retreat of mountain glaciers around the world, loss of mass from the Greenland and West Antarctic ice sheets, and weakening of the oceanic carbon sink.

Biological Diversity. Biological diversity plays a significant role in ecosystem dynamics and functioning and in sustaining a flow of critical ecosystem services. The planetary boundaries work used species extinction rates as a first proxy of diversity loss. Accelerated species loss is likely to compromise the biotic capacity of ecosystems to sustain their current functioning under novel environmental and biotic circumstances. Since the advent of the Anthropocene, humans have increased the rate of species extinction by 100–1,000 times the background rates that were typical over Earth's history. The biodiversity boundary, still under considerable debate, was suggested at 10 extinctions per million species per year. This boundary of biodiversity loss is currently exceeded by two orders of magnitude or more.

Nitrogen and Phosphorus. Nitrogen and phosphorus are critical nutrients for life and are instrumental in enhancing food production through fertilization, but their use also has impacts on forests and landscapes and leads to pollution of waterways and coastal zones. Human activities now convert more nitrogen from the atmosphere into reactive forms than all of Earth's terrestrial processes combined. The nitrogen boundary is tentatively set at 35 million tons of industrially and agriculturally fixed reactive nitrogen per year flowing into the biosphere, which is 25 percent of the total amount now fixed naturally by terrestrial eco-

Eutrophication under way with algal growth in a pond in Lille, France.

systems. This is a first guess only, and new estimates are needed to enable a more informed boundary.

Phosphorus is mined for human use and also added through weathering processes. Inflow of phosphorus to the oceans has been suggested as a key driver behind global-scale ocean anoxic events (depletion of oxygen below the surface). The phosphorus boundary was proposed not to exceed approximately 10 times the natural background rate for human-derived phosphorus inflow to the ocean. New estimates of the phosphorus boundary that incorporate estimates for both freshwater eutrophication and phosphorus flows to the sea conclude that current conditions exceed a proposed planetary boundary for phosphorus in relation to global freshwater eutrophication.[12]

Stratospheric Ozone. Stratospheric ozone filters ultraviolet radiation from the sun and thereby protects humans and other organisms. The suggested ozone boundary is set at a decrease of less than 5 percent in column ozone levels for any particular latitude compared with 1964–80 values. Fortunately, because of the actions taken as a result of the Montreal Protocol and its subsequent amendments, humanity appears to be on a path that avoids exceeding this boundary.

Ocean Acidification. Addition of carbon dioxide to the oceans increases the acidity (lowers the pH) of the surface seawater. The current rate of ocean acidification is much higher than at any other time in the last 20 million years. Many marine organisms are acidity-sensitive, especially those that use calcium carbonate dissolved in the seawater to form shells or skeletal structures (such as corals and marine plankton). Globally, the surface ocean saturation of the aragonite form of carbonate is declining with rising ocean acidity. To avoid possible thresholds, the suggested oceanic acidification boundary is to maintain aragonite saturation in surface waters at a minimum of 80 percent of the average global pre-industrial level.

Global Freshwater Use. Humans alter river flows and the spatial patterns and seasonal timing of other freshwater flows all over the globe. A planetary boundary for freshwater resources needs to secure water flows to regenerate precipitation, support terrestrial ecosystem functioning and services (such as carbon sequestration, biomass growth, food production, and biological diversity), and also ensure the availability of water for aquatic ecosystems. Transgressing a freshwater boundary of roughly 4,000 cubic kilometers per year of consumptive use of runoff may push humanity toward water-induced thresholds at regional to continental scales. Currently, consumptive use is about 2,600 cubic kilometers per year.

Land Use Changes. Land use change, driven primarily by agricultural expansion and intensification, contributes to global environmental change. It is proposed that the boundary for change be set at no more than 15 percent of the global ice-free land surface converted to cropland. Currently that share is about 12 percent. The suggested allowance for expanding agricultural land by three percentage points will likely be used up over the coming decades and includes suitable land that is not currently cultivated or is under forest cover, such as abandoned cropland in Europe, North America, and the former Soviet Union as well as some areas of Africa's savannas and South America's cerrado.

Atmospheric Aerosol Loading. Aerosol loading adds particulates such as dust, soot, and liquid droplets to the atmosphere, and on a regional basis it disrupts monsoon systems and has human health effects. Global threshold behavior is still poorly understood, and no aerosol boundary is yet suggested.

Chemical Pollution. Chemical pollution includes radioactive compounds, heavy metals, and a wide range of organic compounds of human origin that adversely affect human and ecosystem health and are now present in the environment all over the planet. Potential thresholds are largely unknown, and although there is ample scientific evidence on individual chemicals, there is lack of aggregate, global-level analysis, so it is too early to suggest a chemical pollution boundary.

Interdependent Boundaries. Transgressing one or more planetary boundaries may have serious consequences for human well-being due to the risk of crossing thresholds that can trigger non-linear, abrupt environmental change within continental- to planetary-scale systems. Planetary boundaries are interdependent, because crossing one of them may shift the position of other boundaries or cause them to be transgressed. Such interactions between the boundaries are not accounted for in the current estimates. Moreover, the existence of these thresholds in key Earth System processes is independent of peoples' preferences and values or of compromises based on political and socioeconomic feasibility. How far we are willing to move into the uncertainty zones and risk crossing critical thresholds is a reflection of worldviews, choices, and actions—hence the urgent need to reconnect human actions to the biosphere.[13]

Innovation and Transformation for Global Resilience

Humans have changed the way the world works, and now we must change the way we think about it too. Society must seriously consider new ways to support Earth System resilience and explore options for the deliberate transformation of unsustainable trends and practices that undermine it. The future is uncertain, with surprises and shocks in store—and also opportunities. Incremental tweaking is not likely to be sufficient for the new Anthropocene era to remain in a state as favorable for humans as the Holocene. Preventing dangerous transitions at the regional and global levels will require innovation and novelty. It is increasingly clear that development goals and efforts need to relate to the safe operating spaces and create opportunities for prosperous societal development within those dynamic limits.[14]

Large-scale developments in information technology, nano- and biotechnology, and new energy systems have the potential to significantly improve our lives. But if, in framing them, society fails to consider the adaptive capacity of the biosphere and the safe operating spaces for humanity, there is a risk that unsustainable development may be reinforced by technological innovations and policies that are successful in the short term.

Can we innovate sufficiently rapidly and with sufficient intelligence to steer our system out of a destructive pathway and onto one that leads to long-term social and ecological resilience? Whatever forms a transition to sustainability might take, it implies finding the institutional frameworks to stimulate the kinds of innovation that solve rather than aggravate our environmental challenges.[15]

The environment has for too long been looked on as an externality for economic progress—a handy and limitless stock of resources for human economic exploitation. Many even continue to view it as a sector of society rather than the other way around and are truly ignorant about its dynamics and significance.

But it has become crystal clear that people and societies are integral components of the biosphere, depending on the functioning and services of life-supporting ecosystems. It is urgent to start accounting for and governing natural capital and ecosystem services, not just for saving the environment but for the sake of our own development. The question is about responsibility—whether humanity has the understanding, wisdom, and maturity as a species to become wise stewards of the living planet, instead of treating it as an inexhaustible collection of raw materials.

At the core of the global sustainability challenge is extending the period of relative stability of the last 10,000 years that has allowed our species to flourish and create civilizations. It represents a globally desirable social-ecological state. A significant part of this challenge is to make the work of the biosphere visible in the minds of people, in financial and economic transactions and in society as a whole.

In a globalized society, there are no ecosystems without people and no people who do not depend on ecosystem functioning. They are inextricably intertwined. Ecosystem services therefore are not really generated by nature but by social-ecological systems. Social-ecological systems are dynamic and connected, from the local to the global, in complex webs of interactions subject to both gradual and abrupt changes. Dynamic and complex social-ecological systems require strategies that build resilience rather than attempt to control for optimal production and short-term gain in environments assumed to be relatively stable.

The planetary boundaries approach sheds light on the crucial significance of a functioning Earth and its biosphere for human well-being. It inspires stewardship of our critical natural capital at all levels. The shift from perceiving people and nature as separate actors to seeing them as interdependent social-ecological systems creates exciting opportunities for societal development in tune with the biosphere: a global sustainability agenda for humanity.

CHAPTER 3

Defining a Safe and Just Space for Humanity

Kate Raworth

Every pilot knows the importance of flying with a compass: without one, they would be in danger of straying far from course. No wonder that modern airplane cockpits are equipped with an array of dials and indicators—from compass and fuel gauge to altimeter and speedometer. Pity, then, that economic policymakers have used nothing close to that for charting the course of the whole economy.

The excessive attention given to gross domestic product (GDP) in recent decades as an indicator of a nation's economic performance is like trying to fly a plane by its altimeter alone: it tells you if you are going up or down, but nothing of where you are headed or how much fuel you have left in the tank. Such a focus on monetized economic output has failed to reflect the growing degradation of natural resources, the invaluable but unpaid work of carers and volunteers, and the inequalities of income that leave people in every society facing poverty and social exclusion. GDP's dominance has long passed its legitimacy: it is clearly time to create a better dashboard for navigating the twenty-first century's journey toward equity and sustainability. The good news is that better metrics are on the way.

In 2009, Nobel prize–winning economists Joseph Stiglitz and Amartya Sen led a commission of economic thinkers to reassess how best to measure economic performance and social progress. They concluded, "We are almost blind when the metrics on which action is based are ill-designed or when they are not well understood. For many purposes, we need better metrics. Fortunately, research in recent years has enabled us to improve our metrics, and it is time to incorporate in our measurement system some of these advances."[1]

Metrics for assessing environmental sustainability are under development—from calculating ecological footprints (see Chapter 4) to quantifying natural capital. But a new measurement framework that focused only on bringing environmental sustainability into the picture would fail to reflect social outcomes and would overlook the equity implications of pursuing

Kate Raworth is a senior researcher at Oxfam and teaches at Oxford University's Environmental Change Institute. This chapter is written in her personal capacity. Lisa Dittmar provided research assistance.

www.sustainabilitypossible.org

sustainability. For where there is a limit on resource availability, there is always a question of how those limited resources are to be distributed and used. If that question is left unspoken, it can lead to political stalemate, injustice, and suffering. So in any discussion of what it will take to achieve global environmental sustainability, it is crucial to bring the issue of international social justice in resource distribution explicitly into the framework, including into the metrics to be used. The concept of *planetary boundaries* offers a powerful starting point for doing just that.

Between Social Boundaries and Planetary Boundaries

In 2009, a group of leading Earth-system scientists brought together by Johan Rockström of the Stockholm Resilience Centre put forward the concept of planetary boundaries. (See Chapter 2.) They proposed a set of nine interrelated Earth System processes—such as climate regulation, the freshwater cycle, and the nitrogen cycle—that are critical for keeping the planet in the relatively stable state known as the Holocene, a state that has been so beneficial to humanity over the past 10,000 years. Under too much pressure from human activity, these processes could be pushed over biophysical thresholds—some on global scales, others on regional scales—into abrupt and even irreversible change, dangerously undermining the natural resource base on which humanity depends for well-being. To avoid this, the scientists made a first proposition of a set of boundaries below these danger zones, such as a boundary of 350 parts per million of carbon dioxide (CO_2) in the atmosphere to prevent dangerous climate change.[2]

Together the nine boundaries can be depicted as forming a circle, and Rockström's group called the area within it "a safe operating space for humanity." Their first estimates indicated that at least three of the nine boundaries have already been crossed—for climate change, the nitrogen cycle, and biodiversity loss—and that resource pressures are moving rapidly toward the estimated global boundary for several others too.[3]

The concept of nine planetary boundaries powerfully communicates complex scientific issues to a broad audience, and it challenges traditional understandings of economy and environment. While mainstream economics treats environmental degradation as an "externality" that largely falls outside of the monetized economy, natural scientists have effectively turned that approach on its head and proposed a quantified set of resource-use boundaries within which the global economy should operate if we are to avoid critical Earth System tipping points. These boundaries are described not in monetary metrics but in natural metrics fundamental to ensuring the planet's resilience for remaining in a Holocene-like state.

Further work is needed—and is under way—to refine the planetary boundaries approach, both in terms of clarifying the different scales (from local to global) of the critical biophysical thresholds and in terms of understanding their dynamic interactions. Yet even while the nuances of defining the nature and scale of boundaries are being debated, a critical part of the picture is still missing.[4]

Yes, human well-being depends on keeping total resource use below critical natural thresholds, but it equally depends upon every person having a claim on the resources they need to lead a life of dignity and opportunity. International human rights norms have long asserted the fundamental moral claim each person has to life's essentials—such as food, water, basic health care, education, freedom of expression, political participation, and personal security—no matter how much or how little money or power they have. Just as there is an outer boundary of resource use, an "environmental ceiling" beyond which lies unacceptable environmental degradation, so too there is an inner boundary of resource use, a "social foundation" below which lies unacceptable human deprivation.

Of course, a social foundation of this kind provides only for the minimum of every human's needs. But given the current extent of poverty and extreme inequality in the world, ensuring that this social foundation of human rights is achieved for all must be a first focus.

Since 2000, the Millennium Development Goals (MDGs) have provided an important international focus for social priorities in development and have addressed many deprivations—of income, nutrition, gender equality, health, education, and water and sanitation—whose urgency has not receded. The emerging international debate about what should follow the MDGs after 2015, and simultaneously what should underpin a set of Sustainable Development Goals, is bringing attention to additional social concerns such as resilience, access to energy, and social equity.

These major initiatives to generate a new set of global development goals could result in an international consensus about priority social issues to be tackled in coming decades, effectively setting an internationally agreed-upon social foundation. In advance of such agreement, one indication of shared international concerns comes from the social priorities most raised by governments in the run-up to the Rio+20 Conference, as set out in their national and regional submissions before the meeting. Analysis of these submissions reveals that 11 social priorities were raised in over half of them: deprivations in food, water, health care, income, education, energy, jobs, voice, gender equality, social equity, and resilience to shocks. These 11 are taken here as an illustrative social foundation.[5]

Between the social foundation of human rights and the environmental ceiling of planetary boundaries lies a space—shaped like a doughnut—that is both an environmentally safe and a socially just space for humanity. (See Figure 3–1.)[6]

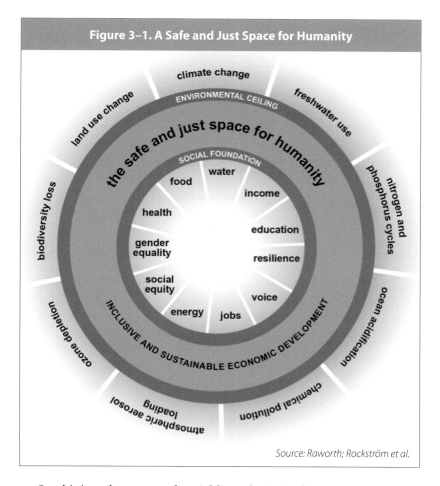

Figure 3–1. A Safe and Just Space for Humanity

climate change

ENVIRONMENTAL CEILING

freshwater use

land use change

the safe and just space for humanity

SOCIAL FOUNDATION

water

food

income

health

education

gender equality

resilience

social equity

voice

energy   jobs

biodiversity loss

nitrogen and phosphorus cycles

INCLUSIVE AND SUSTAINABLE ECONOMIC DEVELOPMENT

ozone depletion

ocean acidification

atmospheric aerosol loading

chemical pollution

Source: Raworth; Rockström et al.

Combining planetary and social boundaries in this way creates a new perspective on sustainable development. Human-rights advocates have long highlighted the imperative of ensuring every person's claim to life's essentials, while ecological economists have emphasized the need to situate the global economy within environmental limits. This framework brings the two together, creating a space that is bounded by both human rights and environmental sustainability, while acknowledging that there are many complex and dynamic interactions across and between the multiple boundaries.[7]

Just as Rockström and the other scientists in 2009 estimated that humanity has already transgressed at least three planetary boundaries, so too it is possible to quantify human outcomes against the social foundation. A first assessment, based on international data, indicates that humanity is falling far below the social foundation on eight dimensions for which comparable indicators are available. Around 13 percent of the world's population is undernourished, for example, 19 percent of people have no access to electricity, and 21 percent live in extreme income poverty. (See Table 3–1.)[8]

Quantifying social boundaries alongside planetary boundaries in this way makes plain humanity's extraordinary situation. (See Figure 3–2.) Many millions of people still live in appalling deprivation, far below the social foundation. Yet collectively humanity has already transgressed several of the planetary boundaries. This is a powerful indication of just how deeply unequal and unsustainable the path of global development has been to date.[9]

Table 3–1. How Far Below the Social Foundation Is Humanity?			
Social Foundation	Illustrative Indicators of Global Deprivation	Share of Population	Year
		(percent)	
Food security	Population undernourished	13	2010–12
Income	Population living below $1.25 (purchasing power parity) per day	21	2005
Water and sanitation	Population without access to an improved drinking water source	13	2008
	Population without access to improved sanitation	39	2008
Health care	Population without regular access to essential medicines	30	2004
Education	Children not enrolled in primary school	10	2009
	Illiteracy among 15–24 year olds	11	2009
Energy	Population lacking access to electricity	19	2009
	Population lacking access to clean cooking facilities	39	2009
Gender equality	Employment gap between women and men in waged work (excluding agriculture)	34	2009
	Representation gap between women and men in national parliaments	77	2011
Social equity	Population living in countries with significant income inequality	33	1995–2009
Voice	Population living in countries perceived (in surveys) not to permit political participation or freedom of expression	To be determined	
Jobs	Labor force not employed in decent work	To be determined	
Resilience	Population facing multiple dimensions of poverty	To be determined	

Source: See endnote 8.

Figure 3–2. Falling Far Below the Social Foundation While Exceeding Planetary Boundaries

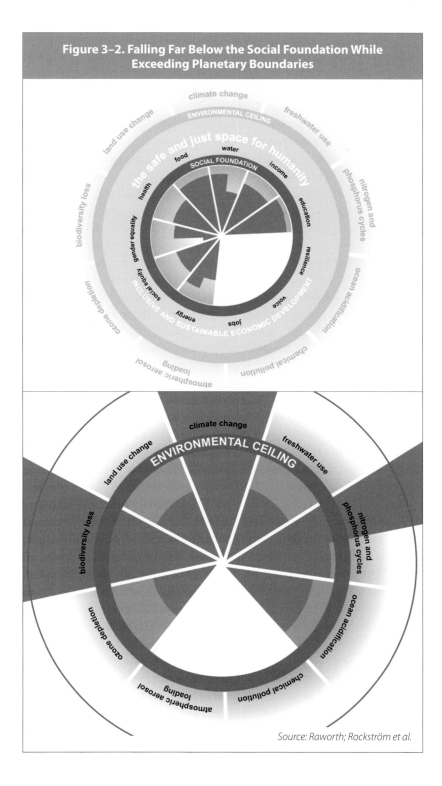

Source: Raworth; Rockström et al.

Dynamics and Distribution between the Boundaries

One striking implication from this initial attempt to quantify both social and planetary boundaries is that ending poverty for all 7 billion people alive today need not be a source of significant stress on planetary boundaries. According to data from the U.N. Food and Agriculture Organization, providing the additional calories needed by the 13 percent of the world who are facing hunger would require just 3 percent of the current global food supply. Consider that against the fact that around 30 percent of the world's food supply is lost in post-harvest processing, wasted in retail supply chains, or thrown away by consumers. Likewise, according to the International Energy Agency, bringing electricity to the 19 percent of the world who currently lack it could be achieved, using a mix of technologies, for as little as a 1 percent increase in global CO_2 emissions—making it clear that tackling climate change and ending energy poverty are essentially distinct challenges. And according to researchers at the Brookings Institute, ending extreme income poverty for the 21 percent of people who live on less than $1.25 a day would require just 0.2 percent of current global income.[10]

What, then, is the biggest source of stress on planetary boundaries today? It is the excessive consumption levels of roughly the wealthiest 10 percent of people in the world and the resource-intensive production patterns of companies producing the goods and services that they buy. The richest 10 percent of people in the world hold 57 percent of global income. Just 11 percent of the global population generates about half of global CO_2 emissions. And one third of the world's "sustainable budget" for reactive nitrogen use is used to produce meat for people in the European Union, just 7 percent of the world's population.[11]

Cutting the resource intensity of the most affluent lifestyles is essential for both equity of and sustainability in global resource use. The global middle class is projected to grow from 2 billion today to nearly 5 billion by 2030, with global demand for water expected to rise by 30 percent, and demand for food and energy each by 50 percent. Families moving into the lower end of the global middle class (spending around $10 per person a day) will be able to afford meat in their diets, electric power at home, and the use of public or private motor transport. As a result, lifelong prospects for many of these families will be transformed. Production patterns that are far more resource-efficient—including resource-saving technologies, investments, and infrastructure in key sectors—are essential to make this possible.[12]

As other families move up to the higher-income end of the global middle class, however, spending $50–100 per person a day, their expectations, aspirations, and hence resource use will be strongly influenced by the consumption and production patterns underpinning the lifestyles of today's most affluent consumers. Achieving more-equitable and more-efficient resource use within and between countries and transforming today's resource-intensive lifestyles will clearly be crucial if humanity is to move onto development pathways that operate in the space between social and planetary boundaries.

Creating Metrics for a New Economic Dashboard

There is wide agreement that it is time to get beyond GDP and toward a far richer conception of what constitutes economic development. The global crises of environmental degradation and extreme human deprivation, coupled with the projected growth of the global middle class, urgently demand a better tool kit for economic policymaking.

What are the implications, then, of this framework of social and planetary boundaries for rethinking the metrics needed to govern economies? The overriding aim of global economic development must surely be to enable humanity to thrive in the safe and just space, ending human deprivation while keeping within safe boundaries of natural resource use locally, regionally, and globally. Traditional economic growth policies have largely failed to deliver on both accounts: far too few benefits of economic growth have gone to people living in poverty, and far too much of GDP's rise has been at the cost of degrading natural resources. And the focus on monetized exchange in the economy overlooks the enormous value for human well-being of unpaid work in terms of both caring for and nurturing others and stewarding natural resources.

Imagine if the doughnut-shaped diagram of social and planetary boundaries found its way onto the opening page of every macroeconomics textbook. So you want to be an economist? Then first, there are a few facts you should know about this planet, how it sustains us, how it responds to excessive pressure from human activity, and how that undermines our own well-being. You should also know about the human rights of its people and about the human, social, and natural resources that it will take to fulfill those. With these fundamental concepts of planetary and social boundaries in place, your task as an economist is clear and crucial: to design economic policies and regulations that help bring humanity into the safe and just space between the boundaries and that enable us all to thrive there.

Of course, redefining the economist's mandate cannot get us there alone. We also need deeper knowledge of Earth System processes at multiple scales and far wider use of resource-efficient technologies and techniques. We need breakthroughs in understanding consumer psychology, in promoting empathy and long-term decisionmaking, and in governing for collective interests. But given that economics is the dominant language and currency of policymaking, we stand little chance of getting there without having that discipline on our side.

Under this framing of what successful economic policymaking looks like, the metrics for assessing the journey toward sustainable and equitable development must widen significantly. In line with the recommendations of the Commission on the Measurement of Economic Performance and Social Progress, at least four broad shifts are needed—and are under way (see Box 3–1)—for creating a better dashboard of economic and social progress.[13]

The first shift is from measuring just what is sold to what is provided for free too. Many of the goods and services that are essential for well-being are provided for free—by parents, by volunteers, and by nature—and have significant value. One 2003 study of the unpaid care economy in Basel, Switzerland, found that the imputed value of housework, unpaid care, and volunteer services was 50 percent greater than the city's public spending on hospitals and schools. Likewise, a recent U.S. study found that accounting for unpaid household production, such as housework, child care, and cooking, effectively increased the country's GDP by 26 percent in 2010.[14]

Assessments of the contribution made by unpriced ecosystem functions are also under way. The United Kingdom's National Ecosystem Assessment in 2011 found that 30 percent of the country's ecosystems were in decline but that ecosystem functioning—such as inland wetlands and pollination by bees—was of high economic value to the economy. Measures such as these that better reflect the value of the unpaid care economy and unpriced ecosystem functions are essential for broadening concepts of what contributes to economic and social development.[15]

Second, we need to shift from a focus on the flow of goods and services to monitoring changes in underlying stocks as well. The flow of goods and services is only half the economic story, as any company knows. Indeed, companies that only published their profit and loss accounts would be laughed off the stock exchange. It is also critical to know what is happening to a company's assets and liabilities. And nations should be held to the same standard.

The physical and financial assets of countries have been measured for some time, but attention is now turning to better accounting of every nation's fundamental wealth: its natural, human, and social assets. Creating metrics that help to assess, value, restore, and expand these assets is at the heart of creating long-term prosperity. The Inclusive Wealth Index (IWI) prepared by the United Nations sets out to do just that, assessing changes in countries' manufactured, human, and natural capital stocks—with the initial finding that 6 out of 20 countries assessed have seen their IWI per capita fall since 1990.[16]

The third shift needed is from a focus on aggregates and averages to monitoring distribution too. Many economic indicators are either aggregates (national GDP, for example) or averages (GDP per capita). But it is the

Box 3–1. Moving Beyond GDP

Beginning in the early 1970s, and initially focusing on the problem of pollution costs and other environmental externalities, economists have been working to develop alternatives to GDP that better capture the full scope of our economy. These include the Measure of Economic Welfare developed by William Nordhaus and James Tobin and a later, better-known derivation, the Genuine Progress Indicator.

More recently, and particularly in the wake of the recession, interest among policymakers has surged and we are now in the early phases of major implementation efforts in multilateral institutions and government. The Beyond GDP movement has entered a new phase, toward the goal of widespread implementation of alternative measurement frameworks in national accounting systems, other levels of governance, and concrete policy settings. Identifiable, large-scale impacts on policy and social outcomes, however, remain a good way off in the face of many technical, institutional, and political challenges.

One major stepping-stone was France's high-profile Commission on the Measurement of Economic Performance and Social Progress. With the widely noted release of its groundbreaking report in 2009, the Commission set a high bar for national implementation of comprehensive accounting reforms, incorporating principles of equity, quality of life, and sustainability. Other important institutional developments include a 2011 U.N. resolution calling for member states to reform national accounting systems based on the principles of well-being and sustainability. Led by Bhutan, the resolution was affirmed by more than 60 countries, including most of Europe as well as India and Brazil.

Government efforts to implement alternative indicators are multiplying. The World Bank's WAVES partnership—Wealth Accounting and the Valuation of Ecosystem Services—is currently developing implementation plans for environmental accounting in Botswana, Colombia, Costa Rica, Madagascar, and the Philippines. Twenty-four countries, mostly in the developing world, are engaged in some form of environmental accounting, particularly around resource management, according to a recent World Bank study.

Industrial countries are also moving forward in certain areas. The United Kingdom has adopted "happiness accounting," incorporating measures of subjective well-being into its national accounts, and Australia and Canada are developing alternative dashboards of well-being indicators. There is also progress in the United States, including high-level federal research programs on nonmarket accounting and happiness measures, a programmatic blueprint for GDP and Beyond measures issued by the Bureau of Economic Analysis in the Department of Commerce, and adoption of the Genuine Progress Indicator in the states of Maryland and Vermont.

—*Lew Daly*
Director, Sustainable Progress Initiative, Demos
Source: See endnote 13.

---

actual distribution of incomes, wealth, and outcomes across a society that determines how inclusive its path of development is. In 17 out of 22 countries in the Organisation for Economic Co-operation and Development (OECD), income inequality has risen since 1985. In OECD countries today, the richest 10 percent of people have, on average, nine times the income of the poorest 10 percent.[17]

Just as there are striking inequalities of income, there are striking inequalities of resource use as well. In the United Kingdom, the richest 10 percent of people produce twice the carbon emissions of the poorest 10 percent; in Sweden, it's four times as much; in China, 18 times as much. Data on income distribution and resource use also need to be disaggregated by sex and by ethnicity in order to ensure that economic policies and their social outcomes are equitable.[18]

The final shift to create a better dashboard of economic and social progress is from monetary metrics to natural and social metrics too. Not everything that matters can be monetized, nor should it be. "Social metrics," such as the number of hours of unpaid caring work provided by women and by men, and "natural metrics," such as per capita footprint calculations for carbon, water, nitrogen, and land, must be given more visibility and weight in policy assessments.

Natural metrics such as these are relatively new but fast improving. More and better data of this kind are essential, most urgently in high-income and resource-intensive countries, for assessing whether a nation's GDP growth is being decoupled from natural resource use—and not just in relative terms (with GDP rising faster than resource use) but in absolute terms (with GDP rising while total resource use falls), since this reveals whether or not "green growth" is taking place and, ultimately, whether it is possible.

What difference are these four shifts making? Gone are the days of GDP as the lone altimeter guiding the economic journey. The interest and progress in creating new metrics is starting to generate a dashboard of indicators that places the monetized economy in a much broader context of what constitutes, and contributes to, equitable and sustainable development. For sure, the direction of GDP still matters—indeed, its growth is absolutely crucial in low-income countries—but it matters alongside other important dimensions of development.

This creation of metrics beyond GDP is crucial, but of course it brings new complexities and controversies. There is an ongoing dance (or a battle) back and forth between the metrics of economics and ecology to determine whose language, concepts, and measurements will define the emerging paradigm of development. Will economics subsume ecology, assigning a monetary value to all natural resources, complete with assumptions of shadow prices, substitutability, and market exchange? Will ecology predominate, proscribing a space for economic activity within safe boundaries designed to avoid critical natural thresholds, expressed and governed only through the evolving natural metrics of the planet? Or will it be possible to create a dashboard of indicators that incorporates the realities and insights brought by both approaches?

If such holistic metrics can be created, they must be compiled and reported in ways that empower people around the world to hold policymakers to account. This change alone would provide governments, civil society, citizens, and companies alike with a far better dashboard for navigating humanity into a safe and just space in which we all can thrive.

CHAPTER 4

Getting to One-Planet Living

Jennie Moore and William E. Rees

In *Collapse: How Societies Choose to Fail or Succeed,* Jared Diamond asks the obvious question of a forest-dependent society: "What was the Easter Islander who cut down the last tree thinking?" For those familiar with the human tendency to habituate to virtually any conditions, the answer might very well be "nothing much." The individual who cut down Easter Island's last significant tree probably did not noticeably alter a familiar landscape. True, that person was likely standing in a scrubby woodland with vastly diminished biodiversity compared with the dense forest of earlier generations. Nevertheless, the incremental encroachments that eventually precipitated the collapse of Easter Island society were likely insufficient in the course of any one islander's life to raise general alarm. Some of the tribal elders might have worried about the shrinking forest, but there is no evidence that they did—or could have done—much to reverse the inexorable decline of the island's ecosystem.[1]

Too bad. With the felling of the last "old-growth" trees on the island, the forest passed a no-return threshold beyond which collapse of the entire socio-ecosystem was inevitable. No doubt several factors contributed to this tragic implosion—perhaps a combination of natural stresses coupled with rat predation of palm nuts, human "predation" of adult trees, overpopulation of both rats and humans, the misallocation of resources to an intertribal competition to construct ever bigger *moai* (the famous sacred monolithic stone heads), or perhaps even some tribal invincibility myth. But there is little doubt that human overexploitation of the limited resources of a finite island was a major driver. The wiser members of the community probably saw what was coming. In slightly different circumstances the islanders could conceivably have responded to reverse the decline, but in the end Easter Island society was unable to organize effectively to save itself.

Fast forward. We might well ask ourselves what the Canadian govern-

Jennie Moore is the director of sustainable development and environmental stewardship in the School of Construction and the Environment at the British Columbia Institute of Technology. **William E. Rees** is Professor Emeritus in the School of Community and Regional Planning at the University of British Columbia.

www.sustainabilitypossible.org

ment was thinking in the early 1990s when it ignored scientists' warnings and a well-documented 30-year decline in spawning stock biomass and allowed commercial fishers to drive the Atlantic Cod stock to collapse. What are North Americans thinking today as they strip the boreal forest to get at tar-sands crude or jeopardize already shrinking water supplies by "fracking" oil-shales for natural gas and petroleum, even as burning the stuff threatens to push the global climate system over the brink? And what are Brazilians, Congolese, Malaysians, and Indonesians thinking as they harvest the world's great rainforests for short-term economic gain (through rare tropical hardwoods, cattle farms, soy production, or oil-palm plantations, for instance)?

Certainly the governments and corporate leaders of these nations know that their actions are destroying the world's greatest deposits of biodiversity, increasing the atmosphere's carbon burden, and accelerating long-term climate change. Nevertheless, as the U.N. Department of Economic and Social Affairs notes, because "so many of the components of existing economic systems are 'locked into' the use of non-green and non-sustainable technologies, much is at stake in terms of the high cost of moving out of those technologies." Result? A world in policy paralysis. [2]

System collapse is a complicated process. Ecosystem thresholds are not marked with signs warning of impending danger. We may actually pass through a tipping point unaware because nothing much happens at first. However, positive feedback ensures that accelerating changes in key variables eventually trigger a chain reaction: critical functions fail and the system can implode like a house of cards. Complexity theory and ecosystems dynamics warn of the dangers of overexploitation and explain observed cycles of climax and collapse. Yet the world community is in effect running a massive unplanned experiment on the only planet we have to see how far we can push the ecosphere before it "flips" into an alternative stability domain that may not be amenable to human civilization. Examples of inexorable trends include the loss of topsoil, atmospheric greenhouse gas accumulation, acidification of oceans with negative impacts on fisheries, coastal erosion, and the flooding of cities. [3]

We can illustrate the human pressure on nature using Ecological Footprint accounting. (See Box 4–1.) Ecological Footprints estimate the productive ecosystem area required, on a continuous basis, by any specified population to produce the renewable resources it consumes and to assimilate its (mostly carbon) wastes. There are only 11.9 billion hectares of productive ecosystem area on the planet. If this area were distributed equally among the 7 billion people on Earth today, each person would be allocated just 1.7 global hectares (gha) per capita. (A global hectare represents a hectare of global average biological productivity.) [4]

Box 4–1. What Is the Ecological Footprint?

The Ecological Footprint compares a population's demand on productive ecosystems—its footprint—with biocapacity, the ability of those ecosystems to keep up with this demand. The Global Footprint Network's National Footprint Accounts tracks the footprints of countries by measuring the area of cropland, grazing land, forest, and fisheries required to produce the food, fiber, and timber resources being consumed and to absorb the carbon dioxide (CO_2) waste emitted when burning fossil fuels. When humanity's Ecological Footprint exceeds the planet's biocapacity, harvests are exceeding yields, causing a depletion of existing stocks or the accumulation of carbon dioxide in the atmosphere and oceans. Such overuse potentially damages ecosystems' regenerative capacity. Locally, demand can exceed biocapacity without depletion if resources can be imported.

In 1961, humanity's Ecological Footprint was at about two thirds of global biocapacity; today humanity is in ecological overshoot—requiring the equivalent of 1.5 planets to provide the renewable resources we use and to absorb our carbon waste. Local overshoot has occurred all through history, but global overshoot only began in the mid-1970s. Overshoot cannot continue indefinitely; ultimately, productive ecosystems will become depleted. Global productivity is further at risk because of potential climate change, ocean acidification, and other consequences of the buildup of CO_2 in the biosphere.

Most nations demand more biocapacity than they have available within their own borders. This means they are liquidating their national ecological wealth, relying through trade on the biocapacity of others, or using the global commons as a carbon sink. This increases the risk of volatile costs or supply disruptions. For example, the Mediterranean region has a rapidly widening ecological deficit: in less than 50 years, demand for ecological resources and services has nearly tripled, expanding its ecological deficit by 230 percent. But it is not just high-income countries where Ecological Footprints exceed biocapacity. The Philippines has been in ecological deficit since the 1960s. In 2008, people there demanded from nature twice the country's capacity to provide biological resources and sequester carbon emissions.

The United Arab Emirates, Qatar, Kuwait, Denmark, and the United States have the largest per capita footprints among countries with populations over 1 million. If everybody consumed like residents of these countries, we would need more than four Earths. Other nations, such as China, have lower per capita footprints but are rapidly pursuing consumption habits that are trending in the direction of high-income, high-footprint nations. And although China's footprint per person is low, we would still need slightly more than one Earth if everyone in the world consumed at that level. Despite relatively small per capita Ecological Footprints, countries with large populations, like India and China, have significant biocapacity deficits and large total Ecological Footprints, similar to that of the United States.

—*Global Footprint Network*
Source: See endnote 4.

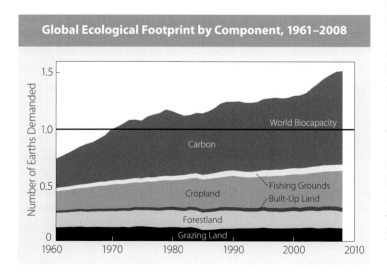

Global Ecological Footprint by Component, 1961–2008

World Biocapacity

Carbon

Fishing Grounds

Built-Up Land

Cropland

Forestland

Grazing Land

Comparing Fair Earth-Share and High-Consumption Societies

Ecological Footprint studies reveal that the world is in ecological overshoot by as much as 50 percent. The growth of the human enterprise today is fueled in large part by the liquidation of natural capital, including essential ecosystems, and the overfilling of waste sinks. In short, the human enterprise is exploiting natural systems faster than they can regenerate. Would a truly intelligent species risk permanently disabling the very ecosystems that sustain it for the increasingly questionable benefits of unequal growth?[5]

Ironically, the main perpetrators of this global experiment are the relatively well educated 20 percent of the human population who live in high-income consumer societies, including most of North America, Europe, Japan, and Australia, along with consumer elites of low-income countries. Densely populated, high-income countries typically exceed their domestic carrying capacities by a factor of three to six or more and thus impose a growing burden on other countries and the global commons. This wealthy minority of the human family appropriates almost 80 percent of the world's resources and generates most of its carbon emissions from fossil fuels.[6]

To achieve sustainability—that is, to live within the ecological carrying capacity of Earth—on average, people would have to live on the biologically productive and assimilative capacity of just 1.7 gha per capita. (If, as good stewards, we reserved more biocapacity solely for wild species, our Earth-shares per person would be even smaller.) In this chapter we use this amount of globally available per capita biocapacity as a starting point to consider the implications of living with a more equitable distribution of Earth's resources. In short, for policy and planning purposes, we consider 1.7 gha/per capita to be each person's equitable or "fair Earth-share" of global biocapacity.

More than half the world's population lives at or below a fair Earth-share. These people are mostly in Latin America, Asia, and Africa. As Table 4–1 shows, such fair Earth-share societies enjoy comparable longevity but have somewhat larger households and lower per capita calorie intake, meat consumption, household energy use, vehicle ownership, and carbon dioxide emissions than average world citizens. The differences between people living at a fair Earth-share and those in high-income countries (which typically need three planets) are much greater.[7]

The data for fair Earth-share societies used in this analysis are based on Cuba, Ecuador, Ethiopia, Guatemala, Haiti, India, Mali, the Philippines, Uzbekistan, and Vietnam. While some of these countries stay within the one-planet parameter due to low socioeconomic development (which also explains lower life expectancy than in the high-consumption societies), others—like Cuba and Ecuador—have high levels of development even with

Table 4–1. Comparing Fair Earth-Share, World Average, and High-Consumption Countries

Consumption Measures	Fair Earth-Share: 1 Planet	World Average: 1.5 Planets	High-Consumption: 3 Planets
		(per person)	
Daily calorie supply	2,424	2,809	3,383
Meat consumption (kilograms per year)	20	40	100
Living space (square meters)	8	10	34
People per household	5	4	3
Home energy use in gigajoules (per year)	8.4	12.6	33.5
Home energy use in kilowatt-hours (per year)	2,300	3,500	9,300
Motor vehicle ownership	0.004	0.1	0.5
Motor vehicle travel (kilometers per year)	582	2,600	6,600
Air travel (kilometers per year)	125	564	2,943
Carbon dioxide emissions (tons per year)	2	4	14
Life expectancy (years)	66	67	79

Source: See endnote 7.

their modest incomes and ecological footprints. In fact, an average Cuban's life expectancy is equivalent to that of an average American (at 78 years). (See Chapter 30.)[8]

The high-consumption societies used in this analysis are Australia, Canada, Germany, Israel, Italy, Japan, Kuwait, New Zealand, Norway, Russia, Spain, Sweden, the United Kingdom, and the United States. While these countries enjoy comparable levels of longevity, education, and quality of life, people in North America, Australia, and the oil-producing states in the Middle East tend to consume twice as much as their three-planet counterparts in other parts of the world. These comparisons show that beyond a certain point, income and consumption have little effect on quality-of-life outcomes compared with other sociocultural factors.

Learning to Live within the (Natural) Law

What might life look like for a high-income consumer society that decided to get serious about sustainability and implement strategies to live on its equitable share of Earth's resources? While this answer will depend on specific geographic, climatic, and cultural realities, a sense of the magnitude of change is available by looking at how one city could make this transition—Vancouver, Canada, which has aspirations to be the "world's greenest city."

The City of Vancouver proper (not the broader metropolitan area), in

British Columbia, is home to approximately 600,000 people and covers 11,467 hectares. Using data compiled by the city, by the Metro-Vancouver region, and by provincial, national, and international statistical agencies, the city's Ecological Footprint is conservatively estimated at 2,352,627 global hectares, or 4.2 gha per person.[9]

The average Vancouver Ecological Footprint can be attributed to various sectors as follows (see Figure 4–1): food (2.13 gha per person) accounts for 51 percent of the footprint, buildings (0.67 gha per person) account for 16 percent, transportation (0.81 gha per person) is 19 percent, consumables (0.58 gha per person) are 14 percent of the footprint, and water use is less than 1 percent.[10]

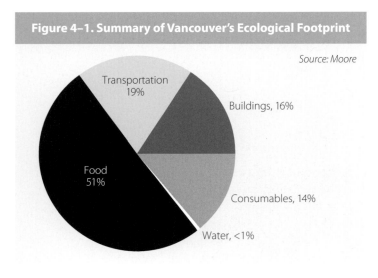

Figure 4–1. Summary of Vancouver's Ecological Footprint

Source: Moore

Transportation 19%

Buildings, 16%

Food 51%

Consumables, 14%

Water, <1%

These data do not include contributions from provincial and national government public services (such as the treasury and military) that take place outside the city for the benefit of all Canadians. Vancouver city staff estimate that these services add an additional 18 percent to the per person eco-footprint. This would be equivalent to approximately 0.76 gha per person, bringing Vancouver's total Ecological Footprint per person to 4.96 global hectares. To achieve one-planet living, the average Vancouverite would need to reduce his or her Ecological Footprint by 66 percent. Note, however, that this is still a minimum number. Ecological Footprint estimates err on the side of caution because they cannot account for elements of consumption and waste assimilation for which data are unavailable or for such things as the fact that much "appropriated" ecosystem area is being degraded.[11]

Food represents half the footprint and includes cropland as well as carbon-sink land associated with processing, distribution, retailing, and consumption. Although many people are concerned about the carbon emissions associated with "food miles" (transporting food from farm to plate), this accounts for less than 3 percent of the food-footprint component and is mostly associated with imported fruits and vegetables. Animal protein production, however, constitutes most of the food footprint (see Figure 4–2), due mostly to cropland used to produce livestock feed.[12]

Transportation is the next largest contributor to the average Vancouverite's Ecolocial Footprint at 19 percent; personal automobile use accounts for 55 percent of this, followed by air travel at 17 percent. Buildings contribute 16 percent to the total Ecological Footprint. Operating energy (mostly natural gas used for water heating and space conditioning) accounts for 80 percent of the buildings footprint and is split equally between the residential and commercial-institutional sectors. The buildings component is smaller than might be expected because 80 percent of Vancouver's electricity is hydroelectric. Moreover, British Columbia was the first jurisdiction in North America to introduce a carbon tax and require all public institutions to be greenhouse-gas neutral in their operations.[13]

Fourteen percent of the Vancouver Ecological Footprint is attributable to consumer products, with paper accounting for 53 percent of this. Fortunately, Vancouverites recycle most of the paper they use (78 percent), reducing its potential Ecological Footprint by almost half. The material content of consumer goods accounts for only 7 percent of the total quantity of energy and material used to produce them; 91 percent of the Ecological Footprint of consumer goods is associated with the manufacturing process and another 2 percent with managing the products as wastes at the end of their life cycle.[14]

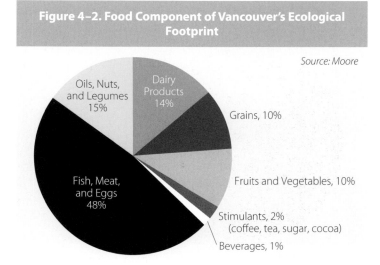

Figure 4–2. Food Component of Vancouver's Ecological Footprint

Source: Moore

Clearly, lifestyle choices have a significant impact on our Ecological Footprint. However, even if average Vancouverites followed a vegan diet; avoided driving or flying and only walked, cycled, or used public transit; lived in a passive solar house that used almost no fossil-based energy; and cut their personal consumption by half, they could only reduce their per capita Ecological Footprint by 44 percent (from 4.96 to 2.8 gha per capita). That seems like an impossible challenge already—and yet it is still a full global hectare beyond the one-planet threshold.[15]

That said, the City of Vancouver is willing to wrestle with this challenge, and in 2011 it launched its *Greenest City 2020 Action Plan*, including a goal to reduce the city's Ecological Footprint 33 percent by 2020 and 66 percent by 2050. Actions in the plan span 10 areas: food, transportation, buildings, economy, waste, climate change, water, access to nature, clean air, and the Ecological Footprint. Indeed, almost all the planned actions contribute to the lighter footprint objective. Nevertheless, the plan falls short of what would be required to achieve stated Ecological Footprint reduction targets.[16]

Through the planning process, city staff explored various approaches, including reducing consumption of high-impact foods (such as meat and dairy products) by up to 20 percent, lowering consumption of new products by up to 30 percent, and cutting the amount of waste sent to landfills and incinerators in half. Note that Vancouver already recycles more than 50 percent of its wastes, so *Greenest City 2020* would achieve a total waste diversion rate of up to 75 percent. Vehicle kilometers travelled would be reduced by up to 20 percent and air travel by up to 30 percent. Building energy efficiency would be improved by up to 30 percent, and all new construction would be zero emissions starting in 2020.[17]

Implementation of these actions is estimated to reduce Vancouverites' Ecological Footprints by 20 percent. Even though the changes in consumption and waste production are substantial (ranging from 20 to 50 percent), this does not directly translate into equivalent reductions in Ecological Footprint. Take the following comparison, for example. Meat and dairy consumption accounts for nearly 23 percent of Vancouver's Ecological Footprint (and 21 percent of food consumed by weight). Reducing that by 20 percent translates into an approximate 4.5 percent reduction in the total Ecological Footprint. Indeed, this is one of the most effective actions that could be taken to achieve one-planet living. Municipal solid waste, on the other hand, only accounts for 1 percent of Vancouver's total Ecological Footprint. So cutting the total tonnage of municipal waste in half has an almost insignificant impact on the Ecological Footprint (assuming there are no upstream impacts on the supply chain of energy and materials used to produce consumer products).[18]

Getting to one-planet living therefore requires strategic consideration of which lifestyle changes can have the most significant impacts. Unfortunately, in the final *Action Plan* some of the actions that would have the greatest impact—such as reducing meat and dairy consumption—were omitted, largely because their implementation relied on people's voluntary actions

Jennie Moore

Bicycling infrastructure on Clark Street in Vancouver.

that could not, and perhaps should not, be regulated by government.[19]

The question remains: even if citizens were willing to do all they could, how would Vancouver shave another global hectare off the average Ecological Footprint? Recall that senior government services from which all Canadians benefit account for an estimated 0.76 gha per capita of Vancouver's Ecological Footprint. Changes in senior government policy and practice are therefore also needed and could include efforts toward demilitarization, an emphasis on population health through disease prevention, and a careful public examination of existing rules, regulations, tax incentives, and assumptions about whether the current administration of public funds is aligned with the goals of a sustainable society.

These are bold measures that move past the current emphasis on efficiency gains across society. The latter would, of course, still be needed—indeed, there is considerable room for additional energy/material efficiency gains across the entire building stock and in manufacturing; farmers and food processors could also greatly reduce their reliance on fossil fuels and inputs (fertilizers and pesticides, for instance). One way to induce efficiency gains is to eliminate "perverse subsidies" (including tax breaks to highly profitable oil and gas producers and subsidies to farmers to produce certain food products, such as corn) that facilitate unsustainable industrial practices and generate false price signals in consumer markets. If necessary, this should be accompanied by pollution charges or taxes to address market failures (that is, to internalize negative externalities) and to ensure that market prices reflect the true social costs of production. Policy alignment at the national and provincial government levels to support all such initiatives is essential.[20]

A second challenge involves engaging civil society with political leaders to advance a paradigm of sufficiency, meaning a shared social commitment to consuming enough for a good life but not so much that total throughput exceeds critical biophysical limits. Such a new consumer paradigm is also necessary to avoid the "rebound effect," in which people spend savings from efficiency on other things—canceling the gains. A survey of 65 studies in North America found that this rebound is responsible for 10–30 percent of expenditures in sectors that account for most energy and material consumption: food, transportation, and buildings. Indeed, total resource and energy demand in most of the world's industrial countries has increased in absolute terms over the past 40 years despite efficiency gains of 50 percent in materials and 30 percent in energy use.[21]

Different people will make different lifestyle choices and changes as required. If one-planet living is the goal, these choices will obviously have to entail more than recycling programs and stay-at-home vacations. For success, the world's nations will have to commit to whole new development strategies with elements ranging from public re-education to ecological fiscal reform, all within a negotiated global sustainability treaty.[22]

While it is beyond the scope of this chapter to detail elements of such an economic transformation, others have tried. In *Factor Five*, for example, Ernst von Weizsäcker and colleagues attempt numerous sector studies to demonstrate how an 80 percent reduction in resource intensity could be achieved in agriculture, transportation, buildings, and selected manufacturing industries. They show that many of the technologies needed for one-planet living already exist, but in the absence of global agreements and enforceable regulations, there is insufficient incentive for corporate, government, and consumer uptake. In a global economy, states will not act alone for fear of losing competitive ground. And even international cooperation or agreements do not ensure success: although some global initiatives (such as the Montreal Protocol on ozone depletion) have succeeded, others (such as the Kyoto Protocol on climate change) have succumbed to shorter-term economic considerations.[23]

A parking lot adapted for use as an urban farm, Vancouver.

What Lies Ahead

Despite the pressing need for cultural transformation, prospects for real progress toward socially just ecological sustainability are not encouraging. Global society remains committed to the progress myth and to unconstrained economic growth. Indeed, the international community views sheer material growth rather than income redistribution as the only feasible solution to chronic poverty.

In *Our Common Future*, the World Commission on Environment and Development recognized peoples' reticence to contemplate serious measures for wealth redistribution. Such an approach might follow a strategy of contraction and convergence, during which industrial countries reduced their energy and material throughput to allow room for developing countries to grow. Instead, the Commission advocated for "more rapid economic growth in both industrial and developing countries," albeit predicated on global cooperation to develop more equitable trade relationships and noting that "rapid growth combined with deteriorating income distribution may be worse than slower growth combined with redistribution in favour of the poor."[24]

Since that report came out in 1987, economic growth has far outpaced population growth, so there are more dollars per person circulating in the world today than ever before. But while some developing states have prospered in the increasingly global economy—such as Singapore, South Korea, China, and India—others have not. Moreover, income disparity is increasing both among and within countries; even in the richest nations, lower-income groups have seen real wages stagnate or decline. It is now apparent that growth alone is failing as a solution to poverty. Most of the human family is still materially deprived, consuming less than its just share of economic output. This has led to renewed recognition—at least in progressive circles—that policy measures explicitly designed to spread the benefits of economic prosperity are more effective than increasing gross domestic product for alleviating material poverty.[25]

Overall, the combined evidence of widening income gaps and accelerating ecological change suggests that the mainstream global community still pays little more than lip service to the sustainability ideal. The growth economy, now dressed in green, remains the dominant social construct. Rio+20, the latest U.N. conference on economy and development, essentially equated sustainable development with sustained economic growth and produced no binding commitments for anyone to do anything. So it is that 40 years after the first global conference on humanity and the environment (Stockholm in 1972) and 20 years after the first world summit on the environment and development (Rio in 1992), the policy focus remains on economic growth—while ecological decline accelerates and social disparity worsens.

Discouraging, yes, but let us recognize that the notion of perpetual growth is just a social construct, initiated as a transition strategy to reboot the economy after World War II. It has now run its course. What society has constructed it can theoretically deconstruct and replace. The time has come for a new social contract that recognizes humanity's collective interest in designing a better form of prosperity for a world in which ecological limits are all too apparent and the growing gap between rich and poor is morally unconscionable. Our individual interests have converged with our collective interests. What more motivation should civil society need to get on with the task at hand?[26]

The major challenges to sustainability are in the social and cultural domains. The global task requires nothing less than a rewrite of our prevailing growth-oriented cultural narrative. As Jared Diamond emphasized in *Collapse*, societies can consciously "choose to fail or succeed," and global society today is in the unique position of knowing the dismal fates of earlier cultures that made unfortunate choices. We can also consider the prospects of those who acted differently. Indeed, in contrast to the fate of Easter Islanders, the people of Tikopia—living on a small South Pacific island—made successful choices to reduce their livestock populations when confronted with signs of ecological deterioration. Today the Tikopian culture serves as an example of conscious self-management in the face of limited resources. Of course, Tikopia has the advantage of being a small population with a homogenous culture on a tiny island where the crises were evident to all and affected everyone. Contrast that with today's heterogeneous global culture characterized by various disparities (tribal, national, linguistic, religious, political, and so on) and the anticipation of uneven impacts.[27]

Meanwhile, our best science is telling us that we are doing no better than previous failures: staying our present course means potential catastrophe. The (un)sustainability conundrum therefore creates a clear choice for people to exercise their remaining democratic freedoms in the name of societal survival. Difficult though it may be, ordinary citizens owe it to themselves and the future to engage with their leaders and insist that they begin the national planning processes and draft the international accords needed to implement options and choices for an economically secure, ecologically stable, socially just future.

CHAPTER 5

Sustaining Freshwater and Its Dependents

Sandra Postel

Access to water is essential for human survival, much less human advancement. The great early human civilizations—from the ancient Egyptians to the Mesopotamians to the early Chinese—sprung up and flourished alongside rivers. Without sufficient water to drink and to grow food, no society—however advanced—can last.

So here's the conundrum. Water is finite. The volume of freshwater on Earth today is the same as when Caesar ruled ancient Rome. Yet in those intervening 2,000 years, the human population has risen from 250 million to more than 7 billion. The annual production of global goods and services, now valued at $70 trillion, has expanded even faster.[1]

Water is needed to produce nearly everything—from electricity and paper to burgers and blue jeans. As consumer demands have risen, the limits of accessible water supplies have become increasingly apparent. An unsettling number of large rivers are now so overtapped that they discharge little or no water to the sea for months, or years, at a time. Lakes and wetlands are shrinking, and crucial aquifers are being depleted. Some 10 percent of the global food supply today depends on the unsustainable use of groundwater.[2]

At the same time, basic human needs for water continue to go unmet. Nearly 800 million people—about 11 percent of humanity—lack access to safe drinking water. An even larger number of people are hungry and malnourished. Many live on farms but lack access to water to irrigate their crops during droughts and yearly dry seasons, which keeps them mired in poverty and chronically malnourished.[3]

Is there hope of achieving a sustainable balance with freshwater? The answer is yes. But to envision how it can be achieved we must dig a little deeper into what sustainable water use means, assess where we stand today, and then develop a vision and a set of practical actions for moving toward it.

Sandra Postel is director of the Global Water Policy Project and Freshwater Fellow of the National Geographic Society.

www.sustainabilitypossible.org

Freshwater by the Numbers

Images from space show Earth to be a strikingly blue planet harboring great stores of water. Some 97.5 percent of that water is ocean, which provides a vast array of benefits but is too salty to drink or irrigate crops. Nor is there great scope to tap this salty water for human use through desalination. (See Box 5–1.) Most of the remaining 2.5 percent is locked up in glaciers and ice caps or resides deep under the surface. Only a tiny fraction of all the water on Earth—less than one one-hundredth of 1 percent—is fresh and renewed each year by the sun-powered hydrological cycle.[4]

At first glance, even this small share of the planet's water—the renewable freshwater supply—would seem to be more than ample to satisfy human needs now and for generations to come. Each year, the global water cycle delivers 110,000 cubic kilometers of water over land in the form of rain, sleet, and snow. (These values are approximate; various models have produced different estimates.) About 64 percent of that precipitation returns to the atmosphere through evaporation or transpiration (the use of water by plants, crops, grasses, and trees). The remaining 36 percent flows toward the sea in rivers, streams, or underground aquifers. This "runoff" is the water supply we tap for irrigation, drinking water, electricity production, and manufacturing.[5]

But when we account for the share of runoff that is too remote to get to (about 19 percent) or that runs off in floods (about 42 percent), the picture darkens a bit. Even taking into account the floodwaters captured by dams, only about 15,600 cubic kilometers of global runoff—39 percent of the total—is accessible. Today, worldwide water demands use about 30 percent of that accessible supply. (Agriculture accounts for 70 percent of the global demand, industries for 20 percent, cities and towns for about 10 percent.)[6]

Humanity's impact on Earth's water, however, is greater than these figures would suggest. First, we not only use water, we often pollute it. For example, many rivers, streams, aquifers, and coastal zones receive harmful levels of nitrogen fertilizers and chemical pesticides carried by runoff from farms and suburban lawns.

Second, we not only tap into rivers, lakes, and aquifers, we also rely on

---

Box 5–1. Desalination

With so much water in the oceans, desalting seawater would seem to provide the ultimate solution to the world's water problems. Desalination is indeed a viable water-supply option, and the process has steadily improved. With new membrane technologies and other developments, the energy required to desalinate seawater has fallen by 60–80 percent over the last two decades.

Nevertheless, the process remains energy-intensive, expensive, and potentially harmful to coastal marine environments. Currently, the roughly 15,000 desalination plants worldwide have the capacity to produce 15.3 cubic kilometers of water per year—less than half of 1 percent of global water demand. Moreover, most de-salting plants run on fossil fuels, which means they contribute to climate change while attempting to "solve" water shortage problems—a Faustian bargain at best. While it provides a lifeline for some island nations and desert regions, desalination is no silver bullet for solving the world's water problems.

Source: See endnote 4.

natural rainfall, especially to grow crops. Some 82 percent of cropland worldwide is watered solely by natural precipitation; it gets no supplemental irrigation. This direct use of precipitation is typically excluded from estimates of water demand.[7]

One other distinction is important. About half of the water we use is "consumed" (or depleted) through evaporation or transpiration, which means it returns to the atmosphere and resides there as vapor until it falls to the earth again. Since it may not return as rain during the same season or in the same location, water is effectively "depleted" from any particular watershed. On the other hand, water used but not consumed is available to use again. The water we use to shower or flush toilets, for example, typically returns to a local river or aquifer, where it can be reused. This distinction between use and consumption is crucial for assessing how much water is actually available to meet the demands in a given watershed.

Researchers Arjen Hoekstra and Mesfin Mekonnen of the University of Twente in the Netherlands have made the most detailed estimates to date of the scale and patterns of humanity's water consumption. They tabulated all the water from both rainfall and irrigation that is consumed in making goods and services for everyone in the world. To complete the picture, they added in the volume of water needed to assimilate the pollution generated along the way. They then calculated the annual average global water "footprint" for 1996–2005, the most recent 10-year period with the data they needed. The upshot: humanity's water footprint totals an estimated 9,087 cubic kilometers per year—a volume of water equivalent to the annual flow of 500 Colorado Rivers.[8]

Whether looking at use, consumption, or "footprints," these global numbers tell only a small part of the story. A large share of the world's people and irrigated farms are located where renewable water is not very abundant. (See Figure 5–1.) China, for instance, has nearly 20 percent of the world's population and 21 percent of total irrigated area but only 6.5 percent of the world's renewable freshwater—and most of that supply is in the southern part of the country. The United States, by contrast, has 4.5 percent of the world's people and 7 percent of the renewable water supply. But most of that nation's irrigated land and recent population growth are found in the drier West: hence the depletion of rivers and aquifers in that region.[9]

In addition to this geographic mismatch there is a timing mismatch: nature does not deliver water evenly or predictably throughout the year. Much of India, for instance, gets most of its water during the summer monsoons, often in just a few intense storms. In much of sub-Saharan Africa, rainfall is highly variable and unreliable. Fourteen countries in that region each experienced at least 10 droughts between 1970 and 2004.[10]

With human-induced climate change likely to make many dry areas drier

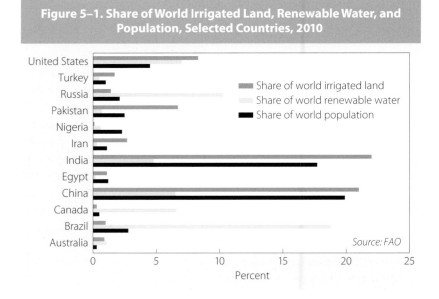

Figure 5–1. Share of World Irrigated Land, Renewable Water, and Population, Selected Countries, 2010

and wet areas wetter, hydrologic variability will become more extreme. In 2008, seven top water scientists argued persuasively in *Science* that "stationarity"—the concept that natural variability remains within a known set of boundaries—is no longer valid. We have moved outside that known envelope of variability into new territory. When it comes to water, in other words, the past is no longer a reliable guide to the future.[11]

How Sustainable Is Our Water Use Today?

Three critical attributes distinguish freshwater from other "resources": it is essential to life, there are no substitutes for it, and because we cannot ship it around the world in large quantities, how it is used and managed locally or regionally is what matters. A working definition of freshwater sustainability that is true to these attributes might be this: in any watershed, ensure that basic water needs are met for all people; preserve ecological infrastructure so as to provide the quantity, quality, and timing of water flows needed to sustain ecosystem services; and where groundwater is tapped, ensure that extraction does not deplete the water in storage or degrade connected ecosystems. Judged according to these criteria, our use and management of water fails the sustainability test on multiple fronts.[12]

Drinking Water for All. The failure to provide universal access to safe drinking water ranks among the greatest shortcomings of human development. As of 2010, some 780 million people—more than 1 in 10—lacked access to a safe supply of water to meet their basic needs for drinking, cooking, and washing. Most live in poor parts of Asia and sub-Saharan Africa, where women and girls often spend hours each day trekking to a water source, lifting what water they can carry home to their families, and then hoping it does not sicken or kill themselves or a family member. The issue is not a lack of water: providing 20 liters per person per day for 780 million people would require only 0.1 percent of current global water withdrawals. There is sufficient water but thus far insufficient political will and financing to provide universal access to safe water.[13]

The good news is that substantial progress has been made over the last two decades in this area: more than 2 billion people acquired access to safe drinking water during this time. The Millennium Development Goal of halving the proportion of the population without access to safe drinking water by 2015 (compared with the 1990 level) was actually met in 2010, five years early. Still, a backlog of need remains. Many dedicated groups—including U.N. agencies, country ministries, grassroots groups, and nongovernmental organizations—are working diligently to complete the job. [14]

Ecosystem Needs for Water. Unfortunately, the progress in meeting basic human needs for water is not paralleled by progress in meeting ecosystem needs. Indeed, when it comes to preserving ecosystem health and services, most of the trends are going in the wrong direction.

Over the decades, water management has largely aimed at getting water to people and farms where and when they need it. Since 1950, the number of large dams has climbed from 5,000 to more than 45,000. Dams and reservoirs are now able to hold 26 percent of annual global runoff all at once, causing immense changes to the flow of rivers.[15]

At the same time, large diversions by canal or pipeline move water hundreds of kilometers. Around the world, 364 transfer schemes move approximately 400 cubic kilometers of water annually from one river basin to another—equivalent to transferring the yearly flow of 22 Colorado Rivers. China is proceeding with a massive $60-billion project to transfer 41.3 cubic kilometers a year from the Yangtze River basin in the south to the water-scarce north. If completed, it will be the largest construction project on Earth.[16]

Many more projects divert water from one location to another within the same river basin. Phoenix, Arizona, for example, gets 40 percent of its supply through the Central Arizona Project (CAP), which transfers water from the Colorado River 300 kilometers to the east. The CAP is just one of many diversion schemes in the Colorado Basin that contribute to that river's drying up before it reaches its final destination, the Sea of Cortez in Mexico.[17]

Dams to store water and diversion schemes to move it around have allowed burgeoning oasis cities in the desert, from Phoenix and Los Angeles to Cairo and Karachi. They have enabled the desert to bloom and food production to keep pace with population growth. Dams have also added to the world's energy supply: hydropower facilities now generate 16 percent of the world's gross electricity, and many large new hydroelectric schemes are planned or under construction in Brazil, Canada, China, India, Turkey, Southeast Asian nations, and elsewhere.[18]

In short, control over water has allowed the human enterprise to grow and prosper as conventionally measured by the number of hectares irrigated, kilowatt-hours generated, and people served. Yet those benefits have come at great expense to some 470 million river-dependent people downstream of large dams, as well as to the health and productivity of freshwater ecosystems that deliver services of great value. (See Box 5–2.) Healthy rivers, for example, supply fish to eat, recreational opportunities, and riparian habitats for birds and wildlife; wetlands mitigate floods, recharge groundwater, and filter out pollutants; and forested watersheds increase the reliability and quality of drinking water supplies. Maintaining these services for this and future generations is part of the sustainability challenge—yet we have paid little attention to them.[19]

Although dams and reservoirs do the important work of storing freshwater for human use, they often result in rivers being turned on and off like plumbing works. Instead of rivers flowing to their own natural rhythms, which create the cues and habitats that fish and wildlife need, they now often flow to suit human demands for electricity, irrigation, water supply, and flood control. This flow alteration is a dominant factor in the loss of freshwater life: extinction rates for freshwater species are estimated to be four to six times higher than for terrestrial or marine species. In North America, 700 freshwater fish species (39 percent of the total) are imperiled, nearly double the number in 1989; of this total, 61 are presumed locally or globally extinct.[20]

Dams and reservoirs worldwide also trap more than 100 billion tons of sediment that would otherwise replenish deltas and nourish coastal habitats crucial to commercial fisheries. From the Colorado to the Indus to the Nile, the depletion of river flows and trapping of nutrient-rich sediments are causing deltas—among the most productive ecosystems on Earth—to shrink and degrade. The Colorado Delta—a crucial stopover for migratory birds on the western Pacific flyway—has lost more than 90 percent of its wetlands; the Nile Delta, which provides about one third of Egypt's crops, is losing ground to the Mediter-

---

Box 5–2. Services Provided by Rivers, Wetlands, Floodplains, and Other Freshwater Ecosystems

- Water supplies for irrigation, industries, cities, and homes
- Fish, waterfowl, mussels, and other foods for people and wildlife
- Water purification and filtration of pollutants
- Flood mitigation
- Drought mitigation
- Groundwater recharge
- Water storage
- Wildlife habitat and nursery grounds
- Soil fertility maintenance
- Nutrient delivery to deltas and estuaries
- Delivery of freshwater flows to maintain estuarine salinity balances
- Aesthetic, cultural, and spiritual values
- Recreational opportunities
- Conservation of biodiversity, which preserves resilience and options for the future

Source: See endnote 19.

ranean Sea as 100 million tons of sediment per year get trapped behind the Aswan Dam.[21]

The services provided by ecological infrastructure run on free energy from the sun, while all the technological replacements for these services—from river levees to treatment plants—require increasingly expensive human-created energy to build, operate, and maintain. As a result, the economic costs of these lost ecological services, though untallied, are high and rising. Scientists participating in the Millennium Ecosystem Assessment estimated in 2005 that wetlands alone provide water purification, flood mitigation, and other services worth $200–940 billion per year. Worldwide, we have filled or drained up to half of the planet's original wetland area.[22]

Ecological infrastructure will be increasingly important as climate change further alters the global water cycle and as droughts, floods, and other extreme events become more common and severe. In the spring of 2011, as floodwaters raged through the Mississippi River, forcing the federal Army Corps of Engineers to breach a levee to save Cairo, Illinois, an important piece of ecological infrastructure was missing: 14 million hectares of wetlands in the upper Mississippi Basin that over time had been drained and filled to make way for farms and homes. Those wetlands—an area the size of Illinois—had worked like a giant sponge, absorbing rainwater and then releasing it slowly to nearby streams or underground aquifers. With those natural protections gone, and with more people and farms in harm's way, flood risks grew. According to ecologists Donald Hey and Nancy Philippi, despite the massive construction of levees throughout the upper Mississippi Basin during the twentieth century, annual average flood damage over the century more than doubled.[23]

Groundwater Trends. Some of the most troubling signs of unsustainable water use come from underground, where we are building up a sizable water debt in the form of aquifer depletion. Just as bank accounts shrink when withdrawals exceed deposits, so do groundwater accounts. Most of the depletion is occurring in some of the world's most crucial farming regions.

Using data from a U.S. National Aeronautics and Space Administration satellite mission called GRACE (for Gravity Recovery and Climate Experiment), scientists have estimated that northern India, which includes that nation's breadbasket, is depleting groundwater at a rate of 54 cubic kilometers per year, a volume that could support a subsistence-level diet for some 180 million people. In another study, led by Jay Famiglietti at the University of California in Irvine and also using data from GRACE, researchers found that between October 2003 and March 2010, California's Central Valley—the fruit and vegetable bowl of the United States—lost a volume of groundwater equivalent to two thirds of the capacity of Lake Mead, the nation's largest human-made reservoir.[24]

Indeed, all four of the world's top irrigators—China, India, Pakistan, and the United States—are pumping groundwater faster than it is being replenished in crucial crop-producing areas. The problem is most serious in India, where 60 percent of irrigated farming depends on groundwater. Water tables are falling extensively in Andhra Pradesh, Gujarat, Maharashtra, Rajasthan, and Tamil Nadu in addition to the breadbasket states of Punjab and Haryana in the northwest. At least 15 percent of India's food is being produced by mining groundwater.[25]

In addition to the problem in California's Central Valley, U.S. groundwaters are being heavily depleted in the western Great Plains, where parts of eight states are above the Ogallala Aquifer. The Ogallala supplies water to 27 percent of U.S. irrigated land, sustaining wheat, corn, and cotton production. According to the U.S. Geological Survey, depletion of the Ogallala—or more precisely, the High Plains Aquifer, most of which is made up of the Ogallala—over the last six decades totals some 328 cubic kilometers, a volume of water sufficient to sustain the U.S. wheat harvest for about six years.[26]

Using state-of-the-art hydrological models and estimates of groundwater withdrawals, Yoshihide Wada of Utrecht University in the Netherlands and his colleagues estimated in a 2010 study that some 283 cubic kilometers of groundwater were depleted from aquifers around the world in 2000. While some of the depletion occurs for urban or industrial purposes, the vast majority is for crop irrigation. Since it takes about 1,500 cubic meters of water to grow one ton of grain (an approximate average for rice, wheat, and corn), that volume of depleted groundwater could have produced 189 million tons of grain, equal to 10 percent of global grain output in 2000.[27]

While many countries are depleting aquifers locally or regionally, five are mining groundwater faster than replenishment rates at the national scale: Saudi Arabia, Libya, Egypt, Pakistan, and Iran. Saudi Arabia's story offers particularly important warnings. This desert nation gets only 59 millimeters (2.3 inches) of rainfall a year, and its renewable groundwater supply is a meager 2.4 cubic kilometers per year. To meet their water demands, the Saudis draw heavily upon nonrenewable or "fossil" aquifers that formed some 20,000 years ago. These aquifers were heavily pumped during several decades of massive desert farming aimed at making the nation self-sufficient in wheat. The Saudis were so successful that for a time Saudi Arabia even exported wheat.[28]

Between 1980 and 2006, the volume of water used for Saudi irrigation more than tripled, and nearly all of it was groundwater. As of 2006, Saudi farmers were pumping nearly 10 times more groundwater than was being replenished by nature. In January 2008, aquifer depletion and the rising costs of pumping from ever greater depths led the Saudis to announce a gradual phaseout of irrigated wheat production. In addition to importing grain, they are now buying or leasing farmland in Ethiopia and elsewhere to try to ensure some degree of food self-sufficiency.[29]

Looking ahead, the prospect of longer and deeper droughts due to human-induced climate change will hasten the depletion of groundwater. The High Plains Water District based in Lubbock, Texas, found that during the severe 2011 drought farmers in their district, who rely on the Ogallala Aquifer, stepped up their groundwater pumping to compensate for the lack of rain. Groundwater levels across the 16-county service area fell an average of 0.78 meters—the largest annual decline recorded in the last quarter-century and more than triple the annual average for the last decade.[30]

Moving Toward Sustainability

Given this snapshot of water use around the world, achieving sustainability might seem like an impossible dream. But here and there, farms, villages, businesses, cities, states, provinces, and nations are taking actions that move communities toward a more secure and sustainable water future. What these examples have in common is an effort to use and manage water in ways that preserve or restore rivers, lakes, aquifers, and watersheds. They place ecosystem health and sustainability principles at the core of water management instead of at the periphery. When this is done, water productivity and the range of benefits derived from water climb upward.

A handful of places around the world are beginning to address groundwater depletion. In the Indian state of Andhra Pradesh, for example, village-level farmer groups are measuring and monitoring rainfall and aquifer levels and then collectively developing water budgets for their crop production in an effort to arrest depletion of the aquifers they depend on. Participation is voluntary and driven by data, education, capacity building, and cooperation. Farmers engaged in the effort have shifted to less-thirsty crops and adopted water-saving irrigation methods, all with an aim of aligning their water use with the sustainable groundwater supply. Farm profitability has increased: surveys indicate that the net value of farm outputs has nearly doubled. The project, which has reached some 1 million farmers, appears to be the first success worldwide in community groundwater management aimed at sustainability. Similar projects are now under way in Maharashtra and are being considered in several other Indian states.[31]

Similarly, the High Plains Water District in Lubbock, Texas, has taken steps to slow depletion of the Ogallala Aquifer. In January 2012, the district declared it illegal to pump groundwater in excess of a pumping limit it established, called an "allowable production rate." Since the Texas portion of the Ogallala gets very little recharge (its water was put in place thousands of years ago), any significant pumping drains the aquifer. The district's goal is to slow the depletion so that at least 50 percent of its Ogallala water is still there in 50 years. As the pumping limits get more stringent, farmers will need to choose crops and irrigation methods that allow them to get more value per drop. And engineers, agronomists, and entrepreneurs will have an incentive to develop new technologies and agricultural practices that help them do this.[32]

Although farmer outcry and the threat of legal action have led the High Plains Water District to delay enforcing the new rule until 2014, both Texas law and a February 2012 state Supreme Court opinion affirm that although farmers do indeed own the groundwater beneath their property, conservation districts can regulate pumping rates.[33]

With crop production accounting for the lion's share of world water consumption, measures to raise irrigation efficiency and get more nutritional value per drop are crucial. Drip irrigation, which delivers water directly to the roots of plants in just the right amounts, can double or triple water productivity, and it appears to be on a rapidly rising growth curve.

Courtesy iDE

A farmer in Nepal uses a low-cost drip system.

Over the last two decades, the area under drip and other "micro" irrigation methods has risen more than sixfold, from 1.6 million hectares to more than 10.3 million. The most dramatic gains have occurred in China and India, the top two irrigators, where over the last two decades the area under micro-irrigation expanded 88-fold and 111-fold, respectively. Anil Jain, managing director of Jain Irrigation—the second largest global micro-irrigation company—expects the drip irrigation market in India to expand by 1 million hectares annually during the coming years and to soon become a $1 billion market in India alone.[34]

Despite recent growth, less than 4 percent of global irrigated area is equipped with micro-irrigation, so its potential has barely been tapped. Markets are widening, however, with the development of low-cost drip systems tailored to the needs of poor farmers. The nongovernmental group iDE (formerly International Development Enterprises), which successfully introduced the human-powered treadle pump to Bangladeshi farmers, has developed a suite of drip systems ranging from $5 bucket kits for home gardens to $25 drum kits for 100-square-meter plots (about 400 plants) and $100 shiftable drip systems that can irrigate 0.2 hectares, including plots on terraced hillsides. More than 600,000 of iDE's low-cost drip systems have been sold in India, Nepal, Zambia, and Zimbabwe, helping farmers raise their land productivity and move out of poverty.[35]

After a decade of drought, Australia's Murray-Darling Basin has engaged in one of the boldest efforts anywhere to return flows to depleted rivers and wetlands while at the same time sustaining its vibrant agricultural economy. The basin spans 14 percent of Australia's territory and supports 39 percent of its agricultural production. It is also home to 30,000 unique wetlands, many internationally recognized, as well as a rich diversity of freshwater species, including the prized Murray cod.[36]

The proposed plan, released in November 2011, would set "sustainable diversion limits" that reduce basin-wide consumption so as to restore 2.75 cubic kilometers of water to the river system. To achieve the savings, the Australian government would spend billions of dollars over 10 years to improve irrigation efficiency and purchase water entitlements from willing sellers. While irrigators assert that the proposed water cuts are too severe (though they have been eased from earlier proposed levels) and threaten their livelihoods, scientists maintain that the cuts are insufficient to meet critical targets for ecosystem health. Though far from resolved, the societal debate in Australia about rebalancing water use between people and nature is crucial and will need to occur in many more river basins around the world. The experiment there will no doubt yield important lessons, particularly for other drought-prone, agriculturally vital regions. [37]

Although cities and towns account for only about 10 percent of global water demand, their concentrated water use can severely strain local and regional water sources. As a result, conservation and efficiency improvements have a crucial role to play in urban areas, too. In the mid-1980s, as Boston, Massachusetts, approached the safe yield of its water supply, the city began considering a large new diversion from the Connecticut River, the largest river in New England. Citizen concern about the effects on Atlantic salmon restoration and the overall health of the river forced Boston water officials to consider aggressive conservation measures instead—including finding and fixing leaks in the distribution system, retrofitting homes with efficient fixtures, conducting industrial water audits, and providing pricing incentives and consumer education. From its 1980 peak, greater Boston's water use has fallen 43 percent, dropping back to levels not seen in 50 years.[38]

Cities are also investing in watershed protection to safeguard the reliability and quality of their drinking water supplies. A healthy watershed can filter out pollutants, often at lower cost than a water treatment plant can, while also saving on energy and chemicals. New York City, which has pioneered good watershed protection for decades, is now investing $1.9 billion to restore and further protect the Catskills-Delaware watershed (which supplies 90 percent of the city's drinking water) in lieu of constructing a $10-billion filtration plant that would cost $100 million a year to operate.[39]

Likewise, Quito in Ecuador partnered with The Nature Conservancy (TNC) to start a watershed protection fund that receives nearly $1 million a year from municipal water utilities and hydroelectric companies that benefit from the clean, reliable water supplies. Launched in 2000, Quito's water fund has become a model for many other Latin American cities, including Bogotá in Colombia and Lima in Peru. By 2015, TNC aims to have helped initiate 32 watershed funds in South America, protecting 3.6 million hectares of land that filter and supply drinking water for some 50 million people.[40]

Increasingly, corporations recognize that water shortages present risks to their bottom lines and reputations, and they are beginning to set their own sustainability goals. The brewing company MillerCoors, for instance, aims by 2014 to reduce the water required to make a pint of beer by 15 percent from 2008 levels (not counting the water used to produce the grain that goes into the beer). The London-based conglomerate Unilever, recognizing that agriculture accounts for half of its water impact (to grow the raw materials for its products), works with farmers to install drip irrigation and improve irrigation practices. On tomato farms in Brazil, these efforts led to a 30 percent reduction in farm water use and higher yields; as more farm suppliers switch to drip, the water footprint of the company's tomato sauce shrinks.[41]

Individuals can make a difference as well, by shrinking their personal water footprints and by consuming less overall. A single cotton shirt takes 2,500 liters of water to make; a pair of blue jeans, 8,000 liters. Most of this water is consumed in growing the cotton, so more-efficient irrigation can shrink the footprint as well. But if 1 billion consumers each bought two fewer new cotton shirts a year, the water savings would be sufficient to meet the annual dietary needs of 4.6 million people. And every day we "eat" about a thousand times more water than we drink, so making more water-conscious choices about our diets could save a great deal of water. Likewise, filling up automobiles takes about 13 liters of water per liter of fuel, so carpooling, biking, taking public transportation, and choosing fuel-efficient vehicles saves not only energy but also water.[42]

If the world is to have any hope of sustainably meeting everyone's water-related needs, these kinds of policy, technology, and consumer shifts must become mainstream. The good news is that we have barely begun to apply our human ingenuity and inventiveness to meeting this challenge. It is time to let the solutions flow.

CHAPTER 6

Sustainable Fisheries and Seas: Preventing Ecological Collapse

Antonia Sohns and Larry Crowder

Over 50 years ago, Rachel Carson noted that "it is a curious situation that the sea, from which life first arose, should now be threatened by the activities of one form of that life. But the sea, though changed in a sinister way, will continue to exist: the threat is rather to life itself."[1]

Carson depicts the relationship between humans and the sea as one of both dependence and conflict. Despite our profound dependence on the sea for survival, improper management of the atmosphere, the seas, and fisheries has brought the ocean to the verge of unprecedented ecological change. This crisis differs from earlier changes, as it was brought on by the actions of a single species. While ocean ecosystems are resilient and have some capacity to adapt, the rate and magnitude of change rivals previous periods of marine mass extinctions. In order to mitigate additional damage to the seas, all stakeholders must be engaged and implement collaborative policies that drastically reduce carbon dioxide (CO_2) emissions and curb population growth.

The ocean has always been vast and mysterious, first captured in the epic poems of the Ancient Greeks through voyages and celestial navigation. It is Homer's wine-dark sea that soaks Earth in those early days. Aristotle is said to be the first to record marine life; hundreds of years later, expeditions sailed across the sea transforming society forever. In 1728, Captain James Cook voyaged into the unknown, collecting specimens and stories as he circumnavigated the globe. Cook encountered island empires where the sea was a mighty god. The expedition of Charles Darwin and the HMS *Beagle* from 1831 to 1836, as well as that of Sir Charles Wyville Thomson and the HMS *Challenger* from 1873 to 1876, enriched the study of marine biology and oceanography—seeding theories on coral reef formation and natural selection and detailing the first systemic plots of ocean currents and temperature.[2]

From scientific studies to great exploration, much has been learned about the intricacy of the seas and the life that has evolved there. The ocean

Antonia Sohns was the Sustainable Prosperity Project Fellow at the Worldwatch Institute. Larry Crowder is the science director at the Center for Ocean Solutions and a professor of biology and a senior fellow at the Stanford Woods Institute for the Environment, both part of Stanford University.

controls climate, absorbs carbon dioxide, generates oxygen, and determines weather patterns through heat exchange. The stability of life on Earth depends on healthy seas.

Despite the ocean's critical role and immense value, policymakers have done little to ensure the future health of the seas. Perhaps the lack of action stems from the perceived distant nature of these problems in place and time—whether it is ice disappearing from the Arctic, invisible changes in sea surface temperatures, ocean acidification and hypoxia (oxygen deficiency), or the consequences of climate change decades from now. While these issues may be challenging to understand and address, they are among the greatest concerns of our time.

Value of the Sea

The ocean's expanse covers 71 percent of Earth's surface and supports 50 percent of its species. Worldwide, 1 billion people depend on fish for their primary source of protein. Approximately 500 million people depend on coral reefs for food resources or supplementary income from fishing or tourism, and 30 million people are wholly dependent on coral reefs for their livelihood and the land they live on, such as atolls. Societies have developed whole economies around the ocean's resources.[3]

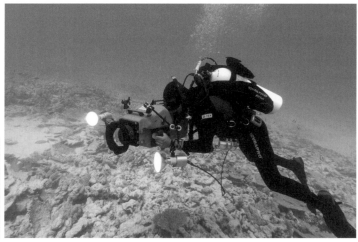

Filming on Rapture Reef in the Northwestern Hawaiian Islands Marine National Monument.

In the United States, the coastal and ocean economies are critical sources of employment and revenue. The U.S. ocean economy generated 2.6 million jobs through tourism and recreation, transportation, and construction and added $223 billion to the U.S. economy in 2009. One of every six U.S. jobs is marine-related.[4]

However, these statistics say nothing of the intrinsic value of the ocean and the importance of ecosystem services to everyday life. The nonmarket value of the sea includes the ecosystem and biodiversity benefits, the value of water quality for nearby communities, and carbon storage. Coral reefs and mangroves are two of the most valuable ecosystems to humankind; for example, reefs are valued at $100,000–600,000 per square kilometer (km²) and mangroves at $200,000–900,000. In Hawaii, the direct economic benefits of coral reefs, based on the values such as recreation, fishing, and biodiversity, are an estimated $360 million per year. Valuing ecosystem services can embolden efforts to protect the environment, as governments use economic tools to influence policymaking, though it is critical that governments not encourage the commodification of nature.[5]

Troubled Seas, Oceanic Threats

Although society derives great economic and intrinsic value from the ocean, ever-increasing demand and the effects of climate change will alter the ocean's biological and chemical properties, making the ocean less resilient at the very time it is subject to heightened threats. Marine life and community structure is determined by salinity and temperature profiles throughout the ocean, blooming in places rich with nutrients, such as nitrate and phosphate, and with adequate dissolved oxygen.[6]

Peru, for example, is home to the world's most productive fishing grounds due to coastal upwelling. Peru's coast plunges dramatically into the sea, and southeasterly trade winds result in wind-driven coastal upwelling. As the surface waters are pulled westward, cold, nutrient-rich water is transported to the surface from the deep ocean. This influx of nutrients, such as nitrogen, phosphorus, and silicic acid, stimulates phytoplankton blooms, which may extend hundreds of kilometers offshore. The phytoplankton feed marine life, making Peru the second largest fishing nation after China.[7]

Peru's fisheries reveal how variables such as water temperature, salinity, and nutrient concentration dictate the productivity of fisheries and how a country's stability depends on relatively consistent ocean conditions. As rising carbon emissions fuel climate change, however, the ocean's biochemical environment is significantly altered, threatening marine ecosystems.

Ocean Acidification. In 2010, global carbon dioxide emissions accumulated to 30.6 gigatons (Gt), with industrial countries logging CO_2 emissions per person 10 times higher than those of developing countries. Globally, CO_2 emissions increased by approximately 45 percent between 1992 and 2010.[8]

These record years of carbon emissions endure in the environment. Over the past 200 years, the ocean has absorbed 525 Gt of carbon dioxide from the atmosphere, approximately half of what was emitted through fossil fuel use over this period. The ocean stores carbon in surface waters, in intermediate and deep ocean, and in marine sediments. The residence time of carbon in each reservoir varies, with surface waters capable of storing 600 Gt of carbon, which can still be there six years later, and marine sediments, which can store 30 million Gt for 100 million years. Though some carbon may remain in the reservoir for the residence time, carbon is exchanged readily between reservoirs every year.[9]

As chemistry dictates, rising CO_2 concentrations in the atmosphere increase the rate at which carbon dioxide is assimilated into the ocean. This absorption of carbon buffers global climate change, yet it also changes ocean chemistry by lowering the pH and reducing the number of carbonate ions available.

Between 1992 and 2007, the ocean's pH declined from 8.11 to 8.01. This rate of acidification may be faster than at any time within the last 300 million years. One consequence of a lower pH is an uncertain future for reef structures because acidic seawater stresses reef-building corals and the photosynthetic algae (zooxanthellae), which have a mutual dependence relationship with the coral. Zooxanthellae supply coral with critical nutrients, such as glucose, glycerol, and amino acids, that are products of the photosynthesis. The coral incorporates 90 percent of the organic material generated by the zooxanthellae into its tissue, producing proteins, fats, calcium carbonate, and carbohydrates. When corals are stressed in acidic waters they expel the zooxanthellae from the reef-structure, crippling their ability to grow. This is the phenomenon known as coral bleaching.[10]

Today coral reefs are experiencing the lowest pH and warmest ocean temperatures of the last 400,000 years, endangering 75 percent of these reefs worldwide. If carbon emissions continue to go unabated, by mid-century nearly all coral reefs will be threatened by such stresses as acidification, overfishing, shipping, and agricultural runoff.[11]

In addition, more acidic waters will adversely affect phytoplankton, which are responsible for nearly half of primary production on Earth. As seawater pH decreases, there is reduced availability of essential minerals such as calcium carbonate. Lower concentrations of these essential compounds will slow calcification, thus weakening skeletons of many phytoplankton species. Reduced rates of calcification will further disrupt carbon cycling because phytoplankton absorb carbon dioxide from surface waters and transform the carbon into sugar during photosynthesis. When phytoplankton die they sink, removing CO_2 from the surface waters and storing it in the deep ocean. This allows further absorption of atmospheric carbon dioxide by phytoplankton in surface waters.[12]

Ocean Warming and Hypoxia. Increased atmospheric concentrations of CO_2 will not only lower seawater's pH, it will also warm the ocean as a result of warmer air. When comparing the last 20 years to the average ocean temperature of the past century, it is apparent there has been a steady increase in seawater temperature—from 0.22 degrees Celsius above the long-term average in 1992 to 0.5 degrees Celsius above it in 2010. Ocean warming not only stresses marine organisms, it also stimulates bacterial activity, consequently expanding larger low-oxygen regions, known as dead zones.[13]

As climate change is predicted to enlarge dead zones, marine life and fish such as the blue marlin will lose critical habitat. Dissolved oxygen concentrations determine the habitat of the blue marlin because it is an energetic fish that requires large amounts of dissolved oxygen. When levels are high, marlin swim deeper, but when hypoxic zones encroach on their habitat from depth, the deep oxygen minimum layer becomes less deep, restricting the blue marlin to a habitat within a narrow surface layer. This further exposes the overfished marlin and other pelagic open-sea predators to surface fishing gear.[14]

Climate change is expected to alter ocean circulation in the Pacific and thus locations of critical marine habitats and migratory pathways, which will have uncertain impacts on large pelagic predators.[15]

Loss of Sea Ice. The habitat of ice-dependent species will be threatened by increased atmospheric CO_2 concentrations as well. Arctic sea ice extent shows a pronounced yearly cycle, with approximately 15 million km^2 in March and 5 million km^2 in September. Yet in 2012, Arctic sea ice reached a new low point at 3.41 million km^2. This was the lowest summer minimum extent in the 33-year satellite record. In fact, scientists estimate the sea ice extent might have been at its lowest in 8,000 years.[16]

At the current rate of ice loss, the Arctic may be completely ice-free during summer months 30 years from now. The last time that happened was at the height of the last major interglacial period, 125,000 years ago. The disappearance of ice threatens critical habitats for organisms at the base of the food web, such as algae or krill, which in turn feed larger animals. As shrimp populations are reduced, ice-dependent ecosystems will be threatened by disappearing habitats and a loss of species fundamental to their food web.[17]

In the Arctic, the ice layer restricts winds and wave action near the coastlines, buffering the force of storms and reducing erosion. As ice disappears and sea level rises, the impact of storms will be compounded in Arctic communities. Worldwide, melting ice and water expansion due to warming temperatures also mean rising sea levels, which threatens coastal communities, island nations, and critical habitats, such as coral reefs, mangroves, and wetlands.

Unexpected Sea Changes. The effects of climate change have already manifested themselves globally and are happening more rapidly than at any time in history, outpacing many species' ability to adapt to the new environment. The broad consequences of climate change have not been fully anticipated, with certain environments changing at a rate greater than the global average. The Arctic is experiencing rapid transformation, with a rate of temperature increase much higher than the global rate and with extensive ice loss. One unexpected consequence of melting sea ice is likely to be more-productive phytoplankton blooms earlier in the season. This may alter marine food webs, such that benthic, deep-sea communities are favored over pelagic, open-ocean communities.[18]

In addition, depending on the severity of climate change, ocean circulation may be transformed entirely. It has been hypothesized that ocean circulation may be altered due to melting of the Greenland ice sheet and Arctic sea ice. As the Gulf Stream current flows northeastward toward Europe, the warm, salty water releases heat to the atmosphere. As the water cools, it becomes very dense compared with the surrounding waters, sinking to the bottom of the ocean. Thus the North Atlantic is an area of "deep-water formation," driving thermohaline circulation in the ocean.

As large volumes of ice are lost from the Arctic, the ocean's salinity is lowered, decreasing its density. The influx of freshwater would inhibit the formation of deep water in the North Atlantic. This would dramatically alter the climate and reduce the oceanic sequestration of carbon dioxide in these regions, thereby leading to a positive feedback mechanism that would increase atmospheric CO_2 concentrations and do more to melt polar ice. The increase in temperature and change in salinity regime due to climate change will drastically affect fisheries and the marine ecosystems upon which geopolitical stability depends.

Devastated Fisheries. More than 500 million people rely on fisheries and aquaculture for their livelihood, and 3 billion people consume fish for 15 percent of their protein intake. The increasing human population will place additional pressure on already stressed fish populations and marine ecosystems as a result of biogeochemical regime shifts and warming sea surface temperatures.[19]

The total fish catch has stabilized at around 80 million tons over the last several years—up from approximately 60 million tons in 1970. Pressure on marine ecosystems due to exploitation of commercial fish species has led to the depletion and overexploitation of 70 percent of the world's fisheries. This trend is cause for significant concern. Between 1992 and 2008, the proportion of fish stocks considered overexploited, depleted, or recovering increased by 33 percent, reaching 52 percent of all fish stocks, while the share of fully exploited stocks rose by 13 percent, reaching 33 percent of all fish stocks.[20]

Valuable fish species such as tuna have been especially targeted by commercial fishing operations. Bluefin tuna species are susceptible to collapse under continued fishing pressure. Tuna catches, for example, reached 4.2 million tons in 2008, up from 600,000 tons in the 1950s. With the massive reduction of top predators like sharks and tuna, marine food webs may be functionally changed, adversely affecting the remaining marine ecosystem by altering how productivity is expressed. Trophic cascades that reflect changes down the food web from predator removals have been increasingly documented in marine ecosystems. Overfishing may therefore not only reflect on target species, it may also cascade throughout the food web. Additional pressure on fisheries and habitat will come not only from increased demand on ocean resources but also from coastal development and pollution.[21]

Impacts of Human Activities. In October 2011 world population reached 7 billion people, with 60 percent of people living within 100 kilometers of a coastline. Of the world's 39 cities with populations over 5 million, 60 percent are within 100 kilometers of a coast—including 12 of the 16 cities that have more than 10 million people in them. Development of

Bluefin Tuna for sale at the Tsujiki Fish Market in Tokyo.

Stewart Butterfield

the world's coasts alters watershed hydrology due to changes in upstream vegetation and the installation of roads and other impervious surfaces, which increase runoff into the sea. Coastal development additionally results in nutrient and sedimentation loading due to human activity, such as agriculture and use of road salts. Such alterations to the hydrological and chemical environment imperil fisheries and critical habitats such as wetlands, mangroves, and estuaries.[22]

Furthermore, pollution has lasting consequences on marine life. Plastic debris particularly affects marine ecosystems through entanglement and ingestion. In the North Pacific gyre (a giant circular ocean surface current), approximately 35 percent of the plankton-eating fish studied had ingested plastic, and they averaged 2.1 plastic items per fish. Plastic debris degrades very slowly and therefore has an enduring legacy on marine life. The ocean's role as a repository for plastic debris must end as its costs rise ecologically and economically. In the Asia-Pacific region alone, the estimated cost of marine debris on activities such as boat repairs is more than $1 billion per year.[23]

Solutions for Sustaining the Seas

In order to protect oceans and fisheries, governments and all stakeholders must implement a variety of strategies domestically and internationally. To mitigate the effects of climate change and ensure global stability, it is critical that action plans that engage an inclusive and broad-based governance approach are used effectively as soon as possible.

The critical first step toward sustainable fishery operations and a healthy ocean is international collaboration. Globally, governments must commit to a far-reaching climate change agreement to reduce atmospheric CO_2, protect marine life, and mitigate acidification, ocean warming, and the disappearance of the world's ice sheets. The impact of population growth on ocean resources must also be considered in global climate discussions in order to prepare for sustainable management of the seas.

In order to reduce demand on fisheries and the oceans, governments can enact policies that implement coastal and marine spatial planning (CMSP) frameworks and can establish catch shares. Coastal and marine spatial planning can greatly help marine ecosystems as it emphasizes comprehensive, adaptive, ecosystem-based management systems. CMSP identifies areas of the coasts and seas that are most suitable for various classes of activity in order to reduce environmental impacts, preserve critical ecosystem services, and meet economic objectives. CMSP facilitates compatible uses, maximizing benefits for all.[24]

In recent years, CMSP has gained popularity because it provides a multifaceted perspective on demands from different sectors, which provides a more complete evaluation of cumulative effects. Thus, coasts and marine areas are planned to simultaneously preserve resilient ecosystems and biodiversity and support a range of human uses.

In the United States, CMSP enabled the National Oceanic and Atmospheric Administration, the U.S. Coast Guard, and several other stakeholders to examine a range of demands in the Boston coastal area in order to decrease whale mortality from ship traffic in the Stellwagen Bank National Marine Sanctuary. The stakeholders reconfigured the Boston Traffic Separation Scheme (TSS) and succeeded in reducing whale mortality from risk of collision with a ship by 81 percent for baleen whales and 58 percent for engendered right whales. The new TSS increased shipping time by only 9–22 minutes and eliminated conflict with deepwater liquefied natural gas port locations. Furthermore, the TSS increased marine safety by separating shipping traffic from areas traveled by commercial and recreational vessels.[25]

Catch shares provide communities with a strategy to combat overfishing. Catch shares allocate shares of fish to individual fishers, communities, or fishery associations. These dedicated access privileges allocate shares of the fish stock to each group or individual, encouraging sustainable practices. Well-designed catch shares not only reward fishers for innovation, lowering their costs and delivering quality products to the market, they may also prevent fishery collapse across a range of ecosystems. Catch share programs must be carefully designed to avoid aggregation of the shares by a few individuals or entities; they also require strong institutions to create and enforce appropriate arrangements.[26]

While implementation of fishery management programs such as catch shares can reduce destructive fishing practices and fishery collapse, pressure on fish stocks remains high. In order to minimize bycatch and destructive fishing practices, governments must elevate the role of small-scale and artisanal fisheries, which have largely been overlooked thus far.

Although small-scale fishing and large-scale fishing operations catch about the same amount of fish for human consumption each year, large-scale operations receive government subsidies. This leads to overcapacity and overfishing and so should end, as large fishing operations consume approximately seven times more fuel and cost 10 times as much as small-scale fishing. They also employ 11.5 million fewer people and hire fewer people for each $1 million invested in fishing vessels, and they discard 8–20 million tons of fish and marine life at sea, whereas small-scale fishing wastes very little sealife.[27]

In order to reduce the volume of discarded fish and sealife, governments and communities could develop markets for bycatch, such as tradable bycatch credits. These aim to create a market for marine life so that it is not wasted, while protecting conservation goals by preventing exploitation of the system and sales of valuable species.

Governments and scientists are working to establish sustainable aquaculture to further diminish pressure on wild fisheries. Although aquaculture is a relatively new contributor to global food production, it has become increasingly important over the last several decades. Global production of food fish from aquaculture increased from 1 million tons in 1950 to 52.5 million tons in 2008. Between 1992 and 2009, aquaculture increased by 260 percent—growing primarily in Asia, including by 315 percent in China alone.[28]

Sustainable aquaculture holistically farms marine life. In the 1980s, John Ryther of the Woods Hole Oceanographic Institute developed an oyster farming approach that raised oysters in the sewage water generated by 50,000 people. The oysters fed on algae that grew in the nutrient-rich environment. To manage the waste produced by the oysters, Ryther introduced polychaete worms that would feed and then be harvested and sold as fish bait. Thus, properly managed aquaculture can decrease pressure on wild fisheries and supply commercial species for the world's market.[29]

Yet if aquaculture is poorly governed, it can have devastating effects on the surrounding environment. Shrimp and salmon aquaculture operations can be particularly damaging. Salmon and shrimp require large quantities of fishmeal and fish oil in their diet. The fish caught to supply this would otherwise support wild fish species. Globally, shrimp and prawn aquaculture has increased approximately 400 percent between 1992 and 2009. In many regions, such as Southeast Asia, highly productive coastal regions are developed and valuable mangroves are cleared for aquaculture. Between 1990 and 2010, some 3 percent of mangrove extent (approximately 500,000 hectares of mangrove forest) was lost to coastal development and conversions to agriculture and aquaculture.[30]

Furthermore, aquaculture operations can hurt the surrounding environment through poor management of high volumes of fish waste, an influx of antibiotics or pesticides, and competition between wild fish species and escaped farm fish. It is estimated that for every ton of fish raised in aquaculture operations, 42–66 kilograms of nitrogen waste and 7.2–10.5 kilograms of phosphorus waste are produced annually. Such organic loading of the seabed and nutrient enrichment of the water column can cause eutrophication, the creation of dead zones that are inhospitable to marine life. Shifting targets of aquaculture from top predators toward lower trophic levels, particularly filter-feeders such as oysters and other bivalves, may make aquaculture more sustainable.[31]

Consumers could decrease their demand on ocean resources by eating less seafood and eating lower on the food chain, preferring anchovies to tuna, for example. Seafood guides from the Monterey Bay Aquarium and Blue Ocean Institute, among others, help consumers purchase more-sustainable seafood options.

To address another problem, fishers can modify their fishing gear in order to decrease bycatch. For example, changing the type of hook used on long-lines from J-hooks to circle hooks can reduce leatherback turtle catch by up to 90 percent.[32]

As the impacts of climate change intensify and as national and global policies are delayed by a dearth of political leadership, the ocean is becoming irreparably damaged. In order to prevent a convergence of changes in the ocean through acidification, ocean warming, sea level rise, pollution, hypoxia, and exploitation of marine resources, solutions must be implemented immediately. If action is not taken, stressors will combine to create an outcome more extreme than any individual change currently projected.

Individuals can press their political leaders to collaborate internationally in order to address these global threats. Catch shares, tradable bycatch credits, and well-managed aquaculture are a few solutions available to governments. Through a broad-based governance approach, resources can be managed at multiple levels, and all stakeholders can cooperate to advance initiatives that protect a common future.

The ocean is Earth's greatest resource. Future planetary and geopolitical stability will depend on managing the seas sustainably and protecting the global environment. If governments fail to do so, the ocean and its fisheries will be further degraded, leading to an ecological collapse and unraveling the ecosystems that humans depend on for so much.

CHAPTER 7

Energy as Master Resource

Eric Zencey

On a spring morning in 1890, the German chemist Wilhelm Ostwald arose early in a Berlin hotel room, preoccupied by a conversation of the previous evening. He had come to Berlin to meet with physicists to discuss his work developing a new theoretical foundation for chemistry, one consistent with the first and second laws of thermodynamics. The first law holds that matter and energy can be neither created nor destroyed, only transformed. The second law states that in any such transformation, the capacity of the energy to do useful work is diminished. The energy does not disappear—the first law—but some of it has become "bound" energy, energy incapable of being useful. In 1865, Rudolf Clausius coined the term *entropy* as a label for this degraded energy, and it allowed him to state the law succinctly: within any thermodynamically closed system, energy is conserved but entropy must increase.[1]

Ostwald was finding these laws enormously useful in developing a rigorous understanding of chemical transformations—work that would eventually win him a Nobel Prize. He had come to the conclusion that the science of energy was not merely a subfield within physics but its very foundation. While in Berlin, he told the physicists that their discipline, too, needed to undergo a "radical reorientation" to accommodate these fundamental truths. Because matter is indestructible and energy degrades, energy must be the key: "From now on . . . the whole of physics had to be represented as a theory of energies."[2]

The group did not give him a warm reception. Ostwald wrote later that they found his idea "so absurd that they refused to take it seriously at all" and instead offered just "ridicule and abuse." He spent a fitful, nearly sleepless night and arose early to walk the still-dark streets, mulling over how best to proceed. Sunrise found him in the *Tiergarten*, surrounded by the budding life of a spring morning in the park. And there he had an insight that he later described in religious terms, calling it a "personal Pentecost" that came to him

Eric Zencey is a fellow of the Gund Institute for Ecological Economics at the University of Vermont and a visiting lecturer in the Sam Fox School of Visual Design and Arts at Washington University in St. Louis.

www.sustainabilitypossible.org

with a force and clarity he had never experienced: "All," he saw, "is energy." And if energy cannot be created and cannot be recycled, then the energy budget of the planet, and of the human economy on the planet, must be finite.[3]

Energy and the Transformation of Science

Ostwald developed this epiphany into his doctrine of energetics, which he thought should revolutionize all human understanding: natural and earth sciences, of course, but also history, economics, sociology, politics, even ethics and moral philosophy. (This, because to Ostwald the laws of thermodynamics implied a new categorical imperative: "Waste no energy!")[4]

Thermodynamics did indeed begin to reshape many disciplines. Solutions to three of the outstanding thermodynamic problems in the Newtonian physics of the day—the photoelectric effect, Brownian motion, and black-box radiation—led a young Swiss patent clerk, Albert Einstein, to his overthrow of the discipline's mechanistic foundations with his general and special theories of relativity. Biology was reconstructed on thermodynamic grounds in the 1920s through the work of A. G. Tansley, Edgar Transeau, Max Kleiber, and others who began conceiving of organisms as energy fixers or consumers and of natural systems as complex webs of energy flows and transformations, thereby developing the modern science of ecology. Alfred Lotka and Howard Odum extended the approach, pointing to the role that energy appropriation plays in evolution: individuals and species that have the largest net energy surplus can dedicate more of their life energy to reproduction, outcompeting their rivals.[5]

At the turn of the nineteenth century, the American historian Henry Adams, having read Ostwald and others on the subject of energy, toyed with a thermodynamic interpretation of history, perhaps merely as metaphor, perhaps as a parodic dissent from the scientific progressivism of the day, perhaps as a literal modeling based on the figures for coal consumption in which he briefly immersed himself. In the mid-1950s William Frederick Cottrell, an American sociologist, linked social and economic change to changes in energy sources and the technologies they power. And in his 1970 *Pentagon of Power*, historian Lewis Mumford took up the theme.[6]

Increased interest in ecological and environmental history late in the twentieth century led to sustained inquiries that focused on the energy history of the human economy, such as Alfred Crosby's *Children of the Sun: A History of Humanity's Unappeasable Appetite for Energy* in 2006. Seen through the thermodynamic lens, what has been called the Industrial Revolution is, more properly, the Hydrocarbon Revolution, a once-in-planetary-history drawdown of stored sunlight to do work and make wealth in the present. The petroleum era will most likely depart as suddenly as it came; in the grand sweep of geologic time, our use of petroleum is just an instant, a brief burst of frantic activity that has produced exponential growth in wealth and human population—and in humanity's impact on planetary ecosystems. (See Figure 15–1 in Chapter 15.)[7]

Economics: The Failed Revolution

Alone among disciplines that aspire to the status of rigorous science, economics remains relatively unaffected by the reconstructive impulse of thermodynamics. Most of the discipline retains its roots in the Newtonian mechanism, in which every action has an equal and opposite reaction and there are no irreversible flows. Nowhere is this clearer than in the circular flow model of production and consumption that lies at the heart of standard economics modeling, in which the economy is seen as a closed system of exchange between households (which supply factors of production and buy goods and services) and firms (which use factors of production to make goods and services for sale to households). As Lester Thurow and Robert Heilbroner describe it in *The Economic Problem*, "the flow of output is circular, self-renewing and self-feeding," because "outputs of the system are returned as fresh inputs." This is patent nonsense. Anything that can take as input what it excretes as output is a perpetual motion machine, a violation of the second law of thermodynamics.[8]

In reality, an economy—like any living thing or any machine—sucks low entropy from its environment and excretes a high-entropy wake of degraded matter and energy. Matter

The flow of output: a B-29 assembly line in 1944.

USAFHRA

can be recycled; once extracted from the planet, much of it could be kept within the circular flow of the monetary economy instead of being discarded back into the environment. But recycling matter takes energy, which cannot be recycled. Thus energy is ultimately the limiting factor on the generative side of the human economy. (There are also limits on the waste side, in the finite capacity of the planet to absorb our effluents.) This is why Romanian-born American economist Nicholas Georgescu-Roegen described the entropy process as "the taproot of economic scarcity"—and why energy is the master resource.[9]

Over the years, conventional economics has been critiqued several times in light of thermodynamics. One critique came from another Nobel-laureate chemist, the Englishman Frederick Soddy. In the 1920s and 1930s he produced a series of books developing the idea that an economy is, at bottom, a system of energy use. The chief mechanism by which the economy denies this physical truth, Soddy believed, was its monetary system.[10]

Soddy drew distinctions between wealth, virtual wealth, and debt. Wealth is the stock of physically useful objects the economy has produced; it has an origin in low entropy and is subject to entropic decline. Money is virtual wealth; it symbolizes the bearer's claim on real wealth and resists entropic decay. Debt, held as an asset by those who lend money, is a claim on the future production of real wealth.

Soddy's fundamental insight was that when money is lent at compound interest, claims on the future production of real wealth increase exponentially—but real wealth can only grow incrementally, through an expansion of the economy's matter-and-energy throughput or through achieving greater efficiency. As the monetary system encourages public and private debt to grow faster than the economy can grow the means of paying it back, the system develops an irresistible need for some form of debt repudiation. This comes as inflation, bankruptcy, foreclosure, bond defaults, stock market crashes, bank failure, pension fund wipeouts, collapse of pyramid schemes, and loss of paper assets and expected investment income of any form.

Aggressive expansion of the economy's matter-and-energy throughput raises hopes and expectations along with output of real wealth. Those hopes and expectations make growth-through-debt seem normal, which can stave off the inevitable financial reconciliation for a time. Eventually, however, expansion of throughput hits a local or absolute limit, confidence falters, and the system rapidly "de-leverages" into collapse. Staving off debt repudiation simply ensures that when it comes it will come hard and fast, as a crisis—as it did in the Great Depression, as it has in every other downturn the global economy has experienced since then.[11]

A few economists gave Soddy's ideas serious attention and found merit in them. The discipline as a whole, however, closed ranks against him, ignoring his ideas and dismissing him as a crank, a scientist who had overstepped his expertise—much as the physicists in Berlin had responded to Ostwald.[12]

Another thermodynamics-based critique of economics was offered in the 1970s by Georgescu-Roegen and his student, Herman Daly. Georgescu-Roegen's masterwork, *The Entropy Law and the Economic Process*, serves as the foundation of ecological economics—an emergent school that combines an appreciation of the laws of thermodynamics with a recognition that humans receive economically valuable but generally nonmarket, unpriced ecosystem services from nature.[13]

In purely physical terms, Georgescu-Roegen noted, an economy consists of nothing more than a set of institutions and processes by which we turn valuable low-entropy inputs into valueless, high-entropy waste. Production of waste is, of course, hardly the point. What we seek is psychological: the "augmentation of an immaterial flux, the enjoyment of life." If that is the ultimate purpose, then it is foolish and ultimately dysfunctional to judge the economy by any other measure. Appreciation of energy as a master resource thus leads directly to use of alternative economic indicators, metrics that assess the economy's capacity to provide sustainable well-being, happiness, or life satisfaction to its participants. (See Chapter 11.)[14]

The thermodynamic revolution in economics also suggests a different conceptual slicing of human productive activity, an alternative to the triumvirate of land, labor, and capital that is offered by neoclassical theory. All economic value is produced by intelligence operating on matter using energy. Capital—the tools and equipment we use to increase labor productivity—is matter embodying both energy (the energy used to extract, refine, shape, and assemble the materials from which it is made) and intelligence (the accumulated inventions and innovations that have gone into its design). Labor is discretionary intelligent energy that participates in production. Land—nature—is the source of all matter and energy, and its systems also embody billions of years of trial-and-error design intelligence encoded into genes, evolution's information storage system. Energy as master resource thus offers a continuity of explanation and understanding between economics and ecology, a necessary step in establishing our economies on an ecologically sound foundation.[15]

In this model, it is easier to see that under conditions of maximum sustainable uptake of matter and energy from the environment, any further increase in the sum total of human well-being has to come from the development of intelligence—from innovation, from intelligent distribution of the products of the economy to achieve maximum well-being, from the application of what we know and can learn about wringing greater efficiency from matter and energy throughput. However inventive humans turn out to be, they will never invent their way around the laws of thermodynamics. That fundamental truth is denied by standard infinite-growth theory, which blithely projects productivity gains from technological innovation indefinitely into the future.

We can continue to seek and enjoy greater life satisfaction while maintaining a constant, steady-state, sustainable throughput of matter and energy in the economy. Our ability to raise our standard of living in a steady-state economy is limited only by our intelligence and our imagination—and the laws of thermodynamics.[16]

Net Energy Analysis and Energy Return on Energy Invested

An appreciation of energy as master resource leads directly to an appreciation of a key economic indicator that is more fundamental than the monetary price of energy or even an economy's gross energy throughput: its net energy uptake, the energy available to an economy after the energy costs of obtaining that energy are paid. Crucial to this figure is the energy return on energy invested, or EROI, of energy sources, a calculation pioneered by researchers Cutler Cleveland, Charles Hall, Robert Herendeen, and Randall Plant. It takes energy to acquire energy: to make economic use of a barrel of oil requires not only drilling the well but also transporting the oil to a refinery, converting it to a variety of petroleum products, and shipping them to end users—as well as expending energy to make the drilling rig, the steel in the refinery equipment, the tank trucks that take gasoline to service stations, the automobiles that burn the fuel, and so on. Only the net that is left after all this energy expense has been paid is available to augment that "immaterial flux, the enjoyment of life," as Georgescu-Roegen put it.[17]

The EROI of fuels can rise with technical efficiencies but tends to decline over time. For instance, according to a 1981 paper exploring this idea, the petroleum energy obtained per foot of drilling effort declined from about 50 barrels of oil equivalent in 1946 to about 15 in 1978. While the authors did not calculate EROI specifically, a figure can easily be inferred: the energy return on energy invested in drilling declined from about 50:1 to 8:1 in that period. Direct calculations of EROI for the U.S. oil industry show that it dropped from roughly 24:1 in 1954 to 11:1 in 2007.[18]

Joint Pipeline Office

An oil and gas drilling installation on an artificial island built for the purpose in the Beaufort Sea north of Alaska.

The reason is simple: other things being equal, rational beings will seek the largest increment of benefit for the smallest outlay—the biggest bang for the buck (or calorie). Naturally, high EROI sources were exploited first. Worldwide, and despite aggressive development of more-efficient extraction techniques, the average EROI of petroleum is falling, from a high of 100:1 in the 1920s to about 20:1 today.[19]

In calculating EROI, the boundaries of the analysis are crucial to the result and are the subject of much debate and discussion. If the exploitation of an energy source requires infrastructure (like roads, vehicles, a steel industry) that has other uses, how much of the energy embodied in that infrastructure should be assigned on a per unit basis to the energy source that flows through it? How far should the boundaries of analysis be extended? The answers are by no means clear-cut, and this accounts for some of the confusion, cross talk, and variety of result in this field of study.[20]

An agreed-upon standard for the boundaries of EROI analysis would allow for economically rational decisionmaking between different energy systems. Even without that standard, EROI analysis reveals the irrationality of making those choices according to current market price, which is a human construct, dependent on current demand, subsidies, taxes, and the rates at which a flow of energy is extracted from its global stock. At the macroeconomic level, rational policymakers should be trying to maximize total sustainable delivered well-being, which (other things being equal—which they often are not) would mean maximizing the EROI of a sustainable energy system for the economy. The effort to use price signals to find and promote that outcome requires that the relative monetary prices of different kinds of energy reflect their relative social costs and benefits—a project that must begin with their relative EROIs. (See Table 7–1.)[21]

If we continue to disregard the climate consequences of burning carbon-based fuels, the EROI of oil will decline further, as we drill deeper, transport farther, and bring energetically expensive oil from tar sands and shales (which have EROIs as low as 5:1) online. Is there some minimum EROI

Table 7–1. Energy Return on Energy Invested, Average and High and Low Estimates, Different Energy Sources

Energy Type	Average	High Estimate	Low Estimate
Oil	19:1		5:1
Coal		85:1	50:1
Natural gas	10:1		
Hydroelectric		267:1	11:1
Nuclear		15:1	1.1:1
Wind	18:1		
Solar photovoltaic		10:1	3.7:1
Geothermal electricity		13:1	2:1
Geothermal heat pump		5:1	3:1
U.S. corn ethanol		1.8:1	< 1:1
Brazilian sugar cane ethanol		10:1	8:1
Soy biodiesel		3.5:1	1.9:1
Palm oil biodiesel	9:1		
Tar sands oil	5:1		
Oil shale		4:1	1.5:1
Wave	15:1		
Tidal	6:1		

Source: See endnote 21.

that an economy or civilization needs in order to be successful? One study postulates that an EROI of 3:1 is "a bare minimum for civilization. It would allow only for energy to run transportation or related systems, but would leave little discretionary surplus for all the things we value about civilization: art, medicine, education and so on." The authors estimate that "we would need something like a 5:1 EROI from our main fuels to maintain anything like what we call civilization."[22]

But a civilization with a 5:1 average EROI cannot support the kind of military investment that can be made by a civilization with a 6:1 or 7:1 EROI—and if military force is useful in securing access to resources, then the minimum EROI a civilization needs to survive is probably some close correlate of the average EROI of its potential enemies and competitors.

If we bracket off such concerns, then the minimum EROI for any particular civilization will depend on a variety of internal factors, some of which are not easily quantified. Appropriation of energy has social, political, and ecological costs and benefits that will depend on factors like the resilience of the host ecosystems, the resilience of the civilization's social systems and social capital, and the expectations its members have for the future, including their expectation of material comfort for themselves and their progeny. It is likely that any definitive answer to the question of a minimum EROI for our civilization can only be derived experimentally—history will reveal it to us when our civilization falls below it.

Can renewables be built out and exploited rapidly enough to avoid making that experimental determination? Perhaps. (See Chapter 8.) If educated guesswork puts the EROI floor at 5:1, a figure that is approached by current petroleum technologies, apparently we can breathe easier knowing that renewables generally do significantly better: photovoltaics (PV) are conservatively estimated at 10:1 and wind at 20:1 or perhaps 50:1.[23]

But some EROI analysts worry that as society is forced to make do with less oil, it will fall into an EROI or Energy Trap. This, according to physicist Tom Murphy, comes about because the energy it takes to build the infrastructure necessary for a sustainable, renewable energy economy must come from current energy consumption. Unlike monetary investments, which can be made on credit and then amortized out of the income stream they produce, the energy investment in energy infrastructure must be made upfront out of a portion of the energy used today: "Nature does not provide an energy financing scheme. You can't build a windmill on *promised* energy."[24]

The arithmetic is daunting. To avoid, for example, a 2-percent annual decline in net energy use, replacing that loss with solar photovoltaic (with an EROI pegged at 10:1) will require giving up 8 percent of the net energy available for the economy. (This is because the EROI of solar PV is calculated over the life of the equipment: a 10:1 return over 40 years means that the break-even point is four years out, and until then most of the energy invested in PV construction is a sunken cost, an incompletely compensated energy expense.) "We cannot," writes Murphy, "build our way out of the problem. If we tried to outsmart the trap by building an eight-unit replacement in year one, it would require 32 units to produce and only dig a deeper hole. The essential point is that up-front infrastructure energy costs mean that one step forward results in four steps back."[25]

The grim truth, Murphy warns, is that on a sheer energetic basis it seems to make more sense to continue to develop oil, even with a 5:1 EROI, than to build wind or solar PV capacity with higher EROIs. While there are plenty of reasons to move to solar and away from oil (climate change prominent among them), EROI, according to Murphy, is not one of them. The problem is rooted in the sunken energy costs of petroleum infrastructure (which makes the continued use of petroleum energetically cheap) and the non-negotiable reality of the energy economy.[26]

The goal of a renewable energy economy is clear, but the path to it seems blocked. The paradox is reminiscent of the one proposed by Zeno, whose logic denied the possibility of all motion: you can never get from point A to point B because first you must go halfway to point B, then halfway again, then halfway again, and so on, never arriving. Legend has it that Diogenes of Sinope refuted Zeno by standing up and walking about. The paradox of the Energy Trap may not be so easily resolved. Refraining from energy expenditure on consumption today in order to use that energy to invest in the infrastructure we need to ensure energy consumption 10, 20, and 50 years into the future, Murphy warns, will require a kind of sacrifice and political will that does not come easily to representative democracies and for which there is scant historical precedent. Politically, the most acceptable path is to finance the energetic investment not by decreasing energy use for consumption today but by maintaining energy use for consumption while increasing the total energy appropriation of the economy—an aggressive expansion of the economy's footprint in paradoxical service to the goal of achieving sustainability.[27]

Eventually, solar and renewables will hit a takeoff point: they will capture enough energy to support the construction of additional solar and renewable infrastructure without requiring us to reallocate energy use away from maintaining the living standards we then enjoy. Achieving this at a high level of energy consumption becomes increasingly difficult as the average EROI of our energy sources declines. If the net energy captured by the economy begins to decline as the peak of fossil fuel production passes, the Energy Trap seems unavoidable.

Can conservation and efficiency save us from the Energy Trap? Maybe. The United States could significantly reduce gasoline use with the simple expedient of carpooling, for instance. Four vehicle occupants instead of one

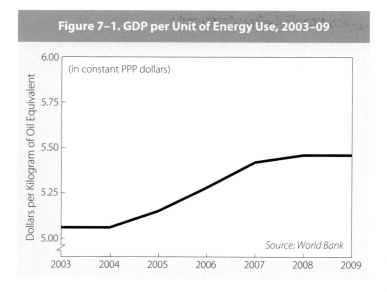

Figure 7–1. GDP per Unit of Energy Use, 2003–09

(in constant PPP dollars)

Dollars per Kilogram of Oil Equivalent

6.00

5.75

5.50

5.25

5.00

2003   2004   2005   2006   2007   2008   2009

Source: World Bank

represents a 75 percent savings, and if the savings were dedicated to building renewable infrastructure (a big "if," but still), this would go a long way toward solving the problem. According to calculations of energy use per constant gross domestic product dollar (see Figure 7–1), current efficiency efforts achieve an annual savings of 1.39 percent, which could be dedicated to building renewable infrastructure with no decrease in the amount of energy going to consumer satisfactions.[28]

But these savings are not sustainable. The low-hanging fruit can be plucked only once, and marginal returns from future conservation and efficiency efforts will necessarily decrease. And whatever savings we achieve, there will be pressure to use them to increase or simply maintain current consumption instead of building solar infrastructure. Yielding to that pressure will condemn future humans to a poorer, stingier, less commodious life.

Sometimes a problem that seems irreducible at the macro scale can, like Zeno's paradox, be solved at the level of individual behavior. Would a rational consumer postpone for a few years some of his or her energy-intensive consumption in order to invest in insulating a house or installing solar panels? Yes—given the right market signals and realistic assumptions about the cost of energy tomorrow. Consumers decide to make this sort of investment every day—and those decisions could cumulate into the macro result that the Energy Trap tells us would be politically difficult to achieve.

This much is clear: sooner or later we will have an economy that runs on its current solar income. The amount of energy that economy will have at its disposal depends on the choices we make today.

Toward a New Worldview

Reality, economic reality included, is sufficiently complex that diametrically opposed idea systems can serve as lenses through which to interpret it, with both systems claiming to be confirmed by what is seen. When an economy is founded on an EROI of 100:1, you can hold almost any economic theory you want and still see an enormous generation of wealth. The decline in average EROI of the world economy brings political challenges—including pressure for austerity in government budgeting—and a kind of evolutionary pressure to get our economic theories right. The incorporation of thermodynamics into economics as a foundational idea system would bring the most influential social science into congruence with physical reality.[29]

Carpool sign on a Maryland Interstate highway.

It would also return economics to its roots in political economy. A steady-state economy will have to face issues of fairness and justice in distribution that were more easily addressed (or postponed to the future) in a high-EROI, supposedly infinite-growth economy. And economically rational, benefit-maximizing choices about energy use will turn on such "externalities" as the social and political costs and benefits of different energy systems, which fall outside of the discipline of economics as currently practiced. Economics will either admit these issues into the discipline or confess its abject impotence to illuminate the most pressing economic issues of our era.

Ultimately, economics will have to recognize that we live on a finite planet and that the laws of thermodynamics apply to economic life as to all other life. This observation from the British physicist Arthur Eddington remains as apt today as when it was written nearly a century ago: "The second law that entropy always increases holds, I think, the supreme position among the laws of Nature. If someone points out to you that your pet theory of the universe is in disagreement with Maxwell's equations—then so much the worse for Maxwell's equations. If it is found to be contradicted by observation—well, these experimentalists do bungle things sometimes. But if it is found to be against the second law of thermodynamics I can give you no hope; there is nothing for it but to collapse in deepest humiliation."[30]

Had economists collapsed in deepest humiliation on being shown in the 1930s or again in the 1970s that their theories fell against the second law, we would have made a great deal more progress toward the goal of establishing our economy and civilization on a sustainable flow of matter-and-energy throughput. Foresters have a saying that is appropriate here. The very best time to plant a tree, like the best time to admit that energy is the master resource, is decades ago. The second best time is today.

CHAPTER 8

Renewable Energy's Natural Resource Impacts

Shakuntala Makhijani and Alexander Ochs

Our fossil-fuel-based economy is environmentally, socially, and economically no longer acceptable. Recent increases in the frequency, severity, and regional spread of heat waves, droughts, wildfires, storms, floods, and other extreme weather events are an early indication of even more damaging climate change impacts sure to come.

Although governments across the world have made a commitment to limit Earth's warming to 1.5–2 degrees Celsius (3.6 degrees Fahrenheit) over pre-industrial levels in order to avoid disastrous climate impacts, current emissions trends put us on a path to much greater warming. Global carbon dioxide emissions from fossil fuel energy combustion, the single largest contributor to greenhouse gases (GHGs), grew by 34 percent from 2000 to 2010. Leading research institutions estimate that global average surface temperatures will increase by between 2 and 11.5 degrees Fahrenheit by 2100, with the most recent estimates projecting that the high end of this warming range is the most probable if no swift action is taken. This warming will affect millions of people through droughts, water stress, decreased agricultural yield, coastal flooding, global species extinctions, heat waves, and the spread of infectious diseases.[1]

In addition to climate impacts, fossil fuel development and emissions cause environmental damage, including altered landscapes, acid rain, freshwater pollution and decline, and polluted soil and rivers, as well as human health impacts such as damage to the brain, heart, kidney, lungs, and immune system. These human and environmental costs are rarely internalized in polluters' fossil fuel energy costs but are instead borne by society as a whole.[2]

Socioeconomic costs are reason enough to question our fossil fuel economy. Today's economies are vulnerable to energy commodity market volatility; price spikes reduce economic output and cause layoffs. Some countries, among them the poorest on the planet, spend more than 10 percent

Shakuntala Makhijani is a research associate with Worldwatch Institute's Climate and Energy Program. **Alexander Ochs** is the director of the program.

www.sustainabilitypossible.org

of their gross domestic product importing fossil fuels. U.S. taxpayers spend $345 billion annually just to pick up the pollution and health bills related to coal use.[3]

Reliance on unsustainable energy sources is no longer necessary; the transition to a sustainable energy system based on high efficiency (see Box 8–1) and renewable sources, as well as smart grid and storage solutions, is under way. Renewable technologies broke all growth records in recent years. In 2011, new investments in renewables topped those in conventional energy technologies for the first time in modern history. U.S. wind power capacity almost tripled and solar energy jumped ninefold since 2007. And 17.1 percent of Germany's electricity comes from renewable sources.[4]

Box 8–1. The Role of Efficiency and Conservation

Current global energy demand is about 14 terawatts, a figure that is expected to double by 2050. Given the rapid acceleration in renewable energy expansion that is needed just to meet energy needs without fossil fuels, energy efficiency measures are essential to ensure that new renewable capacity offsets rather than supplements fossil fuel power production. Energy conservation is especially important in the context of sustainability constraints, as even renewable energy sources can have significant resource and environmental impacts.

Energy efficiency measures work synergistically with renewable energy systems. When an electricity consumer demands one less unit of energy because of efficiency measures, the system typically saves much more than one unit of energy because of avoided losses during transmission and distribution. As a result, efficiency improvements can amplify the benefits of developing utility-scale renewable energy by increasing the impact of added renewable power capacity. Similarly, distributed (as opposed to centralized) renewable generation achieves efficiency gains by producing energy at the point of use, thereby avoiding transmission and distribution losses.

Renewable energy sources such as wind, solar, small hydro, wave, and tidal energy have the additional efficiency advantage of converting natural flows of mechanical energy or sunlight directly into electricity, unlike fossil fuel combustion and nuclear power, which require inherently inefficient thermal energy conversion processes.

Source: See endnote 4.

These promising trends need to be accelerated if global GHG emissions are to peak before 2020, which the consensus among climate scientists deems necessary to avoid climate catastrophes. Numerous studies have shown that renewable energy resources can fully meet global energy demand. But how sustainable would such a system really be? Are resource inputs required that limit the potential of individual renewable technologies? (See Table 8–1.)[5]

Renewable Energy Resources and Constraints

Solar. There are two main categories for solar electricity technologies: photovoltaic (PV) modules that convert light directly into electricity and concentrating solar power systems (CSP) that focus the sun's heat to drive a

Table 8–1. Renewable Energy Potentials, Impacts, and Constraints

Renewable Energy Resource	Estimated Global Potential	Land Needs	Water Needs	Limiting Material Requirements	Other Environmental Impacts
Solar photovoltaics	340 TW	0.29% of global land area to meet 40% of world energy demand in 2030 (both PV and CSP)	Minimal	Crystalline silicon: silver Thin film: tellurium, indium, germanium	Cadmium (heavy metal) emissions—small compared with fossil fuels
Concentrating solar power	240 TW	1.6 to 3.2 hectares per MW	Technology dependent; minimal –3.0 liters per kWh		Possible (desert) ecosystem interference
Wind	40–85 TW	1.17% of global land area to meet half of world energy demand in 2030	Minimal	Neodymium for permanent magnet generators	Possible interference with bird migration routes/bird kills Land use change
Small hydro	1.6 TW (includes large hydro)	Site and technology dependent, can be significant	Site and technology dependent; diversion, pollution, and evaporation of water resources can occur	Neodymium (for some new technologies)	River ecosystem damages, flooded land, methane emissions possible, depending on technology
Geothermal	2.6 TW (excludes EGS)	0.4 to 3.2 hectares per MW	Binary: 1.02 liters per kWh Flash: Minimal, but 10.2 liters per kWh geofluid evaporation EGS: 1.10–2.73 liters per kWh	None	Deforestation (often in protected areas) and interference with sensitive ecosystems possible; seismic activity has been associated with EGS technologies
Wave and tidal	Wave: 500 GW Tidal: 20 GW	Concerns regarding interference with shipping lanes, archaeological sites, pipeline infrastructure, and nature conservation	Minimal	Neodymium (for some technologies)	Sedimentation; biodiversity loss; possible impacts to migratory bird, fish, and mammal populations Positive impacts (artificial reefs) cited for some technologies
Biomass	31.7 TW	Depends on biomass type—can be very significant	Depends on biomass crop / source—can be very significant	None	Deforestation and biodiversity loss, chemical pesticide and fertilizer use, land degradation

Source: See endnote 5.

steam turbine. Solar PV can be used at any scale, from small-scale electronic appliances to decentralized household rooftop systems and from installations that power industrial facilities to utility-scale PV farms. Today's CSP systems are only viable at the utility level.

Solar technology costs are falling rapidly. Crystalline silicon PV module costs fell by 70 percent from 2008 to January 2012 and are forecast to fall by another 30 percent by 2015, without subsidies. PV and CSP installations are now cost-competitive in locations with strong solar potential and relatively expensive alternative power sources—despite distorted prices for fossil fuels that do not reflect their costs to societies. Projections for PV and CSP systems estimate that, averaged over the systems' lifetimes, generation costs in strong resource areas like the southwestern United States will fall to 6–8¢ per kilowatt-hour (kWh) in the near to medium term.[6]

Even when greatly limiting the areas for solar energy development to likely developable resources based on cost and location considerations, the potential capacities are estimated at 340 terawatts (TW) for PV and 240 TW for CSP—much more than projections for energy demand in 2050, even without any efficiency measures.[7]

While land use issues must be considered for individual projects, globally the amount of suitable land area does not pose a significant constraint to installing solar equipment. Existing roof area in the United States alone, excluding areas that are shaded or oriented away from the sun, could support over 600 gigawatts (GW) of PV electricity generation, more than 20 percent of the country's current electricity demand.[8]

Today's utility-scale CSP requires between 1.6 and 3.2 hectares per megawatt (MW) in areas with strong solar resources, depending on the technological specifications. Still, land availability does not pose a significant constraint to CSP either. For example, considering only strong physical resources on uninterrupted available land, the American Southwest has almost 7,500 GW of resource potential and could provide more than four times current U.S. electricity generation. While this estimate does not directly consider desert ecosystem impacts, the potential to supply a large share of electricity demand using just a fraction of this land suggests that harmful effects could be limited. One study found that meeting 40 percent of global energy demand in 2030 with solar PV and CSP would require only 0.29 percent of the world's land area. As a comparison, 11 percent of global land area is used for crop production, and urban areas occupy 3 percent of land area worldwide.[9]

While PV generation requires minimal water (aside from panel cleaning needs in some locations), CSP is the most water-intensive renewable energy technology, requiring 1.9–3.0 liters per kWh. But this is less than or comparable to water needs for coal or nuclear plant cooling. In many areas with the strongest solar potentials, limited water resources rule out this "conventional" form of CSP. However, air-cooled CSP plants offer an alternative mature technology, as they require 90 percent less water and generate only 5 percent less electricity than water-cooled CSP.[10]

The material requirements for PV and CSP are extremely different. CSP plants require an array of fairly unsophisticated mirrors, and the material production needs neither limit the potential for the technology nor impose a significant ecological footprint. Solar energy needs for scarce material resources are mostly limited to module production for the three dominant PV technologies: crystalline and polycrystalline silicon, thin-film PV panels, and concentrator PV cells.

Silver used for electrodes poses a potential limiting factor to crystalline silicon PV cell production. The common and inexpensive use of silver electrodes that are 20–80 micrometers thick would limit potential capacity of silicon PV cells to less than 0.6 TW. Using alternative electrodes that require less silver would reduce—and possibly eliminate—this constraint, however, allowing for about 15 TW of crystalline and/or polycrystalline silicon PV (assuming that no more than 25 percent of the global silver resource is used for PV cell production).[11]

Thin-film PV cells require an indium-tin-oxide conductor layer that includes some materials with resource limits, including tellurium, indium, and germanium. Due in part to greater competition from other uses, germanium and indium pose the greatest constraints to thin-film PV potential. With indium as the limiting factor, thin-film PV potential is limited to 13–22 GW in 2020, 17–106 GW in 2050, and 17–152 GW in 2075. Germanium constraints are even tighter, but alternative silicon- and gallium-based technologies can replace germanium in thin-film modules, removing that limitation. Zinc-oxide alternatives for the conductor layer are currently under development, though future costs and ecological impacts are unknown.[12]

Emissions of cadmium, a heavy metal, from some types of solar thin-film cells have been cited as a concern, but these systems produce only 1 percent of the life cycle cadmium emissions of equivalent fossil fuel generation. The key to limiting cadmium pollution is to ensure high rates of PV cell recovery and materials recycling.[13]

Concentrator PV cells also require germanium. A shift to proven gallium arsenide alternatives would prevent overdependence on this material, but this technology has not reached commercial scale.[14]

Wind. Apart from hydroelectric power, wind has been by far the most successful renewable electricity source to date, with 238 GW installed globally by the end of 2011. Wind power is used mostly for centralized utility-scale generation, though smaller-scale applications are gaining popularity for local and decentralized electricity production.[15]

Wind energy is one of the most economical renewable energy technologies; at attractive locations, it is already fully competitive with fossil fuels. Industry estimates project that the average onshore wind farm will be fully competitive with conventional energy sources by 2016.[16]

Wind potential estimates at land-based and near-shore locations that have strong resources and are practical for energy development range from 40 to 85 TW, far more than is needed to meet future worldwide energy needs even under business-as-usual demand projections. According to one estimate, meeting half of the world's energy needs in 2030 with wind energy would require about 1.17 percent of global land area, almost all of which would be due to the space needed between turbines. The land use impacts of wind energy can be significantly reduced by using wind-farm land for other purposes such as agriculture and by siting some wind turbines offshore. Wind energy is the least water-intensive method of energy production, with operational water use largely limited to what is needed to clean the turbines.[17]

Primary materials in wind turbines include steel and concrete for the basal structures plus copper, glassfiber reinforced plastic, and carbon-filament reinforced plastic for rotor blades. Concrete supplies will remain abundant, as its primary components (sand, gravel, and limestone) are widely available and recycling technology is well established. Steel availability is also of minor concern. At current prices and levels of production, the earth has only about 100–200 years of economically recoverable iron ore remaining if no new major mining areas are discovered. However, recycling technologies for steel used in wind turbines are well established, and recycling rates for construction plates and beams in the United States are close to 100 percent. According to one estimate, adding over 300 GW of wind capacity in the United States by 2030 would require less than 2 percent of the country's 2008 steel use. These bulk materials are therefore not expected to impose a serious constraint on meeting global energy demand with wind energy.[18]

The greatest future supply risk for the wind industry will be the availability of rare earth metals. As permanent magnet generators used in the newest commercial-scale wind turbines are increasingly replacing gear-based generators due to their greater efficiency, a rapid scale-up in production of neodymium, their primary element, is required to keep pace with needs of wind turbine manufacturers and the increasing demand for permanent magnets in other sectors. China is currently the overwhelmingly dominant producer of neodymium and other rare earth materials, despite considerable reserves in other countries, including the United States. Significant expansion of rare earth availability is not expected before 2015, however, as countries other than China work to establish environmentally sound mining and production practices.[19]

One study estimates that meeting 50 percent of global energy demand in 2030 with permanent magnet wind generators would require a more than fivefold increase in annual neodymium production. Current economically available reserves could meet this level of production for about 100 years; thereafter, neodymium recycling (which has been proved possible, although at unknown cost) will be necessary to sustain wind generation. The wind industry will also be able to adapt to future neodymium shortages due to viable alternative technologies that do not require permanent magnet generators.[20]

Small Hydropower. Hydropower is the world's best-established renewable energy resource, providing over 15 percent of global electricity production in 2011, mostly from large hydropower dams. Due to the significant environmental and human impacts of large-scale hydroelectric dams, however—including often devastating effects on river ecosystems, flooding of land ecosystems and human settlements, methane emissions from submerged and decaying vegetation, and consumption of scarce water resources—the discussion of hydropower is limited here to small-scale generation, including both micro hydro (0.1 MW or less) and mini hydro (greater than 0.1 MW but less than 10 MW). Global hydropower technical potential in likely developable locations is estimated at 1.6 TW. But hydropower resource estimates do not typically differentiate between large and small generation facilities, so it is difficult to judge the sustainability of developing small hydro's full technical potential.[21]

Some new small hydro models call for permanent magnet generators, requiring rare earth inputs equivalent to those described for wind turbines. Still, the availability of developable resources, much more than material limitations for hydro generator manufacturing and installation, is the main constraint on significant global expansion of small hydropower.

A widespread scale-up of small hydro facilities could have large cumulative impacts. These effects include disturbance of aquatic ecosystems, upstream and downstream flooding, and reduced water quality and supply. In some cases, the impacts—especially siltation (sediment buildup) and eutrophication (depletion of oxygen in the water)—can be even greater for small hydro than for large hydro on a per-kilowatt basis. Sound environmental management can mitigate some of these impacts, but implementation of best practices should not be taken for granted, especially with widespread proliferation in countries with limited capacities for monitoring and enforcement. Less damaging applications, such as small-scale run-of-the-river hydro to power remote locations, should be the focal point of small hydro development.[22]

Geothermal. Geothermal energy, or thermal energy extracted from rock beneath Earth's surface, can be used to generate electricity or to provide heating and cooling services. A major advantage of geothermal power over intermittent renewable sources like the sun and the wind is that it can be used as a baseload source of energy. The main limitation for geothermal electricity generation is the need for reservoirs with very high temperatures (over 100 degrees Celsius) near Earth's surface.[23]

Heat pumps and geothermal electricity generation are well-developed and mature technologies that are cost-competitive in locations with viable resources. Dry steam and flash steam geothermal technologies use extracted hot liquid or vapor directly to drive a turbine. Binary cycle power plants use extracted fluids to heat a secondary fluid, which in turn drives a turbine.[24]

The share of geothermal in electricity generation currently stands at only 0.3 percent worldwide, but it is much higher in countries with large potential. Nicaragua, for example, already generates more than 12 percent of its power from geothermal sources, with additional sites currently planned.[25]

The Palinpinon Geothermal Power Plant in Negros Oriental, Philippines.

The use of enhanced geothermal systems (EGS), a technology that is still in the demonstration phase, has the potential to greatly expand the feasible area for electricity generation. EGS allows for the use of geothermal resources even where there is not a permeable reservoir of high-temperature water by injecting high-pressure water into a well to open and extend fractures, freeing up thermal energy previously trapped in the rock. Including the resources accessible through EGS increases the technically exploitable geothermal electricity generation potential in the United States alone to nearly 3 TW, although seismicity concerns could limit the areas considered safe and viable for EGS development.[26]

Geothermal electricity generation requires relatively high water consumption compared with other renewable energy sources. EGS and binary generation consume 1.10–2.73 and 1.02 liters per kWh generated, respectively. These levels are comparable to the 0.49–3.94 liters per kWh of water consumption from conventional thermal (coal, natural gas, and nuclear) electricity generation. While flash generation consumes minimal water, it makes direct use of water in the hydrothermal reservoir. Evaporation rates of this "geofluid" average 10.2 liters per kWh, raising questions regarding the long-term viability of geothermal flash generation as liquid volume and pressure in the reservoir decline.[27]

Hot water suitable for geothermal energy is also produced at many oil and gas wells. This "produced water" is typically discarded as waste, but it can provide a cheap and efficient source of geothermal power. One study calculates that over 70 GW of geothermal capacity could be established at existing oil and gas wells within the United States by 2030.[28]

Wave and Tidal Power. Marine energy in the form of waves and tidal patterns can be captured to generate electricity. Wave power generators capture the energy from the rising and falling of waves on the ocean surface, and tidal generators on the ocean floor harness energy from the ebb and flow of tides. The costs of these technologies remain prohibitively high for commercial development, but they are expected to come down as technologies mature and more demonstration projects are implemented.[29]

Wave and tidal energy potentials in likely developable locations are estimated at 500 GW and 20 GW, respectively. Marine energy constraints include the need to avoid offshore areas with competing uses such as shipping lanes, marine archaeological sites, sites of pipeline infrastructures, and nature conservation. Some more recent wave and tidal power models also use permanent magnet generators, requiring the same rare earth inputs as described for wind turbines.[30]

In some cases, marine energy ecosystem impacts could actually be positive. For example, wave and tidal power infrastructure is expected to help fish populations recover in some areas by preventing commercial fishing and providing artificial reef structures for marine organisms. Negative impacts are also possible, such as increased sedimentation around wave energy buoys, which can lead to euthrophication and biodiversity loss. Additional studies are needed on the impacts on migratory bird, fish, and mammal populations, including on spawning areas, from the physical infrastructure as well as from noise and electromagnetism.[31]

Biomass. Biomass energy covers a range of resources that can be combusted for electricity generation, including wood and wood wastes, agricultural crops and residues, municipal solid waste, animal wastes, waste from food processing, and aquatic plants and algae. Biomass has the advantage of providing reliable baseload renewable power, and many biomass projects are already cost-competitive with conventional power sources.[32]

Several studies have attempted to estimate the contribution that biomass energy can sustainably make to energy needs, with wide-ranging results. One study cited by the European Commission found that taking food, water, and biodiversity sustainability constraints into account, biomass could meet up to one third of energy demand in 2050, with up to half of this from residue and wastes alone.[33]

The estimates on biomass energy vary widely due to different assumptions regarding food production and consumption, agricultural techniques, and other variables. Land use is one of the primary concerns, as biomass energy can result in high net GHG emissions in cases where forests or other carbon sinks are destroyed to clear agricultural land for energy crops. This activity can also contribute to significant biodiversity loss. Other environmental impacts of intensive farming include the use of chemical pesticides and fertilizers, land degradation, and unsustainable rates of water consumption.

The potential of biomass to provide sustainable energy therefore depends largely on whether sustainable agriculture techniques are implemented on a global scale. Furthermore, widespread concerns have been raised about diverting crops or cropland from food to energy uses, exacerbating the food price increases of recent years. In order to mitigate environmental and food price impacts, biomass electricity should be produced from the widely available supply of different waste resources (although even this approach has drawbacks, as removing agricultural waste can deprive soil of nutrients, especially in sustainable agricultural systems with limited external inputs).

Addressing the Intermittency and Variability of Renewable Energy

One of the major remaining barriers to meeting energy needs with renewables is the reliability of intermittent renewable energy resources, notably wind and solar. A number of technological solutions already exist for storing surplus renewable energy generated during periods when production exceeds demand and then dispatching this energy at times of low renewable generation. As with renewable energy technologies, advanced storage and grid options have sustainability constraints of their own. (See Table 8–2.)[34]

Batteries. Several battery technologies that can be paired with renewable energy systems are currently available or in development. Lead-acid and nickel-cadmium batteries are both mature technologies with widespread applications, including in hybrid and electric vehicles as well as for standby power storage. Lead-acid batteries are already commonly used to store energy for PV systems. These systems are not considered suitable for bulk or utility-scale storage due to high costs per unit of storage, but they work well for stand-alone decentralized renewable energy storage, particularly at the household level.[35]

The major sustainability limitation for these battery technologies is that both lead and cadmium are toxic heavy metals. Lead-acid batteries enjoy high recycling rates in many countries due to their predominant use as engine starting batteries in automobiles. The toxicity risks of nickel-cadmium batteries, on the other hand, have led the European Union to ban their use except for limited applications. Use of these batteries should therefore be limited to small-scale rural energy storage, in locations with robust battery recycling programs and regulations. Limited lead and nickel reserves

Table 8–2. Energy Storage and Transmission Technologies and Constraints

Storage or Transmission Technology	Technology Status	Limiting Material Needs	Other Environmental Impacts
Lead-acid batteries	Mature	Lead	Lead toxicity
Nickel-cadmium batteries	Mature	Nickel	Cadmium toxicity
Lithium ion batteries	Mature	Lithium	
Liquid metal batteries	Demonstration	None	
Vanadium redox flow batteries	Demonstration	None	
Pumped hydropower	Mature	None	Same as hydropower: land use and ecosystem impacts
Compressed air energy storage	Mature	None	
Molten salt thermal storage	Demonstration	Sodium and potassium nitrates (can be synthetically produced)	
Hydrogen	Demonstration	None	Natural gas (for reformation) and water needs
High-voltage direct current transmission lines	Mature	Copper	Land use needs for transmission lines
High-temperature superconducting cables	Demonstration	None	Land use needs for transmission lines

Source: See endnote 34.

(especially if the use of nickel-cadmium batteries for hybrid and electric vehicles is greatly expanded) further constrain this technology's viability as a widely implementable storage solution.[36]

Lithium ion batteries (LIB) can provide storage capacity of up to 5 MW, and they have higher energy density (and are thus lighter) than lead-acid and nickel-cadmium batteries. LIBs are also free of heavy metal toxicity risks. These batteries have multiple applications, including in hybrid and electric vehicles. Costs are projected to decline from current levels but will likely still make these batteries most suitable for small-scale decentralized capacity rather than utility-scale renewable generation. The availability of lithium resources is a frequently cited concern regarding the viability of LIBs for widespread future use. However, economically exploitable lithium reserve estimates are rapidly increasing, from 4.1 million tons (Mt) in 2009 to 13 Mt in 2012. Additionally, the global resource base of 39 Mt of lithium compares favorably to projected demand from 2010 to 2100, which is estimated at less than 20 Mt even in the highest demand scenario.[37]

Emerging battery technologies, including liquid-metal (sodium-sulfur) batteries and vanadium redox flow batteries, are not yet commercialized but hold promise for future renewable energy storage systems, including for utility-scale generation up to 35 MW, a viable size for a wind farm, especially if low-end cost estimates prove realistic.

Pumped Hydropower. Pumped hydropower uses excess electricity to pump water from a lower to a higher reservoir during low-demand and high-generation periods and then releases the stored water through a hydropower turbine during peak demand periods, in effect turning intermittent resources like wind and solar into on-demand baseload hydro energy sources. Pumped hydro is a mature technology and can be used for utility generation up to the GW scale for several hours of storage potential. Costs vary widely, depending on the size and location of the plant.[38]

Pumped hydro systems are limited in their geographic scope to mountainous landscapes with hydro resources. Furthermore, the sustainability constraints of pumped hydropower are much the same as those for hydropower dams, including land use changes as well as human and ecosystem impacts, especially in the case of large-scale systems. While pumped hydropower can provide sustainable energy storage on a case-by-case basis, its potential for widespread implementation is limited.

Compressed Air Energy and Biogas Storage. Compressed air energy storage is a mature technology that compresses air in tight underground reservoirs during periods of low demand and releases and heats the air with natural gas during peak demand periods, causing it to expand and drive turbines to generate electricity. Like pumped hydro, it can provide storage at the GW scale, but its potential is limited by the low availability of suitable natural storage sites. Costs depend on location and are higher per unit of storage for smaller systems. A number of projects are currently under way that analyze the commercial feasibility of the use of gas (including biogas) in specially designed appliances.[39]

Molten Salt Thermal Storage. Molten salt thermal storage systems are used in conjunction with concentrating solar power generating facilities. Molten salt absorbs and stores heat, which can be released to drive the CSP system's steam turbine during cloudy days or at night. Thermal storage can be used for megawatt-scale CSP facilities and can store energy for up to two days. Although molten salt storage is still in the demonstration stage, it has the potential to be one of the more cost-effective storage options. Its storage potential is largely limited to locations where CSP is a viable energy option.[40]

Molten salt storage requires large amounts of sodium and potassium nitrates. There is currently only one commercially exploited nitrate resource in the world, in Chile, and the estimated reserve is insufficient to provide 12-hour storage to meet a significant share of global energy demand with

CSP. This resource constraint can be eliminated through synthetic nitrate production, although this would reduce the power output of CSP facilities, as some energy would be reallocated for the production process.[41]

Hydrogen. Hydrogen is a potential energy storage option in the long-term future, with applications for powering vehicles as well as storing variable renewable generation up to the megawatt scale. Hydrogen can be produced by the electrolysis of water or by reforming natural gas with steam. Both processes require significant energy inputs. Hydrogen can be produced with excess renewable generation, dispatching stored energy at peak demand periods. Significant barriers remain to be addressed, however, including high costs, safety concerns, and issues relating to storage: while hydrogen has high energy content by weight, it has a low energy density by volume.[42]

Electrolysis and reformation to produce hydrogen consume water (0.27 and 0.56 liters per kWh respectively), both at or below the low range of water consumption levels for conventional thermal power production. From a sustainability perspective, electrolysis is the preferable technology due to its lower water consumption and the requirement of natural gas for reformation.[43]

Electricity Transmission and Distribution. Reliable integration of renewable energy generation into electricity grids is an essential aspect of a future sustainable energy system, especially for utility-scale facilities. Extending the grid will result in environmental disturbance in the areas surrounding new transmission lines. Much of this impact can be mitigated by burying transmission cables, although this option is not as viable for high-voltage lines.[44]

High-voltage direct current (HVDC) lines are considered one of the most efficient means of long-distance transmission for moving electricity from areas of strong renewable generation potentials to end users. HVDC lines require large copper inputs, making copper availability a significant challenge for a future efficient grid system. Even with copper recycling, the need for new copper resources for HVDC lines, wind turbines, CSP facilities, and grid connections for a renewable-powered world will require an estimated 40 percent of total copper reserves, or the equivalent of 14 years of global production at current levels. Aluminum requirements are not expected to add to the resource constraint, as only an estimated 1 percent of global reserves are required for the necessary HVDC lines.[45]

High-temperature superconducting (HTS) cables provide another efficient alternative and can transmit 10 times as much power over long distances as conventional copper transmission lines. Although HTS cable material requirements include the rare earth element yttrium, this component is not expected to pose a constraint to expanded use of HTS transmission. Yttrium reserves are sufficient to meet current production levels, and world yttrium resources, although not yet quantified, are expected to be very large.[46]

Outlook for a Sustainable Renewable Energy System

As with all energy and infrastructure projects, renewable-energy development must take environmental, resource, economic, and social constraints into account in order to be truly sustainable. While material resource and environmental constraints pose a challenge to developing specific renewable energy systems in specific locations, these limitations can be overcome through integrated energy planning, responsible environmental management, and the implementation of clean and widely available substitute technologies.

The analysis in this chapter leads to three key conclusions. First, sustainable renewable-energy planning should be integrated. A strong and efficient electricity grid can connect multiple generation sources over a broad geographic area, which enables the integration of complementary renewable facilities.

For example, certain wind farms generate more energy during the morning and others generate more during the afternoon; likewise, different wind resources have higher generation at different times of year. Different renewable resources such as wind and solar, which are each variable but often have different times of peak production, can also be integrated. Combining these complementary resources can go a long way to resolving renewable intermittency and can create relatively consistent energy supply. Integration with conventional energy technologies during the transition to a fully renewable system is equally important. Natural gas, in particular, can act as an ally of renewables due to its flexibility in dispatch, an advantage over coal and nuclear energy.

Second, sustainable renewable-energy planning should be local. Decisions for siting energy projects must be fully integrated with sustainable and just land policies that ensure protection of ecologically sensitive areas, take into account alternative land uses and environmental services, and fully respect the rights of people living on or close to those lands. (See Box 8–2.) Renewable energy projects that would seriously compromise the surrounding environment or threaten local communities should be abandoned or re-sited.[47]

Renewable energy developments should also be in complete accord with priorities for sustainable water use to avoid large diversions of water from natural systems and to preserve scarce resources for human needs. Water scarcity already affects around 1.2 billion people globally, almost one fifth of the world, and an additional 500 million people are at risk of scarcity. In cases where renewable-resource strength is strong enough to justify project development in water-scarce locations, alternative technologies (such as air cooling) should be used to minimize water consumption.[48]

And third, sustainable renewable-energy planning should at the same

Box 8–2. Land Use Priorities and Land Rights Considerations

While globally the land area required to power the world with renewable energy sources is minimal, local land use impacts of individual projects can be significant. Areas with strong renewable resources can overlap with ecologically rare or sensitive areas or with private or indigenous land rights. Some of the strongest geothermal resources in the United States, for example, are located on public land, but regulations are in place to protect national parks and wilderness areas from development. Clearing cropland for biomass energy resources has caused devastating deforestation in some rainforest nations, including Malaysia and Indonesia. Transporting energy from new renewable facilities can also have negative land use impacts if transmission lines pass through forests or other sensitive ecosystems.

With regard to local and indigenous land rights, hydropower dams have flooded millions of homes in China, Latin America, and elsewhere. As large wind, solar, and other renewable generation expands, increased land rights disputes can be expected that are similar to existing conflicts over the siting of conventional power plants and their transmission lines.

The extent to which environmental impacts of renewable energy projects are mitigated and land rights are respected depends on the strength and effectiveness of the regulatory regime in place.

Source: See endnote 47.

time be global. This is certainly true for the climate crisis, which in the long run can only be solved if all countries contribute to reducing energy-related GHG emissions. But it is also true with regard to the worldwide availability of scarce resources and the extensive environmental damage that can result from material production. Rare earth mining and processing in China, for example, demonstrates the need for strict regulations as extraction of these materials increases around the world for renewable energy, grid, and storage technologies. Robust environmental protections are needed to prevent further soil erosion, damage to vegetation and cropland, surface and groundwater pollution, landslides, and clogged rivers. Governments must not abandon unsustainable practices at home while accepting similar or worse procedures elsewhere.[49]

Recycling regimes should be implemented and strengthened for the materials required for sustainable energy development that are already in wide use today. These include bulk materials such as cement, copper, and steel as well as rarer or toxic materials such as neodymium and cadmium.

The technical, economic, and resource challenges to transitioning to a fully sustainable global energy system are enormous, but they can be fully addressed with solutions that exist today. Rapidly declining renewable energy costs and the need to replace aging fossil fuel infrastructure present an opportunity to rapidly usher in a new era of truly sustainable energy.

CHAPTER 9

Conserving
Nonrenewable Resources

Gary Gardner

A 2012 study by researchers at the Massachusetts Institute of Technology (MIT) cast a long shadow across the otherwise bright future of clean technologies like wind power and electric cars. The study warned that global supplies of neodymium, which is used in the magnets in wind turbines, and dysprosium, used in electric vehicles, could soon be scarce in markets worldwide as demand for clean technologies skyrockets. Demand for neodymium could increase by 700 percent and demand for dysprosium by 2,600 percent over the next 25 years, they calculated, if serious goals for reductions in greenhouse gas (GHG) emissions are adopted. But it may be beyond the capacity of markets to meet these levels of demand. These "rare earth elements" are currently mined almost exclusively in China, which restricts mining licenses and exports in an effort to conserve supplies.[1]

The challenge of sufficient market supply in the decades ahead is not confined to little-known elements. It extends to more common resources, such as phosphorus, a mineral critical to agriculture, and metals like copper and gold. Because these resources are nonrenewable, a growing chorus of analysts worries that whereas minerals and metals in the twentieth century were easy to reach and cheap to extract, nonrenewables this century may be increasingly scarce and costly to bring to market.[2]

Neodymium and dysprosium are not geologically scarce, it should be noted, and as with many minerals, new sources are regularly identified. (Greenland emerged as a possible new source of rare earth elements after the 2012 MIT study appeared.) The issue instead is the accessibility of metals and minerals and whether their extraction can continue to be profitable. Indeed, nonrenewable resources could become increasingly market-scarce this century as a perfect storm of constraints—from declining resource quality to rising prices for water, energy, and other inputs to extraction—begin to kick in. Together, these constraints create a markedly more worrisome environment for nonrenewable resources than the one that existed just a decade ago.[3]

Gary Gardner is a senior fellow at the Worldwatch Institute.

www.sustainabilitypossible.org

Increasing Dependence on Nonrenewables

Nonrenewable materials are the blood and bones of industrial economies. High-speed roads, multistory buildings, electronic gadgets, high-yield agriculture—these and myriad other achievements of industrial economies are built on massive quantities of nonrenewable resources. Indeed, most materials flowing through industrial economies—in the United States the share is 95 percent; in China, 88 percent—are nonrenewables, a stark contrast to pre-industrial societies whose economies were dominated by wood, water, plant fibers, animal skins, and other renewable resources.[4]

The rise of industrial economies in the twentieth century marked an exponential increase in the extraction of nonrenewable resources, from construction gravel and agricultural minerals to base metals, precious metals, and fuels. (See Figure 9–1.) Note in particular the very rapid rise in global output since 2000, as economic growth in emerging economies in Asia and Latin America has accelerated. Note, too, the minimal impact of the global recession of 2009: it slowed but did not reverse the use of nonrenewables, and the pace quickly resumed once global economic output picked up. Supply optimists are quick to note, correctly, that the trend over the last century was one of rising output and falling prices—surely conclusive evidence of plentiful supply. But because of galloping demand and emerging constraints on supply, that run of abundance could be coming to an end.[5]

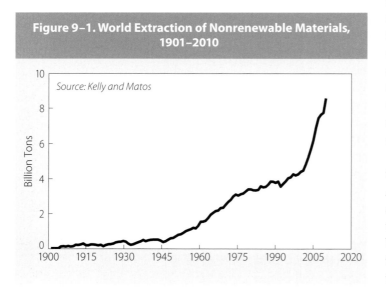

Figure 9–1. World Extraction of Nonrenewable Materials, 1901–2010

Source: Kelly and Matos

Today many emerging industrial economies in Asia and Latin America are moving into a resource-intensive phase of industrialization, as they build roads, buildings, water and sewage systems, airports, power grids, irrigation canals, railroads, and a host of other works of infrastructure that require enormous volumes of energy, metals, minerals, and other nonrenewables. The increase in demand is huge: analysts at the McKinsey Global Institute note that China and India "are experiencing roughly ten times the economic acceleration of the Industrial Revolution, on 100 times the scale"—because of their far larger populations—"resulting in an economic force that is 1000 times as big."[6]

Despite the run-up in resource demand, industrial nations continue to build throw-away economies. Advances in recycling over the past 40 years have been modest at best, as data for metals show. Whether measuring the share of discarded metal that gets recycled (the end-of-life recycling rate) or the share of newly manufactured metal that is recycled metal (recycled content), recycling levels are mostly poor. More than half of 60 metals studied by the U.N. Environment Programme have an end-of-life recycling rate of less than 1 percent, and fewer than a third of the 60 are recycled at 50 percent or more.[7]

In sum, the voracious materials appetite of industrial countries, the rapid expansion of emerging industrial economies, and the ingrained modern habit of using materials only once before they are discarded raise an urgent question: Will the market supply of nonrenewable resources be plentiful and affordable enough to meet human needs in the decades ahead?

Suggestions of Scarcity

Several signals suggest that market scarcity could increasingly become the norm for nonrenewable resources. The indicators include rapidly rising prices for nonrenewables, the declining quality of resources and difficulty accessing them, the rising cost of inputs to mining and oil drilling, the growing environmental burden of extractive activity, and the possibility that "net energy" will be insufficient to support mining and pumping.

In this chapter, *scarcity* refers to market scarcity. (See Box 9–1.) While sometimes exacerbated by declining geological supplies, market scarcity is generally driven by economic, political, or other constraining factors. Some of these are temporary obstacles, but others are intractable and can render resources as unavailable as if they were physically depleted.

Rising Prices. The first worrisome development suggestive of scarcity is the sharp upward trend in the prices of nonrenewable resources starting in 2002. This is best appreciated in contrast to the overall decline in prices during the last century. U.S. Geological Survey (USGS) data for 86 metals and minerals show an average price decline of 0.9 percent annually between 1900 and 2001; for metals, a subset of the 86, the average annual decline was 1.4 percent. But between 2002 and 2010, prices of the 86 resources increased annually by 6.4 percent and those for metals

Box 9–1. What is Scarcity?

The term *scarcity* brings up images of physical insufficiency and raises the specter of "running out." But a range of issues can limit supply long before a resource is exhausted. Often the tightest constraint on supply is cost: if the energy needed to extract a resource becomes too expensive, or if environmental regulations prohibit cheap extraction methods, or if low-quality minerals require extensive processing to be economically useful, the resources may become too expensive to tap. Political considerations may affect supply as well. Some nations prohibit exploitation of key nonrenewable resources, preferring to tap overseas supplies and treating their own endowment as a strategic reserve. In either case, market supply is constrained and resources can be described as scarce, even if they remain geologically abundant.

On the other hand, resource availability can increase even as a resource is being depleted. Advances in drilling or mineral processing, for example, can lower the cost of extraction and increase supplies. Similarly, recycling can increase resource supplies and reduce market scarcity.

went up 11 percent. So great was the change of fortune that rising prices over the eight-year period entirely canceled the price declines of the twentieth century. Although some prices softened in 2012 because of a slowing Chinese economy, this is likely temporary; the pressure on prices could well resume with renewed demand.[8]

Supply optimists argue that the recent run-up in prices is merely an anomaly in the century-long trend of downward prices and that the run-up is driven by speculation and hoarding. But Jeremy Grantham, chief strategist at the investment firm GMO and a student of resource trends, uses statistical analysis to counter this argument. He has found that for 27 of the 33 commodities he studied, there is less than a 3-percent probability that their sharp increases in price over the past decade are an extension of the twentieth-century trend of declining prices. For the 11 commodities with the greatest price rises, the odds are less than one tenth of 1 percent that they are part of the old trend. He concludes that humanity has entered a new era of global resource use in which commodities will no longer be cheap and abundant.[9]

The drop in prices during the last century was largely the result of productivity gains that outpaced the rise in extraction costs. But these costs have recently risen as metals and minerals have become more difficult to get to and as their quality has declined. Lower-quality and less-accessible ores often require more processing to coax out smaller quantities of metal, which adds to costs. And contrary to the expectations of supply optimists, increasing prices are not prompting similar increases in output. In Australia between 1989–90 and 2005–06, for example, prices in the mining sector increased by an average 9 percent annually (with the greatest increases occurring since 2000), whereas the tonnage of materials increased by only 3 percent.[10]

Ore Grade Declines. A second indication of growing scarcity, at least with regard to metals, is the decline in ore grade—that is, the shrinking share of desired metals in mined rock. The downward trend in ore quality is not new; it extends back decades for many metals, and more than a century for some. But ore grade attracted little attention among policymakers during the past century when metals extraction was robust and prices were falling.

No publicly available dataset exists to document a decline in ore grades for all metals across the entire world, but leading research demonstrates that the problem is widespread. Gavin Mudd of Monash University in Australia, whose research on mining covers a broad range of metals, documents long-term ore grade declines for gold in the United States, South Africa, Brazil, and Canada (see Figure 9–2) and for nickel in Canada and Russia. He finds similar declining ore values in Australia for copper, nickel, uranium, lead, zinc, silver, gold, iron, diamonds, and bauxite. While ore grades can increase as new discoveries, new technologies, or new techniques open access

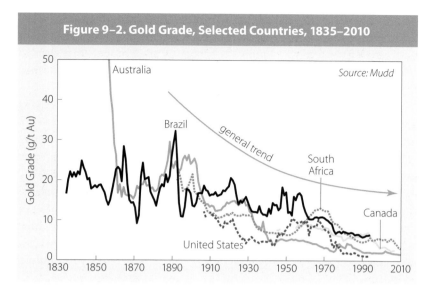

Figure 9–2. Gold Grade, Selected Countries, 1835–2010

to high-grade ores, increases in ore grade are fewer and smaller as mining matures in each nation—and the long-term trend over several decades is almost always a decline in ore grades. Mudd concludes that "based on known deposits, it is hard to envisage new discoveries or mining techniques leading to ore grades rising in the future."[11]

Environmental Costs. Lower-grade minerals can have greater environmental impact, in terms of both inputs and pollutants. Consider water, which is often needed in greater supply as ore grades decline, although the particular characteristics of a mine—open pit versus underground, for example, or the chemistry of the particular metal and even water quality and climate—also affect the quantity of water needed. The inverse relationship between declining ore grades and increased water use has been documented in Mudd's research for a number of metals. (See Table 9–1.)[12]

As long as the energy used in mining is fossil in origin, greater energy use will typically drive greater emissions of greenhouse gases—and all the more as ore grades decline. Gavin Mudd uses a rough-cut analysis to show that a decline in ore grade of copper from 0.95 percent in 2008 to 0.40 percent in 2050 would be associated with easily a doubling (and very possibly much more) of GHG

Table 9–1. Relationship between Ore Grade and Water Use

Metal	Ore Grade	Embodied Water
	(percent)	(cubic meters per ton of metal)
Lead-zinc	10–15	29
Copper	1–2	172
Uranium	0.04–0.3	505

Source: See endnote 12.

emissions from copper mining just when policymakers are struggling to reduce emissions by 50–80 percent below 2000 levels. Meeting these ambitious GHG goals would require emissions cuts per ton of copper of at least 75 percent. Unless these reductions are made through much greater efficiency, they will depend on a scaling back of mining.[13]

Declining ore grades and increasingly inaccessible minerals are driving a trend toward ever-larger mines in which greater tonnages of waste rock are generated per ton of metal extracted. At the Rossing uranium mine in Namibia, expansion of the open pit mine to maintain production led to an increase in the annual generation of waste rock from 7.5 tons in 2005 to 42 tons in 2010. Today waste rock tonnage can often be at least as great as the tonnage of ore mined, and sometimes it is several times greater—3.5 times greater in the case of Rossing—which can mean more remediation after a mine is closed. Indeed, the growing environmental cost of operating ever-larger mines is yet another factor that could constrain mining output in the future.[14]

Scarce and Expensive Inputs. Tight supplies of inputs to the extraction of nonrenewable resources could hamper mining and pumping activities. Energy is the input of greatest concern, particularly as awareness increases of "peak oil" and the finite nature of fossil fuels. Materials analyst Andre Diederen notes that while "the absolute amount of various metal minerals in the earth's crust are large beyond imagination," the bulk of these minerals "might as well not be there" because of the energy required to extract them. Because minerals extraction is so directly tied to availability of cheap energy, Diederen expects that the global peak in net energy production by the mid-2020s will also bring about the peak of global minerals production, as many minerals simply become too energy-intensive to get access to.[15]

The problem is made worse by declining ore grades, which increase the energy needed to find, extract, and process minerals. In Australia, for example, the energy intensity of mining—the amount of energy needed to produce a ton of metal or mineral—increased by 3.7 percent annually between 1989–90 and 2005–06, largely because of the shift to lower-grade and more-remote resources that require more energy-intensive technologies, according to government officials.[16]

Metals output faces an ore-grade governed "mineralogical barrier"—the grade at which the energy needed to continue mining becomes prohibitively expensive. For copper (Cu), a long-standing estimate of the mineralogical barrier is 0.1 percent Cu. This is below the global average ore grade for copper of 0.62 percent Cu. But economic impacts begin to kick in well before the mineralogical barrier is reached. The energy intensity of copper production begins to increase as ore grades approach 1 percent Cu (that is, 10 times higher than the mineralogical barrier) and grows exponentially

below 0.25 Cu. Reaching the mineralogical barrier for copper may be decades
away, but the economic consequences
should appear sooner.[17]

Thus, two reinforcing trends are rac-
ing toward a collision that could translate to declining market availability of
minerals in the near to medium term:
energy scarcity could well limit minerals
output even as declining ore values require ever-greater inputs of energy.

But a third compounding trend is in
play as well, known as the "energy return
on energy invested" (EROI). (See Chapter 7.) The power of the EROI argument
lies in its compelling logic: drilling for

Pit of the Prominent Hill copper, silver, and gold mine in South Australia.

oil or digging for coal makes little sense if the energy required for extraction
is greater than the energy extracted—that is, if the energy return on energy
invested is negative.

Indeed, analysts suggest that the energy invested in pumping and drilling is growing rapidly while the yields of wells and mines decline: the EROI
is dipping to worrisome levels. Cutler Cleveland of Boston University has
found that the EROI of oil and gas in the United States declined from 100:1
in 1930 (which means that the energy in 1 barrel of oil could pump out 100
barrels) to 30:1 in 1970 and 11:1 in 2000. In other words, more and more energy is needed to extract the same amount of energy content as companies
drill or dig deeper or as they extract lower-quality resources that need to be
processed more extensively.[18]

The implications are sobering. The surplus, or net, energy—the energy
liberated from mines or wells after an energy investment of a barrel of oil
or a ton of coal—was the life force for the extraordinary economic, technological, social, and other advances of the last two centuries. Without exaggeration, that surplus energy is the foundation of our civilization. Now, as
a growing share of extracted energy is needed to extract even more energy,
less surplus energy is available for all other economic activity—including
mining and other extractive activities.

Worse still, the break-even EROI may actually be much higher than 1:1.
Charles Hall of the State University of New York calculates the minimum
EROI for transportation fuels as 3:1, after accounting for the energy needed to process the fuel, build the machinery to use it (say, a car), and build
and maintain the infrastructure (highways) needed by the machinery. But
economic disruptions could arrive well before the 3:1 threshold is reached.

Hall's modeling suggests that price increases associated with a declining EROI start to accelerate when EROI reaches roughly 10:1—very close to the 11:1 EROI that Cleveland calculated for 2000. Once the price-acceleration threshold of various fossil fuels is reached, the viability of every process that uses fossil energy is called into serious question.[19]

Another little-recognized dynamic that could affect extractive activity is the growing tendency of price increases in one resource to spread to others. The McKinsey Global Institute reports that prices across four commodity categories—energy, metals, agricultural raw materials, and food—are more closely connected than at any time in the past century. This means that the price of inputs, such as water and energy, can move together to drive up mining costs.[20]

Creating a Circular Economy

Emerging indications of tight resource supplies require a comprehensive societal effort to conserve remaining stocks and be smarter about resource use. The challenge is to increase resource productivity markedly, similar to the increases in labor productivity over the past 100 years—about 1 percent annually in the first half of the last century, then 2–3 percent a year after 1950. This may well be achievable: analysts have long asserted that fivefold increases in material productivity are possible in industrial economies—if policymakers make this a priority. The key is to decouple resource use from economic growth.[21]

One conceptual framework for large and steady increases in resource productivity, known as a "circular economy," emphasizes meeting economic needs using a minimum of natural resources. By eliminating the wasteful one-way flow of resources that characterizes industrial economies today, a circular economy reduces the need for virgin resources and the environmental degradation associated with extractive activities. Creating a circular economy requires resource policies designed to conserve nonrenewables, as well as policies that generate more-intelligent patterns of production and consumption.

A circular economy features intelligent policies that treat nonrenewable resources for what they are: scarce and finite assets. Removal of public subsidies for nonrenewable minerals and fuels, such as the $600 million to $1 trillion in public subsidies paid by governments to fossil fuel companies, is a logical place to start, because such subsidies encourage use of nonrenewable resources and the environmental problems created by extractive activities. The European Commission has set a goal of eliminating environmentally harmful subsidies by 2020, and in 2009 and 2010 the Group of 20 industrial nations and Asia-Pacific Economic Cooperation announced that they would end fossil fuel subsidies. Steps like these are helpful, and if expanded to cover all nonrenewable resources, they would help create an ethic of conservation.[22]

Indeed, far from being subsidized, nonrenewable resources arguably should be taxed at their source—at the mine shaft and the oil well—to encourage conservation. Many countries already tax mining—but not at levels that discourage use of virgin nonrenewables and encourage development of a sophisticated infrastructure for materials recycling and product remanufacturing (including, perhaps, landfill mining—see Box 9–2). High taxes, along with programs that help mining companies convert to recycling activities, would help create employment (recycling is more labor-intensive than mining) and would husband virgin mineral stocks for the future.[23]

Beyond the mining sector, governments can take steps to create an ethic of resource conservation throughout their economies. In 2011 the European Commission released *Roadmap to a Resource Efficient Europe*, which seeks to ensure that by 2020 "waste" is essentially an obsolete concept, with discarded material fed back into the economy as raw materials. One tool to this end is "takeback" laws under which producers re-assume responsibility for products at the end of their useful lives. Such laws create a strong incentive for companies to reduce the materials used in products and packaging and to make them recyclable or remanufacturable. These practices typically save materials and energy: a 2009 report noted that studies at the Massachusetts Institute of Technology and in Germany have found that some 85 percent of the energy and materials embodied in a product are preserved in remanufacturing.[24]

> ### Box 9–2. Can Landfills Be Mined?
>
> The need to conserve virgin nonrenewable resources and tap existing resources raises an intriguing question: can landfills be mined? The potential appears to be enormous—the USGS reported in 2005 that landfills in the United States alone contain enough steel to build 11,000 Golden Gate bridges. Landfill mining has been proposed periodically since the 1950s, but it is regularly rejected for reasons of cost.
>
> Yet it is already happening. A waste management firm in Belgium has begun excavation of the Remo Milieubeheer landfill some 80 kilometers east of Brussels. Its aim is to recycle 45 percent of the site's 16.5 million tons of content, convert residues into construction materials, and siphon off methane from the landfill to generate electricity—enough to power 200,000 homes over the 20-year life of the project, according to the firm. It will then return the land to nature.
>
> A number of factors make the Belgian landfill project viable, including the high price of metals and other materials, the fact that the landfill is well mapped (so that they know the location of various types of refuse), rising demand for recycled products, and government subsidies in the form of renewable energy credits. But the firm believes the Belgian project is the way of the future, and it is working to interest other authorities worldwide in landfill mining.
>
> *Source: See endnote 23.*

Take-back laws and other reuse and recycling initiatives require proper infrastructure in order to collect, separate, recycle, and reuse materials. San Francisco built a waste collection infrastructure that accommodates recyclables, compostables, and trash as an essential step to achieving its "zero waste to landfills" goal by 2020. As of 2012, some 78 percent of materials collected in that city are recovered for composting or recycling—compared with 34 percent for the United States as a whole. Next, products must be designed for recycling, like the parts on BMW automobiles that are bar-coded with information about metal content and recycling possibilities. Finally, technologies for materials separation and recycling must be improved to make recycling more economical.[25]

But building a circular economy also requires attention to production and consumption patterns. Businesspeople, policymakers, and analysts have come up with an array of creative ideas for giving consumers what they need at reduced levels of materials use. Table 9–2 summarizes many of these initiatives.[26]

Because consumerism is a strong driver of resource use, policies are needed to steer consumption in resource-light directions. These could include taxing consumption rather than income (with a design that protects consumption of basics such as food and shelter), subsidizing solar panels

Table 9–2. Innovative Practices That Reduce Consumption of Materials and Energy

Innovation	Description	Example
Services in place of goods	Focus is on the service a consumer needs, rather than a good	Car sharing gives participants access to a private automobile without requiring them to own one. A survey of more than 6,000 car-sharing participants in North America found that cars per household fell from 0.47 to 0.24 after signing up for car sharing.
Eco-industrial parks	Discards from one production process become inputs to another	China is particularly ambitious, having created more than 50 eco-industrial parks. In Guigang City, wastes from a sugar refinery, paper plant, cement mill, thermo-electric plant, and local farms are used as inputs to other industrial operations.
Whole system design	One process serves multiple purposes	Cogeneration uses the waste heat from electricity generation to heat and cool buildings and to heat water, achieving energy efficiencies of 65–75 percent compared with 45 percent when electricity generation and heating/cooling are provided separately.
Intelligent design	Advantages are sought wherever possible	Bus rapid transit (BRT) systems, conceived in Brazil, offer the high-speed advantages of a subway system at the lower cost of surface transportation. Passengers prepay and can board quickly, and buses have dedicated lanes and driver control of stoplights. By making public transport more attractive and affordable, BRT reduces the demand for material-intensive private cars.
Shared use	Goods serve multiple users	Dozens of tool libraries, toy libraries, and other sharing institutions give people access to infrequently used goods. Portland, Oregon, has three tool libraries, for example.
Competitive efficiency	Efficiency improvements are benchmarked and ratcheted upward	A Japanese government program designates the most energy-efficient consumer products as Top Runners and challenges all products to meet the Top Runner standard within five years. Goals for 21 major energy-using consumer products have been met—and often exceeded.

Source: See endnote 26.

and other technologies that shift consumption away from nonrenewables, and using government procurement power to expand the market for goods with high recycled content or with other sustainability advantages. Conservation of nonrenewables will not happen without rethinking the dominant model of consumerist-driven economies.

The Krupp Bagger 288 is the world's biggest bucket wheel excavator and one of the biggest vehicles ever built.

The challenge of conserving nonrenewable resources is great, and it will require long-term thinking and a new conservation ethic among policymakers and publics. Whether people in the twenty-first century are up to the task remains to be seen. Jeremy Grantham of the investment firm GMO observes, with sadness and deep irony, that investing in increasingly market-scarce nonrenewable resources could prove profitable in the decades ahead, even as the prospects for human civilization decline. The challenge is to reverse incentives, rules, and other structures that cause us to be myopic users of resources and replace them with principles and practices that would make our children, and their children, grateful and proud.[27]

RIO PLUS 20 ENVIRONMENT SUMMIT

Victor Ndula/Cartoon Movement

Getting to True Sustainability

Despite scattered attempts to impute progress on climate change to the U.N. summit in Rio de Janeiro in June 2012, the consensus persists that it produced lots of gaseous talk and no significant action—leaving, according to one cartoonist, Rio's statue of Christ the Redeemer gasping for purer air.

Climate change is only the most prominent environmental trend that threatens sustainability; the first section of this book details several other areas where humanity seems to be overdrawing its accounts with nature. Yet we are hardly helpless. This section samples a variety of measures that, if pursued vigorously, could set us on a sustainable path. Indeed, had we done so after the first Rio summit 20 years ago, we would be well down that path now.

A long first stride would be jettisoning consumer cultures. As Erik Assadourian writes, consumerism has turned out to undermine both human well-being and the planet's life-support functions. But it is a willfully engineered way of living, supported by enormous sums spent annually on advertising, subsidies, tax breaks, and public relations. We can, and must, replace it with a culture of sustainability.

Many cultural options might qualify as sustainable, but certain attributes seem critical. Robert Costanza and his coauthors argue for an economy that focuses on human well-being rather than on economic growth as an end in itself. Pavan Sukhdev urges sharp reforms of corporations—the main agents of the "brown economy"—which account for 60 percent of global gross domestic product but also generate trillions of dollars of externalities and exert pernicious influence on national policies. Jeff Hohensee describes the efforts of international accounting agencies to build externality disclosure into routine corporate reporting—an important step in the right direction.

Energy is perhaps the most daunting challenge before us. In a real sense, fossil energy is the author of modern civilization—but now threatens to destroy it. The only solution, say Thomas Princen and his colleagues, is to take a true precautionary approach and leave fossil fuels in the ground by "delegitimizing" them, as happened with slavery and smoking. In their place, we must rapidly transition to renewables, and T. W. Murphy tallies the pros and cons of solar, wind, biomass, and other alternatives. He notes, however, that they are inferior in many respects to fossil fuels and warns against delaying the renewable transition so long that it diverts too much energy from other uses. In any case, such a transition will falter absent serious efficiency efforts, and Phillip Saieg reminds us that buildings remain a neglected but highly promising sector for those.

Like energy, global agriculture is at a turning point. Danielle Nierenberg notes that 1.5 billion people are overweight while billions of others are hungry or malnourished, all while the system wastes staggering amounts of food. Agriculture can help solve multiple problems through reducing food waste, promoting agroecological approaches to farming, and focusing on nutrient-rich, indigenous foods rather than high-calorie commodified foods. Those indigenous foods are stewarded by native peoples all over the globe, and in separate chapters Melissa Nelson and Rebecca Adamson (with her coauthors) make the case that the ongoing mistreatment of native peoples is not only unjust but shortsighted, as it threatens loss of valuable knowledge of key biodiversity habitats and ways of living sustainably in them.

Finally, how to achieve these changes? If civilizational survival is not motivation enough, Kathleen Moore and Michael Nelson believe that eco-disasters are violations of human rights and principles of justice. Dwight Collins and his coauthors suggest that an appreciation of humanity's place in the universe, through the teaching of Big History, can support effective planet-wide action.

In the end it boils down to politics. Melissa Leach offers strategies for bridging and connecting top-down and bottom-up approaches and stresses deliberation, citizen mobilization, network building, and the shrewd exploitation of political openings. Creating such a movement, says Annie Leonard, requires the realization that individual actions are "a fine place to start" but "a terrible place to stop." They must be linked to organized political action, to "bigger visions and bolder campaigns" for broad change.

—*Tom Prugh*

CHAPTER 10

Re-engineering Cultures to Create a Sustainable Civilization

Erik Assadourian

At the heart of how humans live their lives are the cultures they are part of. These cultures—and the norms, stories, rituals, values, symbols, and traditions that they incorporate—guide nearly all of our choices, from what we eat and how we raise our children to how we work, move, play, and celebrate. Unfortunately, consumerism—a cultural pattern that was nurtured by a nexus of business and government leaders over the past few centuries—has now spread around the globe, becoming the dominant paradigm across most cultures. More people are defining themselves first and foremost through how they consume and are striving to own or use ever more stuff, whether in fashion, food, travel, electronics, or countless other products and services.[1]

But consumerism is not a viable cultural paradigm on a planet whose systems are deeply stressed and that is currently home to 7 billion people, let alone on a planet of 8–10.6 billion people, the population the United Nations projects for 2050. Ultimately, to create a sustainable human civilization—one that can thrive for millennia without degrading the planet on which we all depend—consumer cultures will have to be re-engineered into cultures of sustainability, so that living sustainably feels as natural as living as a consumer does today.[2]

Granted, this is no easy task. It will and is being resisted by myriad interests that have a huge stake in sustaining the global consumer culture—from the fossil fuel industry and big agribusiness to food processors, car manufacturers, advertisers, and so on. But given that consumerism and the consumption patterns it fuels are not compatible with the flourishing of a living planetary system, either we find ways to wrestle our cultural patterns out of the grip of those with a vested interest in maintaining consumerism or Earth's ecosystems decline and bring down the consumer culture for the vast majority of humanity in a much crueler way.

Erik Assadourian is a senior fellow at Worldwatch Institute and director of the Transforming Cultures Project. He is codirector of *State of the World 2013*.

www.sustainabilitypossible.org

Consuming the Planet

In 2008, people around the world used 68 billion tons of materials, including metals and minerals, fossil fuels, and biomass. That is an average of 10 tons per person—or 27 kilograms each and every day. That same year, humanity used the biocapacity of 1.5 planets, consuming far beyond what the Earth can sustainably provide.[3]

Of course, not every human consumes at the same level. While the average Southeast Asian used 3.3 tons of materials in 2008, the average North American used 27.5 tons—eight times as much. And the spread of consumerism has driven many regions to dramatically accelerate material consumption. Asia used 21.1 billion tons of materials in 2008, up 450 percent from the 4.7 billion tons that the region used in 1980.[4]

This vast differentiation in consumption is often explained as simply a difference in development levels—with growth in consumption trends routinely celebrated by leading newspapers, policymakers, and economists, regardless of the current size of the host economy. In reality, however, such high levels of consumption often undermine the well-being of high-income consumers themselves, while also deeply undermining humanity's long-term well-being and security.

The United States, for example, now suffers from an obesity epidemic in which two thirds of Americans are overweight or obese. This leads to significant increases in mortality and morbidity from a variety of chronic, diet-related diseases like diabetes, heart disease, and several forms of cancer. Worse, obesity has reached a point that it is affecting children and even shortening the average American life span, not to mention costing the United States $270 billion a year in additional health care costs and lost productivity.[5]

Beyond the personal impact, this obesity epidemic—which has spread around the world, with 1.9 billion people now overweight or obese globally and suffering similar health impacts—adds significantly to the demands humanity puts on Earth. Obesity has added an extra 5.4 percent of human biomass to the planet—15.5 million tons of human flesh—which means that people are eating enough extra food each year to feed an additional 242 million people of healthy weight. And obesity is just one manifestation of the ills of overconsumption, to which we could add urban sprawl, traffic, air pollution from automobiles and factories, and dependence on a growing number of pharmaceutical drugs like anti-depressants.[6]

Consuming at such high levels is depleting the capacity of Earth to provide vital ecosystem services—from a stable climate, due to the profligate use of fossil fuels and consumption of meat, to provision of freshwater and fish, through pollution by chemicals and plastics. And as high consumption levels are promoted as ways to increase well-being, development, and economic growth, these pressures only increase. Indeed, if all humans consumed like Americans, the earth could sustain only about one quarter of the human population without undermining the planet's biocapacity. But even if everyone only consumed like the average Chinese, the planet could sustain just 84 percent of today's population.[7]

Why are people consuming so much? The answer cannot be simply because they can afford to. In short, it stems from decades of engineering of a set of cultural norms, values, traditions, symbols, and stories that make it feel natural to consume ever larger amounts—of food, of energy, of stuff. Policymakers changed laws, marketers and the media cultivated desire, businesses created and aggressively pushed new products, and over time "consumers" deeply internalized this new way of living.[8]

In a majority of societies today, consumerism feels so natural that it is hard to even imagine a different cultural model. Certain goods and services—from air conditioning and large homes to cars, vacation travel, and pets—are seen as a right, even an entitlement. Yet it is these and countless other lifestyle choices that in the aggregate are undermining the well-being of countless humans, today and for centuries into the future.[9]

Moving away from consumerism—now propped up by more than $500 billion in annual advertising expenditures, by hundreds of billions of dollars in government subsidies and tax breaks, billions more in lobbying and public relations spending, and the momentum of generations of living the consumer dream—will undoubtedly be the most difficult part of the transition to a sustainable society. Especially if, as analysts predict, an additional 1 billion consumers join the global consumer class by 2025.[10]

But ultimately consumerism will decline whether people act proactively or not, as human society has far transcended Earth's limits. Our profligate use of fossil fuels has all but guaranteed an increase in average global temperatures of 2 degrees Celsius, and current projections suggest that unless a dramatic shift in policies and behaviors occurs, an increase of 4 degrees Celsius or more by the end of this century, or even mid-century, is possible.[11]

These vast climatic changes will bring unprecedented heat waves, megastorms, massive droughts, dramatic floods, population displacements, and the deaths of tens, even hundreds of millions of people—not to mention political instability. (See Chapter 31.) None of these are conducive to the perpetuation of a global consumer culture, though surely a small elite will still be able to maintain the materialistic version of "the good life." Ideally, however, we will not accept this as our likely future but instead will grapple with the main challenge of our times: re-engineering human cultures to be inherently sustainable. (See Box 10–1.)[12]

Box 10–1. What Would a Culture of Sustainability Look Like?

When discussing the transition beyond consumerism, opponents often conjure up a return to hunting and gathering and living in caves. In reality, if proactive—that is, if we do not wait until Earth's systems are irrevocably degraded—humanity can maintain a decent quality of life for all (and not just current consumers) at a much lower level of impact.

Roland Stulz and Tanja Lütolf of Novatlantis looked at what an equitable and sustainable consumption level would look like. They found that from an energy perspective—with a commitment to move to a sustainable energy paradigm based on renewables (admittedly a big qualifier)—the average human could continuously use 2,000 watts of energy (or 17,520 kilowatt-hours per year) for all of his or her needs, including, food, transportation, water, services, and possessions.

This is the current global average energy use—but it is unequally divided, with people in industrial countries using far more, such as in the United States, which uses six times this amount per person. What does living off this amount of energy look like?

One Australian researcher and inventor, Saul Griffith, analyzed a 2,000-watt lifestyle at a personal level and found that he would need to own one tenth as much stuff and make it last 10 times as long, that he would have to fly rarely, drive infrequently (and mostly in

efficient vehicles fully loaded with passengers), and become six-sevenths vegetarian.

Put simply, a 2,000-watt lifestyle looks like the way much of the world lives today, or better, but gone are the celebrated entitlements of the high-income lifestyle—79 kilograms of meat a year (2.5 servings a day), nearly daily access to a private car (often with only one passenger), air-conditioned homes, family pets, and unfettered access to flights around the world. In truth, these luxuries will no longer be routinely accessible to the vast majority of people in a truly sustainable society, though they may be available as rarer treats, like the once-every-three-years flight to visit his parents that Saul Griffith factored in to his new energy allowance.

Sometimes these lost consumer luxuries will be difficult sacrifices to accept after a lifetime with free access to them, though rarer consumption of luxuries may actually make them more enjoyable, like escaping to a cool café on a very hot day or enjoying meat on special occasions. But offsetting these lost consumer luxuries will in all likelihood be improved health, more free time, less stress, a strengthening of community ties (as people rely on each other instead of on privatized services), and—most important—a stop to the decline of major ecosystems on which a stable human civilization depends.

Source: See endnote 12.

Learning from Past Greatness

Keep in mind that cultures are always changing in large ways and small—sometimes organically and other times intentionally with a push in certain directions, whether driven by religious, political, technological, or other forces. There have been many spectacular beneficial cultural shifts in recent history: slavery was abolished in the United States, apartheid disappeared in South Africa, women have equal representation in many societies, fascism was defeated in Western Europe. Of course, some of these shifts required military power, not just "people power," and none of the victories is guaranteed to stay with us indefinitely without vigilance. But perhaps the biggest cultural transformation of all—one often overlooked but in reality one to draw inspiration from—was the initial engineering of consumerism.

At first there was resistance to the introduction of some elements of consumerism. For example, the first generation of factory workers typically chose to work fewer hours when receiving raises, not buy more stuff. The purpose of life, after all, was not to spend most of a person's waking hours in hot, dangerous conditions, away from family and community. This resistance could be seen over and over: to disposable goods that were introduced in the 1950s, which went against the cultural norm of thrift that had been so important to family survival; even to the switch from oil lamps to gas lights, which to some seemed unnaturally bright and "glaring." But over time people got used to new products, some of which did indeed improve life quality and many of which were at least marketed as such by clever entrepreneurs and a new advertising industry. Eventually we could hardly imagine life without an abundance of products. Three sectors deserve special recognition for so effectively shifting (and continuing to shift) cultural norms around transportation, food, and even relationships—and in turn, even if unintentionally, helping to engineer a global consumer culture.[13]

The automobile industry offers an excellent case study on how to change cultural norms. Car companies used nearly every societal institution to shift transportation norms and even our understanding of the street, which before cars came along was understood as multimodal—shared by humans, horses, carts, and trolleys. A combination of tactics shifted this norm.

Automobile companies bought up city trolley systems and dismantled them. They distributed propaganda (disguised as safety educational materials) in schools, teaching children from an early age that the street was built for cars, not them. Companies helped create and finance citizen groups to oppose people who were concerned with the spread of cars and the accidents they were causing. They even helped local police forces fine, arrest, or shame pedestrians who crossed streets wherever they wanted to (known today as "jaywalkers"—a word that was intentionally spread by car companies and their allies), helping to further establish the car as the dominant user of streets. And of course they spent huge sums marketing cars as sexy, fun, and liberating. Today the car industry spends $31 billion a year just in the United States on advertising and has effectively exported car culture to developing countries—like China, where the automobile fleet has grown from less than 10 million to 73 million in just 11 years—using lessons learned in earlier successes.[14]

The fast-food industry provides another good example. Serving over 69 million people around the world every day, McDonald's is a global power. So it may come as a surprise that less than a century ago the hamburger—today's iconic American meal—was a taboo food, unsafe, unclean, and eaten only by the poor. But technological changes, including the assembly line and the automobile, helped make the conditions right for a transformation in how we eat: quickly, on the go, and out of the home. McDonald's not only seized on this, it accelerated the transformation, retraining the palates of entire generations of Americans and now the 119 countries in which the company operates.[15]

McDonald's did not just create a cheap and tasty food, it effectively targeted children to get them to eat at McDonald's early on—shaping their palate for both the company's food and the high-sugar, high-salt, high-fat consumer diet. McDonald's was one of the earliest companies to market to children. It created cartoon characters to appeal to kids, including the globally recognized clown, Ronald McDonald. The company built playgrounds in its restaurants and offered toys in its kids' meals to get children excited to go to McDonald's (and to pressure their parents to bring them), even before they had acquired a taste for the food. Add to that the more than $2 billion in global advertising the company spends each year, and the sheer economic and political power today to keep its prices low (through lobbying and commodity purchasing power), and you have a powerful shaper of cultural and dietary norms that has a global and even generational reach.[16]

The third relevant case study is the pet industry. In India, dog ownership has grown significantly in recent years. In part this has been driven by demographic changes that include later marriages and increasing social isolation, but the obvious solution to this did not have to be pet ownership. Yet a global pet industry, recognizing an opportunity to grow, worked to stoke this enormous potential new market. It is part of the larger industry effort to transform pets into family members so that more people will buy pets and that owners will spend more on them (which industry and many owners call their "children").[17]

And it has worked. People spend more than $58 billion on pet food each year around the world. Americans spend another $11.8 billion on pet supplies annually—with nearly $2 billion of that on just cat litter, adding up to billions of pounds of litter annually diverted to landfills—and $13.4 billion on veterinarian care that is often more sophisticated than most humans have access to. Considering the ecological impact of the millions of dogs and cats (133 million dogs and 162 million cats in just the top five dog- and cat-owning countries in the world), this is not just another curious consumer trend. Two German Shepherds have a larger ecological footprint from their food requirements alone than a person in Bangladesh does in total. And unfortunately it is Bangladeshis—whose country is one of the most vulnerable to climate change—not wealthier people's pets, who will bear the brunt of climate change.[18]

These products and countless others—from doughnuts to disposable diapers—are all being spread to new consumer populations, supported by $16,000 of advertising every single second somewhere in the world. So how do we transform the world's cultures so that living sustainably becomes as natural as living as a consumer has been made to feel today? Just as consumer interests learned over the decades as they worked to stimulate markets and, intentionally or inadvertently, engineer cultural norms, it will be essential to use the full complement of societal institutions to shift cultural norms—business, media and marketing, government, education, social movements, even traditions.[19]

First Attempts to Pioneer Cultures of Sustainability

While consumerism is being spread more aggressively every year, many cultural pioneers are working to spread a culture of sustainability, in both bold and subtle ways, locally and globally, and often in ways they may not even recognize as culture changing. The most effective of these pioneers tend to use dominant societal institutions to normalize an alternative set of practices, values, beliefs, stories, and symbols.[20]

Within the business sector, a handful of executives are using their companies to transform broader consumption norms. The clothing company Patagonia, for instance, recognizing that its continued success depends on the earth and that "the environmental cost of everything we make is astonishing," has taken the bold step of encouraging its customers to not even buy its products unless truly needed, encouraging them to instead either buy used Patagonia products or do without. The company even worked with eBay to create a ready supply of used Patagonia gear.[21]

While some change will be driven by large corporations—which have significant capital and influence at their disposal—the real drivers of a culture of sustainability in the business sector are entrepreneurs and business leaders working to transform the sector's mission altogether, with a positive social purpose being first and foremost and with revenue generation simply being the means to achieve that. The good news is that an increasing number of business leaders, when creating new businesses, are establishing these "social enterprises" with the specific goal of using their businesses, and the profits they generate, to improve society. In Thailand, the restaurant Cabbages & Condoms has for decades helped to normalize safe sex to prevent sexually transmitted diseases and unwanted pregnancies—using a clever mix of décor, events, and information. It donates its profits to the Population and Community Development Association (its parent organization) to promote family planning projects in Thai communities.[22]

And today, more social enterprises like these are flourishing and even locking their beneficial missions directly into their corporate charters. Many businesses are now incorporating or getting certified as "B" or "benefit" corporations. Twelve states in the United States have set up laws that allow businesses to incorporate as benefit corporations, which requires them to work toward having an overall positive effect on society and the environment. And the company must take into account the impact of its decisions on not just shareholders but all stakeholders, including workers, local communities, and the planet. Where laws do not allow incorporation as a benefit corporation, many businesses have worked with B Lab, a nonprofit organization, to be certified as B corporations. As of fall 2012, there were 650 certified B corporations in 18 countries and 60 industries, with annual revenues of more than $4.2 billion.[23]

Within government, more policymakers are recognizing the need to use this institution to help steer citizens toward consuming less and living more sustainably, editing out unsustainable options like supersized sodas in New York City and plastic bags in San Francisco. (See Box 10–2.) And some are supporting sustainable choices like mass transit, bicycle lanes, even super accessible libraries, as with the series of library kiosks that Madrid placed in its subway system.[24]

A few governments are starting to lead even bolder transformations—such as expanding fundamental rights to the planet itself. Just as the introduction of human rights transformed the legal realm and was a catalyst for social change around the world, Earth's rights could have the same potential. In recent years, Ecuador and Bolivia have both incorporated Earth's rights into their constitutions, in turn empowering people to legally defend Earth's interest even when no humans are directly harmed—for example, by stopping mining projects in an uninhabited area.[25]

Beyond governance, local communities are organizing themselves to both reinforce sustainability norms locally and inspire others to do the same. There are now hundreds of ecovillages around the world modeling sustainable and low-consumption lifestyles. And hundreds of Transition Towns are working to transform existing communities to be both more sustainable and more resilient. While all these efforts are small in scale and scope, their potential to inspire and experiment with new cultural norms is exponentially larger.[26]

A number of schools and universities are also working to embed sustainability directly into their school cultures, including integrating environmental science, media literacy, and critical thinking into their curricula. In Europe, 39,500 schools have now been awarded a "Green Flag" for greening their curricula, empowering students to make their schools more sustainable, and articulating the schools' ecological values alongside their educational values. Some schools are also modeling a sustainable way of living, from integrating gardening programs and renewable energy production onto school grounds to changing what is served in the cafeteria. In Rome, a leader in school food reform, two thirds of food served in cafeterias is organic, one quarter is locally sourced, and 14 percent is certified Fair Trade.[27]

Like education, cultural and religious traditions play a central role in

Box 10–2. Shifting Norms with Choice Editing

On September 13, 2012, after months of debate, stacks of scientific reports, several City Hall press events, and a $1-million counter-campaign by the soda industry, the New York City Board of Health banned the sale of large cups of sodas and other sugary drinks. For Mayor Michael Bloomberg, the ban was the "the single biggest step any city has taken to curb obesity." But some people are not so sure. Fearing that the ban will spread to other cities (Richmond, California, and Philadelphia, Pennsylvania, are considering similar action), the soda industry promises to fight on. Many New Yorkers are also skeptical—60 percent view the ban as infringing on their consumer freedom. And yet the science is clear: large portion sizes, defined as 32 ounces or more for soda and sugary drinks, increase consumption, often beyond the point of any additional satisfaction, and are a major driver of the obesity crisis.

With this ban, Mayor Bloomberg joins the swelling ranks of policymakers, scientists, public interest groups, and communities that are re-engineering the norms of consumerism through a frontal assault on the fabric of choice. Colleges and universities are removing trays from their cafeterias, making it more difficult for students to pile on food as they move down the line. This simple "choice edit" has reduced food waste by 30 percent on many campuses. A plastic bag tax in Washington, DC, and a ban in San Francisco have produced striking reductions in plastic-bag pollution; more important, it has begun to foster a culture of reuse (in this case, of cloth shopping bags) that could spill over into other consumer venues.

The construction of bicycle superhighways in Denmark and the focus on better bike paths, joined with financial incentives to bicycle to work in the United States, promise to make the choice of riding a bike over driving a car more attractive. And communities like Albert Lea in Minnesota are enjoying better health, longer life spans, and greater happiness by subtly changing everything from the size of plates in restaurants and the choice of snacks in vending machines to the configuration of sidewalks and the availability of walking paths.

Successful choice editors tend to focus on small aspects of choice that produce big outcomes, like the food trays in cafeterias or the 5¢-per-bag tax in Washington. They foster choices that clearly deliver benefits to health and happiness. They also strive to preserve choice, or at least the illusion of choice. The ban on incandescent lightbulbs soon to take effect in the United States will succeed in part because of the expanding choice of acceptable lighting alternatives. The best choice editors, moreover, resist reacting too quickly to initial public objections to choice edits. They know that people frequently become habituated to their new choices and forget their initial objections.

Scores of choice-editing strategies for sustainability are hiding in plain sight. They remain largely untapped in part because of qualms about the manipulative quality of choice editing. It is easy to forget, though, that existing patterns of choice are often no less manipulative than the more-sustainable patterns that choice editors advocate. After all, 32-ounce drink cups were created to drive consumers to buy more, while the lack of good sidewalks and bicycle paths subtly but firmly pushed people to motorized transport. Reconfiguring cultural norms will mean, in part, overcoming the aversion to choice editing while simultaneously engaging the public in a conversation about the growing costs of a consumer society.

—Michael Maniates
Professor, Allegheny College
Source: See endnote 24.

shaping our understandings of the world. Fortunately, more religious communities are drawing attention to practices and teachings that reinforce our sustainable stewardship of Creation. These initiatives include everything from promoting carbon fasts for Lent to reclaiming *shemitah*—the seven-year sabbath cycle in Judaism—to encourage sustainability. Perhaps most important is the greening of life's rites of passages—births, coming-of-age celebrations, weddings, and funerals—which, while infrequent, have disproportionate impacts both on the planet and on shaping cultural norms.[28]

In many cultures, funeral traditions reinforce an idea that humans are separate from nature, with humans being embalmed and hermetically sealed in coffins to delay the decaying process. If, on the other hand, funerals celebrated our return to the natural cycle of life and reinforced our place as part of a larger living Earth system, this ritual could play an important role in nurturing a culture of sustainability. Instead, the current form uses significant ecological resources. In the United States, 3.1 million liters of embalming fluid, 1.5 million tons of concrete, 90,000 tons of steel, and more than 45 million board feet of lumber are used each year in burials, costing the average family about $10,000, often a significant financial burden at a distressing time. Groups like The Green Burial Council are helping to shift this tradition, promoting natural burial—free of chemicals and of expensive coffins or vaults and in natural cemeteries that provide parkland for people to enjoy, space for biodiversity, and trees to absorb carbon dioxide.[29]

Storytelling and myth building also have tremendous potential to help transform cultures, from efforts like Big History, which is working to incorporate sustainability into cultural creation stories (see Chapter 20), to a plethora of documentaries and films that wrestle with sustainability themes. Two examples are worth noting for their similarity: the documentary *Crude* and the blockbuster science-fiction film *Avatar*. These films, each produced in 2009, are essentially the same story, both about indigenous peoples fighting to protect their land from those pursuing the resource wealth underneath. *Avatar*—with its global reach and $2.8 billion in sales so far—in particular has the potential to deeply shift beliefs and raise awareness that our current consumptive path will lead to the future of Earth described by the protagonist Jake Sully in the final moments of the film: "There's no green there. They killed their Mother."[30]

Finally, given that media—and the marketing now embedded at its every level—play such a powerful role in shaping modern cultures, social marketing and "ad jamming" will be a powerful means to harness marketing energy for positive ends. Examples include social marketing efforts like The Story of Stuff project, which uses short, catchy videos to build political support for reduced consumption (see Chapter 23), and ad jamming efforts by Adbusters, the Billboard Liberation Front, and The Yes Men. The Yes Men, for example, uses fake ads and press conferences to draw attention to hypocritical positions of businesses and global institutions, such as their subversive effort to pose as Dow Chemical representatives and announce that the company would pay reparations for the 1984 Bhopal disaster (leading to a stock plunge of 4.2 percent in 23 minutes and the company's temporary loss of $2 billion in market value) and their efforts to jam the multimillion-dollar "We Agree" advertising campaign by the oil company Chevron. With few resources—leveraged in aikido-like fashion—these efforts garner significant

Chevron ad from its "We Agree" advertising campaign.

Spoof ad of Chevron's "We Agree" advertising campaign, Inspired by The Yes Men's ad jamming campaign, by Jonathan McIntosh.

attention and undermine the public relations efforts of those spending millions on advertising to shape the public's view of the company, their products, and, more generally, progress.[31]

Just as water can erode rock into a grand canyon, the continuing pursuit of culture-changing efforts can add up to much more than their constituent parts. And the seeds that pioneers like these sow today, even if they fail to take root while consumerism is dominant, may sprout as humanity desperately reaches for a new set of norms, symbols, rituals, and stories to rebuild a semblance of normality once Earth's systems unravel under the unbearable burden of sustaining a global consumer economy.

Tilting at Cultural Norms?

When the dominant institutions of most societies are primarily still promoting consumerism, and probably will not stop anytime soon, how will upstart efforts to engineer cultures of sustainability have any chance of success? Ultimately, if Don Quixote had just waited long enough, the passage of time would have brought down his windmill giants. The same is true for the consumer culture giants, which depend completely on the bounty of the energy embedded in fossil fuels, abundant resources, and a stable planetary system provided to humanity at this stage in its development. (See Box 10–3.)[32]

But given Earth's weakening capacity to absorb greenhouse gases and other wastes generated in pursuit of the consumer dream, the end of the consumer culture will come—willingly or unwillingly, proactively chosen or not—and sooner than we would like to believe. The only question is whether we greet it with a series of alternative ways of orienting our lives and our cultures to maintain a good life, even as we consume much less. Every culture-changing effort, whether small or large, will help facilitate this transition and lay the foundation for a new set of cultural norms—quite possibly only implemented when humanity has no other choice.

While some will argue until the bitter end that letting go of certain consumer luxuries is a step backwards, as North Face apparel company co-founder and environmentalist Doug Tompkins notes, "What happens if you get to the cliff and you take one step forward or you do a 180-degree turn and take one step forward? Which way are you going? Which is progress?" Patagonia founder Yvon Chouinard answered that the solution for a lot of the world's problems may be "to turn around and take a forward step. You can't just keep trying to make a flawed system work."[33]

The challenge will be convincing more individuals that further efforts to spread a consumer culture are truly a step in the wrong direction and that the faster we use our talents and energies to promote a culture of sustainability, the better off all of humanity will be.

Box 10–3. Development and Decline

Since 1990, *development* has been added to the rubbish heap of dismantled ideas in history. The development age lasted 40 years, from President Truman's announced intention at the onset of the Cold War to raise the living standards of poor nations through to the Washington Consensus in 1989 that paved the way for the end of Keynesianism and the ascent of market fundamentalism.

The epoch of development was then replaced by the age of globalization. It was not the nation-state developing but the purchasing power of consumer classes worldwide. Cold War divisions faded away, corporations relocated freely across borders, politicians and many others pinned their hopes on the model of a western-style consumer economy. In a rapid—even meteoric—advance, a number of newly industrializing countries acquired a larger share of economic activity. For them, it was as if President Truman's promise—that poor nations would catch up with the rich—had finally come true. But this success was paid for by destruction of local and global ecosystems. Development-as-growth turns out to be mortally dangerous.

Since the outbreak of the financial crisis in 2007, the age of security is on the rise. States line up to bolster the failing confidence of the economy, and in turn the economy burdens the state with an insurmountable pile of debt. The newcomers are preoccupied with the fossil and biotic raw materials needed for growth: the resource imperialism of China, India, and Brazil is similar to that of the rich countries, albeit in fast motion. Above all, the age of security is an era when human security of the poor and powerless is being violated on a large scale. Freeways cut through neighborhoods, high-rise buildings displace traditional housing, dams drive tribal groups from their homelands, trawlers marginalize local fishers, supermarkets undercut small shopkeepers. As development proceeds, the land and the living spaces of indigenous peoples, small farmers, and the urban poor are put under ever more pressure.

Economic growth is of a cannibalistic nature; it feeds on both nature and communities, and shifts unpaid costs back onto them as well. The shiny side of development is often accompanied by a dark side of displacement and dispossession; this is why economic growth has time and again produced impoverishment next to enrichment.

In hindsight, the consumptive Euro-Atlantic development path turns out to be a special case; it cannot be repeated everywhere and at any time. Access to biotic resources from colonies and fossil raw materials from the crust of the earth were essential to the rise of the Euro-Atlantic civilization. There would have been no industrial or consumer society without the mobilization of resources from both the expanse of geographical space and the depth of geological time. Climate chaos as well as the limits to growth suggest that the past 200 years of Euro-Atlantic development will remain a parenthesis in world history.

Indeed, it is difficult to see how, for example, the automobile society, chemical agriculture, or a meat-based food system could be spread completely across the globe. In other words, pursuing the resource-intensive Euro-Atlantic model requires social exclusion by its very structure; it is unfit to underpin equity on a global scale. Development-as-growth cannot continue to be a guiding concept of international politics unless global apartheid is taken for granted. Politics, therefore, is at a crossroads. The choice is for either affluence with persistent disparity or moderation with prospects for equity. If there is to be some kind of prosperity for all world citizens, the Euro-Atlantic model needs to be superseded, making room for ways of living, producing, and consuming that leave only a light footprint on the earth.

—Wolfgang Sachs
Senior Fellow, Wuppertal Institute
Source: See endnote 32.

CHAPTER 11

Building a Sustainable and Desirable Economy-in-Society-in-Nature

Robert Costanza, Gar Alperovitz, Herman Daly, Joshua Farley, Carol Franco, Tim Jackson, Ida Kubiszewski, Juliet Schor, and Peter Victor

Robert Costanza is a visiting fellow in the Crawford School of Public Policy, Australian National University. **Gar Alperovitz** is Lionel R. Bauman Professor of Political Economy at the University of Maryland. **Herman Daly** is professor emeritus in the School of Public Policy at the University of Maryland. **Joshua Farley** is an associate professor in the Department of Community Development & Applied Economics and Public Administration at the University of Vermont. **Carol Franco** is a project administrator at the Woods Hole Research Center. **Tim Jackson** is a professor of sustainable development at the University of Surrey, United Kingdom. **Ida Kubiszewski** is a visiting fellow in the Crawford School of Public Policy, Australian National University. **Juliet Schor** is a professor of sociology at Boston College. **Peter Victor** is a professor in the Faculty of Environmental Studies at York University.

www.sustainabilitypossible.org

The current mainstream model of the global economy is based on a number of assumptions about the way the world works, what the economy is, and what the economy is for. (See Table 11–1.) These assumptions arose in an earlier period, when the world was relatively empty of humans and their artifacts. Built capital was the limiting factor, while natural capital was abundant. It made sense not to worry too much about environmental "externalities," since they could be assumed to be relatively small and ultimately solvable. It also made sense to focus on the growth of the market economy, as measured by gross domestic product (GDP), as a primary means to improve human welfare. And it made sense to think of the economy as only marketed goods and services and to think of the goal as increasing the amount of these that were produced and consumed.[1]

Now, however, we live in a radically different world—one that is relatively full of humans and their built capital infrastructure. We need to reconceptualize what the economy is and what it is for. We have to first remember that the goal of any economy should be to sustainably improve human well-being and quality of life and that material consumption and GDP are merely means to that end. We have to recognize, as both ancient wisdom and new psychological research tell us, that too much of a focus on material consumption can actually reduce human well-being. We have to understand better what really does contribute to sustainable human well-being and recognize the substantial contributions of natural and social capital, which are now the limiting factors to improving well-being in many countries. We have to be able to distinguish between real poverty, in terms of low quality of life, and low monetary income. Ultimately we have to create a new model of the economy that acknowledges this new "full-world" context and vision.[2]

Some people argue that relatively minor adjustments to the current

	Current Economic Model	Green Economy Model	Ecological Economics Model

Table 11–1. Basic Characteristics of Current Economic Model, Green Economy Model, and Ecological Economics Model

	Current Economic Model	Green Economy Model	Ecological Economics Model
Primary policy goal	**More:** Economic growth in the conventional sense, as measured by GDP. The assumption is that growth will ultimately allow the solution of all other problems. More is always better.	More but with lower environmental impact: GDP growth decoupled from carbon and from other material and energy impacts.	**Better:** Focus must shift from merely growth to "development" in the real sense of improvement in sustainable human well-being, recognizing that growth has significant negative by-products.
Primary measure of progress	GDP	Still GDP, but recognizing impacts on natural capital.	Index of Sustainable Economic Welfare, Genuine Progress Indicator, or other improved measures of real welfare.
Scale/carrying capacity/role of environment	Not an issue, since markets are assumed to be able to overcome any resource limits via new technology, and substitutes for resources are always available.	Recognized, but assumed to be solvable via decoupling.	A primary concern as a determinant of ecological sustainability. Natural capital and ecosystem services are not infinitely substitutable, and real limits exist.
Distribution/ poverty	Given lip service, but relegated to "politics" and a "trickle-down" policy: a rising tide lifts all boats.	Recognized as important, assumes greening the economy will reduce poverty via enhanced agriculture and employment in green sectors.	A primary concern, since it directly affects quality of life and social capital and is often exacerbated by growth: a too rapidly rising tide only lifts yachts, while swamping small boats.
Economic efficiency/ allocation	The primary concern, but generally including only marketed goods and services (GDP) and market institutions.	Recognized to include natural capital and the need to incorporate its value into market incentives.	A primary concern, but including both market and nonmarket goods and services and the effects. Emphasis on the need to incorporate the value of natural and social capital to achieve true allocative efficiency.
Property rights	Emphasis on private property and conventional markets.	Recognition of the need for instruments beyond the market.	Emphasis on a balance of property rights regimes appropriate to the nature and scale of the system, and a linking of rights with responsibilities. Includes larger role for common-property institutions.
Role of government	Government intervention to be minimized and replaced with private and market institutions.	Recognition of the need for government intervention to internalize natural capital.	Government plays a central role, including new functions as referee, facilitator, and broker in a new suite of common-asset institutions.
Principles of governance	Laissez-faire market capitalism.	Recognition of the need for government.	Lisbon principles of sustainable governance.

Source: See endnote 1.

economic model will produce the desired results. For example, they maintain that by adequately pricing the depletion of natural capital (such as putting a price on carbon emissions) we can address many of the problems of the current economy while still allowing growth to continue. This approach can be called the "green economy" model. Some of the areas of intervention promoted by its advocates, such as investing in natural capital, are necessary and should be pursued. But they are not sufficient to achieve sustainable human well-being. We need a more fundamental change, a change of our goals and paradigm.[3]

Both the shortcomings and the critics of the current model are abundant—and many of them are described in this book. A coherent and viable alternative is sorely needed. This chapter aims to sketch a framework for a new model of the economy based on the worldview and following principles of ecological economics:[4]

• Our material economy is embedded in society, which is embedded in our ecological life-support system, and we cannot understand or manage our economy without understanding the whole interconnected system.

• Growth and development are not always linked, and true development must be defined in terms of the improvement of sustainable human well-being, not merely improvement in material consumption.

• A balance of four basic types of assets is necessary for sustainable human well-being: built, human, social, and natural capital (financial capital is merely a marker for real capital and must be managed as such).

• Growth in material consumption is ultimately unsustainable because of fundamental planetary boundaries, and such growth is or eventually becomes counterproductive (uneconomic) in that it has negative effects on well-being and on social and natural capital.

There is a substantial and growing body of new research on what actually contributes to human well-being and quality of life. Although there is still much ongoing debate, this new science clearly demonstrates the limits of conventional economic income and consumption's contribution to well-being. For example, economist Richard Easterlin has shown that well-being tends to correlate well with health, level of education, and marital status and shows sharply diminishing returns to income beyond a fairly low threshold. Economist Richard Layard argues that current economic policies are not improving well-being and happiness and that "happiness should become the goal of policy, and the progress of national happiness should be measured and analyzed as closely as the growth of GNP (gross national product)."[5]

In fact, if we want to assess the "real" economy—all the things that contribute to real, sustainable, human well-being—as opposed to only the "market" economy, we have to measure and include the nonmarketed contributions to human well-being from nature, from family, friends, and other social relationships at many scales, and from health and education. Doing so often yields a very different picture of the state of well-being than may be implied by growth in per capita GDP. Surveys, for instance, have found people's life satisfaction to be relatively flat in the United States (see Figure 11–1) and many other industrial countries since about 1975, in spite of a near doubling in per capita income.[6]

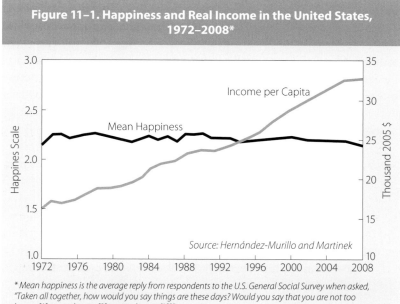

Figure 11–1. Happiness and Real Income in the United States, 1972–2008*

Source: Hernández-Murillo and Martinek

* Mean happiness is the average reply from respondents to the U.S. General Social Survey when asked, "Taken all together, how would you say things are these days? Would you say that you are not too happy [1], pretty happy [2], or very happy [3]?"

A second approach is an aggregate measure of the real economy that has been developed as an alternative to GDP, called the Index of Sustainable Economic Well-Being, or a variation called the Genuine Progress Indicator (GPI). The GPI attempts to correct for the many shortcomings of GDP as a measure of true human well-being. For example, GDP is not just limited—measuring only marketed economic activity or gross income—it also counts all activity as positive. It does not separate desirable, well-being-enhancing activity from undesirable, well-being-reducing activity. An oil spill increases GDP because someone has to clean it up, but it obviously detracts from society's well-being. From the perspective of GDP, more crime, sickness, war, pollution, fires, storms, and pestilence are all potentially good things because they can increase marketed activity in the economy.[7]

GDP also leaves out many things that actually do enhance well-being but that are outside the market, such as the unpaid work of parents caring for their children at home or the nonmarketed work of natural capital in providing clean air and water, food, natural resources, and other ecosystem services. And GDP takes no account of the distribution of income among individuals, even though it is well known that an additional dollar of income produces more well-being if a person is poor rather than rich.

The GPI addresses these problems by separating the positive from the negative components of marketed economic activity, adding in estimates of the value of nonmarketed goods and services provided by natural, human, and social capital and adjusting for income-distribution effects. Comparing GDP and GPI for the United States, Figure 11–2 shows that while GDP has steadily increased since 1950, with the occasional dip or recession, the GPI peaked in about 1975 and has been flat or gradually decreasing ever since. The United States and several other industrial countries are now in a period of what might be called uneconomic growth, in which further growth in marketed economic activity (GDP) is actually reducing well-being, on balance, rather than enhancing it.[8]

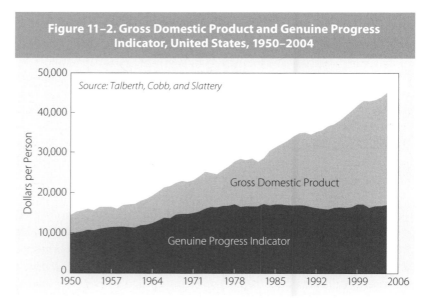

Figure 11–2. Gross Domestic Product and Genuine Progress Indicator, United States, 1950–2004

A new model of the economy consistent with our new full-world context would be based clearly on the goal of sustainable human well-being. It would use measures of progress that openly acknowledge this goal (for example, GPI instead of GDP). It would acknowledge the importance of ecological sustainability, social fairness, and real economic efficiency.

One way to interrelate the goals of the new economy is by combining planetary boundaries as the "environmental ceiling" with basic human needs as the "social foundation." This creates an environmentally sustainable, socially desirable and just space within which humanity can thrive. (See Chapter 3.)[9]

A Framework for a New Economy

A report prepared for the United Nations Rio+20 Conference described in detail what a new economy-in-society-in-nature might look like. A number of other groups—for example, the Great Transition initiative and the Future We Want—have performed similar exercises. All are meant to reflect the essential broad features of a better, more-sustainable world, but it is unlikely that any particular one of these will emerge wholly intact from efforts to reach that goal. For that reason, and because of space limitations, those visions will not be described here. This chapter instead lays out the changes in policy, governance, and institutional design that are needed in order to achieve any of these sustainable and desirable futures.[10]

The key to achieving sustainable governance in the new, full-world context is an integrated approach—across disciplines, stakeholder groups, and generations—whereby policymaking is an iterative experiment acknowledging uncertainty, rather than a static "answer." Within this paradigm, six core principles—known as the Lisbon principles following a 1997 conference in Lisbon and originally developed for sustainable governance of the oceans—embody the essential criteria for sustainable governance and the use of common natural and social capital assets:[11]

- *Responsibility.* Access to common asset resources carries attendant responsibilities to use them in an ecologically sustainable, economically efficient, and socially fair manner. Individual and corporate responsibilities and incentives should be aligned with each other and with broad social and ecological goals.
- *Scale-matching.* Problems of managing natural and social capital assets are rarely confined to a single scale. Decisionmaking should be assigned to institutional levels that maximize ecological input, ensure the flow of information between institutional levels, take ownership and actors into account, and internalize social costs and benefits. Appropriate scales of governance will be those that have the most relevant information, can respond quickly and efficiently, and are able to integrate across scale boundaries.
- *Precaution.* In the face of uncertainty about potentially irreversible impacts on natural and social capital assets, decisions concerning their use should err on the side of caution. The burden of proof should shift to those whose activities potentially damage natural and social capital.
- *Adaptive management.* Given that some level of uncertainty always exists in common asset management, decisionmakers should continuously gather and integrate appropriate ecological, social, and economic information with the goal of adaptive improvement.

- *Full-cost allocation.* All of the internal and external costs and benefits, including social and ecological, of alternative decisions concerning the use of natural and social capital should be identified and allocated, to the extent possible. When appropriate, markets should be adjusted to reflect full costs.
- *Participation.* All stakeholders should be engaged in the formulation and implementation of decisions concerning natural and social capital assets. Full stakeholder awareness and participation contributes to credible, accepted rules that identify and assign the corresponding responsibilities appropriately.

This section describes examples of worldviews, institutions and institutional instruments, and technologies that can help the world move toward the new economic paradigm.[12]

Respecting Ecological Limits. Once society has accepted the worldview that the economic system is sustained and contained by our finite global ecosystem, it becomes obvious that we must respect ecological limits. This requires that we understand precisely what these limits entail and where economic activity currently stands in relation to them.

A key category of ecological limit is dangerous waste emissions, including nuclear waste, particulates, toxic chemicals, heavy metals, greenhouse gases (GHGs), and excess nutrients. The poster child for dangerous wastes is greenhouse gases, as excessive stocks of them in the atmosphere are disrupting the climate. Since most of the energy currently used for economic production comes from fossil fuels, economic activity inevitably generates flows of GHGs into the atmosphere. Ecosystem processes such as plant growth, soil formation, and dissolution of carbon dioxide (CO_2) in the ocean can sequester CO_2 from the atmosphere. But when flows into the atmosphere exceed flows out of the atmosphere, atmospheric stocks accumulate. This represents a critical ecological threshold, and exceeding it risks runaway climate change with disastrous consequences. At a minimum, then, for any type of waste where accumulated stocks are the main problem, emissions must be reduced below absorption capacity.

Current atmospheric CO_2 stocks are well over 390 parts per million, and there is already clear evidence of global climate change in current weather patterns. Moreover, the oceans are beginning to acidify as they sequester more CO_2. Acidification threatens the numerous forms of oceanic life that form carbon-based shells or skeletons, such as mollusks, corals, and diatoms. In short, the weight of evidence suggests that we have already exceeded the critical ecological threshold for atmospheric GHG stocks. (See Chapter 2.) This means that we must reduce flows by more than 80 percent or increase sequestration until atmospheric stocks are reduced to acceptable levels. If we accept that all individuals are entitled to an equal share of CO_2

absorption capacity, then the wealthy nations need to reduce net emissions by 95 percent or more.[13]

Another category of ecological limit entails renewable-resource stocks, flows, and services. All economic production requires the transformation of raw materials provided by nature, including renewable resources (for example, trees). To a large extent, society can choose the rate at which it harvests these raw materials—that is, cuts down trees. Whenever extraction rates of renewable resources exceed their regeneration rates, however, stocks decline. Eventually, the stock of trees (the forest) will no longer be able to regenerate. So the first rule for renewable-resource stocks is that extraction rates must not exceed regeneration rates, thus maintaining the stocks to provide appropriate levels of raw materials at an acceptable cost.

But a forest is not just a warehouse of trees; it is an ecosystem that generates critical services, including life support for its inhabitants. These services are diminished when the structure is depleted or its configuration is changed. So another rule guiding resource extraction and land use conversion is that they must not threaten the capacity of the ecosystem stock or fund to provide essential services. Our limited understanding of ecosystem structure and function and the dynamic nature of ecological and economic systems mean that this precise point may be difficult to determine. However, it is increasingly obvious that the extraction of many resources to drive growth has already gone far beyond this point. Rates of resource extraction must therefore be reduced to below regeneration rates in order to restore ecosystem funds to desirable levels.

Protecting Capabilities for Flourishing. In a zero-growth or contracting economy, working-time policies that enable equitable sharing of the available work are essential to achieve economic stability and to protect people's jobs and livelihoods. Reduced working hours can also increase people's ability to flourish by improving the work/life balance, and there is evidence that working fewer hours can reduce consumption-related environmental impacts. Specific policies should include greater choice for employees about working time; measures to combat discrimination against part-time work as regards grading, promotion, training, security of employment, rate of pay, health insurance, and so on; and better incentives to employees (and flexibility for employers) for family time, parental leave, and sabbatical breaks.[14]

Systemic social inequality can likewise undermine the capacity to flourish. It expresses itself in many forms besides income inequality, such as life expectancy, poverty, malnourishment, and infant mortality. Inequality can also drive other social problems (such as overconsumption), increase anxiety, undermine social capital, and expose lower-income households to higher morbidity and lower life satisfaction.[15]

The degree of inequality varies widely from one sector or country to another. In the U.S. civil service, military, and university sectors, for example, income inequality ranges within a factor of 15 or 20 between the highest and lowest paying jobs. Corporate America has a range of 500 or more. Many industrial nations are below 25.[16]

A sense of community—which is necessary for democracy—is hard to maintain across such vast income differences. The main justification for such differences has been that they stimulate growth, which will one day filter down, making everyone rich. But in today's full world, with its steady-state or contracting economy, this is unrealistic. And without aggregate growth, poverty reduction requires redistribution.

Fair limits to the range of inequality need to be determined—that is, a minimum and a maximum income. Studies have shown that most adults would be willing to give up personal gain in return for reducing inequality they see as unfair. Redistributive mechanisms and policies could include revising income tax structures, improving access to high-quality education, introducing anti-discrimination legislation, implementing anti-crime measures and improving the local environment in deprived areas, and addressing the impact of immigration on urban and rural poverty. New forms of cooperative ownership (as in the Mondragón model) or public ownership, as is common in many European nations, can also help lower internal pay ratios.[17]

The dominance of markets and property rights in allocating resources also can impair communities' capacity to flourish. Private property rights are established when resources can be made "excludable"—that is, when one person or group can use a resource while denying access to others. But many resources essential to human welfare are "non-excludable," meaning that it is difficult or impossible to exclude others from access to them. Examples include oceanic fisheries, timber from unprotected forests, and numerous ecosystem services, including waste absorption capacity for unregulated pollutants.

Absent property rights, resources are "open access"—anyone may use them, whether or not they pay. However, individual owners of property rights are likely to overexploit or underprovide the resource, imposing costs on others, which is unsustainable, unjust, and inefficient. Private property rights also favor the conversion of ecosystem stocks into market products regardless of the difference in contributions that ecosystems and market products have to human welfare. The incentives are to privatize benefits and socialize costs.

One solution to these problems, at least for some resources, is common ownership. A commons sector, separate from the public or private sector, can hold property rights to resources created by nature or society as a whole and manage them for the equal benefit of all citizens, present and future. Contrary to wide belief, the misleadingly labeled "tragedy of the commons" results from no ownership or open access to resources, not common ownership. Abundant research shows that resources owned in common can be effectively managed through collective institutions that assure cooperative compliance with established rules.[18]

Finally, flourishing communities will be supported and maintained by the social capital built by a strong democracy. A strong democracy is most easily understood at the level of community governance, where all citizens are free (and expected) to participate in all political decisions affecting the community. Broad participation requires the removal of distorting influences like special interest lobbying and funding of political campaigns. The process itself helps to satisfy myriad human needs, such as enhancing people's understanding of relevant issues, affirming their sense of belonging and commitment to the community, offering opportunity for expression and cooperation, and strengthening the sense of rights and responsibilities. Historical examples (though participation was restricted to elites) include the town meetings of New England and the system of ancient Athenians.[19]

Building a Sustainable Macroeconomy. The central focus of macroeconomic policies is typically to maximize economic growth; lesser goals include price stabilization and full employment. If society instead adopts the central economic goal of sustainable human well-being, macroeconomic policy will change radically. The goals will be to create an economy that offers meaningful employment to all and that balances investments across the four types of capital to maximize well-being. Such an approach would lead to fundamentally different macroeconomic policies and rules.

A key leverage point is the current monetary system, which is inherently unsustainable. Most of the money supply is a result of what is known as fractional reserve banking. (See Box 11–1.) Banks are required by law to retain a percentage of every deposit they receive; the rest they loan at interest. However, loans are then deposited in other banks, which in turn can lend out all but the reserve requirement. The net result is that the new money issued by banks, plus the initial deposit, will be equal to the initial deposit divided by the fractional reserve. For example, if a government credits $1 million to a bank and the fractional reserve requirement is 10 percent, banks can create $9 million in new money, for a total money supply of $10 million. In this way, most money is today created as interest-bearing debt. Total debt in the United States—adding together consumers, businesses, and the government—is about $50 trillion. This is the source of the national money supply.[20]

There are several serious problems with this system. First, it is highly destabilizing. When the economy is booming, banks will be eager to loan money and investors will be eager to borrow, which leads to a rapid increase

Box 11–1. The Social Costs of the U.S. Banking System

In recent decades the United States has seen the eclipse of banking regulations, leading to a radical concentration of money power in too-big-to-fail banks and Wall Street generally. In 1994, the five largest U.S. banks held 12 percent of total U.S. deposits. By 2009 they held nearly 40 percent. The country's 20 largest banks control almost 60 percent of bank assets. Market concentration is even higher in other banking-type businesses, such as credit cards, debt and equity underwriting, and derivatives trading. Many of America's earlier leaders warned against such concentration of power in the hands of a financial elite. As Thomas Greco notes in *The End of Money and the Future of Civilization*, "Thomas Jefferson said, 'I sincerely believe . . . that banking establishments are more dangerous than standing armies.'"

Today banks are required to hold deposits that are only a small fraction—less than 10 percent—of the loans they make. Anyone who takes on debt is creating new money. Banks do not actually lend money; they create promises to supply money they do not possess. Mary Mellor has summed up the resulting situation: "The most important outcome of the dominance of bank issued money is that the supply of money is largely in private hands determined by commercial decisions, while the state retains responsibility for managing and supporting the system, as has become clear through the [2008] financial crisis." In the United States, the Federal Reserve can powerfully influence the supply and hence the price of money, but private banks decide how much to lend and where to lend it. The capital allocation process has become far removed from institutions that serve the public interest and is instead dominated by institutions and individuals seeking only to maximize profits.

The evidence is already abundant that today's system of money and finance cannot deliver a fair and sustaining economy. Its transformation is an integral, essential aspect of the overall transition to a new economy. Otto Scharmer of the Massachusetts Institute of Technology explains why: "Today we have a system that accumulates an oversupply of money and capital in areas that produce high financial and low environmental and social returns, while at the same time we have an undersupply of money and capital in areas that serve important societal and community investment needs (high social and low financial returns, such as the education of children in low-income communities)." Among other urgently needed reforms, economist Herman Daly has recommended returning the power to create money to government by abandoning today's fractional-reserve banking and moving to a 100 percent reserve requirement on demand deposits. Banks would lend time deposits, and the depositor would not have access to the money for the period of the deposit. The lending bank would have to count on new and renewing short-term time deposits or on long-term time deposits. These requirements would eliminate the bank's ability to create new money. As needed, government would create new money instead. As Daly explains, "This would put control of the money supply and seigniorage (profit made by the issuer of fiat money) in hands of the government rather than private banks, which would no longer be able to live the alchemist's dream by creating money out of nothing and lending it at interest."

—James Gustave Speth
Professor of Law, Vermont Law School
Source: See endnote 20.

in money supply. This stimulates further growth, encouraging more lending and borrowing, in a positive feedback loop. A booming economy stimulates firms and households to take on more debt relative to the income flows they use to repay the loans. This means that any slowdown in the economy makes it very difficult for borrowers to meet their debt obligations. Eventually some borrowers are forced to default. Widespread default eventually creates a self-reinforcing downward economic spiral, leading to recession or worse.

Second, the current system steadily transfers resources to the financial sector. Borrowers must always pay back more than they borrowed. At 5.5 percent interest, homeowners will be forced to pay back twice what they borrowed on a 30-year mortgage. Conservatively speaking, interest on the $50 trillion total debt (in 2009) of the United States must be at least $2.5 trillion a year, one sixth of national output.[21]

Third, the banking system will only create money to finance market activities that can generate the revenue required to repay the debt plus interest. Since the banking system currently creates far more money than the government, this system prioritizes investments in market goods over public goods, regardless of the relative rates of return to human well-being.

Fourth, and most important, the system is ecologically unsustainable. Debt, which is a claim on future production, grows exponentially, obeying the abstract laws of mathematics. Future production, in contrast, confronts ecological limits and cannot possibly keep pace. Interest rates exceed economic growth rates even in good times. Eventually, the exponentially increasing debt must exceed the value of current real wealth and potential future wealth, and the system collapses.

To address this problem, the public sector must reclaim the power to create money, a constitutional right in the United States and most other countries, and at the same time take away from the banks the right to do so by gradually moving toward 100-percent fractional-reserve requirements.

A second key lever for macroeconomic reform is tax policy. Conventional economists generally look at taxes as a necessary but significant drag on economic growth. However, taxes are an effective tool for internalizing negative externalities into market prices and for improving income distribution.

A shift in the burden of taxation from value added (economic "goods," such as income earned by labor and capital) to throughput flow (ecological "bads," such as resource extraction and pollution) is critical for shifting toward sustainability. Such a reform would internalize external costs, thus increasing efficiency. Taxing the origin and narrowest point in the throughput flow—for example, oil wells rather than sources of CO_2 emissions—induces more-efficient resource use in production as well as consumption and facilitates monitoring and collection. Such taxes could be introduced in a revenue-neutral way, for example by phasing in resource severance taxes while phasing out regressive taxes such as those on payrolls or sales.[22]

Taxes should also be used to capture unearned income (rent, in economic parlance). Green taxes are a form of rent capture, since they charge for the private use of resources created by nature. But there are many other sources of unearned income in society. For example, if a government builds a light rail or subway system—more-sustainable alternatives to private cars—adjacent land values typically skyrocket, providing a windfall profit for landowners. New technologies also increase the value of land, due to its role as an essential input into all production. Because the supply of land is fixed, any increase in demand results in an increase in price. Landowners therefore automatically grow wealthier independent of any investments in the land. High taxes on land values (but not on improvements, such as buildings) allow the public sector to capture this unearned income. Public ownership through land trusts and other means also allows for public capture of the unearned income and eliminates any reward from land speculation, thus stabilizing the economy.[23]

Tax policy can also be used to reduce income inequality. (See Figure 11–3.) Taxing the highest incomes at high marginal rates has been shown to significantly reduce income inequality. There is also a strong correlation between tax rates and social justice. (See Figure 11–4.) High tax rates that contribute to income equality appear to be closely related to human well-being. This suggests that tax rates should be highly progressive, perhaps asymptotically approaching 100 percent on marginal income. The measure of tax justice should not be how much is taxed away but rather how much income remains after taxes. For example, hedge fund manager John Paulson earned $4.9 billion in 2010. If Paulson had to pay a flat tax of 99 percent, he would still retain nearly $1 million per week in income.[24]

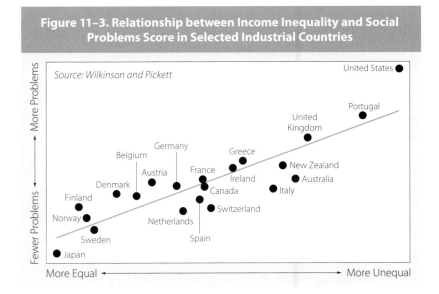

Figure 11–3. Relationship between Income Inequality and Social Problems Score in Selected Industrial Countries

Other policies for achieving financial and fiscal prudence will almost certainly be required as well. Our relentless pursuit of debt-driven growth has contributed to the global economic crisis. A new era of financial and fiscal prudence needs to increase the regulation of national and international

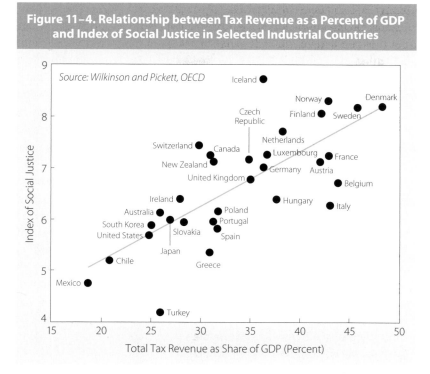

Figure 11–4. Relationship between Tax Revenue as a Percent of GDP and Index of Social Justice in Selected Industrial Countries

financial markets; incentivize domestic savings, for example through secure (green) national or community-based bonds; outlaw unscrupulous and destabilizing market practices (such as "short selling," in which borrowed securities are sold with the intention of repurchasing them later at a lower price); and provide greater protection against consumer debt. Governments must pass laws that restrict the size of financial sector institutions, eliminating any that impose systemic risks for the economy.[25]

Finally, as indicated earlier, we need to improve macroeconomic accounting, replacing or supplementing GDP as the prime economic indicator. GDP does, however, belong as an indicator of economic efficiency. The more efficient we are, the less economic activity, raw materials, energy, and work are required to provide satisfying lives. When GDP rises faster than life satisfaction, efficiency declines. The goal should be to minimize GDP, subject to maintaining a high and sustainable quality of life.

Is a Sustainable Civilization Possible?

The brief sketch presented here of a sustainable and desirable "ecological economy," along with some of the policies required to achieve it, begs the important question of whether these policies taken together are consistent and whether they are sufficient to achieve the goals articulated. Can we have a global economy that is not growing in material terms but that is sustainable and provides a high quality of life for most, if not all, people? Several lines of evidence suggest that the answer is yes.

The first comes from history. Achieving long-lasting zero- or low-growth desirable societies has been difficult—but not unheard of. While many societies have collapsed in the past and many of them were not what would be called "desirable," there have been a few successful historical cases in which decline did not occur, as these examples indicate:[26]

• Tikopia Islanders have maintained a sustainable food supply and non-increasing population with a bottom-up social organization.

• New Guinea features a silviculture system that is more than 7,000 years old with an extremely democratic, bottom-up decisionmaking structure.

• Japan's top-down forest and population policies in the Tokugawa era arose as a response to an environmental and population crisis, bringing an era of stable population, peace, and prosperity.

A second line of evidence comes from the many groups and communities around the world that are involved in building a new economic vision and testing solutions. Here are a few examples:

• Transition Initiative movement (www.transitionnetwork.org)

• Global EcoVillage Network (gen.ecovillage.org)

• Co-Housing Network (www.cohousing.org/)

• Wiser Earth (www.wiserearth.org)

• Sustainable Cities International (www.sustainablecities.net)

• Center for a New American Dream (www.newdream.org)

• Democracy Collaborative (www.community-wealth.org)

• Portland, Oregon, Bureau of Planning and Sustainability (www.portland online.com/bps/)

All these examples to some extent embody the vision, worldview, and policies elaborated in this chapter. Their experiences collectively provide evidence that the policies are feasible at a smaller scale. The challenge is to scale up some of these models to society as a whole. Several cities, states, regions, and countries have made significant progress along that path, including Portland in Oregon; Stockholm and Malmö in Sweden; London; the states of Vermont, Washington, and Oregon in the United States; Germany; Sweden; Iceland; Denmark; Costa Rica; and Bhutan.[27]

A third line of evidence for the feasibility of this vision is based on integrated modeling studies that suggest a sustainable, non-growing economy is both possible and desirable. These include studies using such well-established models as World3, the subject of *The Limits to Growth* in 1972 and other more recent books, and the Global Unified Metamodel of the BiOsphere (GUMBO).[28]

A recent addition to this suite of modeling tools is LowGrow, a model of the Canadian economy that has been used to assess the possibility of constructing an economy that is not growing in GDP terms but that is stable, with high employment, low carbon emissions, and a high quality of life. LowGrow was explicitly constructed as a fairly conventional macroeconomic model calibrated for the Canadian economy, with added features to simulate the effects on natural and social capital.[29]

LowGrow includes features that are particularly relevant for exploring a low-/no-growth economy, such as emissions of carbon dioxide and other greenhouse gases, a carbon tax, a forestry submodel, and provisions for redistributing incomes. It measures poverty using the Human Poverty Index of the United Nations. LowGrow allows additional funds to be spent on health care and on programs for reducing adult illiteracy and estimates their impacts on longevity and adult literacy.

A wide range of low- and no-growth scenarios can be examined with LowGrow, and some (including the one shown in Figure 11–5) offer considerable promise. Compared with the business-as-usual scenario, in this scenario GDP per capita grows more slowly, leveling off around 2028, at which time the rate of unemployment is 5.7 percent. The unemployment rate declines to 4 percent by 2035. By 2020 the poverty index declines from 10.7 to an internationally unprecedented level of 4.9, where it remains, and the debt-to-GDP ratio declines to about 30 percent and is maintained at that level to 2035. GHG emissions are 41 percent lower at the start of 2035 than in 2010.[30]

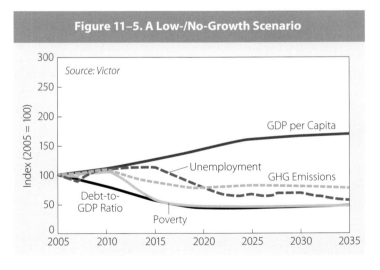

Figure 11–5. A Low-/No-Growth Scenario

Source: Victor

These results are obtained by slower growth in overall government expenditures, net investment, and productivity; a positive net trade balance; cessation of growth in population; a reduced workweek; a revenue-neutral carbon tax; and increased government investment in public goods, on anti-poverty programs, adult literacy programs, and health care. In addition, there are more public goods and fewer status goods through changes in taxation and marketing; there are limits on throughput and the use of space through better land use planning and habitat protection and ecological fiscal reform; and fiscal and trade policies strengthen local economies.

No model results can be taken as definitive, since models are only as good as the assumptions that go into them. But what World3, GUMBO, and

LowGrow have provided is some evidence for the consistency and feasibility of these policies, taken together, to produce an economy that is not growing in GDP terms but that is sustainable and desirable.

This chapter offers a vision of the structure of an "ecological economics" option and how to achieve it—an economy that can provide nearly full employment and a high quality of life for everyone into the indefinite future while staying within the safe environmental operating space for humanity on Earth. The policies laid out here are mutually supportive and the resulting system is feasible. Due to their privileged position, industrial countries have a special responsibility for achieving these goals. Yet this is not a utopian fantasy; to the contrary, it is business as usual that is the utopian fantasy. Humanity will have to create something different and better—or risk collapse into something far worse.

CHAPTER 12

Transforming the Corporation into a Driver of Sustainability

Pavan Sukhdev

There is an emerging consensus among government and business leaders that all is not well with the market-centric economic model that dominates today's world. Although it has delivered wealth in most economies over the last half-century and pulled millions out of poverty, it is recession-prone, leaves too many people unemployed, widens the gap between the rich and the poor, creates ecological scarcities that affect water and food, and generates environmental risks such as climate change.

Planetary boundaries are now being approached—and in some domains, have been breached—across many critical axes, including greenhouse gas emissions, the nitrogen cycle, freshwater use, land use and food security, ocean fisheries, and coral reefs. Within the next decade, significant changes are needed in the way we deal with Earth's resources. The failure of intergovernmental efforts points to the need to recognize the vital role of the private sector in determining economic direction and resource use globally. The corporate world must be brought to the table as planetary stewards rather than value-neutral agents that are free-riding their way to global resource depletion.[1]

The rationale for engaging with the private sector is compelling: corporations produce almost everything we consume, generating 60 percent of global gross domestic product (GDP) and providing a comparable share of global employment. Their advertising creates and drives consumer demand. Their production feeds this demand and drives economic growth.[2]

Corporations thus drive our economic system, but the way they have been operating also threatens the system's very survival. Externalities—the unaccounted costs to society of doing "business as usual"—by just the top 3,000 public corporations cost an estimated $2.15 trillion, or 3.5 percent of GDP, every year. Corporate lobbying frequently influences national policies and politics to the detriment of the public good. Advertising often converts human insecurities into wants, wants into needs, and needs into exces-

Pavan Sukhdev is founder-director of Corporation 2020. This chapter is based on *Corporation 2020: Transforming Business for Tomorrow's World* (Washington, DC: Island Press, 2012), as well as on the e-chapter "Why Corporation 2020? The Case for a New Corporation in the Next Decade."

www.sustainabilitypossible.org

sive consumer demand. Corporate production rises to meet such demand, which has already made humanity's ecological footprint exceed the planet's biocapacity by over 50 percent. We are now living by consuming Earth's capital, not its interest.[3]

We can blame consumerism, but consumerism was created by the corporation and its marketing and advertising. We can blame the free market, and indeed the free market has been the rallying cry of many in the private sector. But what they usually mean by "free market" is the "status-quo market." Around $1 trillion per year in harmful subsidies—including $650 billion in fossil fuel subsidies—promote "business-as-usual" while obscuring its associated environmental and societal costs. The finger of blame must finally point to the main agent of our "brown economy": today's corporation and the rules that govern its operations and behaviors.[4]

To break free of this system, the rules of the game need to be changed, so that corporations are enabled to truly compete on the basis of innovation, resource conservation, and satisfaction of multiple stakeholder demands—rather than on the basis of who can best influence government regulation, avoid taxes, and obtain subsidies for harmful activities in order to optimize shareholder returns. These rules of the game include policies regarding accounting practices, taxation, financial leverage, and advertising that can result in a new corporate model, an agent for tomorrow's green economy.

This new model can be called Corporation 2020 because the pace at which we are approaching planetary boundaries suggests that 2020 is the date by which it needs to be in place in order for us not to cross these boundaries. Like a biological species that evolves in response to its environment, and in turn influences it, today's corporation can evolve into Corporation 2020 in response to a changed environment of prices, institutions, and regulations. Its success can lead to a green economy. Achieving such an environment requires that four important change drivers be in place:

• First, taxes and subsidies have to be transformed to tax the "bads" more (such as resource extraction and fossil fuel use) and tax the "goods" (such as wages and profits) less, rather than the other way around, as is the case today.

- Second, we must introduce rules and limits to govern financial leverage, especially if the borrower is considered "too big to fail."
- Third, advertising norms and standards must be introduced so that advertising is much more responsible and accountable.
- Fourth, all major corporate externalities—both positive and negative—must be measured, audited, and reported as disclosures in the annual financial statements of companies.

These four reforms will together ensure that the new corporate model evolves from the old and does so profitably. As it increasingly wins business away from the old model, its net impacts on society will be positive because it is now hard-wired to create positive externalities, not negative ones. Collectively, its activities will bring us closer to a green economy, one that increases human well-being and social equity and decreases environmental risks and ecological losses.

Using Taxes as Incentives

The end of the twentieth century saw global consumption of almost every principal industrial commodity increase dramatically, fueling the 242 percent economic expansion of the last four decades. Between 1973 and 2009, world energy consumption nearly doubled from the equivalent of 4.6 billion to 8.4 billion tons of oil. Fossil fuels—coal, petroleum, and natural gas—represented over 80 percent of global energy consumption during this period.[5]

This practice of fueling our economic activity using nonrenewable resources has been very effective at increasing GDP, but it is ultimately not sustainable. Most of the increase in energy use has occurred, and will continue to occur, in the developing world. If the material living standards there were equal to those of the average American, the natural resource inputs required for this consumption would exceed five Earths' worth of global ecological capacity.[6]

Taxing the resource base of our predominantly "brown" economy—coal, petroleum, and many other minerals—can steer the market away from resource-intensive growth and toward smart-technology industries in renewable energy, clean water, new and better materials, and waste management. Taxing resources and removing all resource subsidies would force a revaluation of resources, in turn allowing us to manage, not simply extract, natural assets. Resource taxation will not only reduce the resource intensity of consumption, it has the potential to generate new revenues and additional financing that can be used for high-priority areas such as education and health care—or it can be applied against the rising cost of nature's remaining resources.

The philosophy of free markets and small government has long demonized taxation as a job-killing, "socialist" redistribution tool that robs the rich in order to feed the inefficiencies of "big government." Like any tool, however, taxes are either good or bad depending on how they are used. Using taxes to revalue natural resources positions an innovative Corporation 2020 as the successful protagonist of twenty-first century capitalism.

"Too Big to Fail" is Too Big

Over the last few decades, "sustainability" has become nearly synonymous with environmental initiatives. But as has become evident over the last few years, businesses have not even succeeded in being financially sustainable, let alone environmentally sustainable. In general, it should not be worrying if a business is not financially sustainable, because bankruptcy is a normal element of a functioning market. But governments have increasingly viewed a diverse group of companies as "too big to fail"—a term that refers not only to big banks (which provide clearing and settlement, which if disrupted can have far-reaching economic consequences) but now includes giant insurers, airlines, and auto manufacturers. These companies are hotbeds of "moral hazard"—they are inherently incentivized to take risks that push the entire economic system toward instability because they are comforted by the confidence that governments will socialize their losses when meltdowns happen.

The problem with having so many such companies is that it adds to systemic risks. Financial leverage has played a large role in each of the last four major economic crises in the world—the Latin American debt crisis, the savings and loan crisis in the United States, the Asian debt crisis, and the recent global financial crisis led by the housing sector. The wisdom of allowing a growing population of "too big to fail" companies yet more financial leverage to grow even larger is highly questionable, even when cloaked in the garb of promoting growth or aiding development.[7]

At present, most regulators are shying away from addressing the risks that excess leverage imposes. Even when they do engage the idea of more or better controls, they focus mainly on further capital requirements for banks and financial intermediaries. We know this cannot be the answer because out of the four financial crises just mentioned, the last two happened while sophisticated capital adequacy regulations by the Bank for International Settlements and the European Community were in place for the banks involved. They are in effect leaving the better interests of society to be met by the "invisible hand" of markets. In other words, investor behavior is expected to determine how much leverage is appropriate, with fund managers becoming the unlikely conscience-keepers of society. Unfettered markets were never meant to solve social problems, yet the system today is set up as if they were.

It is essential that we re-evaluate and rebuild the financial sector's regulatory infrastructure to better monitor systemic risk and control of leverage. In addition, we must explore regulatory options for nonbanking corporations that include reasonable limitations on leverage.

The most widespread tools to control leverage of financial institutions are reserve requirements and capital adequacy ratios:

- *Reserve requirements.* These represent the fraction of deposits that banks are required to retain either as cash in the vault, as a balance directly with the central bank, or as government and other high-quality liquid securities. Reserve requirements help limit the leverage in the banking system as a whole and also help reduce the risk of liquidity problems.
- *Capital adequacy ratios.* Whereas reserve ratios are akin to using "brute force" to preempt bank liquidity away from markets, capital adequacy ratios are a more subtle device in that they use the economic disincentive of raising the capital costs of leverage to achieve similar ends. A capital adequacy ratio limits an institution's financial leverage by requiring the financial firm to have a minimum amount of capital—including ownership equity and other forms of long-term capital—based on a specified percentage of the firm's assets.

For nonfinancial corporations, other tools are available:

- *Consortium banking.* An interesting case of nonfinancial corporations' leverage being monitored actively is India's "consortium banking" arrangements. Under these schemes, banks form lending groups that share key financial information about their corporate borrowers, including information about their credit ratings, financial exposure, securities outstanding, and compliance with financial covenants. This enables the group to minimize the possibility that a borrowing firm can play banks off against one another in order to take on more leverage than is advisable.[8]
- *Eliminating the tax deductibility of interest.* One significant incentive for corporations to increase their use of debt is the tax deductibility of interest expenses. This creates a clear inducement for companies to lever up, with governments effectively subsidizing a portion of the cost of debt. A simple solution would be to impose limits on the tax deductibility of interest expense for nonfinancial corporations by phasing out or capping the total amount of interest deductible.
- *Strengthening disclosure requirements.* Improved disclosure requirements need to be enacted for off-balance-sheet obligations and derivative transactions. Proper measurement and reporting of leverage is critical to the effective control of leverage at nonfinancial firms.
- *Constraining leverage from acquisitions.* Mergers and acquisitions (M&A) represent an important source of leverage around the world, especially when they take the form of leveraged-buyout (LBO) transactions, which

involve the heavy use of debt. Approximately 14,000 LBOs took place in 2007, up from 5,000 in 2000. LBOs often have leverage ratios of at least 4 or 5 and higher, meaning that the majority of the funds they use to acquire the new company consist of loans that must be paid back. M&A transactions that exceed a given transaction amount—such as $10 billion— should be subject to review and approval by that country's central bank (in the United States, the Federal Reserve) in order to ensure the amount of leverage used is not likely to sink the company in debt and create downstream economic ripples.[9]

Breaking the Cycle of Advertising and Consumption

In addition to making taxes more effective and placing limits on leverage, we must look closely at the demand side of the equation and ask what is driving today's unsustainable level of consumption. This brings us to the issue of corporate advertising. (See also Chapter 10.) Global advertising turnover is estimated to be around $500 billion, which is less than Walmart and Carrefour combined are worth. So while advertising is a relatively small global business, it has an inordinately high share of voice: it impinges on us more than any other communication, every day of the week, every week of the year. And every commercial message that enters our conscious or unconscious mind was placed there by marketing and advertising companies.[10]

David Evers

Billboard landscape in Alexandria, Egypt.

Marketing and advertising convert wants into needs, sometimes creating new wants out of human insecurities, which are then skillfully transformed into new consumer needs that must be met. It would not be an exaggeration to say that advertising is the single biggest force driving consumer demand today.

But for many consumers, advertising has become the bane of modern existence. So there are opposing forces at play. Consumer resistance has built up, and in some cases vocal consumer resistance has resulted in legislation to control advertising, if not ban it entirely. Consumers increasingly want to shut down the cacophony—or at least "talk back." A delightful example of this two-way interaction is the Bubble Project, in which communication designer Ji Lee pasted 50,000 blank stickers that looked like "speech bubbles" on advertisements throughout New York City, allowing passers-by to write in their reactions, thoughts, and witticisms.[11]

In other parts of the world, legislation has intervened to keep public spaces "public." In 2007, São Paulo became the first major city outside the communist world to ban almost all outdoor advertising. In a city with two conflicting identities—it is both the commercial capital of Brazil and the epicenter of gang violence and extensive slums—São Paulo's *Lei Cidade Limpa* (Clean City Law) is now considered an unexpected success. Nearly all outdoor advertisements—including billboards, outdoor video screens, and ads on buses—were torn down, and the size of storefront signage was regulated. The law was enforced with nearly $8 million in fines. Despite protests and legal challenges, more than 70 percent of city residents welcomed the move. In fact, even Nizan Guanaes, head of Grupo ABC, Brazil's largest advertising group, said "I think it's a good law. It was a challenge for us because, of course, it's easier to simply throw garbage advertising all over your city."[12]

Apart from legislative actions, consumers are also increasingly unwilling to put up with advertising that is misleading. The ability to talk back empowers two-way communication and co-creation. Bob Garfield, journalist and advertising commentator, coined the term "listenomics" to describe the trend toward businesses using open-source techniques to find ideas for product development, marketing, production, and many other activities that have traditionally been controlled by isolated corporate departments. These companies can be viewed as either encouraging or co-opting these forces, depending on your viewpoint.[13]

Regardless, it is clear that a certain degree of serious change in advertising is going to come endogenously—through the changing balance of power between consumer and producer. However, this is an evolutionary process and will take time—several decades perhaps. But what can be done over the next decade, given the urgency of reform in the corporate world?

Two basic principles underlie the movement for change during this decade. The first principle for advertising goes beyond what industry self-regulation and governmental standards generally require: corporate advertisers need to treat all consumers as equal, no matter where they live—whether in an industrial country or the developing world.

Second, transparency and disclosure are key elements of accountable advertising. A robust practice of disclosure around advertising can improve the comparison between corporate bodies and also push them to be more accountable. An annual Accountable Advertising Report would reveal which relevant industry standards have been used, provide a place to share newly created corporate principles on responsible advertising, and, most important, be a vehicle for companies to differentiate themselves from—and be better than—their competitors.

Therefore in addition to following the two principles just described, four strategies can bring us closer to a more accountable system of advertising:

- *Disclose life span on products and in all advertisements.* This would drive individuals to question whether they really need a new version of an item or whether they should purchase an item that has such a short life span in the first place.

- *Disclose countries of origin on the product.* On the product itself, this should be a simple visual that highlights all the countries in which any part of the product was produced. While this simplifies a more formal life-cycle analysis process, its simplicity is what makes it effective in getting people to avoid products that have too many "miles" in their assembly or that come from countries where human rights are disrespected or nature is excessively exploited.

- *Recommend on the product itself how to dispose of it.* Advertisers should communicate how to dispose of a product when advertising it, so that consumers recognize the residual or waste value of the product and the responsibility they have to dispose of it properly.

- *Voluntarily commit a "10 percent development donation" on total advertising spent in developing countries.* This recommendation is specific to the developing world: to offset "footprint" expansion in local economies, advertisers could support local sustainability projects through a 10 percent "ad dollars to development dollars" commitment. The benefit of a proportion like this is that companies might have an incentive to spend less on advertising, which in some cases may reduce consumption.

These principles and strategies are not the only tools available to move us toward a more acceptable form of "accountable advertising," but they would be a start—and they would be especially effective if coupled with additional taxes and bans on the most pernicious forms of advertising, such as ads that promote social ills like smoking and those that target vulnerable populations like children. As companies begin to think more seriously about both the unintended consequences of their production as well as the potential good they could do with their advertising, new tools and strategies will surely emerge.

Taking Externalities into Account

The modern corporation is responsible for immense negative externalities, the largest of which is most likely its impact on the environment. Many corporations undertake processes that have negative impacts on the environment, such as air pollution or deforestation. Sometimes these impacts are rare, catastrophic events, like BP's oil spill in the Gulf of Mexico. But they can also be so ubiquitous that people do not even notice them anymore. One recent study estimates that 3,000 of the biggest public companies alone cause $1.44 trillion in damages from their greenhouse gas emissions.[14]

On the other hand, corporations can also create positive externalities. One leader in creating human capital has been the Indian software giant Infosys. Its primary training campus in Mysore is the largest corporate university in the world, with the capacity to train 14,000 employees at a time. Simply due to the sheer scale of its training initiatives, Infosys is probably one of the largest generators of positive human-capital externalities in the world. The reason is that Infosys's training programs enhance the earning potential of thousands of people, some of whom leave to work in organizations elsewhere. Thus these people represent a positive externality for society for which the company receives no economic gain—an externality estimated to be worth over $1.4 billion in 2012.[15]

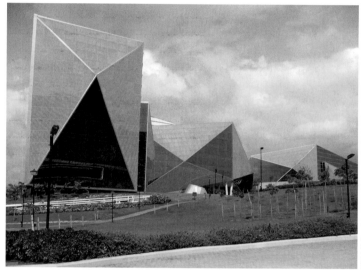

Part of the Infosys training campus in Mysore.

It is clearly to companies' own benefit to measure their positive externalities, but it is essential for the survival of the economy as a whole that they start measuring and disclosing their negative externalities as well. Our current understanding of the extent to which corporations cause externalities is fuzzy at best. There is a common aphorism in business management that "you cannot manage what you do not measure." Most corporations only measure financial performance, not their externalities—the third-party effects of doing "business as usual." The same problem is seen at the country level as well: governments are fixated on measuring only GDP and targeting its growth, forgoing many more holistic and relevant macroeconomic indicators such as Green GDP, Inclusive Wealth, and so on, which subtract negative environmental externalities from overall economic performance.

We need a better accounting framework, one that reflects both positive and negative externalities in a corporation's financial statements and thus makes transparent not only its holistic impact on the economy, society, and the environment, but also its exposure to risks of resource constraints and regulation. Furthermore, the external impacts of companies must be standardized. (See also Chapter 13.) Although there may be a dozen ways, for instance, to calculate the freshwater externalities of a cement plant—across locations, ecosystem types, and types of cement plants—there should not be a dozen accounting standards. On the contrary, there should be just one— with clear parameters and simple enough for the industry to use.

The recently formed TEEB for Business Coalition (The Economics of Ecosystems & Biodiversity) has as its primary task to standardize the methodologies for calculating exactly these types of corporate externalities. As of November 2012, an ambitious program of first establishing priorities and then quantifying the top 100 global externalities has been launched. A mechanism of this type ensures that investors are adequately aware of the broad set of relevant risks faced by any corporation with large externalities, as opposed to the narrower risks that are currently measured and reported.[16]

A uniform reporting mechanism established by combining leading current research on externalities valuation and risk assessment would ensure awareness of the current and projected magnitude of a corporation's operations, supply chain, and investments' external impact on society, economy, and capital stock. It would also allow corporations to identify sources of negative and positive impacts that they can target to improve.

Moving toward a More Responsible Corporation

If the recommendations of this chapter are implemented—taxation shifted toward resource extraction, corporate leverage limited for those "too big to fail," advertising made more accountable, and externalities measured and disclosed—the new corporations will likely look quite different from those of today. They will be more responsible, with goals aligned to the communities and societies that host them.

First of all, tomorrow's corporation will be a "capital factory," not just a goods-and-services factory. It will create financial capital for its shareholders through its operations, but without depleting (and ideally, while growing) natural capital, social capital, and human capital for society at large— the stakeholders of the corporation.

Second, Corporation 2020 will be a community. The loss of community around the world is a palpable result of the dominant economic model. Corporation 2020 can be a modern-day community, tied by a shared culture created by its values, mission, goals, objectives, and governance. It can (and in the best of today's companies, it already does) recreate the sense of belonging that has been lost due to the forces of modernization and globalization.

Third, the corporation of tomorrow must be an institute of learning and skills training, providing employees with an increasing base of knowledge and skills with which to add value to the corporation and also add to each individual's earnings profile.

Finally, the goals of Corporation 2020 should be the goals of human society: increased human well-being, increased social equity, improved social and communal harmony, reduced ecological scarcities, and reduced environmental risks. Profitability is undoubtedly a key objective for Corporation 2020, which ensures its financial sustainability while pursuing these goals, but it is not the only objective. There are other important goals—not just those determined by the corporation's shareholders but also those determined by its stakeholders: the public, those who are affected by the corporation.

If the ideas presented here seem complex, that is because they have to be. Complex problems merit complex solutions, and there are no elegant or easy ways to transform corporate purpose and behavior to create a sustainable economy. Too many people still underestimate the urgency, extent, and complexity of the challenge ahead. No one institution, be it government or civil society or the market or the corporation itself, can succeed alone. And the challenge is too often presented as solely about the environment, or social justice, or economics. But it is in fact a challenge of survival for the corporation itself, for the modern economies that corporations constitute and operate, and for human civilization as we know it.

CHAPTER 13

Corporate Reporting and Externalities

Jeff Hohensee

A bag of groceries carried into the house is a snapshot of the global economy. The raspberries could have come from Chile. The plastic container they came in could have been manufactured in Mexico from oil extracted in the Middle East. Even things that seem local might not be: bread baked in Los Angeles, for instance, could be made with wheat from the San Joaquin Valley in California, water from Colorado, and salt from Pakistan.

The world's economy is a web of activities that span the globe. The commercial activities of the world's largest corporations extend around the world, collectively touching almost everywhere, and their gross revenues often exceed those of many national economies. This economic web supplies the labor, materials, and resources that make the products and services we enjoy. The environmental and social impacts of these activities are broad and deep but rarely counted in corporate reporting, forcing business leaders to make decisions with partial information. But some promising recent trends in corporate reporting could provide regulators, investors, corporate decisionmakers, and community leaders with more accurate views of the activities that affect their companies and communities.

Most notable in this trend is "integrated reporting," a new form of corporate disclosure that integrates financial data with the environmental and social challenges that affect a company's health. In the mid-1990s, the Prince of Wales initiated the Accounting for Sustainability (A4S) project in the United Kingdom. A4S proposed that reporting regimes integrate strategy, governance, and financial performance into the social, environmental, and economic contexts of a company. A4S's work draws from concepts like ecological economics, natural capitalism, and full-cost accounting.[1]

Numerous attempts have been made to build these concepts into corporate reporting, including Baxter's *Environmental Financial Statement*, Puma's *Environmental Profit & Loss Statement*, Wilhelm's *Return on Sustainability*, Willard's *Sustainability Advantage*, and Krzus and Eccles' *One Report*.

Jeff Hohensee is an associate of the Colorado-based nonprofit Natural Capitalism Solutions.

www.sustainabilitypossible.org

Building on this work, A4S collaborated with the Global Reporting Initiative (GRI) to spearhead the creation of the International Integrated Reporting Council (IIRC). The IIRC is currently piloting an integrated report as an alternative to traditional corporate reports, as described later in this chapter. The integrated report has the potential to help corporate directors and officers make better decisions as well as to help investors and other stakeholders understand better how a company is really performing and its impacts in local communities.[2]

Externalities

The global economy provides people with the food on their tables, the shelter over their heads, and many of the routine supplies of daily life. Over the past 30 years, in great part because of the expansion of the global economy, more than a half-billion people have been brought out of abject poverty. This has increased life expectancy and improved the quality of life. As documented throughout this book, however, the rapid expansion of globalism has also increased water and air pollution and the production of hazardous wastes, and it has brought most major ecosystems to the brink of collapse. We have lost many of the services these ecosystems provide, such as clean air, clean water, and arable soil. As global population has increased, so has per capita natural resource use, energy consumption, and demands on the environment to provide raw materials and other natural resources.[3]

From a corporate reporting standpoint, these negative impacts are often considered externalities, which are costs (such as air pollution) or benefits from an economic activity that are not fully reflected in the price of the good or service involved. Externalities disguise the actual cost of goods by leaving these costs or benefits unaccounted for in product pricing and in corporate reporting. To take just one example, the emissions from the long-haul trucks used to bring goods to market cause air pollution that imposes health care costs that are not included in the price of the goods sold. In California's San Joaquin Valley, it is estimated that meeting federal clean air standards would save the region air pollution–related health care costs of more than $1,600 per person a year—for $6 billion in annual savings for the region's economy. Yet the cost of air pollution from long-haul trucks is not included in the cost of San Joaquin Valley farm goods.[4]

Unreported externalities can also hide business risks. "Chemicals of concern" is a case in point. According to the California Department of Toxic

Substances Control, "because of the many chemicals in commerce . . . and the increased interest by scientists and the public in understanding the types of toxicity that chemicals may pose, more and more scientists and toxicologists are identifying 'emerging chemicals of concern,' or ECCs. . . . Some examples of ECCs include bisphenol A, phthalates, arsenic, perchlorate, nonylphenols, synthetic musks and other personal care product ingredients."[5]

Bisphenol A (BPA) illustrates the potential of an ECC to have a measurable financial impact on business profit even before such chemicals are regulated. BPA is an endocrine disruptor that has been associated with a wide variety of health problems, including miscarriages, retarded infant brain development, obesity, heart disease, and cancer. Evidence that BPA was a chemical of concern first emerged in 1931. Transparent corporate reporting would have notified all involved that a risk existed. Instead, more than half a century went by until the adverse effects of low-dose exposure on laboratory animals were widely reported in 1997. It was another decade until several government agencies questioned the safety of BPA, which prompted a public outcry. Shortly thereafter, several retailers pulled drinking bottles containing BPA from their shelves. In 2010, Canada became the first country to declare BPA a toxic substance.[6]

The worst consequence of this delay in reporting, of course, is that millions of people were fed a hazardous chemical. But the cost to companies of quickly transitioning away from BPA was also huge. They carried an unreported risk that was entirely predictable. The cost of a quick, expensive transition to BPA-free products could have been avoided if companies had acknowledged the risk and avoided using BPA in the first place. Unreported externalities like these burden the public, which is forced to shoulder the costs that companies should bear, but they also pose risks to companies and investors.

The omission of externalities in corporate reporting is not a small problem. The global value of externalities is staggering: it was estimated at nearly $7 trillion—11 percent of the value of the global economy—in 2008. And just 3,000 companies were responsible for 35 percent of these costs. Among the largest of the unreported externalities are those related to oil. Since the Industrial Revolution, this form of natural capital has been extracted and used in the production of electricity, gasoline, diesel fuel, fertilizer, herbicides, plastics, and explosives. The negative impacts of the petrochemical industries on ecosystems, cultural heritage, and economic equity—chief among them, climate change—have been well documented. According to the Principles for Responsible Investment Initiative, the environmental externalities of the oil and gas industries are valued at more than $300 billion a year. The exclusion of undisclosed externalities is a significant oversight that needs to be addressed.[7]

Corporate Reporting of Externalities

Efforts to promote corporate reporting of externalities have taken both mandatory and voluntary forms. A variety of regulatory bodies are responsible for what is included in mandatory corporate reporting as well as what is left out. In the United States, the private, nonprofit Financial Accounting Standards Board sets Generally Accepted Accounting Principles (GAAP) for financial reporting. The International Accounting Standards Board sets International Financial Reporting Standards (IFRS), the measure used for financial reporting in most other countries. GAAP and IFRS are being combined into one system through a convergence process that will eventually produce a unified accounting standard. This process, however, has not addressed externalities, nor is financial reporting likely to do so soon.

Limited progress has been made in other mandatory corporate reports. In the United States, the Securities and Exchange Commission (SEC) oversees mandatory reporting for companies whose stock is traded in that country. Beginning in 1982, SEC Regulation S-K required disclosure of the cost of compliance with environmental regulations and the potential cost of legal proceedings for environmental liabilities when those costs were large enough to affect earnings. While not explicitly mentioning environmental externalities, the SEC also requires companies to disclose trends, events, and uncertainties that may materially (measurably) affect the company's financial position.[8]

After years of pressure from groups like Ceres (a nonprofit founded in 1989 as the Coalition for Environmentally Responsible Economies), in 2010 the SEC took the unprecedented action of issuing an interpretative release on disclosure of externalities related to climate change. This guidance explains the SEC's position on how climate change risk should be addressed under existing reporting requirements by expanding the requirement for environmental externalities to include the indirect consequences of climate change.[9]

Numerous voluntary corporate reporting initiatives encourage disclosure of environmental, social, and governance externalities to varying degrees. In the 1980s, many corporations began reporting voluntarily on sustainability metrics. These are commonly called corporate social responsibility (CSR) reports. The pioneers of CSR reporting gave shareholders, regulators, and other interested parties previously unavailable views into the positive and negative impacts of corporate activities. While the early CSR reports were better than nothing, the points made there were often incomplete and, in some cases, intentionally misleading. By the 1990s Ceres, the World Resources Institute, and several other groups launched an initiative to create standards for CSR reporting. This effort led to the Global Reporting Initiative.

GRI has become what many call the gold standard for CSR reporting.

By 2008, some 80 percent of the world's 250 largest corporations were producing GRI-based CSR reports. This movement dramatically improved the quality and transparency of corporate sustainability reporting. Yet companies can exclude vast portions of their sustainability impacts from CSR reporting and still achieve a high GRI reporting standard.[10]

Flotilla of vessels working to stop the flow of oil at the site of the Deepwater Horizon spill.

U.S. Coast Guard

BP's 2009 CSR report described its environmental controls and approach to risk by stating that its "commitment to competence is through having the right people with the right skills doing the right thing supported by our leadership framework." The report included an entire section on deep-water drilling that touted BP's technical expertise. BP's 2010 CSR report then had to address the Deepwater Horizon oil spill in the Gulf of Mexico. In a sea of carefully worded narrative, it reported that almost all of the environmental metrics in the report improved over the previous several years. The metrics in BP's 2010 CSR report prove, to the point of absurdity, that CSR reporting, as currently configured is insufficient to guarantee reporting of externalities.[11]

Integrated Reporting

The most promising move to include externalities in corporate reporting is the integrated report proposed by the International Integrated Reporting Council. The IIRC's participants include the Global Reporting Initiative, WWF, and the World Resources Institute. The world's largest accounting firms are involved, as are the key regulatory agencies responsible for corporate reporting. Diverse multinational corporations have been involved in establishing the IIRC framework and piloting the integrated reports, including AB Volvo, the Clorox Company, the Coca-Cola Company, Deloitte LLP, Deutsche Bank, Jones Lang Lasalle, Microsoft, Sainsbury's, Tata Steel, and Unilever.[12]

An integrated report, as described by the IIRC, is "a principles-based approach that requires senior management and those charged with governance to apply considerable judgement to determine which matters are material and to ensure they are appropriately disclosed given the specific circumstances of the organization and, where appropriate, the application of generally accepted measurement and disclosure methods." The conceptual foundation of an IIRC integrated report specifically mentions the disclosure of externalities:

- *Capital (resources and relationships)*: The IIRC's 2011 Discussion Paper noted that "Integrated Reporting . . . makes visible an organization's use of and dependence on . . . 'capitals' (financial, manufactured, human, intellectual, natural and social), and the organization's access to and impact on them."
- *External factors*: A July 2011 draft outline notes that the framework for integrated reporting "is expected to discuss how external factors affect the organization both directly and indirectly, for example, how they affect the availability, affordability, and quality of capitals that the organization depends upon and impacts in creating and preserving value. External factors include macro and micro economic conditions, market forces, the speed and impact of technological change, societal issues, environmental challenges, and the legislative and regulatory environment in which the organization operates."[13]

The integrated report framework includes a description of a company's business model that describes the forms of capital it relies on, strategic objectives to add value to these capitals, and delivery of products or services to achieve these objectives.

The IIRC challenges corporate reporting requirements, stating that policymakers should "question capital market orthodoxy and challenge traditional accounting practices, business models and value creation methods. One concern is whether capital is being allocated in the most effective way to achieve sustainable returns over the short, medium and long term." This has far-reaching implications for sustainability. An integrated report discloses the external environmental and social factors that directly and indirectly affect a company. In doing this, IIRC's integrated report effectively eliminates the concept of externalities by bringing indirect environmental and social costs inside corporate reporting.[14]

The U.S. nonprofit Sustainability Accounting Standards Board (SASB) is establishing sustainability standards that can be used in integrated reporting and other forms of corporate disclosure. The SASB views the SEC as one of its primary stakeholders, and the Board's objective is to create sustainability accounting standards for use by SEC registrants (publicly held companies) using the definition of materiality found in U.S. securities law (essentially, "big enough to matter"—which is sensitive to context, in that a sum that is material to a hot dog vendor would be immaterial to a transnational corporation).[15]

Ultimately, more-relevant, transparent, and useful disclosures of nonfinancial information by SEC registrants will affect the quality of disclosures made by companies around the world. In using the SEC definition of materiality, SASB is creating a de facto mandatory reporting environment for disclosure of material nonfinancial information. Even if the integrated report does not become mandatory, the SASB key performance indicators have the potential to credibly include externalities in mandatory investor reporting.

The Future of Corporate Reporting of Externalities

The persistent disconnect between "business as usual" and the need to report externalities is made clear by an Association of Chartered Certified Accountants 2012 survey. Forty-nine percent of the respondents identified natural capital as measurably valuable for businesses. The same survey notes that few companies include the value of natural capital in their financial reports. This omission puts investors, companies, and communities at risk.[16]

At the Rio+20 Conference in 2012, a group of 37 investment companies—including the International Finance Corporation, part of the World Bank Group—released an official Natural Capital Declaration that states the need to accurately calculate and disclose the value of externalities in corporate reporting: "It is becoming ever clearer that natural capital can have an impact on specific financial products, as well as on long-term growth. Endorsement of the Declaration represents an opportunity to understand how natural capital, as part of a range of other material [environmental, social, and governance] issues, can affect your institution's bottom line." Signatories of the Natural Capital Declaration specifically committed to collaborating with the IIRC and other stakeholders to "build a global consensus around the development of Integrated Reporting, which includes Natural Capital as part of the wider definition of resources and relationships key to an organization's success." More than at any other time in history, the value of reporting externalities is being recognized as a critical part of corporate reporting.[17]

The IIRC's work aims to fix the problem altogether. The integrated report has the potential to transform the nature of corporate reporting. By the beginning of December 2012, some 80 companies were piloting the IIRC's integrated report. The results of these pilots will be reviewed by the IIRC, which includes the accounting standards boards that oversee all financial reporting. The groundwork has been laid for a significant change in corporate reporting. The right players to mandate the change are at the table. The IIRC's work thus has the potential to create a level playing field that requires all corporations to include externalities in the information they disclose about performance.[18]

CHAPTER 14

Keep Them in the Ground: Ending the Fossil Fuel Era

Thomas Princen, Jack P. Manno, and Pamela Martin

Coal, oil, and gas—fossil fuels: we can't do without them. They are the life-blood of modern industrial civilization. These highly concentrated, widely available stores of energy have unleashed modern civilization's astonishing productivity, liberating billions of people from drudgery and insecurity. Finding more fossil fuels and getting them to markets around the world is the challenge of our times.

Fossil fuels: we must do without them. They feed the fire in the oven destined to bake civilization beyond recognition. When these hydrocarbons from the concentrated, pressurized remains of ancient organisms are burned, they overwhelm the Earth's ecosystems and condemn billions of people to climate-induced misery. Shifting to renewable energy sources and alternative ways of life is the challenge of our time.

Two existential positions, poles apart. Both may be accurate. The contradiction is the crux of the contemporary energy and environment dilemma and one reason governments have done so little in the face of obvious and ramifying threats.

Is there a way out? Not as long as technological optimism and trust in the magic of "the market" sustain the belief that the growth-dependent, consumerist, debt-laden, risk-accumulating world is the best of all possible worlds. Not when those who live in this world and those who aspire to join it see no reason to exchange the current model for an uncertain new model. Not when leaders and citizens alike cannot imagine replacing the current, fossil-fuel-dependent economic and social system. Why? Because too many people believe that the next energy transition, like previous transitions—from human power to animal power, animal to wood, wood to coal, and coal to oil—will make life better for all. As happened before, they believe, the next energy source will spur convenience, higher speeds, greater labor productivity, and more consumer choice—material progress forever. The bridge, this view has it, is new technologies to extract and burn every last bit

Thomas Princen is a professor of natural resources and environment at the University of Michigan. Jack P. Manno is a professor of environmental studies at SUNY College of Environmental Science and Forestry. Pamela Martin is a professor of politics at Coastal Carolina University.

www.sustainabilitypossible.org

of affordable oil, coal, and gas. This is the dominant worldview, what we dub the "industrial progressive" view.[1]

It is time to choose a different view and a different future. A first step is to recognize that Earth's "gift" to humanity of high-quality fossil fuels, those that pack the greatest energy wallop, is a one-time nonrenewing and diminishing reserve. There was a "before" (before the late nineteenth century), when fossil fuels powered only a tiny proportion of the world's work, and there will be an "after," when fossil fuels are reserved for tasks for which they, and they alone, are best suited. The question humanity faces at this historic juncture is how to navigate the transition, and how to do so given that the fossil fuel era will end and these fuels will be rationed—although on the current path, not soon enough to avert catastrophic climate and other environmental and social impacts.

The choice to keep fossil fuels in the ground in the face of otherwise overwhelming pressures to exploit them to the end is, we have come to believe, the only way to ensure that greenhouse gases and other pollutants remain out of the atmosphere and out of our bodies. The power and momentum of the fossil fuel complex is simply too great. And the predominant approach to ground-level air pollution, high-level climate change, and petrochemical contamination of human beings and nonhumans—that is, to manage fossil fuel emissions—is too ineffectual, too much of an accommodating, end-of-pipe approach. All too often, such an approach reduces a couple of centuries of history to one chemical element, carbon, when the real problem is upstream, in a global infrastructure and power structure that is extremely adept at drilling holes, blasting mountains, and laying pipes.

The Problem: Extraction . . . Not Emissions, Fossil Fuels . . . Not Carbon

The central problem is not emissions, but extraction. Put differently, it is not about carbon dioxide but fossil fuels—not about what comes out of the exhaust pipes and smokestacks but what comes out of the ground. To direct political attention away from end-of-pipe management to extraction is to be precautionary, a widely accepted approach for known toxic and ozone-depleting substances but not, as yet, for fossil fuels.

A carbon focus is reductionist, possibly the greatest and most dangerous reductionism of all time: a 150-year history of complex geologic, political, economic, and military security issues all reduced to one element—carbon. This framing implies that the problem only arises once fuels are burned. It effectively absolves of responsibility all those who organize to extract, process, and distribute. It leaves unquestioned the legal requirement to extract created by the selling of fossil-fuel reserves in futures markets and the widespread use of reserves for collateral in financial transactions. So constructed, extraction is called "production," and the burden of harm and of responsibility for amelioration falls on governments and consumers rather than on extractors. Inside the carbon logic, extraction is presumed to be a given—normal, inevitable, even desirable. What is more, the carbon lens portrays the global ecological predicament as one-dimensional: deal with carbon emissions, and everything else will follow.

To focus on fossil fuel extraction, in contrast, is to ask how and why removal of these fuels is deemed inevitable and net beneficial. A fossil fuel focus does not take such how-and-why questions as self-evident (people want the energy, producers get it). It directs analytic and political attention upstream to a whole set of decisions, incentives, and institutions that conspire to bring to the surface hydrocarbons that otherwise sit safely and permanently in the ground. It forces us to consider that once fossil fuels are extracted, their by-products—ground-level pollution, atmospheric greenhouse gases, petrochemical endocrine disruptors—inevitably and unavoidably move into people's bloodstream, into ecosystems, and into the atmosphere and oceans.

To question extraction is to consider deliberately limiting an otherwise valuable resource, rationing and setting priorities for its uses. It is to take renewable energy, conservation, equity, and environmental justice seriously and to create the institutions, local to global, capable of doing so. It is to ask what the prior ethics of fossil fuel allocation have been and what, given the imperative to reverse course and build a sustainable society, they must be. It is to ask what a politics of fossil fuel resistance and abolition would be and to imagine a deliberately chosen post–fossil fuel world.

All this leads to the conclusion—unthinkable for fossil fuel proponents and business-as-usual-only-greener proponents—that the only realistic means of stopping fossil fuel emissions is to keep the fuels in the ground. The only safe place for fossil fuels is in place, where they lie, where they are solid or liquid (or, for natural gas, geologically well contained already), where their chemistry is mostly of complex chains, not simple molecules like carbon dioxide, that find their way out of the tiniest crevices, that lubricate tectonic plates perpetually under stress, that react readily with water to acidify the oceans, and that float into high places filtering and reflecting sunlight, heating beyond livability the habitats below.

And yet the fossil fuel complex is extremely powerful. That power is at once energetic, economic, and political. Its weakness is ultimately geologic and ethical.

Fossil Fuel Influence

One measure of the industry's influence is the fact that 88 percent of the world's energy comes from fossil fuels. (See Box 14–1.) Sixty-one percent of that is produced by national oil companies—created, subsidized, and defend-

Box 14–1. Fossil Fuels by the Numbers

- Fossil fuels provide 88 percent of the world's energy.

- Fossil fuel infrastructure occupies an area the size of Belgium.

- Biofuel infrastructure roughly the size of the United States and India would be needed if biofuels were to replace fossil fuels.

- To meet industry and agency projections of increased energy demands, $38 trillion in oil and gas infrastructure is needed by 2035.

- It takes 7.3–10 calories of energy input to produce 1 calorie of food energy.

- Direct fuel subsidies to agriculture in the United States total $2.4 billion.

- Proven fossil fuel reserves, owned by private companies, state companies, and governments, exceed the planet's remaining carbon budget (in order to keep within a 2 degree Celsius temperature increase) by a factor of five.

- Occupationally related fatalities among workers in the oil and gas extraction process are higher than deaths for workers from all other U.S. industries combined.

Source: See endnote 2.

ed by national governments. Another is that the petroleum industry is the world's largest, capitalized at $2.3 trillion and accounting for 14.2 percent of all commodity trade. What's more, it is by far the most capital-intensive industry—$3.2 million is invested for every person employed. By comparison, the textile industry is capitalized at $13,000 per worker, the computer industry at $100,000, and the chemical industry at $200,000. And the petroleum industry is among the most profitable. In 2008, for example, ExxonMobil made $11.68 billion in second-quarter profits, amounting to profits of some $1,400 per second, and it ranked forty-fifth on a list of the top 100 economic entities in the world, a list that includes national governments. In 2010, ExxonMobil jumped to thirty-fifth on the list, just behind Royal Dutch Shell.[2]

Yet another indication of the influence of the fossil fuel complex is the flow of tax dollars to and from the industry. Worldwide, governments subsidize the fossil fuel industry to the tune of some $300–500 billion per year. In the United States in 2008, the petroleum industry paid $23 billion in royalties to the U.S. Treasury. In Saudi Arabia, the world's largest oil producer, oil and gas account for 90 percent of the gross domestic product while employing only 1.6 percent of the active labor force.[3]

Perhaps the industry's greatest source of influence is its ability to advance a vision, one of abundant and cheap energy, of powering and defending nations, of feeding and sheltering billions of people. It is a vision with appeal to nearly every sector of a modern industrial society—manufacturers, investors, military and political leaders, consumers. But its appeal has begun to erode.

For one, under the rubric of the "resource curse" (broadly construed), the social and economic costs have become well established. "The irony of oil wealth," writes political scientist Michael Ross in *The Oil Curse*, is that "the greater a country's need for additional income—because it is poor and has a weak economy—the more likely its oil wealth will be misused or squandered. . . . Since the oil nationalizations of the 1970s, the oil-producing countries have had less democracy, fewer opportunities for women, more frequent civil wars, and more volatile economic growth than the rest of the world, especially in the developing world." In addition, Ross finds, "by 2005, at least half of the OPEC countries were poorer than they had been thirty years earlier."[4]

From a national security perspective, former CIA director Jim Woolsey says: "It was obvious that oil was dominant in a lot of places that generated trouble. There's almost nothing that doesn't get better if you move away from dependence on oil." Even industry insiders have taken stock and are trying to imagine a different world. "The resources are there," writes John Hofmeister, former president of the Shell Oil Company in the United States. "The question is: do we *want* to continue to use these fossil fuels at current—or increasing—rates until they are eventually exhausted? The answer, unequivocally, is no. The economic, social, and environmental costs of such an approach are becoming ever clearer and ever higher." Or, as the German Advisory Council on Global Change put it, "The 'fossil-nuclear metabolism' of the industrialized society has no future. The longer we cling to it, the higher the prices will be for future generations."[5]

In short, for all the power of the fossil fuel players, their deliberate construction of fossil fuels' net beneficence and inevitable use is beginning to crumble.

A Politics of Urgent Transition

To limit extraction, not just manage emissions, requires a particular kind of politics. Its thrust is accelerating the transition out of fossil fuels, confronting extremely powerful actors, and creating a norm of the good life, life without endless expansion and extraction.

The politics of this transition is ultimately moral, and so the ultimate strategy is delegitimization. This does not mean a vilification of the fossil fuel industry. The industry has a century and more of vilification, starting with charges against Rockefeller's Standard Oil (the "Octopus") and continuing through to today (Hofmeister entitled his book *Why We Hate the Oil Companies*). Nor does this mean simply a repudiation of the industry's anti-democratic, anti-environmental tactics. Rather, delegitimization means the reconceptualizing and revaluing of fossil fuels—or, to be precise, of humans' relationship with fossil fuels. It means a shift in understanding of fossil fuels from constructive substances to destructive substances, from necessity to indulgence or even addiction, from a "good" to a "bad," from lifeblood (of modern society) to poison (of a potentially sustainable society).[6]

In other words, fossil fuels will make a moral transition in parallel to the material transition. Much as slavery went from universal institution to universal abomination and as tobacco went from medicinal and cool to lethal and disgusting, the delegitimization of fossil fuels will flip the valence of these otherwise wondrous, free-for-the-taking complex hydrocarbons. And rather than pin blame on "big bad oil (and coal) companies" or, even worse, on "all of us" because everyone uses fossil fuels, delegitimization simply recognizes that a substance once deemed net beneficial can become net detrimental. As in abolition and the delegitimization of smoking, what it takes is some compelling examples (begin with climate disruption and smog, add acid rain and oil slicks, include carbon monoxide and scores of other air pollutants), incisive critics, effective communication, and—for the moral entrepreneurs—a whole lot of persistence and willingness to be vilified.[7]

Delegitimization of fossil fuels would start with the simple observation that there are some things humans cannot handle. And for these things, humans can decide not to use them, just as they have with respect to ozone-depleting substances, lead in paint and gasoline, drift nets, land mines, rhino horns, and someday, perhaps, nuclear power plants and nuclear weapons.

Fortunately, some bold and clever people, North and South, are already saying no to fossil fuels and other mined materials. Their experiments, indeed their courage, suggest that such delegitimization has begun. This is particularly true among otherwise marginalized peoples. Their politics is not parochial protectionism, not localism. It is simultaneously protecting livelihood and the planet. Every new act of local resistance contributes to a new normative belief, one that says that the game is illegitimate, that it benefits a powerful few and their clients while fobbing the costs off on others in space and time. While such local acts of resistance are quickly dismissed as NIMBY (Not in My Back Yard) by defenders of the fossil fuel order, from the perspective of global threat and globalization from below, they are part of a larger project of delegitimization.

And so, what the climate scientists and others started yet cannot finish with their top-down, expert-led, apolitical, managerialist schemes and technological fixes is being augmented and accelerated by moral commitments in small pockets all over the world. But clearly fossil-fuel-dependent societies cannot stop cold. They can, however, start stopping now. One ethical justification for continued fossil-fuel-consumption is to facilitate a future without fossil fuels. Others are self-preservation and self-defense. What is more, because the transition away from current high-energy patterns will require considerable energy, those societies and communities deliberately living on little energy will have an advantage. Local action matters most in part because a top-down, centralized phaseout of fossil fuels by those with the most to lose is highly unlikely.[8]

Finally, delegitimizing a substance (or a process like exploring and drilling), as opposed to condemning an actor or all of humanity, puts the focus on the offending substance or, more specifically, on its use. Fossil fuels are perfectly "natural"; traditional uses of petroleum (rock oil) for pitch, lighting, and medicinal purposes were, for all we can tell, only harmful locally if at all. In a strategy of delegitimization, the burden shifts from the contest of interest groups (environmentalists versus industrialists, for example) to a contest over the politics of the good life. Industrialists have enacted one vision of the good life. Its efficacy in the twentieth century can be debated, but the politics of delegitimization are about now and the future, including the distant future. It is an affirmative politics, about creating a different vision of the good life given the biophysical trends under way.

Early Efforts to Keep Fossil Fuels in the Ground

On the face of it, keep them in the ground, for all its environmental and ethical justifications, is just an idea. The world is happily (some might say madly) pumping oil, devouring coal, and capturing natural gas—all at record levels. Everyone wants in the game for reasons of profit and power (or both), everyone from private energy companies to petrostates to investors. The juggernaut is rolling across the landscape; it cannot be stopped.

Except in some places, including some of the unlikeliest of places—major oil-producing countries, for instance—where key actors have begun stopping this monstrous vehicle. None of these exceptions are successful in the sense of a complete shutdown of fossil fuel extraction. None are large-scale. But all are significant in that these actors have had the temerity to challenge an established order that is local, national, and international as well as hugely powerful. What is more, these efforts are occurring largely peacefully and through democratic means. And perhaps most significant, they are doing so at a time when the world as a whole sees no crisis, no existential threat, just the odd pollutant to clean up, emissions to be managed, and efficiencies to be realized.

In the global South, for example, coalitions of indigenous peoples, nongovernmental organizations, and government agencies in Ecuador and Bolivia have rewritten their constitutions to enshrine the right of nature and define a new model of sustainable development, one that excludes fossil fuels. In Ecuador, it is called *sumak kawsay* in Quichua, *buen vivir* in Spanish, and the *good life* in English. The leaders there recognize that petroleum production will eventually decline, that there have been long-term costs to Ecuador, and that costs to the planet are becoming increasingly dangerous.[9]

As a first step, the Yasuní-ITT Initiative proposes keeping 20 percent of Ecuador's known oil reserves in the ground. It calls for coresponsibility with the rest of the world in avoiding emissions that the nearly 900 million barrels of oil in the ITT block could produce. The international community would pay for avoided carbon emissions to protect one of the most biodiverse spots on Earth and to limit in a small way global emissions. It would also protect the rights of at least two indigenous groups that live there in voluntary isolation. The $350 million per year that Ecuador seeks for 13 years (half of what they estimate the reserves would earn from oil extraction) would be placed in a U.N. Development Programme Trust Fund with a board of directors that includes Ecuadorans as well as members of the global community. If successful, it would be one of the largest global environmental trust funds of its kind. And it would be created not by burning fossil fuels, but by keeping them in the ground.[10]

Costa Rica, a small Caribbean country with known oil reserves offshore, enacted a moratorium in 2002 on oil extraction, citing ecological and social damage. In his 2002 inaugural address, President Abel Pacheco declared "Costa Rica will become an environmental leader and not an oil or mining enclave." He went on to say, "Costa Rica's real oil and real gold are its waters and the oxygen produced by its forests." Despite a brief encounter with the oil industry in the 1980s and recent considerations of natural gas exploration, Costa Rica has maintained its stance against this industry in favor of ecotourism and alternative energy sources and has achieved high human development indicators.[11]

In the global North, however, fossil fuels once left in the ground as too expensive to retrieve are being revisited. In the United States, federally funded research in the 1980s led to major innovations in imaging and mapping gas-rich shale deep beneath the surface. Blasting the shale with high pressure fracking fluids and drilling horizontally in multiple directions with powerful new diamond-studded drill bits add up to what became known as "slick-water, high-volume horizontal hydraulic fracturing," commonly referred to as *hydrofracking* or just *fracking*.[12]

The machinery of fracking deployed at a site in Texas.

As a result, massive amounts of shale gas can be reached profitably. These shale gas "plays," as the industry refers to them, are spreading rapidly in the traditional coal and oil states: Pennsylvania, Texas, and West Virginia. But when landsmen began knocking on doors in rural New York State enticing homeowners to lease their property for access to the vast Marcellus Shale beneath them, a keep-it-in-the-ground movement came to life. Landowners, environmental activists, artists, and indigenous peoples organized and protested, putting pressure on state and local officials. In 2010, New York Governor David Patterson ordered a moratorium on hydrofracking permits until the state completed an environmental and regulatory review. As of this writing, the latest state proposals would ban hydrofracking in the watersheds from which New York City and Syracuse get their unfiltered municipal supplies; surface drilling would be prohibited on state-owned land, including parks, and on forest areas and wildlife management areas.[13]

In the process, the state Department of Environmental Conservation received more than 13,000 public comments overwhelmingly in opposition to drilling in the remaining areas. Not leaving the decision up to the state, many local municipalities have approved or are considering zoning ordinances and outright bans. These are likely to be challenged in state courts. Concerns focus mostly on the threat to water supplies and aquifers from a process that involves the injection of large volumes of water, industrial fracking chemicals, and sand under high pressure. Water and contaminants are involved in every step of the process: transporting water to the drill site, mixing the chemicals, blasting the shale, recovering the fluids that come back with the gas, and, finally, transporting, treating, and disposing of the wastewater.[14]

Among the most vocal and powerful voices in the hydrofracking uprising have been those of the indigenous peoples of New York State. Representatives of the traditional leadership of the Haudenosaunee (the Iroquois) have pointed out that large-scale industrial drilling would likely disturb burial grounds and other sites of historical and spiritual importance. They have called on the U.S. government to uphold their water and land rights as guaranteed in multiple treaties between the United States and native nations. They remind the state and its citizens that while the gas industry's concern only spans the period of time when the well produces gas, it is everyone's responsibility to protect the land and the water for future generations.[15]

The outcome of the anti-fracking movement in New York State remains inconclusive. Fracking is on hold for now, but the pressures to exploit the resource are great. And conventional environmental arguments do not seem to be enough. What may turn out to be the most significant outcome is a public increasingly open to the possibility of keeping fossil fuels in the ground, an idea largely attributable to the new and powerful influence of Haudenosaunee leaders and the introduction of indigenous perspectives and values into a debate that would otherwise be narrowly technical and economic.

Farther south, long-standing resistance to destructive coal mining practices in Appalachia appear to be taking a new turn, shifting in places from improving practices and cleaning up waste to ending coal extraction entirely. Around the world there are citizen-led actions to keep destructive substances in the ground and stop destructive practices, from uranium in Australia and gold in El Salvador to gold and diamonds in Guyana and oil in the Norwegian Arctic. These examples, though small in the larger scheme of global energy production and consumption, signal a rippling of resistance around the globe against extractivist policies and, simultaneously, support for a good life without fossil fuels.[16]

Envisioning a Post–Fossil Fuel Era

Imagining deliberately keeping fossil fuels in the ground, much less the end of the fossil fuel era, is difficult. No matter how much environmental science is absorbed, how much geologic and ecological perspective is attained, how much ethical commitment is mustered, it is hard to escape industrial progressivism. It just seems like all this modernity will continue, albeit with adjustments—an efficiency here, some greening up there.

In fact, this pervasive impression—that the fossil fuel era has been around for a long, long time and will be for a long, long time to come, indeed that it must be—this impression has been deliberately constructed by the industry and its industrial and governmental enablers. Physical reality, however, speaks otherwise. Unfortunately, for fossil fuel proponents anyway, there is just too much knowledge piled up to believe in the indefinite perpetuation of the fossil fuel era, and not just scientific knowledge but political and strategic knowledge.

So a primary task for those who believe that the fossil fuel era will not continue, and yet will not end soon enough to avoid catastrophic outcomes, is to imagine that end. To facilitate such imagining, arguably a necessary precursor to designing policies and behavior change strategies, we offer two observations as an envisioning exercise.[17]

First, the fossil-fuel era, which began in the 1890s, when fossil fuels surpassed wood as the dominant energy source, is only about six generations old. Many of us alive now have personally known people who lived before the fossil-fuel era. It was not that long ago. The fossil fuel era is not that permanent, nor is its continuation that inevitable. Given that the initial stage of an energy source's use is one where benefits are highlighted and costs unknown or shaded (displaced in time and space), we can expect that fossil fuels have the same quality, only on a far grander scale than anything before. Coal's depredations—from miners' bodies to asthmatics' lungs, from decimated mountains in Appalachia to dug-out deserts in Mongolia —are well known. Coal's early exit is virtually a no-brainer. No wonder the industry's anti-climate-change activism has been so vehement. Oil, arguably the most consequential energy source of all time, is widely deemed essential (and thus the rush for alternative liquid fuels), but it too will eventually fade out.[18]

The costs of fossil fuels, from traffic casualties to climate disruption, will eventually catch up. The fossil-fuel era will come to an end well before conventional analysis and decisionmaking would indicate. And just as global fossil-fuel production will decline as all wells and oil fields do, the industry will decline, too. Just because no one in the industry or anyone dependent on it (virtually everyone) wants to talk about this does not make it otherwise. Fossil fuel production and the fossil fuel industries will most assuredly decline.

Second, one place to start the imagining is, ironically enough, the fossil fuel industry itself. Preliminary evidence suggests that serious people in the oil, gas, and coal industries along with the automobile and petrochemical industries know this game cannot go on. "Energy executives know that the existing supply capacity from traditional sources is about tapped out," writes former Shell president John Hofmeister. They know the easy stuff is effectively gone. Now, they are learning, it is also changing the climate, melting the very tundra their trucks depend on, blowing apart rigs they thought were secure. What they say publicly is different, of course. Their jobs, their way of life, their personal and professional identity, their future is on the line. They seem to pray that a miracle technology will come along to keep the game going a little while longer. This difficulty is perfectly understandable. And yet people in equally entrenched positions (witness slavery and smoking) have made huge shifts in position.[19]

In short, a deliberate policy, state-led or not, of keeping fossil fuels in the ground is at once preposterous and perfectly sensible. Stranger things have happened. How it would happen, at what rate, with what local effects, is still anyone's guess. That fossil fuels will be in the ground and stay there when the fossil fuel era ends is beyond doubt. The only question is whether enough will stay to stabilize climate, reverse degrading trends, and avert social calamity. Bringing about an urgent transition begins with a certain kind of politics, one of delegitimizing fossil fuels and humans' deeply problematic relationship to them. This is a politics that recognizes that once fossil fuels are out of the ground, their by-products will permeate our bodies, the oceans, and the atmosphere and cause catastrophic loss. Those politics and the policies and economies that follow constitute a necessary first step in choosing to end the fossil-fuel era.

CHAPTER 15

Beyond Fossil Fuels: Assessing Energy Alternatives

T. W. Murphy, Jr.

Most discussions of the remarkable trajectory of human development in the past few centuries label the phenomenon the *Industrial Revolution.* This term is apt enough, although it emphasizes the industrious nature of clever humans. An equally important factor—if not more so—has been the abundant supply of cheap surplus energy in the form of fossil fuels. Coal fueled the early stages of the Industrial Revolution, opening the door to accelerated energy-resource discovery and exploitation. Indeed, the first major application of coal was to power steam engines used to pump water out of coal mines in order to gain access to more coal. Perhaps the Coal Revolution would more accurately represent the transformational change marked by the nineteenth century.[1]

Fossil fuel stocks are known to be finite, and by most accounts their extraction rates will peak this century. Thus in the long view it is a near certainty that the current age will be known to history as the *Fossil Fuel Age.* It is the time when humans discovered Earth's battery—solar-charged over millions of years—and depleted it fast enough to effectively constitute a short circuit.

During this epoch, our unprecedented capacity to process materials, manufacture goods, create a "built environment," and revolutionize agricultural productivity has translated into a world of spectacular accomplishments, advanced scientific knowledge, technology that an earlier generation might call magic, sustained economic growth, and a surging population of 7 billion industrially fed human beings. These feats would not have been possible without the bounty of fossil fuels.

In this light, our present state can be seen as a reflection of historically available energy. If depicted in schematic fashion over the course of a civilization-scale timeline, the general history and future of fossil fuel use will very likely appear as a sharp spike. (See Figure 15–1.) Humanity now sits near the apex of the brief fossil fuel energy explosion and prepares to en-

T. W. Murphy, Jr., is an associate professor of physics at the University of California/San Diego.

www.sustainabilitypossible.org

ter an untested regime of unprecedented scale: the loss of a resource that has been unquestionably vital to growth and development.[2]

Bracketing the possible future paths are the optimistic scenario that fossil fuels are merely a kickstart to an ever-growing, ever-improving technological society and the pessimistic view that society will fail to find suitable replacements for fossil fuels and will experience decline to pre-industrial population levels and ways of life. The optimistic view is clearly more appealing, rests on a track record spanning generations, and is closer to mainstream opinion, while the

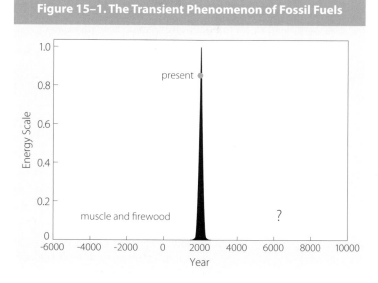

Figure 15–1. The Transient Phenomenon of Fossil Fuels

unpalatable pessimistic perspective seems alarmist and fatalistic. Yet complete dismissal of the pessimistic possibility carries hubristic overtones. It must, after all, be recognized that most of the empirical evidence in support of the optimistic scenario emerged in the context of abundant surplus energy provided by fossil fuels.

In short, recent history has been written in fossil fuels. When production of these fuels declines, the prevailing narrative of growth-based human endeavors may require significant adjustment. Any scientist will affirm that indefinite growth in any physical measure is impossible. Energy use in the world has grown by approximately 3 percent per year for the past few centuries. At this rate, the current 16 terawatts (TW) of global power demand would balloon to equal the entire solar output in about 1,000 years and match all 100 billion stars in our galaxy inside of 2,000 years. Well before this—within 400 years—enough direct heat would be generated on Earth to bring the surface temperature to that of boiling water. Similarly alarming statements can be made about population, resource use, or anything that has seen sustained growth over the past few centuries. Obviously, the "normal" world of growth is a temporary anomaly destined to self-terminate by natural means.[3]

While some current economic activities use little energy or physical resources, no activity can claim zero use. And energy-intensive activities (such as agriculture, transport, and thermal management) will establish a floor below which the economy cannot sink. So an end to energy or resource growth ultimately means an end to economic growth as traditionally defined.[4]

Substitution and the Drumbeat of Improvement

For indoor lighting applications, whale oil replaced beeswax; kerosene derived from coal replaced whale oil; petroleum replaced kerosene; and now we use electricity derived from coal, natural gas, hydropower, nuclear, biomass, and a smattering of renewable sources. The lesson seems clear: new, superior sources come to bear, rendering the prior solutions obsolete. Why should there be any deviation in this recurring storyline as fossil fuels give way in the future? Considering solar, wind, nuclear, geothermal, tidal, wave, and biofuel sources, it appears that the menu of substitutes is full to bursting.

It is worth pointing out, however, that some concepts and technologies find no superior substitute over time; examples include the wheel, metal blades, window glass, and rope. Naturally, refinements accumulate, but the basic concepts are unrivaled and dominate for millennia. And sometimes once-prevalent technologies become unavailable to society without adequate substitutes, such as the recent loss of commercial supersonic transatlantic flight or of U.S. human space launch capability. Perhaps these reversals are temporary setbacks, but the familiar narrative of a constant march toward superior substitutions and "faster, better, cheaper" practices is not an immutable law of nature.

The Alternative Energy Matrix

In exploring potential replacements for fossil energy, it soon becomes apparent that fossil fuels are unparalleled in many respects. Even though viewed as a source of energy from the ground, fossil fuels are perhaps more aptly described as nearly perfect energy storage media, at energy densities that are orders of magnitude higher than anything achieved thus far in the best battery technology today. The storage is nearly perfect because it is reasonably safe, not especially corrosive, easy to transport (via pipelines, often), lightweight yet dense enough to work in airplanes, and indefinitely storable—indeed, for millions of years—without loss of energy. No alternative storage technique can boast all the same benefits, be it batteries, flywheels, hydrogen, or ethanol.

In order to make comparisons, it is helpful to create a matrix of properties of energy sources so that the relative strengths and weaknesses of each are obvious at a glance. (See Figures 15–2 and 15–3.) The matrix is presented as a Figure based on 10 different criteria. White, gray, and black can be loosely interpreted as satisfactory, marginal, and deficient, respectively. Gray boxes are often accompanied by brief reasons for their classification—other extremes often being obvious. While some criteria are quantitative, many are subjective. The following 10 properties are useful for this comparison.

Abundance. Not all ideas, however clever or practical, can scale to meet the needs of modern society. Hydroelectric power cannot expand beyond

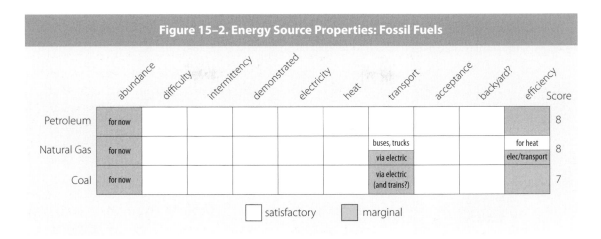

Figure 15–2. Energy Source Properties: Fossil Fuels

	abundance	difficulty	intermittency	demonstrated	electricity	heat	transport	acceptance	backyard?	efficiency	Score
Petroleum	for now										8
Natural Gas	for now						buses, trucks via electric			for heat elec/transport	8
Coal	for now						via electric (and trains?)				7

□ satisfactory   ▨ marginal

about 5 percent of current global demand, while the solar potential reaching Earth's surface is easily calculated to exceed this benchmark by a factor of about 5,000. Abundant sources are coded white, while niche ideas like hydroelectricity that cannot conceivably fulfill one quarter of global demand are colored black. Intermediate players that can satisfy a substantial fraction of demand are coded gray.[5]

Difficulty. This field tries to capture the degree to which a resource brings with it large technical challenges. How many PhDs does it take to run the plant? How intensive is it to maintain an operational state? This one might translate into economic terms: difficult is another term for expensive.

Intermittency. This is colored white if the source is rock-steady or available whenever it is needed. If the availability is beyond our control, then it gets a gray at least. The possibility of substantial underproduction for a few days earns black.

Demonstrated. To be white, a resource has to be commercially available today and providing useful energy. Proof of concept on paper, or prototypes that exhibit some of the technology, do not count as demonstrated.

Electricity. Can the technology produce electricity? For most sources, the answer is yes. Sometimes it would make little sense to try. For other sources, it is impractical.

Heat. Can the resource produce direct heat? This is colored gray if only via electric means.

Transport. Does the technology relieve the looming decline in petroleum production? Anything that makes electricity can power an electric car, earning a gray score. Liquid fuels are white. Bear in mind that a large-scale migration to electric cars is not guaranteed to happen, as the cars may remain too expensive to be widely adopted.

Acceptance. Is public opinion (judging by U.S. attitudes) favorable to this

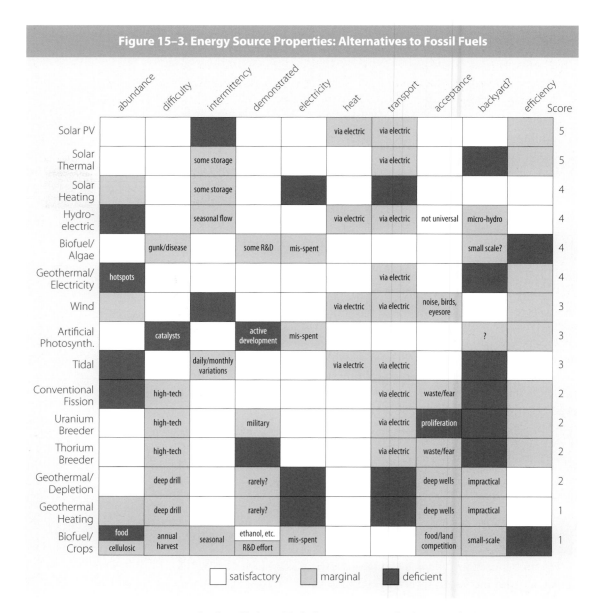

Figure 15–3. Energy Source Properties: Alternatives to Fossil Fuels

method? Will there likely be resistance, whether justified or not?

Backyard. Is this something that can be used domestically, in someone's backyard or small property, managed by the individual? Distributed power adds to system resilience.

Efficiency. Over 50 percent earns white. Below about 10 percent gets black. It is not the most important of criteria, as the property of abundance implicitly incorporates efficiency expectations, but we will always view low efficiency negatively.

Environmental impact has no column in this matrix, although the "acceptance" measure captures some of this. Climate change is an obvious negative for fossil fuels, but not so much as to curtail global demand, in practice. None of the alternatives presented here contributes directly to carbon dioxide emissions, earning an added advantage for all entries.

Each energy source can be assigned a crude numerical score, adding one point for each white box, no points for gray boxes, and deducting a point for each black box. Certainly this is an imperfect scoring scheme, giving each criterion equal weight, but it provides some means of comparing and ranking sources.

The conventional fossil fuels each score 7–8 out of 10 possible points by this scheme, displayed on the right side of Figure 15–2. Natural gas must be divided into heating versus electricity production for a few of the scoring categories.

The overall impression conveyed by this graphic is that fossil fuels perform rather well in almost all criteria. Because fossil fuels account for 81 percent of global energy use, they are each classified as having intermediate abundance. But even this is not a permanent condition—providing significant motivation for exploring alternatives in the first place. Getting energy out of fossil fuels is trivially easy. Being free of intermittency problems, fully demonstrated, and versatile enough to provide heat, electricity, and transportation fuel, fossil fuels have been embraced by society and are frequently used directly in homes. Efficiency for anything but direct heat is middling, typically clocking 15–25 percent for automotive engines and 30–40 percent for power plants.[6]

The commonly discussed alternative energy approaches display a wider range of ratings. Immediately, some overall trends are clear in Figure 15–3. Very few options are both abundant and easy. Solar photovoltaics (PV) and solar thermal are the exceptions. A similar exclusion principle often holds for abundant and demonstrated/available. This uncommon combination plays a large role in the popularity of solar power.

Intermittency mainly plagues solar and wind resources, with mild inconvenience appearing for many of the natural sources.

Electricity is easy to produce, resulting in many options. Since the easiest and cheapest will likely be picked first, the less convenient forms of electricity production are less likely to be exploited (farther down the list, to the extent that the ordering is correlated with economic advantage).

Transportation needs are hard to satisfy. Together with the fact that oil production will peak before natural gas or coal, transportation may appear as the foremost problem to address. Electric cars are an obvious—albeit expensive—solution, but the technology has a number of drawbacks relative to fossil fuels and does not lend itself to air travel or heavy shipping by land or sea.

Few of the options face serious barriers to acceptance, especially when energy scarcity is at stake. Some energy sources are available for individual implementation, allowing distributed power generation as opposed to centralized resources. For example, a passive solar home with PV panels, wind power, and some method to produce liquid fuels on-site would satisfy most domestic energy needs in a self-sufficient manner.

Cost is not directly represented in the matrix, although the difficulty rating may serve as an imperfect proxy. In general, the alternative methods have difficulty competing with cheap fossil fuels. It is not yet clear whether the requisite prosperity needed to afford a more expensive energy future at today's scale will be forthcoming.

The Tally for Individual Alternative Sources

A single chapter cannot adequately detail the myriad complex considerations that went into the matrix in Figure 15–3. Many of the quantitative and qualitative aspects for each were developed at the *Do the Math* website. The key qualities of each resource in relation to the matrix criteria are discussed in this section, focusing especially on less obvious characteristics.[7]

Solar PV. Covering just 0.5 percent of Earth's land area with PV panels that are 15 percent efficient satisfies global annual energy demand, qualifying solar PV as abundant. PV panels are being produced globally at 27 gigawatts (GW) peak capacity per year (translating to about 5 GW of average power added per year), demonstrating a low degree of difficulty. Most people do not object to solar PV on rooftops or over parking areas, or even in open spaces (especially deserts). Solar panels are well suited to individual operation and maintenance. Intermittency is the Achilles' heel of solar PV, requiring storage solutions if adopted on a large scale. To illustrate the difficulty of storage, a lead-acid battery big enough to provide the United States with adequate backup power would require more lead than is estimated to be accessible in the world and would cost approximately $60 trillion at today's price of lead. Lithium or nickel-based batteries fare no better on cost or abundance. The small number of suitable locales limits the potential of pumped storage.[8]

Solar Thermal. Achieving comparable efficiency to PV but using more land area, the process of generating electricity from concentrated solar thermal energy has no problem qualifying as abundant—although somewhat more regionally constrained. It is relatively low-tech: shiny curved mirrors, tracking on (often) one axis, heat oil or a similar fluid to drive a standard heat engine. Intermittency can be mitigated by storing thermal energy, perhaps even for a few days. A number of plants are already in operation, producing cost-competitive electricity. Public acceptance is no worse than for PV, but the technology generally must be implemented in large, centralized facilities.

Solar Heating. On a smaller scale, heat collected directly from the sun can provide domestic hot water and home heating. In the latter case, this can be as simple as a south-facing window. Capturing and using solar heat effectively is not particularly difficult, coming down to plumbing, insulation, and ventilation control. Technically, solar heating potential might be abundant, but since it is usually restricted to building footprints (roof, windows), it gets a gray rating. Solar heating does not lend itself to electricity generation or transport, but it has no difficulty being accepted and almost by definition is a backyard-ready technology.

Hydroelectric. Despite impressive efficiency, hydroelectric potential is already well developed in the world and is destined to remain a small player on the scale of today's energy use. It has seasonal intermittency (a typical hydroelectric plant delivers only 40 percent of its design capacity), does not directly provide heat or transport, and can only rarely be implemented personally, at home. Acceptance is fairly high, although silting and associated dangers—together with habitat destruction and the forced displacement of people—do cause some opposition to expansion.

Biofuels from Algae. Because algae capture solar energy—even at less than 5 percent efficiency—the potential energy scale is enormous. Challenges include keeping the plumbing clean, possible infection (for example, a genetic arms race with evolving viruses), contamination by other species, and so on. At present, no algal sample that secretes the desired fuels has been identified or engineered. No one knows whether genetic engineering will succeed at creating a suitable organism. Otherwise, the ability to provide transportation fuel is the big draw. Heat may also be efficiently produced, but electricity production would represent a misallocation of precious liquid fuel.

Geothermal Electricity. This option makes sense primarily at rare geological hotspots. It will not scale to be a significant part of our entire energy mix. Aside from this, it is relatively easy, steady, and well demonstrated in many locations. It can provide electricity, and obviously direct heat—although usually far from locations demanding heat.

Wind. Wind is neither super-abundant nor scarce, being one of those options that can meet a considerable fraction of present needs under large-scale development. Implementation is relatively straightforward, reasonably efficient, and demonstrated the world over in large wind farms. The biggest downside is intermittency. It is not unusual to have little or no regional input for several days in a row. Objections to wind tend to be more serious than for many other alternatives. Windmills are noisy and tend to be located in prominent places (ridgetops, coastlines) where their high degree of visibility alters scenery. Wind remains viable for small-scale personal use.[9]

Artificial Photosynthesis. Combining the abundance of direct solar with the self-storing flexibility of liquid fuel, artificial photosynthesis is a compelling future possibility. The ability to store the resulting liquid fuel for many months means that intermittency is eliminated to the extent that annual production meets demand. A panel in sunlight dripping liquid fuel could satisfy both heating and transportation needs. Electricity can also be produced, but given an abundance of ways to make electricity, the liquid fuels would be misallocated if used in this way. Unfortunately, an adequate form of artificial photosynthesis has yet to be demonstrated in the laboratory, although the U.S. Department of Energy initiated a large program in 2010 toward this goal.[10]

Dani 7C3

Dam of the tidal power plant on the estuary of the Rance River, Brittany, France. It has been in operation since 1966.

Tidal Power. Restricted to select coastal locations, tidal power will never be a large contributor on the global scale. The resource is intermittent on daily and monthly scales but in a wholly predictable manner. Extracting tidal energy is not terribly hard—it shares technology with similarly efficient hydroelectric installations—and has been demonstrated in a number of locations around the world.

Conventional Fission. Using conventional uranium reactors and conventional mining practices, nuclear fission does not have the legs for a marathon. On the other hand, it is certainly well demonstrated and has no problems with intermittency—except that it cannot easily accommodate intermittency in the face of variable load. Compared with other options, nuclear power qualifies as a high-tech approach—meaning that design, construction, operation, and emergency mitigation require more advanced training and sophistication than the average energy producer.

Acceptance is mixed. Germany and Japan plan to phase out their nuclear programs by 2022 and the 2030s, respectively, despite being serious about carbon reduction. Public unease also contributed to a halt in licensing new reactors in the United States from 1978 to 2012. Some opposition stems from unwarranted—yet no less real—fear, sustained in part by the technical complexity of the subject. But some opposition relates to political difficulty surrounding the onerous waste problem that no country has yet solved to satisfaction.[11]

Uranium Breeder. Extending nuclear fission to use plutonium synthesized from U-238, which is 140 times more abundant than U-235, gives uranium fission the legs to run for at least centuries if not a few millennia, ameliorating abundance issues. Breeding has been practiced in military re-actors, and indeed some significant fraction of the power in conventional uranium reactors comes from incidental synthesis of plutonium (Pu-239) from U-238. But no commercial power plant has been built to deliberately tap the bulk of uranium for power production. Public acceptance of breed-ers will face even higher hurdles because plutonium is more easily separated into bomb material than is U-235, and the trans-uranic radioactive waste from this option is worse than for the conventional reactor.[12]

Thorium Breeder. Thorium is more abundant than uranium and only has one natural isotope, qualifying it as an abundant resource. Like all reac-tors, thorium reactors fall into the high-tech camp and include new chal-lenges (such as liquid sodium) that conventional reactors have not faced. A few small-scale demonstrations have been carried out, but nothing in the commercial realm; bringing thorium reactors online at scale is probably a few decades away. Public reaction will likely be similar to that for conven-tional nuclear: not a show stopper, but some resistance on similar grounds. It is not clear whether the novelty of thorium will be greeted with suspi-cion or enthusiasm. Although thorium also represents a breeding technol-ogy (making fissile U-233 from Th-232), the proliferation aspect is severely diminished for thorium due to a highly radioactive U-232 by-product and virtually no easily separable plutonium.

Geothermal Heating Allowing Depletion. A vast store of thermal ener-gy sits in Earth's crust, permeating the rock and moving slowly outward. Without regard to sustainable practices, boreholes could be drilled a few kilometers down to extract thermal energy out of the rock faster than the geophysical replacement rate, effectively mining heat as a one-time resource. In the absence of water flow to distribute heat, dry rock will deplete its heat within 5–10 meters of the borehole in a matter of a few years, requiring another hole 10 meters away from the previous, in repeated fashion. The re-current large-scale drilling operation across the land qualifies this technique as moderately difficult.

The temperatures are marginal for running heat engines to make elec-tricity with any respectable efficiency (especially given the existence of many easier options for electricity), but at least the thermal resource will not suffer intermittency problems during the time that a given hole is still useful. Kilo-meter-scale drilling hurdles have prevented this technique from being dem-onstrated at geologically normal (inactive) sites. Public acceptance may be less than lukewarm given the scale of drilling involved, dealing with tailings and possibly groundwater contamination issues on a sizable scale. While a backyard might accommodate a borehole, it would be far more practical to use the heat for clusters of buildings rather than for just one—given the ef-fort and lifetime associated with each hole.

Geothermal Heating, Steady State. Sustainable extraction of geothermal heat—replenished by radioactive decay within Earth—offers far less total potential, coming to about 10 TW of flow if summed across all land. And to get to temperatures hot enough to be useful for heating purposes, boreholes at least 1 kilometer deep would be required. It is tremendously challenging to cover any significant fraction of land area with thermal collectors 1 kilometer deep. As a result, a gray score for the abundance factor may be generous. To gather enough steady-flow heat to provide for a normal U.S. home's heating demand, the collection network would have to span a square 200 meters on a side at depth, which is likely unachievable. (Note that ordinary geothermal heat pumps are not accessing an energy resource; they are simply using a large thermal mass against which to regulate temperature.)

Biofuels from Crops. While corn ethanol may not even be net energy-positive, sugarcane and vegetable oils as sources of biofuel fare better. But these sources compete with food production and arable land availability. So biofuels from crops can only graduate from "niche" to moderate scale in the context of plant waste or cellulosic conversion. The abundance and demonstration fields are thus split: food crop energy is demonstrated but severely constrained in scale. Cellulosic matter becomes a potentially larger-scale source but is undemonstrated (perhaps this should even be black). Growing and harvesting annual crops on a relevant scale constitutes a massive, perpetual task and thus scores gray in difficulty.

If exploiting fossil fuels is akin to spending a considerable inheritance, growing and harvesting our energy supply on an annual basis is like getting a manual-labor job: a most difficult transition. The main benefit of biofuels from crops is the liquid fuel aspect. Public acceptance hinges on competition with food or even land in general. Because plants are only about 1–2 percent efficient at harvesting solar energy, this option requires the commandeering of massive tracts of land.[13]

A few other sources not discussed here—ocean thermal, ocean currents, wave energy, and two flavors of fusion—all score 1 point. Notably, the extreme technological challenge of mastering fusion just to provide another avenue for electricity production puts this technique at a disadvantage in the matrix.[14]

The Fossil Fuel Gap

The subjective nature of this exercise certainly allows numerous possibilities for modifying the box rankings in one direction or the other. The matrices embody some biases, but no attempt by anyone would be free from bias. The result, in this case, is dramatic. Even allowing some manipulation, the substantial gap between the fossil fuels and their renewable alternatives would require excessive "cooking" to close.

The lesson is that a transition away from fossil fuels does not appear at this time to involve superior substitutes, as has been characteristic of our energy history. Fossil fuels represent a generous one-time gift from the earth. From our current vantage point, it is not clear that energy—vital to our economic activity—will be as cheap, convenient, and abundant as it has been during our meteoric ascent to the present.

Adding to the hardship is the fact that many alternative energy technologies—solar, wind, nuclear power, hydroelectric, and so on—require substantial up-front energy investments. If society waits until energy scarcity forces large-scale deployment of such alternatives, it risks falling into an Energy Trap in which aggressive use of energy to develop a new energy infrastructure leaves less available to society in general. (See Chapter 7.) If there is to be a transition to a sustainable energy regime, it's best to begin it now.[15]

CHAPTER 16

Energy Efficiency in the Built Environment

Phillip Saieg

The vast majority of carbon emissions into Earth's atmosphere are energy-related, stemming from the combustion of fossil fuels. Curtailing these emissions is crucial to mitigating climate change. The supply-side option for reducing fossil-fuel combustion is renewable energy, and significant efforts are being made in that direction. (See Chapters 8 and 15.) However, there are currently only weak market incentives to develop renewable energy at a scale relative to coal and natural gas. The U.S. Energy Information Administration estimates that the average levelized costs (without accounting for climate change and other externalities) of producing electricity in the United States from natural gas generation plants entering service in 2017 will be $66.10 per megawatt-hour, while the equivalent costs for utility-scale wind power will be $96. So although it is critical to increase the use of renewable sources of energy, the current cost gap between renewables and fossil fuel generation, along with supply integration issues, is impeding large-scale adoption of renewables.[1]

But there is a quicker and more financially feasible way to lessen the amount of carbon being added to the atmosphere. Focusing on the demand side of the energy equation—increasing energy efficiency—can dramatically reduce the relative percentage of emissions created by energy generation, relieve the high demand for increased energy production, and ultimately reduce carbon emissions.

In the United States, the transportation and industrial sectors each use about one quarter of all the energy consumed, while buildings consume nearly half in the course of heating, cooling, ventilating, and lighting their spaces. Worldwide, buildings account for nearly 16 percent of all energy consumption. And with little of the building stock being built new—from 2 percent of U.S. commercial floor space to as much as 10 percent in India—most opportunities to improve efficiency over the next several decades will be in the existing building stock.[2]

Phillip Saieg is an accredited professional under the Leadership in Energy and Environmental Design program of the U.S. Green Building Council and an account executive for McKinstry, a U.S.-based energy services company.

www.sustainabilitypossible.org

This lesson has begun to sink in: many countries, including a number in the developing world, are taking building efficiency seriously. India and China, for example, have begun paying much closer attention to natural gas and electricity consumption as these begin to play a larger role in their growing built environment. Several countries in the Middle East, including Algeria, Egypt, Tunisia, and the United Arab Emirates, have launched efficiency programs. These initiatives, such as energy-conservation building codes and high-performance building standards, are responses to the asymmetrical growth of energy consumption and population growth. In industrial countries such as the United States, energy consumption rises at an annual rate of 1.3 percent while population grows 0.8 percent; in India, energy consumption is exploding by 4.3 percent a year while population grows 1.3 percent.[3]

Energy Efficiency as a Financial Opportunity

Efficiency is an investment opportunity as well as an environmental one. In July 2009, the international consulting firm McKinsey & Company did a comprehensive study of the U.S. building stock and found that if off-the-shelf energy efficiency measures were put in place across the sector, total U.S. energy consumption would decline by 23 percent, yielding more than $1.2 trillion in savings for an investment of $520 billion. These measures included retrofitting existing buildings with more-efficient lighting and updated heating and cooling equipment, as well as insulating walls and roofs, upgrading windows, and optimizing building system controls. Separately, McKinsey also published an analysis and ranking of the most cost-effective strategies for reducing carbon emissions. For example, the report concluded that for a given amount of money, installing building insulation would yield greater net savings than solar photovoltaic panels.[4]

McKinsey's analysis confirmed that energy efficiency strategies routinely yield better emission reduction results than supply-side solutions like solar or wind energy because energy efficiency strategies offer larger carbon savings at lower costs. Energy efficiency, in fact, often wins out as a high-yield financial investment strategy when compared with more traditional investments like stocks or bonds. According to the average of 100 years of U.S. market data, the stock market will return about 10 percent on any given investment (although any given investment in the stock market can result in huge gains or massive losses, of course). But according to the American Council for an Energy-Efficient Economy, the average financial return on investment for efficiency is more like 20 percent. When these energy efficiency projects are guaranteed using a methodology called performance contracting, they become extremely low-risk, high-yield investments—and as one result, the energy performance contracting business has now grown to over a $5-billion-per-year industry.[5]

Though McKinsey and other analysts have identified a vast opportunity for the reduction of carbon emissions and economic development, the idea of addressing buildings as "green buildings" or "energy-efficient buildings" is relatively new to the traditional real estate market. There is, however, a clear business case for renovating buildings to meet high efficiency standards, such as those set by the Energy Star program of the U.S. Environmental Protection Agency (EPA) and by the Leadership in Energy and Environmental Design (LEED) program of the U.S. Green Building Council (USGBC). An Energy Star Leader building is one with an energy efficiency score calculated by EPA to be at least 75, meaning that the building is in the seventy-fifth percentile for efficient buildings. A LEED-certified building has been evaluated under USGBC's nationally accepted third-party assessment program and its construction and operations have been confirmed to be high-performance and sustainable. Despite the recent recession, the number of green buildings in the United States has grown significantly. This trend is predicted to continue and to shift even more toward retrofit and renovation projects.[6]

LEED Certified Gold condos in a renovated and retrofitted building in Hoboken, New Jersey.

Walter Burns

Businesses that invest in a sustainable building and have it certified under either the Energy Star Program or the LEED program are typically differentiated from the market norm by premiums in property value, rental rates, and occupancy rates. They are also more likely to mitigate risks to owners and tenants, such as rising utility costs, new regulations and standards, and a negative reputation. In 2008 the Urban Land Institute had this to say about green buildings: "Green will be measured by the business community, regulators, savvy consumers. . . . stay on top of green or eat everyone's dust. There will be differentiation over the long run, adapt or get crushed." Five years later, the Institute noted that "major tenants willingly pay high rents in return for more efficient design layouts and lower operating costs in LEED rated, green projects. . . . Green buildings with high ratings under the [LEED] program and energy-efficient systems leapfrog the competition."[7]

The numbers bear out these claims. On average, a 10 percent reduction in energy use in certified buildings results in an increase of 1.1–1.2 percent in market value. The aggregate value of the U.S. commercial green real estate market is expected to grow by 18 percent annually, from $35.6 billion in 2010 to $81.8 billion by 2015. And with 185 million square meters of floor space in LEED-certified buildings and another 650 million square meters registered to become certified, sustainability investments are seen to create even larger market differentiation.[8]

Reorienting the Commercial Real Estate Market

The commercial real estate market is beginning to take notice of these evolutionary developments toward sustainable buildings. One milestone reached in 2010 was a concerted effort by the Appraisal Foundation—which is responsible for publishing standards, appraiser qualifications, and guidance regarding valuation methods and techniques—to begin to account for the increased value imparted to a building by its energy efficiency and sustainability features. The foundation and the U.S. Department of Energy signed a memorandum of understanding to promote consistent and fair appraisal standards and practices with respect to energy-efficient and sustainable buildings.[9]

A second development is the emergence of asset rating. Many building operational rating systems are in use today, such as the EPA's Energy Star Portfolio Manager, which is used to rate building energy efficiency in percentile terms compared with other similar buildings. These focus on ongoing energy usage with the intent of improving operations. Asset rating, however, focuses on the energy performance of a building's component parts, enabling direct comparisons of performance among similar buildings regardless of hours of operation, tenant behavior, how well the systems are operated and maintained, and other factors that can have significant impacts on energy consumption. Asset rating of a building's systems (such as lighting, heating and cooling, and insulation) in terms of their energy efficiency offers a new way to objectively value property, creating value for high-performance systems.[10]

These developments have helped to unlock energy efficiency in commercial buildings. Building owners, in response to seeing value beyond the simple payback from spending less on energy, have started changing the way they evaluate building performance upgrades. Traditionally owners have performed straightforward return on investment (ROI) calculations to show how energy efficiency measures can repay an investment, and this has been the tool of choice for evaluating whether to upgrade a building's efficiency.

ROI calculations are a key part of the evaluation process and often help set priorities for upgrades, but they do not give the whole picture. If an ROI calculation yields a payback period longer than an owner plans to hold on to the building, the incentive for the upgrade disappears. Commercial buildings typically change hands every two to four years, which makes the acceptable payback period fairly short. The owner in these cases usually chooses to implement only efficiency measures with short payback times, thus excluding many options that might yield deeper savings over the life of the building.

Since green building has caught on, however, and tenant demand for sustainable buildings has increased, many commercial building owners have broadened their evaluation tools and are using a net present value (NPV) method that takes into account not just payback but total asset value (the sum of the incoming and outgoing cash flows) to help them make efficiency upgrade decisions. Because NPV can be realized before and in the sale of a building, owners are now willing to make strategic efficiency upgrades whose payback times extend beyond their terms of ownership.

Efficiency Policy

In addition to the rise of market demand and the realized financial returns from energy-efficient buildings, a supportive policy framework has grown around the green building movement. In addition to U.S. Department of Energy investments of hundreds of millions of dollars in energy efficiency projects, President Obama's Better Buildings Initiative is partnering with the public and private sectors to invest $4 billion in energy efficiency. And many state and city governments have begun passing energy efficiency legislation.[11]

According to the Institute for Market Transformation, many local jurisdictions—including Austin, Texas; Washington, DC; New York City; Portland, Oregon; San Francisco; Seattle; and the states of California and Washington—now have disclosure policies requiring owners of commercial buildings of a certain size (usually over 5,000 square feet) to report the buildings' annual energy consumption. Thirty-two countries in Europe as well as China and Australia have also adopted disclosure policies. In New York City, commercial buildings over 10,000 square feet are required to undergo an efficiency auditing and evaluation process called retro-commissioning every 10 years to ensure that their owners learn about opportunities for efficiency improvements.[12]

While it is clear that great environmental benefits can result from these policy changes, the justification for most of the policy programs and legislation has been rooted in promoting energy efficiency as a valid tool to drive economic growth. President Obama's Better Buildings Initiative "seeks to tap into job-creation potential with a suite of policies designed to encourage the pursuit of energy efficiency." The administration claims that the initiative has led to the creation of 114,000 jobs. Many local governments have also been using environmental policy as a tool for boosting economic growth, citing job creation and the value of efficiency as an innovative approach to help balance the books in a struggling economy. A good number of them have undertaken efficiency strategies to reach their climate goals as well: at least 141 U.S. cities have registered Climate Action Plans and more than 1,000 have signed on to the U.S. Conference of Mayors' Climate Protection Agreement.[13]

Many nations have also instituted green building codes and standards. Between Australia's Green Star program, Canada's Green Globes, China's Three Star Program, and Britain's BREEAM program, to name just a few, almost every nation has begun requiring some level of sustainable building be incorporated into the their built environment in the last 10 years. Even Sudan has acknowledged this need by reducing fees and customs on liquid petroleum gas stoves in order to promote use of this energy source instead of inefficient biomass, which causes deforestation.[14]

The untapped energy savings waiting to be harvested from existing building stock are vast. And while certain barriers still block this harvest, it is clear from the private and political support for sustainable buildings that an energy-efficient future is good for everyone. While realizing economic savings and improving the world's well-being through a sustainable built environment, the problems of excessive energy consumption and greenhouse gas emissions can be addressed. As Ludwig Wittgenstein once wrote, "The problems of life are insoluble on the surface and can only be solved in depth." Focusing on energy efficiency and creating sustainable buildings is essential to mitigate environmental risk, create long-standing jobs, sustain local governments, and help design a future that leverages waste to prosperity.[15]

EG Focus

One Angel Square in Manchester, England, is planned to be a BREEAM Outstanding building.

CHAPTER 17

Agriculture:
Growing Food—and Solutions

Danielle Nierenberg

In Ahmedabad, India, some women farmers and food processors are changing the way Indians eat. These women belong to the Self-Employed Women's Association (SEWA), a trade union that brings together more than 1 million poor women workers, 54 percent of whom are small and marginal farmers. In India, 93 percent of women working outside the home do not belong to a union, making them nearly invisible—they do not have access to credit, land, or financial services, including bank accounts. But when SEWA involves women in food production and processing, it is helping them improve their livelihoods by becoming more self-sufficient.[1]

SEWA members sort, package, and market rice under their own label. At a SEWA-run farm outside the city, women are growing organic rice and vegetables and producing organic compost on what was once considered unproductive and "marginal" land. "We now earn over 15,000 rupees [$350] per season, an amount we had never dreamed of earning in a lifetime," says Surajben Shankasbhai Rathwa, who has been a member since 2003. These women earn more and eat better than before, and they are providing an important community service by producing healthy, affordable, and sustainably grown food to local consumers, who usually cannot afford high-quality food.[2]

But the women in SEWA are not only interested in what is going on in their own communities—they are interested in what farmers are doing to combat climate change, conserve water, and build soils thousands of miles away, in places like sub-Saharan Africa. During a meeting in early 2011, the women of SEWA made it clear that they wanted to learn from their counterparts elsewhere who face the same challenges—erratic weather events, soil degradation, high food prices, poverty, and malnutrition—throughout India, Africa, and other parts of the developing world. And while SEWA's training farms and agricultural credit services will not change the global food system on their own, they are an important step toward enabling agri-

Danielle Nierenberg is the former director of the Nourishing the Planet program at Worldwatch Institute. This chapter is based on *Eating Planet 2012* by the Barilla Center for Food & Nutrition.

www.sustainabilitypossible.org

culture not only to feed the world but also to nourish livelihoods, environmental sustainability, and vibrant rural and urban economies.[3]

Agriculture is at a turning point. More than 1 billion people in the world remain hungry and 2 billion suffer micronutrient deficiencies. (See Figure 17–1.) Over the last three decades the western food system has been built to promote the overconsumption of a few consolidated commodities—including rice, wheat, and maize—and has neglected nutrient-rich indigenous foods that tend to resist heat, drought, and disease. One result is that 1.5 billion people in the world are obese or overweight and thus at higher risk of diabetes, cardiovascular disease, and other maladies. Moreover, vast amounts of food are wasted in both rich and poor countries, agriculture accounts for one third of global greenhouse gas (GHG) emissions, food-related diseases are on the rise, and the environmental impacts of agriculture—including deforestation, water scarcity, and GHG emissions—are increasing.[4]

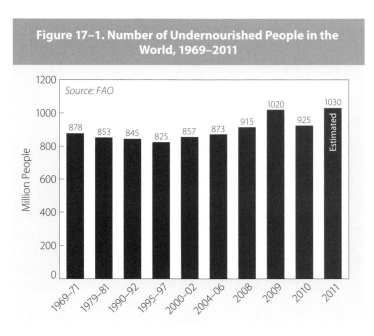

Figure 17–1. Number of Undernourished People in the World, 1969–2011

Source: FAO

The global food system needs a strategy and vision to nourish people and the planet by finding ways to make food production and consumption more socially just and economically and environmentally sustainable.

Food for All

Hunger and malnourishment continue to be a cruel reality for many of the world's poor. More than 239 million people in sub-Saharan Africa are considered undernourished by the U.N. Food and Agriculture Organization (FAO). Asia has the greatest number of undernourished people, with 578 million out of the world's 2010 total of 925 million. In Latin America and the Caribbean, where hunger receded dramatically throughout the 1990s, the number is 53 million.[5]

Food prices also continue to rise. Since 2007, FAO's Food Price Index has recorded a 70 percent jump in international food prices. (See Figure 17–2.) World Bank data show that food prices increased 15 percent for many developing countries between October 2010 and January 2011 alone, which pushed an estimated 44 million people into poverty. In sub-Saharan Africa

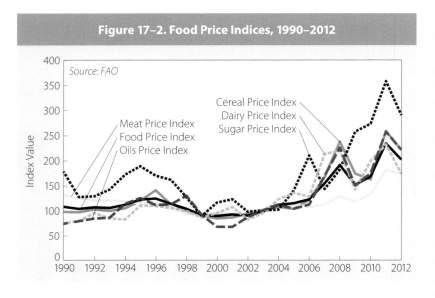

Figure 17–2. Food Price Indices, 1990–2012

Source: FAO

Meat Price Index
Food Price Index
Oils Price Index

Cereal Price Index
Dairy Price Index
Sugar Price Index

and South Asia, many farmers and consumers are earning just $1–2 a day, making any increase in food prices especially painful. Instead of being able to buy nutritious beans, eggs, meat, or vegetables, many households can afford only nutrient-poor staple crops such as rice or cassava.[6]

Governments, development agencies, nongovernmental organizations (NGOs), and funders tend to invest in increasing production and improving yields rather than in more-neglected parts of the food system that have the potential to improve livelihoods, decrease malnutrition, and protect the environment. What is needed is more investment to prevent waste from field to fork and a stronger focus on food aid and local school nutrition programs.[7]

Food waste can total an astonishing 30 percent of yearly harvests. In poorer countries, crop storage remains woefully inadequate, wasting crops in the places that need them the most. Farmers generally do not have access to proper grain stores, drying equipment, fruit crates, refrigeration, or other post-harvest storage and processing technologies.[8]

Even wealthy nations with climate-controlled storage units, refrigeration, drying equipment, chemicals that inhibit fungi and molds, and plant breeds designed to extend shelf life still squander vast amounts of food, throwing away cosmetically imperfect produce, disposing of edible fish at sea, overordering stock at grocery and "big box" stores, and purchasing too much food for home consumption. Much of it ends up in landfills instead of in stomachs.

In 1974, the first World Food Conference called for a 50 percent reduction in post-harvest losses over the following decade. Nearly 40 years later that goal is still not met, and waste prevention efforts remain vastly underfunded. Few donors invest in helping farmers and food processors find better ways to store and manage crops post-harvest, and wealthy consumers remain uninformed about the environmental impact of their (over)buying habits.[9]

But reducing this waste can be simple, inexpensive, and effective. Consider, for example, food contamination by aflatoxin, a toxic fungus that is caused almost exclusively by the consumption of food that has become moldy due to poor storage. The International Institute of Tropical Agriculture is working with farmers to apply a non-toxic, locally occurring strain of the fungus prior to harvest. The new strain, trademarked as Aflasafe, safely outcompetes and virtually eliminates the toxic strain, making it an effective bio-control with the potential to save farmers millions of dollars per year and simultaneously protect human health.[10]

There are also novel and income-generating ways of transforming food so that it does not go to waste. Solar-powered driers and dehydrators are helping farmers around the world preserve abundant harvests of mangoes, papayas, and other fruits, providing important vitamins and nutrients to people all year long.

Some consumers are also changing their eating and buying habits to reduce waste. In the United Kingdom, the campaign Love Food, Hate Waste educates citizens about food waste. The group's work has promoted the recycling of over 1 billion plastic bottles a year and has helped divert 670,000 tons of food from landfills in the last decade, saving consumers over $970 million annually.[11]

Food for Sustainable Growth

Twenty years ago, organic agriculture, conservation farming, and other agroecological practices were considered backward and inadequate ways to feed the world. Today, agriculture is emerging as a solution to the planet's pressing environmental problems—and agroecological approaches are seen as the way forward in a world of declining fossil fuel resources and increasing hunger and poverty. Several major research reports have demonstrated that food production can help address climate change, unemployment, urbanization, desertification, water pollution, and other environmental challenges.[12]

The Green Revolution technologies of the past, although effective at increasing yields in the short term, tended to focus narrowly on yields and very little on biological interaction. Nearly 2 billion hectares and 2.6 billion people have been affected by significant land degradation resulting from large-scale agricultural practices associated with the Green Revolution. Today, 70 percent of freshwater withdrawals are for agricultural irrigation, causing salinization of water in industrial and developing countries alike. The overuse and misuse of artificial fertilizers and pesticides has produced toxic runoff that has created coastal dead zones and reduced biodiversity.[13]

Although the Green Revolution is considered a "success," its benefits are unevenly spread. The most striking results in decreasing poverty and increasing crop yields were seen in South Asia, while people in sub-Saharan Africa have remained poor and undernourished. Many of the poorest of the poor "have gained little or nothing," according to the International Assessment of Agricultural Knowledge, Science and Technology for Development

(IAASTD), a landmark report on global agricultural knowledge released in 2008. Dr. Robert Watson, director of IAASTD, said that "we are putting food that appears cheap on our tables; but it is food that is not always healthy and that costs us dearly in terms of water, soil, and the biological diversity on which all our futures depend."[14]

A return to agroecology, which is a sustainable and environmentally friendly approach to food production, does not mean a return to old-fashioned or outdated practices. On the contrary, such approaches are highly complex, relying on the extensive knowledge of farmers and an understanding of local ecosystems. Agroecology mimics nature and integrates crops and livestock with the environment. For example, crops such as maize, wheat, sorghum, millet, and vegetables are being grown around the world alongside *Acacia, Sesbania, Gliricidia, Tephrosia,* and *Faidherbia* trees. These agro-forested trees provide shade, improve water availability, prevent soil erosion, and add nitrogen—a natural fertilizer—to soils. Integrating trees with crops can double or even triple yields over those obtained when crops are grown without a canopy.

Farmers in Japan are also finding ways to add nutrients to crops without expensive artificial fertilizers or toxic pesticides. By using ducks instead of pesticides for pest control in rice paddies, for instance, farmers have increased their incomes and provided additional protein for their families. The ducks eat weeds, weed seeds, insects, and other pests, and their droppings provide nutrients for the rice plants. In Bangladesh, the International Rice Research Institute reports that these systems have resulted in 20 percent higher crop yields and that farmers using this method have seen their net incomes rise by 80 percent.[15]

Agroecological practices even help farmers cope with natural disasters. A 2001 study compared "conventional" and "sustainable" farms on 880 similar plots of land after Hurricane Mitch devastated Honduras in 1999. The researchers found that the farms engaged in agroecological or sustainable land management practices had higher resistance to the storm.[16]

Food for Health

Hunger and obesity are both tied to inadequate nutrition and poor agricultural infrastructure, and investments in agriculture and hunger relief have often failed to deliver nutritionally. Focusing on agricultural yield and caloric intake has interfered with the actual delivery of vital nutrients, especially in fetuses and children under age three, yet this is what funding agencies, donors, and governments still tend to do. Over the last 20 years, food output in sub-Saharan Africa and Asia has become more concentrated in raw commodities, including maize, wheat, and rice, and less focused on nutritious indigenous foods, like millet, sorghum, and vegetables.[17]

Vegetables are a luxury for many of the world's poor, as many farmers who once grew vegetables have had to focus their attention on staple crops. But vegetable production is the most sustainable and affordable way to alleviate micronutrient deficiencies among the poor. Micronutrient deficiencies lead to poor mental and physical development, blindness, and anemia, especially among children, and they degrade performance in work and in school.[18]

Many low-income and middle-income communities face the double burden of under- and overnutrition. Obesity and malnutrition are the most obvious symptoms of our broken global food system: some 2.5 billion people worldwide suffer from one or the other. While poor nations receive a great deal of attention for high malnutrition rates, researchers and policymakers have paid less attention to the prevalence of noncommunicable diseases (NCDs), such as cardiovascular and respiratory diseases as well as type 2 diabetes, that result from unhealthy and inadequate diets. Sixty-three percent of global deaths are caused by NCDs, and this rate is expected to rise.[19]

Bernard Pollack

Tomatoes growing at the World Vegetable Center in Arusha, Tanzania.

Efforts to make agriculture healthier are being made in laboratories and at numerous conferences but also at the grassroots level in kitchens and backyards all over the world. One successful model is The Food Trust in north Philadelphia in the United States. The Trust runs community-based nutrition and food systems programs that have helped reduce the number of obese children there by half. A more broadly based U.S. program is Food Corps, one of the newest parts of the AmeriCorps program. Food Corps is working to address the country's childhood obesity epidemic by focusing on nutrition education, school gardens, and farm-to-school programs. Food Corps service members partner with local organizations to support community initiatives that are in touch with local needs, while also bringing in new energy and ideas. American children on average receive only 3.4 hours of nutrition education each year, but students in schools working with Food Corps will receive at least 10 hours.[20]

Surprisingly, the lack of nutritious food extends into many hospitals. Even rich-country hospitals can fail on this score: the Texas Children's Hospital in Houston, for instance, is home to a McDonald's restaurant. Hospitals in California, Ohio, Minnesota, and several other states also house fast-food restaurants. Health Care without Harm (HCWH), an international health coalition, is working to leverage the purchasing power of hospitals and health care systems to support food that is more nutritious and environmentally friendly. HCWH member Catholic Healthcare West, a 41-hospital system in Arizona, Nevada, and California, recently announced a partnership with Murray's Chicken, a New York producer, to supply its hospitals with chicken raised without antibiotics or arsenic feed additives. In South Africa, HIV/AIDS patients at the Chris Hani Baragwanath Hospital receive training in permaculture, irrigation, water conservation, food, nutrition, and indigenous medicinal plants. The patients are able to cultivate and harvest a garden at the hospital and are encouraged to bring home nutritious vegetables, fruits, and herbs.[21]

Food for Culture

The disconnection between young people and the global food system is growing. Most young people do not grow up wanting to be farmers. And consumers all over the world have forgotten basic cooking skills because of an overreliance on processed foods. Agricultural diversity is declining: most diets in rich countries consist of just six foods, including maize, wheat, rice, and potatoes. Agriculture is looked down upon as a career and is often viewed as work for the poor or people who have no other options. Farmers also lack access to markets, making it hard for them to earn an income from their work.

In villages outside of Kampala, Uganda, however, something unusual is happening among young people. For the first time, many of them are excited about being involved in agriculture—and instead of moving to the city after they finish primary school, many are choosing to stay in their communities to become involved in raising food.[22]

Betty Nabukalu, a 16-year-old student at Kisoga Secondary School, manages her school's garden. She explained how a project called Developing Innovations in School Cultivation has taught students "new" methods of planting vegetables. Before, she says, "we used to just plant seeds," but now she and the other students know how to fertilize with manure and compost and how to save seeds after harvest. She says they have learned not only that they can produce food but that they can also earn money from its sale. Thanks to their school food program, students no longer see agriculture as an option of last resort but as something that they can enjoy, is intellectually stimulating, and will provide a good income.[23]

Successful programs that turn farming regions into vibrant places where young people want to live and work have led to smarter land use, increased production, and stronger interest in agriculture among the next generation. Another way to help young people become more excited about agriculture is by incorporating information and communications technology into the process of farming.[24]

One obstacle faced by farmers worldwide is the lack of agricultural extension services. In sub-Saharan Africa, extension agents who used to provide information to farmers about weather, new seed varieties, or irrigation technologies have been replaced by agro-dealers who sell artificial fertilizer or pesticides to farmers, often with very little education or training about how to use those inputs.[25]

But in Ghana, farmers are benefitting from better-trained extension officers. At the Department of Agricultural Economics and Extension at Cape Coast University in southern Ghana, learning takes place in classrooms, fields, and farms. Extension officers are working with professors to find context-specific ways to improve food production in their particular communities. "One beauty of the program," says Dr. Ernest Okorley of the School of Agriculture there, "is the on-the-ground research and experimentation. . . . It allows the environment to teach what should be done."[26]

Growing a Better Food System

It is clear that we need a better recipe for ensuring that agriculture contributes to health, environmental sustainability, income generation, and food security. The ingredients will vary by country and region, but there are several key components that will lead to healthier food systems everywhere.

Investing in Agroecological Food Systems. Although many authoritative reports point to the need for more investment in agroecological technologies and practices that alleviate hunger and poverty, little attention is given to ensuring that farmers know about these. In October 2011, philanthropist-farmer Howard G. Buffett called on the agricultural development community to "get loud and get busy" to ensure that sustainable crop production is "back on the table" at the annual climate change meetings, at the 2012 United Nations Conference on Sustainable Development in Rio, and with every major agricultural donor and government in the world.[27]

In March 2012, the Landscapes for People, Food, and Nature (LPFN) initiative brought together farmers, policymakers, food companies, conservation agencies, and grassroots organizations in Nairobi in one of several meetings to develop a long-term strategy to scale up and support agroecological solutions. LPFN is documenting integrated farming landscapes around the world to strengthen policy, investment, capacity building, and research in support of sustainable land management. This sort of research can encourage policymakers to restore investment in agriculture, which has fallen precipitously from $8 billion in 1984 to $3.5 billion in 2005.[28]

Initiatives like Feed the Future and the Global Agriculture and Food Security Program (GAFSP) could have a huge impact on malnutrition, access to markets, and farmer incomes—if they were fully funded. Feed the Future is the U.S. global hunger and food security initiative; GAFSP is a multinational program formed to assist in the monitoring and evaluation of the $1.2 billion in pledges made by the Group of 20 industrial nations in

2009. Unfortunately, these programs have received very little of the funding pledged by donor countries, private businesses, and NGOs.[29]

Recognizing Agriculture's Multiple Benefits. Farmers are business people, educators, and stewards of the land. Finding ways to compensate these women and men for their multiple roles will become increasingly important as agricultural challenges increase.

Women farmers, for example, make up as much as 80 percent of the agricultural labor force in some countries but are often denied basic benefits like land tenure, education, and access to banks. Organizations, policymakers, and community members should recognize women's rights and involve women in decisionmaking processes.[30]

Innovative organizations are also compensating farmers for the ecosystem services their lands provide. And the Rainforest Alliance is working with millions of farmers around the world to ensure that sustainably grown crops get a premium price from consumers in wealthy nations so that the benefits of agroecological practices are recognized. Other projects involve paying farmers for sequestering carbon in their soils.[31]

Cultivating Better Livelihoods. Building a better food system does not mean producing more food—the world can already feed 9–11 billion people with the food grown today. It means addressing poverty. More than 2 billion people live on less than $2 per day, global unemployment is at a record high, and poor households in the developing world spend 70 percent of their income on food.[32]

Financial speculation on the price of food has contributed to volatility in agricultural markets, with grave impacts on the livelihoods of small-scale farmers, many of whom still lack access to the most basic aspects of domestic support, including land, insurance, bargaining power, and credit (despite the expansion of microfinance and other ways of providing financial support; see Box 17–1). Food prices were nearly 20 percent higher in 2011 than in 2010 due to such speculation. Price volatility hurts these farmers, who need stable markets and a fair price for their yields. Clamping down on food price speculation—especially prices for maize, wheat, and rice, the three most heavily traded food commodities, which supply the bulk of dietary calories for 2 billion poor people—would be a major step forward for both farmers and the hungry.[33]

Additionally, farmers need access to markets where they can get a fair price. Institutions such as agricultural cooperatives can help farmers operate more efficiently and earn more money than they can as individuals. By helping farmers come together to grow, distribute, and sell food, cooperatives function as businesses and social groups, enhancing communities' economic powers as well as their social service networks.[34]

Farmers also need better access to information about prices and mar-

Box 17–1. Promoting Sustainable Agriculture through Village Banking

Since Mohammad Yunus launched the Grameen Bank in Bangladesh in 1976, microcredit has become a celebrated tool to help relieve poverty and foster entrepreneurism among the poor. Initially conceived as a purely charitable tool for alleviating poverty, microcredit has become microfinance and now includes loans, insurance, and savings products. Currently there are an estimated 500 million microsavings accounts around the world. As demand for these services grew, many providers aimed to make microfinance profitable, allowing it to attract investor capital and thus achieve greater scale. The microfinance industry has exploded to include over 1,000 institutions serving an estimated 85 million clients.

After an initial burst of wild enthusiasm, there is now a growing debate about the effectiveness of these credit mechanisms as tools for ending poverty. This is especially true where the focus on scalability has caused lending institutions to neglect impoverished rural populations. The farmers who can take out loans sometimes borrow for costly agricultural inputs and then become trapped in a vicious cycle of crop failure and debt. Particularly troubling are the reports of up to 200,000 farmer suicides in India, where farmers have borrowed to buy expensive genetically modified organisms, chemical fertilizers, and pesticides.

But there is another way to help poor farmers gain access to financial services: village savings and loan associations (VSLAs), which were pioneered by CARE in West Africa. VSLAs typically have 20–30 members who meet weekly to pool their savings and create a loan fund. With the help and training of a facilitator, the members draft bylaws and elect leaders. At the beginning of the investment cycle, each member deposits an agreed-upon amount. Then the group meets weekly, and individual members make further deposits as determined by the group's bylaws. After 12 weeks, each member may take out a loan for up to three times the amount he or she has saved.

Groups typically have many more savers than borrowers, which ensures that there are adequate funds for those who wish to borrow. The investment cycle is short, usually 12 months. At the end, members receive back their shares plus a portion of any accrued interest or capital gains from fines and fundraising. The group can then choose whether to initiate a second VSLA cycle.

VSLAs have dramatically improved members' lives and communities. Successful businesses create new jobs, and the interest raised by the bank stays in the local community. The groups also often establish their own charitable funds to help members meet various needs, such as education fees for their children, medical expenses, or emergencies.

The benefits of VSLAs go far beyond economics, however. Weekly meetings strengthen communities and provide opportunities for personal growth, education, and the development of various talents and business skills. Those who succeed in businesses also reach out to help others, so the entire community benefits. In recent impact evaluations of Plant With Purpose's Tanzania VSLA groups, it was found that each group member shared his or her agricultural training with on average more than 20 others.

Plant With Purpose—a nonprofit based in California that works to transform lives in rural areas where poverty is caused by deforestation—is using VSLAs as a vital part of an integrated strategy to address environmental and economic needs. The weekly meetings provide a platform to teach farmers skills that increase agricultural productivity, help gain access to markets, promote crop diversification, reduce deforestation, and help adapt to the challenges of climate change. By offering such training, VSLAs can provide an entirely new skill set of agroecological methods, empowering farmers to make a living in ways that also restore and protect fragile environments.

—Doug Satre
Plant With Purpose, California
Source: See endnote 33.

kets. Information and communication technologies, such as mobile phones, are enabling farmers to obtain real-time data about market prices, which is helping them make better-informed decisions about crop production. Services such as FrontlineSMS allow farmers not only to get real-time food price data but also to connect with one another and with potential consumers, increasing their market size.[35]

The Emergence of Agriculture as a Solution

Governments need to do more to recognize the inherent right of every human being to safe, affordable, and healthy food and back up that right with appropriate policies. Countries such as Ghana and Brazil have already reduced the number of people suffering from hunger through effective government action, such as national school feeding programs and increased support for agricultural extension services.[36]

The projects highlighted in this chapter are exciting because they exemplify how agriculture is emerging as a solution to global problems by reducing public health costs, making communities everywhere more livable, decreasing poverty, creating jobs for young people, and even reducing climate change.

Some innovative programs and individuals are working to ensure that everyone has access to nutritious, healthy, sustainable, and justly grown food. From SEWA in India and villages in rural Uganda to research institutes and governments all over the world, there is a growing realization of the positive impact that agriculture can have on livelihoods, nutrition, and the environment. And these are exactly the sort of innovations that should attract the support of governments, the private sector, and the international funding and donor communities.

CHAPTER 18

Protecting the Sanctity of Native Foods

Melissa K. Nelson

Indigenous peoples are the caretakers of many of the last biodiverse places on Earth. Even though they only constitute 5 percent of world population and occupy 20 percent of the earth's surface, they live in 80 percent of the world's biological diversity hotspots. They are therefore critical to ecosystem health and should be recognized as major stakeholders and leaders in the global sustainability movement.[1]

Through millennium-tested traditional ecological knowledge, land-based lifeways, and a holistic, ethical relationship to the earth, indigenous peoples have a lot of practice in sustainable living. These cultures, with their diverse knowledge systems and integrated life-enhancing practices, can provide relevant and timely examples of how to live sustainably within local ecosystems. They can also provide principles and lessons for the industrial world to relearn how to become native to place.

Indigenous peoples everywhere are critical protectors of biodiversity, more often referred to by native peoples as homelands, territories, sacred lands, or simply "all our relations." One of the most significant ways indigenous peoples have practiced and demonstrated a sustainable relationship to native lands and waters is through the tending, harvesting, hunting, growing, and cultivating of native foods. Living requires eating, eating means taking life, and taking life requires (or at least implies) a philosophy, a process, and a coherent system or "cosmovision" for acquiring an adequate amount of food and nutrition to sustain people and thrive as a culture.

Acquiring food from the earth is both an art and a science. Native foodways traditions are complex, diverse, and beautiful systems that connect nature and culture and that provide both physical survival and cultural meaning to a people. Indigenous foods and lifeways are an ideal example of the profound interface of biological and cultural diversity—or what Yuchi professor and author Dan Wildcat refers to as the nature-culture nexus. Embedded within native food traditions are diverse knowledge systems and

Melissa K. Nelson (Turtle Mountain Chippewa) is an associate professor of American Indian studies at San Francisco State University and president of the Cultural Conservancy.

www.sustainabilitypossible.org

native sciences, languages and distinct cultural heritages, and unique embodied life-affirming practices. They are connected to soils and songs, seeds and stories, ancestors and memory, taste and rain, dance and medicine, and nourishment and place.[2]

According to many native traditions, to live well is the goal of life. And to live well means not only sustaining foods and a lifestyle but actually regenerating the ecological systems people depend on and enhancing their happiness and spirit. So there is an emphasis not only on sustaining basic needs over time but on actually regenerating the resources, or "relatives," that provide the raw materials for food, shelter, clothing, and medicine. In this sense, indigenous livelihoods are about surviving and thriving in a place where all beings in the circle of life thrive together. This philosophy and practice of thriving in place is best demonstrated in the many complex foodways of indigenous peoples.

Sacred Foods of the Americas and the Pacific

Native Americans are the originators and caretakers of many staple foods globally. Many of these foods are considered sacred and have profound teachings and practices associated with them. Corn (*Zea mays*), for example, is one of the staple indigenous foods of the Americas and has become an important food crop globally. Through numerous creation stories, corn is known as "Corn Mother" in many Native American cultures. It is considered a sacred relative and a source of life. Rituals, myths, ceremonies, offerings, dances, and songs all praise the value and sacredness of this native cultivated plant. Whether in the heart of Oaxaca in Mexico, the Rio Grande River valley of New Mexico, the Saint Lawrence river valley of the Haudenosaunee, the Yucatan Peninsula of the Maya, the Canadian Plains of the Cree, or the Andean highlands of the Quechua and Aymara, you can find indigenous peoples cultivating, praising, and eating corn.

Corn has been tragically compromised by genetic engineering, and native farmers are working hard to preserve the heirloom varieties and protect them from genetic contamination and further industrial commodification. As the late Seneca scholar and farmer John Mohawk has said, trouble comes when people start growing food for money rather than for nutrition. He also shared the prophecy that warned of a negative shift in the world when sacred corn was fed to machines rather than to people. And this is happening now, as 40 percent of the corn grown in the United States is converted to ethanol for machines.[3]

For Hawaiians, the origin food taro (*Colocasia esculenta*) or *kalo* is the major staple food of the Pacific. Hawaiians know this food as an elder brother in their origin stories, and they value him as a sacred ancestor. He represents a type of mytho-geneology common among native peoples globally. *Kalo* too has been threatened, as scientists at the University of Hawaii attempt to patent it and genetically alter it. In 2002, three taro varieties were patented, and in 2003, three Hawaiian varieties were genetically modified without any public debate or consultation with Native Hawaiian farmers, who have worked with this plant for thousands of years. Hawaiian farmers and activists as well as other concerned citizens protested, wrote letters, and educated the public about this ethical violation of the sacred *kalo* plant and of Hawaiian lifeways.[4]

Melissa K. Nelson

In 2006, the University of Hawaii withdrew its patents on the three varieties and agreed to stop genetically modifying Hawaiian forms of taro. Researchers continue to experiment with modifying a Chinese form of taro, however. According to the nonprofit group Hawai'i SEED, "Native Hawaiians, taro farmers and Hawaii SEED continue to fight back by supporting legislation that places a moratorium on the cultivation and experimentation of GMO [genetically modified organism] taro in the lab and field." In 2008, Native Hawaiians and allies drafted

Harvested taro root that has been boiled in preparation for being cleaned and mashed into poi.

legislation that banned GMO taro and corn from the Big Island of Hawaii. Bill 361 legally protects taro and coffee from genetic engineering on the Big Island. Due to strong public support, the Hawai'i County Council unanimously passed this bill, although the mayor later vetoed it. Protecting indigenous foods that are considered ancestors and practicing the traditional Hawaiian philosophy of *aloha 'aina*, "to love that which nourishes you," is still a major struggle and challenge for Hawaiians and native peoples globally.[5]

In the heart of North America, members of the Anishinaabeg/Ojibwe/ Chippewa nation are concerned about the sanctity of their food *manoomin*, or wild rice (*Zizania aquatica*). This food was given to the Ojibwe by the Creator in a sacred story of migration and helped the ancestors locate their homelands on the Great Lakes. Winona LaDuke and her nonprofit organization, the White Earth Land Recovery Project & Native Harvest program, are actively protecting this sacred staple food and have secured laws in Minnesota to oversee any research proposed on *manoomin*. This is the only native rice in North America and is a significant food source for many Native American nations in the Great Lakes of the United States and Canada. It is highly nutritious, delicious, and has made it into the mainstream food industry as a unique rice dish.[6]

Most of the commercially available "wild" rice is cultivated, paddy rice, not a true wild rice any more. Industrial agriculture has been cultivating wild rice in nonnative habitats like inland northern California. Cultivating wild rice threatens the native ecosystems in the Great Lakes and other landscapes, where it requires a lot of water to grow well. This *ex-situ* cultivation practice undermines the integrity of the wild rice and its true value as a hand-harvested wild rice and an economic asset to the Ojibwe wild rice gatherers who depend on its sale as part of their seasonal livelihood. As LaDuke has pointed out, "Our *manoomin* grows nowhere else in the world, and our people, the Anishinaabeg from these Great Lakes reservations, intend to keep this tradition alive, vital and nurturing our souls and our bellies."[7]

Freshly hand-harvested wild rice steaming over the coals of an open fire to decrease its moisture content.

These sacred, totem foods—corn, taro, wild rice—have been passed on for generations and traded with other tribes and communities for human and economic well-being for centuries and millennia. They are often seen as intimate relatives. It is tragic that they are now being threatened with life patenting and genetic modification, as Claire Cummings clearly outlines in *Uncertain Peril*. Protecting the foodways of native peoples requires protecting and restoring the sanctity of native seeds and resisting the industrial commodification of these invaluable food sources. We see positive examples of this happening, as just described in Hawaii and Minnesota. Internationally, indigenous farmers in the province of Cuzco, Peru, have been successful in banning GMO potatoes, another global staple food at its center of origin. The larger context of these important bans and laws is the question of who owns and controls these native lands and waters—and their indigenous foods—in the first place.[8]

Environmental Context of Native Foods

Protecting native foods is about territory—land and water rights. It is important to assess the conditions of the native habitats that are the source of these local foods and determine who owns and controls them. Because Native Americans control only 4 percent of U.S. land, they are not likely to have control or access to much of the land that provides their indigenous foods.[9]

Most lands, rivers, and lakes in the United States are under private ownership, in public parklands, or in other federal lands, such as those under the jurisdiction of the Bureau of Land Management or the U.S. military. Creating access agreements and cooperative management plans for the native biodiversity of these lands is an important strategy that many tribes and traditional practitioners are using to reconnect with their ancestral harvesting sites. Federal agencies often benefit as well from the indigenous resource management practices of native peoples, whose land care practices, such as small-scale controlled burning, often enhance biodiversity. Ecological restoration is often incorporated into these practices, helping to clean up toxic landscapes and restore ecosystem health.[10]

There is also a growing native land trust movement in the United States as more and more tribes buy back their ancestral lands for both traditional and contemporary uses. As Slow Food founder and president Carlo Petrini noted in May 2012, "It would be senseless to defend biodiversity without also defending the cultural diversity of peoples and their right to govern their own territories. The right of peoples to have control over their land, to grow food, to hunt, fish and gather according to their own needs and decisions, is inalienable."[11]

Another key factor in assessing access to native lands and waters for traditional food harvesting is environmental quality. An assessment needs to take place to ensure that the foods grown in a specific place have not been contaminated by pesticides, industrial runoff, or other types of pollution. Many native food plants, like watercress and piñon pine, are seen as "weeds" and useless shrubs by government agencies and nonnatives, and they are unfortunately destroyed with herbicides and pesticides. Toxic exposure is thus a very real threat to traditional food gatherers when they do not own the land where they gather.

In addition, many animal food sources bioaccumulate toxins. So when native people eat their traditional meats—whether it is fresh fish or deer, moose, ducks, seal, or caribou—they can be exposed to high levels of such toxins as mercury, lead, polychlorinated biphenyls (PCBs), and other persistent organic pollutants. This exposure has become so extreme that often mothers' breast milk is considered toxic due to the high levels of industrial chemicals in it. As Mohawk midwife and environmental health researcher Katsi Cook says, "women are the first environment," so whatever happens to the environment will happen to women's bodies.[12]

The Arctic peoples of the far north are experiencing this health crisis in a major way because their traditional diet consists primarily of high-protein, high-fat meat foods. Inuit women's breast milk has 5–10 times the level of

PCBs as the breast milk of women in southern Canada. Even with these risks, Arctic people are still hungry for their ancestral foods instead of the imported western diet. As Canadian writer Lisa Charleyboy has noted, "for at least some Inuit, the value of eating the foods of their ancestors is worth the cost. 'Contaminants do not affect our souls,' [Inuit activist Ingmar] Egede said. 'Avoiding our foods from fear does.'"[13]

Without healthy seeds, lands, and waters, native foods will continue to be compromised, damaged, and made scarce, and native health will suffer. Native peoples are seeking to ban GMO foods with legislation and to establish GMO-free zones in local communities, create access and cooperative management agreements with agencies and private landowners, develop ecological restoration plans to clean up contaminated sites, and engage in and purchase back land through native lands trusts. Native peoples and food activists are also exploring unique partnerships and programs to safeguard these foods. The growing need for global food security and food justice has inspired many food groups to partner with native communities and organizations to educate the general public about the true value of native foods and their significance for biodiversity conservation and cultural heritage and health.

New Partnership for Food Security

One major example of new partnerships is Native American and indigenous participation in the international Slow Food movement. Slow Food International and Slow Food USA are interconnected grassroots membership organizations that promote good, clean, fair food for all. There are over 100,000 members globally. Two of their programs are the Ark of Taste and the Presidia. According to the Slow Food USA website, "the Ark is an international catalog of foods that are threatened by industrial standardization, the regulations of large-scale distribution and environmental damage. The US Ark of Taste is a catalogue of over 200 delicious foods in danger of extinction. By promoting and eating Ark products we help ensure they remain in production and on our plates."[14]

One way Slow Food protects the Ark of Taste foods is through the Presidia program. A Presidium in this context is a "garrison" or fort that aims to protect endangered foods. Local projects work to improve the infrastructure of artisan food production. The Presidia aim to guarantee a viable future for traditional foods by stabilizing production techniques, establishing stringent production standards, and promoting local consumption of endangered foods.[15]

Many important Native American foods and beverages are in the US Ark of Taste and are part of the Presidia program, including Anishinaabeg *manoomin*, Navajo-Churro sheep, Arikara yellow bean, greenthread tea,

O'odham pink bean, Tuscarora white corn, Hopi mottled lima beans, traditional Hawaiian poi (*kalo*) and sea salt, and the Ozette potato. Slow Food is highlighting the significance of these foods because they are at risk biologically and as culinary traditions, are sustainably produced, have great cultural or historical significance, and are produced in limited quantities. The main factor also for Slow Food is that they have outstanding taste, even though taste itself is often culturally conditioned and will vary greatly. Some Native American organizations, such as the White Earth Land Recovery Project (with Anishinaabeg *manoomin*) and Diné bé Iiná (with the Navajo-Churro sheep), are working directly with Slow Food to gain support and recognition for their food traditions through that network.[16]

Cultural Heritage and Traditional Ecological Knowledge

A crucial aspect of protecting native foodways is recognizing and honoring the ecological knowledge of elders and traditional food gatherers, because native foods cannot be protected without their hands-on knowledge—how to grow, nurture, harvest, process, cook, and feast on them. This requires intergenerational knowledge transmission. It is the elders who retain an understanding of living off the land before stores and commodity foods dominated native diets. It is the elders who know how to gather and prepare tule bulbs as foods, as the Paiute do. Or how to gather and process the California acorns, as the Pomo do. Or how to hunt and prepare a moose for a feast, as the Cree do. Or how to take an heirloom tepary bean and grow it in a beautiful desert garden, as the Tohono O'odham do.

Navajo chef Walter Whitewater giving a hands-on lesson to Native children about Native American foods and cooking.

Lois Ellen Frank

The keys to cultural health include strong, healthy bodies for all and also healthy elders who feel valued and appreciated. In healthy communities, elders and youth still have a deep relationship and a system of knowledge sharing, often through storytelling, the arts, and hands-on practices like farming. When young people are able to learn the traditions from their elders, their identities are reinforced and invigorated, their sense of pride in their heritage increases, and their overall wellness improves significantly.

Elders, knowledge holders, and traditional practitioners often teach through stories and demonstrations. Through them they impart the importance of the "original instructions"—a tribe or community's enduring values, insights, and worldviews about life-enhancing practices that take care of the gifts of life, of food, of water, of all the relations that make life possible. Intergenerational knowledge transmission and these philosophical and ethical teachings can be seen as parts of the intangible cultural heritage of native foodways.

Combating Health Disparities and Improving Native Wellness

Native foodways are tied to sustainable living in very practical environmental ways and through the revival of cultural memory and heritage. Most directly, native foodways are critical channels for maintaining physical health and solving contemporary health problems such as diabetes and obesity. This topic is beginning to be incorporated into tribal and higher education curricula. As native chef and culinary anthropologist Lois Ellen Frank has noted, "Young, educated Native American activists, such as students at the Institute of American Indian Arts (IAIA) in Santa Fe, New Mexico, are beginning to foster a dialogue about how to decolonize their diets and their bodies by recovering their ancestors' gardens and foodways." IAIA has incorporated an "indigenous concept of Native American food" into its required science class for all students as part of the four-year degree program, regardless of the student's major.[17]

Western doctors are also taking note of how returning to indigenous diets can significantly improve health. The Physicians' Committee for Responsible Medicine has worked with native chefs Lois Ellen Frank (Kiowa) and Walter Whitewater (Navajo) in sponsoring cooking classes at IAIA and the Pueblo Indian Cultural Center, focusing on plant-based ancestral foods of the Southwest. After just eight weeks, native students and participants of these classes lost weight, lowered blood sugar levels, in some instances were able to decrease their diabetes medication while working with medical professionals, and felt much healthier.[18]

This significant correlation between eating native foods, decreasing diabetes, and improving overall wellness has been clearly demonstrated by the Tohono O'odham Community Action organization and its significant work to combat the diabetes epidemic in its tribal community. With Tohono O'odham, Seri, Yaqui, and other tribal and nonnative participants, this relationship was passionately demonstrated in the Desert Walk for Biodiversity, Heritage and Health co-organized and documented by Gary Paul Nabhan in 2000. During this intertribal, multicultural pilgrimage, nearly 200 people walked 240 miles while sustaining themselves only on desert foods and medicines—as well as on songs, stories, and prayers to feed the soul. Again, people lost weight, lowered their blood sugar and cholesterol levels, and felt renewed and reconnected to their ancestral lands and diets.[19]

Native Food Alive and Well

Today indigenous food sovereignty is being reasserted, enacted, and explored in many diverse ways in Native America. Ojibwe and other native professors and students are working on "decolonizing your taste buds" programs in Native Studies classes and in reservation cultural centers. Miwok and Lakota youth are growing intertribal urban gardens in cities like Oakland and Detroit. Western Shoshone environmental directors are building soil and storing water with permaculture rain gardens in the Great Basin desert. Wailaki gatherers harvest kelp, dulse seaweed, and red abalone on the northern California coast for their elders and for ceremonies. Pueblo farmers continue to shape and eat from their desert landscapes with dryland farming methods. Native chefs teach indigenous nutrition and Native American cuisine in tribal colleges and culinary schools.[20]

The native foods movement is alive and well in Turtle Island (as North America is known by some Native Americans) and throughout the world. This movement continues to grow and thrive in a modern context. Native elders, young people, leaders, students, and tribal members are protecting the sanctity of native foods for cultural health and environmental justice, despite continued industrial efforts to marginalize, commodify, and devalue these original foods.

Indigenous peoples are asserting food sovereignty as an indigenous right and responsibility and a human right for all peoples and future generations. They are "re-indigenizing" bodies and minds and lands and communities through native foodways. Native foods are sacred and irreplaceable. They are markers of diversity and are often keystone species for the health of an ecosystem and the health of a people. Indigenous knowledge and foodways are viable and potentially essential alternatives to modernity that remind us all that we can become native to place and serve as regenerative elements in our local foodsheds and ecosystems.[21]

CHAPTER 19

Valuing Indigenous Peoples

Rebecca Adamson, Danielle Nierenberg, and Olivia Arnow

For most of the last century, the Maasai faced the threat of eviction by the Kenyan government and outside corporations eager to profit from Maasai lands. These semi-nomadic pastoralists have lived for centuries on areas that are now part of Kenya and Tanzania. But they have often been denied many basic human rights, including food security, safe drinking water, and adequate sanitation.[1]

In August 2010, things changed for the Maasai. A new Kenyan constitution was passed. It recognized the traditions, customs, languages, and rights of Kenya's indigenous peoples and acknowledged the legitimacy of hunter-gatherer, pastoral, and nomadic ways of life. These policy changes would not have come to pass without the support and strength of indigenous grassroots organizations. Mary Simat, executive director of Maasai Women for Education and Economic Development, embarked on a major initiative, with funding from First Peoples Worldwide, to familiarize Maasai villagers with the new constitution, issuing Maasai-language copies of it and conducting workshops in communities.[2]

The changes to the constitution are having immediate impacts. Land reform initiatives authorize land use according to the Maasai's own customs; by entrusting revenues to county and local authorities, the land reform policies create a channel for regular funding for local priorities. In addition, the Maasai are now recognized, for the first time, as important stewards of the land whose environmental knowledge and practices—including rotational livestock grazing and the fostering of beneficial wildlife habitats—can help build resilience to climate change, improve water conservation, and protect biodiversity. And this shows policymakers and communities the importance of acknowledging the longstanding relationships of indigenous peoples to their lands and their commitment to sustainability. These sorts of victories by indigenous peoples are becoming more common in Asia, Latin America, and North America as well as Africa.

Rebecca Adamson is a Cherokee and the founder of First Peoples Worldwide. Danielle Nierenberg is the former director of the Nourishing the Planet program at Worldwatch Institute. Olivia Arnow is a senior at Vassar College.

www.sustainabilitypossible.org

Indigenous peoples inhabit more than 85 percent of the earth's protected areas. Their territories span most of the last remaining biodiversity-rich conservation priority areas, and they maintain traditional land claims on 18–24 percent of Earth's land surface. But reports from the International Funders for Indigenous Peoples suggest that only about 1 percent of the billions of dollars spent each year on philanthropy goes to indigenous peoples and the ecosystem services they support, including biodiversity protection. The wealth of natural resources preserved within indigenous territories presents an enormous opportunity to expand conservation strategies on a scale that will help alleviate hunger and poverty while also conserving and protecting Earth's natural resources.[3]

Young Maasai herder approaches a mixed herd of cattle and goats.

Andreas Lederer

Forced Evictions

Despite the important role indigenous peoples play in protecting natural resources, their contributions are often overlooked. Even at its best, conventional or science-based conservation can ignore or marginalize the stewardship of indigenous peoples. And at its worst, western approaches to conservation can lead to their violent eviction.[4]

Evictions in the name of nature conservation or preservation are not a new phenomenon. In North America, the Miwok and Awahneeshi people were removed from Yosemite Valley to preserve land for the national park in 1906. Although they used the woods, waters, and plains of Yosemite, they were not considered a part of this wilderness and were evicted or killed.[5]

Governments still use conservation to forcibly relocate and intimidate ethnic groups, including in Central Africa, Asia, and Latin America. Rather than protesting these actions or withdrawing support, conservation groups have ignored them. The Wildlife Conservation Society (WCS), for example, does conservation work in Myanmar—work criticized by human rights advocates. By 2000, Myanmar had designated over 15,000 square kilometers of protected areas in 31 national parks and wildlife sanctuaries. When evidence surfaced that the government was killing and evicting ethnic minorities in the interest of "conservation," WWF and other groups closed down their programs there. WCS, however, continues to manage conservation programs in Myanmar.[6]

Forced evictions devalue not only the importance of indigenous communities but also the traditional ecological and agricultural knowledge these groups possess. It is true that rapid urbanization and the expanding global population over the coming decades will inevitably lead to a scaling up or overhaul of many traditional methods of food production, such as foraging and wild fishing. But evicting people from their native lands and relocating them to urban slums without training, education, or adequate compensation is not a sustainable solution to the problem of feeding the world. Clearing forests and evicting families to grow sugarcane or maize does not necessarily lead to less malnutrition or better incomes for indigenous farmers. Farm families evicted from their land are often forced to rely on imported and processed foods, rather than being able to grow their own nutritious foods, keep livestock, and rely on their local communities for their food and other needs.

By removing indigenous groups from their lands or recklessly exploiting natural resources such as minerals and forests, corporations and governments are effectively erasing thousands of years of practiced traditional ecological knowledge—the cumulative body of experience an indigenous group has collected over generations, encompassing knowledge, practices, and beliefs about their customary lands. (See Table 19–1.)[7]

In 2007 and 2008, food price spikes plunged millions of people into poverty and food insecurity, derailing years of international development work and aid. The World Bank estimates that at least 44 million people were driven into poverty as a result of higher food prices. Helping indigenous communities maintain their traditional knowledge and ways of life can avoid the expenditure of billions of dollars in emergency aid, as well as protect the natural environment that indigenous peoples have cultivated for many generations.[8]

Fighting Back

Indigenous communities all over the world are fighting for the right to free, prior, and informed consent (FPIC) whenever an action may affect their lands, values, or rights. FPIC states that anyone who wishes to use customary land belonging to indigenous communities must enter into open, non-coercive negotiations with them. Private corporations, national governments, nongovernmental organizations (NGOs), and even entire industries have begun enforcing the principle of FPIC for indigenous communities.[9]

The United Nations Declaration on the Rights of Indigenous Peoples, endorsed in 2007, provides an international legal framework and court of public opinion that can be used to slow down commercial development.

Table 19–1. Indigenous Peoples' Resources: What's at Stake?		
Indigenous Group and Endangered Resource	**Why Endangered**	**Why Resource is Valuable**
Mangyan Peoples of the Philippines—Forests of Occidental Mindoro	Large mining corporations are threatening to destroy ancestral lands to profit from gold, natural gas, and minerals worth millions of dollars.	Deforestation threatens the livelihoods of forest-dwelling indigenous communities in the Mindoro region. Without the food and shelter resources traditionally provided by the forest, indigenous communities will be forced to rely on unpredictable markets for their income. Traditional knowledge about agriculture, which is critical to the food security of the communities, may be lost if there is no land to plant indigenous crops.
Ogiek Peoples of Kenya—Mau forest complex	Since 2009, the Kenyan government has evicted thousands of Ogiek people from their ancestral forest, ostensibly to reforest the area. But over the last two decades the government has sold parcels of the forest for agricultural development, both degrading the forest and forcibly evicting Ogiek tribespeople, who have sustainably managed the forest for centuries.	The forest stores and channels rain that is essential for irrigation and hydroelectric power, and it also absorbs and stores carbon dioxide from the atmosphere. The storage of rainwater, as well as the cooler temperatures resulting from forest cover, previously kept malaria outbreaks at bay—but the incidence of malaria is on the rise now that the forest is being degraded and cleared.
Imraguen in Mauritania—Mullet fish	In 2006, Mauritania sold fishing rights to the European Union in exchange for a reduction in public debts. Fishing fleets from western countries often obtain the fishing rights, employ local fishers, and freeze the catch to be sent elsewhere for processing, mainly to North Africa and Europe.	Traditional knowledge of catching and preparing mullet is being lost, resulting in the disappearance of a significant element of Imraguen cultural identity. The waters off Mauritania are among the few left in the world that are still well stocked with fish—demonstrating the ability of the Imraguen to manage their fisheries sustainably over long periods of time. At a time when large-scale and industrial fishing practices have depleted many global fish stocks, preserving and scaling up Imraguen fishing practices can help reverse overfishing and restore sustainable fish populations.
Aboriginal communities of northern Australia—Burn-control strategies/fire management techniques	Throughout the twentieth century, forced removals kicked many aboriginal communities off their lands. Fierce dry-season blazes have destroyed biodiversity and emitted tons of greenhouse gases into the atmosphere.	Aboriginal fire management techniques have been crucial in helping manage habitats and food resources across northern Australia for millennia. If these strategies are not followed, Australia's biodiversity will be seriously threatened.

Source: See endnote 7.

Corporations like BP, ConocoPhillips, ExxonMobil, and Suncor have all announced policies on indigenous peoples recently, and a shareholder meeting of Newmont Mining Corporation voted 96.4 percent in favor of reducing company conflicts with indigenous peoples.[10]

In the northern Pacific island of Sakhalin, over the past 15 years the livelihoods of the Evenk, Nivkh, Nanai, and Uilta peoples have been threatened by companies eager to extract oil. The pipelines, processing facilities, and other industrial sites have degraded the island's biodiversity and decreased food production. In response, Shell International has made efforts to maintain a decent quality of life for the indigenous communities, in keeping with the U.N. Guiding Principles on Business and Human Rights that were adopted in June 2011.[11]

Shell repainted its ships when the Inuit elders told them that red disrupts sea mammal behavior. The company implemented a Sakhalin Indigenous Minorities Development Plan and engages with Sakhalin's indigenous groups to address community grievances, improve health care and education facilities, and preserve and study traditional languages. There are also efforts under way to establish indigenous peoples consultancy services for companies working on native lands, giving indigenous peoples the ability to influence corporate policy and engage in a business relationship with companies that would normally be adversaries.[12]

Protecting People and the Planet

Respecting indigenous peoples and their practices is a potentially invaluable resource in combating climate change. The Accra Caucus on Forests and Climate Change, a network of NGOs representing about 100 civil society and indigenous peoples' organizations from 38 countries, determined that the key to reducing deforestation is to respect the rights and realities of indigenous peoples and forest-dwelling communities. Its 2010 report, *Realizing Rights, Protecting Forests: An Alternative Vision for Reducing Deforestation*, features case studies from Brazil, Cameroon, the Democratic Republic of Congo, Ecuador, Indonesia, Nepal, Papua New Guinea, and Tanzania and concludes that a human rights-based approach should be applied to all policy and development planning, including for agriculture, forests, and the Reducing Emissions from Deforestation and Forest Degradation initiatives of the United Nations.[13]

Well-planned and targeted grants can help indigenous communities preserve their livelihoods. First Peoples Worldwide has developed a progressive and innovative funding model that promotes indigenous-led projects that establish indigenous control of indigenous assets. The Keepers of the Earth Program issues grants of $250 to $20,000 for projects in land conservation, climate change, and food security, and it strives to protect indigenous rights to subsistence hunting and gathering, access to sacred sites, and customary cultural practices while simultaneously protecting biodiversity and sustainable economic production.[14]

Although traditional grants can help alleviate immediate or short-term problems, they can sometimes ignore the values and longer-term needs of the communities receiving the support. A value central to indigenous communities is egalitarian and inclusive development—development that does not benefit some at the expense of others. This was explicitly demonstrated when First Peoples held a roundtable meeting in Kenya in May 2009 to discuss funding available for community stewardship projects throughout Africa. As part of this meeting, the group was offered funding from the Keepers of the Earth Fund, and the participants worked together to decide how to allocate the funding.[15]

A Mangyan village on Mindoro Island, Philippines, where mining interests are threatening to deforest ancestral Mangyan lands.

The deliberations lasted nearly an hour, with ideas ranging from giving money to the one community that needed it the most to dividing it in equal or unequal parts or using it to facilitate regional plans. Representatives of the Mbendjele community in the Democratic Republic of Congo were so adamant about sharing the funds equally that the Mursi representatives from Ethiopia, who thought they needed the funds the most, conceded. The decision was so simple in the end: the Mursi respected the Mbendjele and their beliefs enough to follow their lead to split the funds equally because access to funding was so limited. While many foundations view this kind of funding as the least strategic form of development, indigenous communities measure the success of a development project by its holistic, inclusive results, and they are more willing to work with foundations if they feel these and other values are being heard.[16]

Many other groups are also working to protect indigenous peoples and their assets. The Cultural Conservancy, an organization dedicated to empowering indigenous cultures in the direct application of their traditional knowledge and practices on ancestral lands, works on a variety of projects to help indigenous communities protect and revitalize native lands and cultures. One of its projects, the Native Circle of Food Program, provides educational workshops and creates urban and rural native gardens, in addition to promoting seed saving, coalition building, and public education projects to restore Native American traditional ecological and nutritional knowledge. Through its work, the Cultural Conservancy hopes to restore biodiversity within North America's food supply and to protect and enhance biological diversity in general.[17]

By providing the necessary framework—and taking a hands-off approach—these organizations and many others allow indigenous groups to take charge of projects that protect their assets. With this kind of support, indigenous peoples can work toward maintaining their economic and cultural self-determination in the twenty-first century, all while protecting the environment and preserving cultural identity.

Maintaining indigenous self-determination needs to become a collaborative effort among governments, policymakers, NGOs, private corporations, and indigenous groups themselves. The actors will vary from one country or region to another, but there are some key components that will not only help indigenous economic development but also increase food security, protect biodiversity, and create resilience to climate change.

Policies that Protect Indigenous Peoples. Giving recognition to indigenous groups, respecting their differences, and allowing them all to flourish in a truly democratic spirit can help prevent conflict. In 2010, the Republic of Congo granted indigenous peoples there (10 percent of the total population) access by law to education and health services. This law is the first of its kind on the African continent and marks a significant step in recognizing and protecting the rights of marginalized indigenous peoples worldwide. The law also mandates punishments and fines for anyone who uses indigenous persons as slaves.[18]

Policies like free, prior, and informed consent can help ensure open, noncoercive negotiations between indigenous groups and those interested in using land belonging to indigenous communities. According to Article 10 of the United Nations Declaration on the Rights of Indigenous Peoples, "Indigenous Peoples shall not be forcibly removed from their lands or territories. No relocation shall take place without the free, prior and informed consent of the indigenous peoples concerned and after agreement on just and fair compensation and, where possible, with the option of return."[19]

Corporate Engagement with Indigenous Peoples. While corporate presence in indigenous communities may be inevitable, corporations can work with indigenous groups to ensure mutually beneficial outcomes. Businesses and corporations involved in the use and extraction of natural resources on indigenous peoples' lands should consider their relations with indigenous communities a crucial part of their business practices.

Indigenous peoples are pivotal in changing corporate behavior. Indigenous groups, with the support of NGOs and other organizations, are voicing their opinions on land development and encouraging communities to set and enforce environmental standards. Corporations can reciprocate by partnering with indigenous peoples in project planning, design, and decisionmaking. Mutual benefits can be achieved, such as when corporations build local mapping centers and indigenous groups include local land uses on the maps. This gives companies the information they need for their operations and the indigenous groups the information they need for environmental monitoring, agriculture, hunting, fishing, and other practices.

Unique Grantmaking and Funding Strategies. NGOs and other organizations need to develop funding models that support and suit indigenous needs. Foundations and aid agencies in the United States often lack specific strategies for working with indigenous peoples, but if these funding initiatives can tap into the capacities and resources of the communities they are serving, their work could be far more effective.

By adhering to the cultural values of indigenous communities and adopting more holistic approaches that engage these communities effectively, foundations and aid agencies will be able to ensure that the projects they fund provide the greatest benefit to all. Through small grants, public forums, private discussions, and the transfer of research and information relevant to indigenous peoples, outside groups can change international public opinion, mobilize relevant groups to secure policy reform, and shift the focus of indigenous economic development from income maintenance to a full use and appreciation of indigenous assets and knowledge.

CHAPTER 20

Crafting a New Narrative to Support Sustainability

Dwight E. Collins, Russell M. Genet, and David Christian

In 1968—during the first manned voyage to orbit the moon—Astronaut William Anders took the famous photograph known as *Earthrise*, which graphically depicts Earth as a small oasis in a dark, cold, hostile space. Environmentalists used *Earthrise* to spread their message of the need to care for our fragile planet, and it played a pivotal role in catalyzing the great environmental campaign successes of the 1970s in the United States, such as Earth Day, the Clean Air and Clean Water Acts, and the creation of the Environmental Protection Agency.[1]

There is another more subtle message embedded in the *Earthrise* photograph. It was taken by a species able to travel beyond Earth by building a human-friendly, short-term, artificial environmental system. In both the spaceships we build and Spaceship Earth on which we live, our survival is at stake.

Finding a new set of myths and stories that remind us frequently of our dependence on planet Earth and our role as stewards is essential in this Anthropocene epoch, when humanity is having a severe impact on the biosphere—enough even to disrupt life itself. Many religions are trying to do just that, reminding their adherents of the lessons from their stories about stewardship, protecting the earth. The Judaic concept of a covenant or legal agreement between God and humanity can be extended to all creation. Christianity's focus on sacrament and incarnation can be interpreted as a lens through which one can see the entire natural world as sacred. The Islamic vice-regency concept teaches that the natural world is not owned by humans but rather given to them in trust, implying a responsibility to preserve all of creation. But modern science, too, has much to contribute to people's understanding of our beginning and our future.[2]

One story that is now known globally and understood by billions of people is the story of humanity's evolution—what E. O. Wilson, the Pulitzer Prize–winning Harvard entomologist, calls "probably the best myth we

Dwight E. Collins chairs the MBA Program at the Presidio Graduate School in San Francisco and is president of the Collins Educational Foundation. **Russell M. Genet** is an astronomer and a Research Scholar in Residence at California Polytechnic Institute in San Luis Obispo. **David Christian** is a professor of history at Macquarie University in Sydney, Australia, and primary founder of the academic discipline of Big History.

www.sustainabilitypossible.org

will ever have." This story starts 13 billion years ago with the Big Bang and continues into the future beyond *Homo sapiens* and toward new species into which even humans may evolve. But it also includes much more beyond humans and planet Earth, the "billions and billions" of stars and planets where processes similar to those here on Earth are likely taking place. What is exciting is that there are now efforts around the world to draw on this evolutionary story—which has been incorporated into an academic discipline often called Big History—to help humanity set a course to a sustainable future.[3]

NASA

Teaching Big History

Courses on Big History are now being taught in some 50 colleges and universities around the world—from Harvard University and the University of Amsterdam to the American University in Cairo and the International State University in Moscow. Big History courses offer semester-long or year-long accounts of the history of the cosmos, of life and civilization on planet Earth, and of humanity's place within the universe. These courses, by their very nature, are interdisciplinary, multiscalar, and both global and cosmic in their perspective. Often they take as their central theme the idea of increasing complexity.[4]

These courses typically begin by explaining what Big History is, often comparing it to traditional origin stories. They then launch into a narrative that begins with the Big Bang, explaining the key ideas of Big Bang cosmology in language that nonscientists can grasp. The creation of stars is the next chapter in the story. With the appearance of stars, a universe that was previously both homogenous and quite simple suddenly acquired new chemical elements and energy flows of increased intensity. The narrative then moves on to the dispersal of these new chemical elements from dying stars, a story that helps explain the appearance of chemically complex objects such as planets. Describing the creation of these new chemical elements sets up the story of planets in general and our own solar system in particular, preparing students for the history of Earth and its life.

The emergence of life seems to have been made possible by these chemically complex environments with a liquid solvent (water) and gentle energy flows that allowed the evolution of increasingly sophisticated molecules. The story of life and its evolution on Earth leads to the appearance of our own species some 200,000 years ago. Many Big History courses identify our species as distinct because of our capacity for "collective learning"—the ability to share ideas so efficiently that the information learned by individuals begins to accumulate in the collective memory from generation to generation. This creates a level of technological creativity that no other species has been able to match in the almost 4 billion years that life has existed on Earth.[5]

The final parts of the story describe the results of this collective learning. As humans learned to ever more successfully exploit their environments, they evolved ever larger, complex, populous, and energy-hungry societies. Today, in the Anthropocene epoch, for better or worse humans have acquired the power to transform the biosphere. It is natural, therefore, that Big History courses end by considering where the story is headed—the story of humans and the biosphere, and also the story of the planet, the solar system, and even the Universe as a whole.[6]

There are different schools of thought when teaching Big History. Some focus more on Earth and its origins; others, on life in the universe. But whatever way you slice it, Big History gets to some of the biggest questions of time, space, and our survival.[7]

For example, Big History raises the question of whether the history of our own species is unique. Is it possible that there have been many examples of other species beyond Earth capable of collective learning and able as a result to accumulate new technologies over many generations? Assuming such species exist, we can make some plausible generalizations about the likely shape of their histories. And these generalizations can help place our own predicament into a larger context.

It seems likely that other collective-learning species might pass through similar stages in their histories as their knowledge base and technological resources accumulate. One line of discussion hypothesizes three stages. In Stage 1, childhood, these species accumulate a growing body of knowledge about their environment. This gives them increasing power to extract resources from their environment and support ever larger and more complex communities. Barring extreme events such as asteroid strikes, they eventually reach Stage 2, adolescence. In this stage, they have accumulated so much power over their environment that they can now transform their planet, although it is not yet clear if they have the wisdom needed to use their power well. This potential mismatch of power and wisdom may create a bottleneck, difficult to pass through, and this may explain why we have not heard from other such species although we have been listening for over half a century. Is it possible that all such species are like galactic fireflies, only briefly flashing on and off, here and there? Perhaps our species has reached this adolescent phase.[8]

The primary impediment to making it through our bottleneck is the runaway success of our species. Like other species capable of collective learning, we presumably have not only the ability to fill our own niche but also, because we keep accumulating new technologies, the ability to fill and overexploit almost every niche on Earth. Through our cultural evolution, we have developed powerful machines, tapped fossil fuels, and are now rapidly transforming the biosphere. So far, other species have lacked the power or foresight to restrain us. Our cultural evolution has been too fast for their genetic evolution to counter.

Thanks to our capacity for collective learning, there is a potential pathway through the bottleneck. We can become the first species on Earth to develop the effective planet-wide evolutionary foresight we will need if we are to avoid the dangers of ecological overreach and death as a civilization.

Effective planet-wide action based on foresight is the key to a flourishing future. Science provides the foresight, while long-view narratives such as Big History can energize the public will, enabling politicians to make wise, long-term choices.[9]

The Rocinha Favela in Rio de Janeiro is one of the largest shantytowns in South America, with over 200,000 inhabitants.

In summary, from a cosmic perspective, sustainability can be seen as the requirement for civilizations of species capable of collective learning to safely negotiate their bottlenecks, to pass through their adolescent stages to Stage 3: planet-wide cooperative maturity leading to a flourishing future. The cosmic perspective presented by this Big History narrative places the question of sustainability into a nonconfrontational context. It also provides a foundation of meaning upon which we can unite and align our ethics of exploration and environmental stewardship in pursuit of a common goal: negotiating a way through our cosmic bottleneck to reach Stage 3 of our history.

Can Big History Courses Change Attitudes?

The Big History Project, founded by Bill Gates and David Christian, is bringing this curriculum into high schools by building what will eventually be a free online syllabus in Big History. A two-year pilot offering of the course began in 2011 at individual high schools in the United States. In 2012, schools from Australia, the Netherlands, Scotland, and South Korea joined the pilot.

Eventually, using feedback from these pilot high schools, the syllabus will be revised. In late 2013, it will be made freely available to high schools as well as individual learners. Systematic feedback from high schools will also provide valuable data about the capacity of such courses to change the way students think about issues such as sustainability. The eventual goal of this project is to see Big History taught in a majority of high schools throughout the world. Already Big History is catching on in high schools, colleges—with some, like Dominican University of California, even requiring all undergraduates to take this course—and even science museums.[10]

Adults may react in different ways when exposed to the Big History account. For some, it may generate an awareness that they should change their behavior. But they may need more support for change because, for example, they are caught up in the paradigm of well-being defined by the material things that surround them. Others may react by initiating a change in personal values and priorities for what has meaning out of a heightened awareness of their interconnectedness with all life. Still others may need to connect the contents of the account to their spiritual identity in order to change behavior. They may look to practices like Religious Naturalism, an approach to spirituality with a focus on the religious attributes of the universe and nature.[11]

In any case, a great deal of anecdotal evidence from many Big History courses taught at the college level over the last 20 years suggests the powerful ability of these programs to transform a student's perspectives with respect to the major global challenges of the Anthropocene epoch. Big History has the capacity to expand our vision of humanity and its trajectory just as the *Earthrise* picture changed how the first astronauts and cosmonauts viewed their home planet. Here, for instance, is the reaction of one student to a Big History course taught in the United States:

> When I was first asked to consider my role in the universe four months ago . . . I do not think I fully realized there was even a living community around me, never mind an Earth full of other humans and an entire universe beyond. . . . But after this long, incredible voyage of exploration . . . I have a newfound sense of what the universe is. I have learned . . . that we are all part of the Global Future, and I can make a difference in my life as well as the lives of others. . . . My role is now to change my ways and respect this beautiful planet that granted us life, and to get others to join me.[12]

This anecdotal evidence suggests that students learning the new narrative can change their "reality map," resulting in more-sustainable behavior. This hypothesis can be tested in a rigorous systematic way using before-and-after surveys.

Since 2009, staff from the Alliance for Climate Education (ACE) have been giving presentations on climate science to high school assemblies across the United States. Their presentations incorporate animation, music, and documentary footage of students taking on climate-related projects in their schools. In three years, ACE has engaged over 1.3 million students and won numerous awards for their innovative presentation style. Before-and-after ACE surveys have measured students' knowledge, attitude, behaviors, and intentions related to climate and energy. The results suggest that students have the potential to shift their attitudes and behavior in response to a creatively crafted message about climate science. Before an ACE assembly, 37 percent of 1,388 students surveyed passed a test on climate science; after the assembly, the pass rate rose to 56 percent. And the share of students categorized as concerned or alarmed about climate change rose 43 percent. The key seems to be presenting compelling information in an engaging format that incorporates a sense of hope and empowerment. A course in Big History, given that it is taught over several months, is likely to have an even greater impact on attitudes and behaviors than a one-time high school assembly engagement.[13]

The Future of Big History

As Spaceship Earth speeds toward the brick wall of its own planetary finiteness, Big History has great potential as a teaching vehicle to change the attitudes of its passengers about sustainability. However, a more critical need is to educate its pilots—our leaders in business and government—in Big History.

Graduate schools of management could, for instance, offer a one-semester Big History course at the beginning of their Masters of Business Administration and Public Administration (MBA/MPA) curricula. Knowledge of Big History grounds us in how to live as good citizens of Earth. Hence, this strategy could strengthen MBA/MPA programs by teaching students how to weave Earth citizenship values into the leadership cultures of public and private institutions.

A small number of graduate programs have already made substantial headway in this direction. One is the 10-year-old San Francisco-based Presidio Graduate School, which offers a dual MBA/MPA degree in sustainable management. This program integrates sustainability values and tools for conducting business and managing public institutions throughout every course in its curricula. Addressing the sustainability dimension of businesses and public policies requires students to learn how to think at a global level with a sense of the broadest impacts of decisions. The primary discipline used by the school to teach this skill is "systems thinking," developed and popularized by Jay Forrester, Donella and Dennis Meadows, and others at the Massachusetts Institute of Technology in the 1970s. It was used in connection with the discipline of system dynamics invented by Forrester and found in this team's famous work for the Club of Rome, *Limits to Growth*. Systems thinking is mathematics- and logic-based, with a focus on concepts like feedback loops and leverage points within a system.[14]

The discipline of Big History offers a complementary approach to teaching a student to think globally. The student assimilates a breadth of knowledge that by its very nature requires him or her to think from a global/cosmic perspective. Big History and systems thinking are two very different approaches to achieving similar learning outcomes. A course in Big History—with its broad opportunity for use of both cognitive and affective learning modalities—could augment a student's knowledge of systems thinking, providing the student with an even stronger sense of the interconnectedness of all things in space and time.

It remains to be seen whether or not we Earthlings will safely negotiate Spaceship Earth's bottleneck and advance from our civilization's reckless adolescence to a state of sustainable and flourishing maturity. Anecdotal evidence indicates that teaching people Big History can help on this journey. These courses educate students toward sustainable behavior by enabling them to understand the sustainability challenge in the broadest context and by deepening their understanding of what it means to be a good citizen of Earth. They teach us how to think in terms of multiple time scales and across disciplines. Offering such courses in our high schools and institutions of higher learning can provide the education that both the passengers and the pilots of Spaceship Earth need to steer a safe course through our bottleneck.

The Big History narrative gives new meaning to our journey to a state of true sustainability and flourishing. It anchors the journey's starting point, and its unified perspective serves as a constant reminder of why we are on the journey and why we should not divert from its path. This cosmic narrative was eloquently expressed by Carl Sagan when he ended the thirteenth and final episode of *Cosmos*—"Who Speaks for Earth?"— with these words: "Our loyalties are to the species and to the planet. We speak for Earth. Our obligation to survive and flourish is owed not just to ourselves but also to that Cosmos, ancient and vast, from which we spring!"[15]

CHAPTER 21

Moving Toward a Global Moral Consensus on Environmental Action

Kathleen Dean Moore and Michael P. Nelson

In the summer of 2012, some 10 percent of the earth's land baked under intense heat, a tenfold increase from baseline years. Ninety-seven percent of the surface of the Greenland ice sheet warmed enough to show signs of thawing. The temperature in the state of Kansas broke 115 degrees—an all-time record. And the *U.S. Drought Monitor* reported that 62.3 percent of the United States was suffering from moderate to extreme drought. Hot, dry weather also scorched Moscow, which was cloaked in haze from wildfires. All but 24 percent of the Arctic Ocean was ice-free that summer, the lowest point since measurements began at 50 percent in the late 1970s.[1]

Startling changes, to be sure. But along with the increases in temperature has come an important expansion in the world's understanding of the environmental emergencies that beset the planet. The waves of climate and other environmental change are scientific issues. They are also technological and economic issues. What is new and significant is an increasing awareness that environmental emergencies, especially those caused by rapid climate change, are fundamentally moral issues that call for a moral response.

The call for a response based on justice, compassion, and respect for human rights comes from scientists as well as activists and moral and religious leaders. Averting climate change, NASA scientist James Hansen says, "is a great moral issue" that he compares to the fight against slavery; it is an "injustice of one generation to others." Archbishop Emeritus Desmond Tutu writes, "Climate change is a moral challenge, not simply an economic or technological problem. We are called to honor our duties of justice. . . . We are called to honor our duties of compassion." Environmental issues are human rights issues, former Inuit Circumpolar Council Chair Sheila Watt-Cloutier writes: "We are defending our right to culture. . . . We are defending our right to be cold." And the Dalai Lama says that a "clean environment is a human right like any other. It is therefore part of our responsibility toward others to ensure that the world we pass on is as healthy, if not healthier, than when we found it."[2]

Kathleen Dean Moore is Distinguished Professor of Philosophy, School of History, Philosophy, and Religion, at Oregon State University. Michael P. Nelson is Ruth H. Spaniol Chair of Natural Resources, professor of environmental ethics and philosophy, and lead principal investigator of the H. J. Andrews Long-Term Ecological Research Program at Oregon State University.

www.sustainabilitypossible.org

The emerging global consensus about the moral implications of environmental crises is an important development, given the underlying logic of policymaking. That logic is expressed in the form of the practical moral syllogism: Any argument that reaches a conclusion about what we ought to do must have two premises. The first premise is factual, based on empirical, usually scientific, evidence—*This is the way the world is, and this is the way the world will be if it continues on this path.* But facts alone do not tell us what we ought to do. For that, we need a second premise. The second premise is normative, based on our best judgment of what is right and good, what is of value, what is just, what is worthy of us as moral beings—*This is the way the world ought to be.* From these two premises together, but from neither alone, we can devise policies that empower our values and embody our visions of the world as it ought to be.

This logic helps explain some of the impasses blocking action to avert the emergencies. It helps explain a strategy of climate change deniers, for example. Given the logic of the practical moral syllogism, individuals who would reject climate action and the changes it would require can either deny the science that supports action or deny collected human wisdom about how the world ought to be. Unsurprisingly, they choose to attack the science. It is far easier to pick a fight about, say, whether dramatically increasing levels of carbon dioxide will help or hurt humankind than to quarrel about, say, whether we have a moral obligation to protect children from harm.[3]

The logic also helps explain the frustration of scientists, who see an astonishing decoupling of scientific consensus and public belief, as well as, in some cases, an inverse correlation between the amount people know about climate change and the political will to act. Indeed, scientists have heroically expanded knowledge and explained it to the public on the assumption that if people only knew, if they only knew, then they would act. This, unfortunately, is a fallacy. Better to say, if people only knew the facts about the harmful effects of climate change on the human prospect, and if they affirmed basic principles of justice and compassion, then they would act. It is from the partnership between science and ethics that policies are born. For this reason, university departmentalization and the myriad isolations of expertise, science/religion divides, and other forces that weaken the connection between the realm of the first premise (generally science and technology) and the realm of the second premise (literature, art, religion, indigenous wisdom, ethics, history) have made it harder to devise effective policies.

Shared Moral Principles That Require Action

Hidden behind the well-publicized disagreements about climate change is a body of shared wisdom about fundamental moral principles of human and political action. Just as the world's scientists are achieving a hard-won global consensus about the facts, it is possible to move toward a global consensus about basic principles of morality. This section looks at just a few of the principles fundamental to a global moral response to climate change and other environmental crises.

Everyone has the right to life, liberty, and security of person. This basic moral principle, from the Universal Declaration of Human Rights, is echoed in constitutions around the world. If there is a fundamental, globally shared moral vision, this is it. If we accept what scientists tell us about the effects of environmental assaults, and if we accept this definition of human rights, it follows that the carbon-spewing nations are embarking on the greatest violation of human rights the world has ever seen. The consequences of global warming and widespread environmental degradation—flooding people from their homes, exposing them to new disease vectors, disrupting food supplies, contaminating or exhausting freshwater sources, uprooting the material bases of traditional cultures—are a systematic denial of human rights. By whom? By the wealthy nations and the wealthiest subpopulations of all nations, who cannot or will not stop releasing more than their fair share of carbon into the atmosphere. For what? For the continuing consumption of material goods and the accumulation of wealth. "An environmental human rights movement is the vision under which I labor," writes biologist Sandra Steingraber, "from which I am not free to desist, and which may, if we all work together, become a self-fulfilling prophecy."[4]

Justice, and intergenerational justice in particular, requires an equitable distribution of benefits and burdens. Climate change is not only a violation of rights; it is a violation of the principles of justice. The people who are suffering and will suffer the most severe harms from climate change (at least in the short term, until it engulfs us all) are unlikely ever to see the putative benefits of the profligate use of fossil fuels and natural resources. Moreover, they are the people least responsible for causing the harm. The people who are causing the harm are off-loading its consequences onto those least able to speak in their own defense. Who are the voiceless? They are future people, who do not exist and so cannot defend themselves against the profound destabilization of the world. They are plants and animals and ecosystems, destroyed wholesale to support the lifestyles of the present. They are marginalized people everywhere—economically marginalized and geographically marginalized, in sub-Saharan Africa, in the circumpolar regions, in low-lying islands, in areas of flood or drought or disease or famine. And they are children. That is a violation of distributive justice.

Humans have an absolute obligation to protect children from harm. The suffering of any child is unjust. Small children can never deserve to suffer, because they can never do a wrong that might justify suffering in return. But adults are harming children, even as (especially as) we believe we are acting to provide for them. It is ironic that the amassing of material wealth in the name of very privileged children will harm them in time. Consider the poison in the plastic car seat, the disease in the pesticide-treated fruit, the coal company in the college investment portfolio, the mall where there had been frogs, the carbon load of a distant summer camp. But the harm that adult decisions will do to the children who are not as privileged is not just an irony; it is a violation of our obligation to protect them. The world's less privileged children are the ones who will suffer the most as seas rise, fires scorch cropland, diseases spread north, and famine returns to lands that had been abundant. At this point in history, few can claim the excuse of ignorance. Few can claim they are acting unintentionally. The damage to children's future is a deliberate theft. "This is not the future I want for my daughters," President Barack Obama has said. "It's not the future any of us want for our children."[5]

We have an obligation as moral beings to act with compassion. Of all the virtues that a human being can possess, the greatest may be compassion. Compassion: to "feel with," to imagine ourselves in another's place. Understanding the joys or sufferings of others, the compassionate person is joyous or suffers too. Thus the truly compassionate person strives to create conditions that bring forth joy and to prevent or diminish conditions that create pain. But the price of the accelerating use of fossil fuels and the waste of natural thriving will be paid in human and animal suffering. If virtuous people are compassionate, if compassionate people act to reduce suffering, and if climate change will cause suffering around the world, then we who call ourselves virtuous have a moral obligation to avert the effects of the coming storms.

It is wrong to wreck the world. "A thing is right when it tends to preserve the integrity, stability, and beauty of the biotic community," conservationist and ecologist Aldo Leopold wrote. "It is wrong when it tends otherwise." By this principle, the waste and spoilage that cause climate change are wrong. The timeless unfurling of the universe, or the glory of God, or an unknown mystery, or all of these together have brought the Earth to a glorious fecundity, resilience, and beauty. To let it all slip away because we are too preoccupied to save it? That is wrong. And when the destruction is done knowingly and in exchange for something of far lesser value, this is immorality at its most incomprehensible. A full appreciation of the beauty and wonder of the world calls us to action. If this is the way the world is—beautiful, astonishing, wondrous, awe-inspiring—then this is how we ought to act in that world: with respect, with deep caring and fierce protectiveness, and with a full sense of our obligation to the future, that this world shall remain.[6]

Moral integrity requires us to make decisions that embody our values. It is possible to believe the world is trapped between two unacceptable alternatives. One is the moral complacency that comes from blind hope. The other is the moral abdication that comes from blinding despair. Certainly, there is good reason for despair. Vermont Law School professor Gus Speth wrote, "All we have to do to destroy the planet's climate and ecosystem and leave a ruined world to our children and grandchildren is to keep doing exactly what we are doing today."[7]

But to think that hope and despair are the only two options is a false dichotomy. Between them is a vast and fertile middle ground, which is integrity: a matching between what we believe and what we do. To act justly because we believe in justice. To act lovingly toward children because we love them. To refuse to allow corporations to make us into instruments of destruction because we believe it is wrong to wreck the world. This is moral integrity. This is a fundamental moral obligation—to act in ways that are consistent with our beliefs about what is right. And this is a fundamental moral challenge—to make our lives into works of art that embody our deepest values.

A Competing Moral Value that Blocks Climate Action

Even as consensus grows on the moral necessity of climate action, disagreement grows as to the proper steps to take. A substantial minority of the U.S. populace, for example, believes that the steps required to combat climate change are wrong, primarily because they limit personal freedom. It is surely correct that effective climate action will increase social constraints. It will require limiting the freedom of commerce, limiting the freedom of consumer choices, and, in a variety of ways, limiting the freedom of some to benefit at the expense of others. Climate policy disputes are one manifestation of a division between those who think the primary purpose of government is to bring people to common action, so they can do together what none of them can do alone, and those who think the primary purpose of government is to protect individual freedom of self-development and self-realization.[8]

Either way, freedom has value as a means to the ends people seek. That value raises a paradox of unsurpassed importance: If unfettered freedom unleashes a climate chaos that threatens to undermine the great systems that sustain our lives and nations, then what will be left of freedom? What the world faces is a choice between social constraints democratically chosen and the fierce, uncontrollable, lethally unleashed constraints of flood, fire, and the societal chaos that will accompany rapid ecological changes. (See Box 21–1.)[9]

From Moral Imperative to Moral Action

Work is advancing on many fronts to harness the power of moral conviction in efforts to slow climate destabilization and ecological disruption. Moral arguments about climate change do not have to be abstract and complex; recent scholarship suggests powerful new frames for moral arguments. Ac-

Box 21–1. Ethics at the End of the World

It is possible that planetary civilization will move smoothly into the future through prudence and grace, with all its ethical wisdom intact. But what if we fall hard into a future marked by chaos, scarcity, and calamity? What of ethics then?

Moviemakers like to portray a post-apocalyptic world as post-moral—solitary, poor, nasty, brutish, and short—governed by animal instincts unrestrained by human decency. It is certainly a possible scenario, and even a probable one if we fail to act to prevent global average temperature increases from reaching high-end projections of 6 degrees Celsius. But of course this Hobbesian future is not the only scenario. It is possible that ethics will not disappear but will change. Among the expected casualties of ecological collapse may be those parts of western ethics-as-usual that have not served us well. In a world in which there are few good consequences to be found, for example, we might see the end of utilitarianism, which judges the morality of acts by the desirability of their consequences. We might see as well the end of egoism or radical individualism, as ecological collapse forces us finally to accept that we humans are created and defined by our relation to cultural and ecological communities—that we flourish not as isolated utility-maximizers but as members of communities of interdependent parts.

What will replace the ethics that no longer serve us well? When we study terrible times (concentration camps, wars, the forced relocations of Native Americans, and many more examples), we most often see moral behavior based on personal integrity, by which people choose to do what is right for no other reason than because it is right. To act justly because we believe in justice. To act compassionately because we believe in compassion. "When we are no longer able to change a situation," wrote Austrian psychiatrist and Holocaust survivor Viktor Frankl, "we are challenged to change ourselves." This may be the one choice remaining to us even in the darkest futures we can imagine: "Everything can be taken from a man but one thing: the last of the human freedoms—to choose one's attitude in any given set of circumstances, to choose one's own way," Frankl noted. Making difficult choices, helping others get through the demanding and grim ecological transitions of the future—these may be true acts of moral courage. But the fact is, we have the opportunity to be morally courageous right now, choosing to match our actions to our beliefs about what is right and good, just and beautiful, worthy of us as moral beings.

Source: See endnote 9.

cordingly, the world is now seeing strong, innovative moral climate change initiatives based on moral rights, conscientious objection, and religious conviction, to name a few, and new efforts to reimagine ethics as well as the institutions that embed moral values.[10]

Moral Rights. The Earth Charter in 2000 was the first global effort to expand moral consideration to the earth. It called for "respect for the Earth and life in all its diversity," recognizing that "every form of life has value regardless of its worth to human beings." Since then, many nations have formally granted moral standing and legal rights to the earth. Ecuador declared in 2008 that Nature has the "right to exist, persist, maintain and regenerate its vital cycles, structure, functions and its processes in evolution." In *La Ley de Derechos de la Madre Tierra* (the Law of the Rights of Mother Earth), Bolivia defined 11 rights for the environment in 2011, including "the right to life and to exist; the right to continue vital cycles and processes free from human alteration; the right to pure water and clean air; the right to balance; the right not to be polluted; and the right to not have cellular structure modified or genetically altered."[11]

These laws have the important effect of changing the burden of proof, so that anyone who would do harm to the earth must provide good reasons why this is justified. But efforts to encode obligations to the earth do not stop there. For example, a campaign is under way in Britain to make "ecocide" an international crime comparable to genocide and likewise actionable as a fifth "crime against peace" that can be tried by the International Criminal Court.[12]

Conscientious Action. The world is seeing an increase in direct action or civil disobedience that is guided by moral integrity—the refusal to acquiesce passively in actions believed wrong. For example, 12,000 people surrounded the White House in November 2011 to push President Obama to keep his campaign promise to "end the tyranny of oil." More than 200 were arrested, including event organizer Bill McKibben, who wrote, "This is, at bottom, a moral issue." In Sydney, Australia, a crowd of 10,000 cheered Climate Project coordinator Nell Schofield when she decried the government's lack of action as "not only embarrassing, . . . [but] morally reprehensible." Around the world, thousands have been arrested in demonstrations against fracking, mountaintop removal, open-pit mines, and other particularly destructive industrial practices.[13]

In July 2012, the first-ever nationwide anti-fracking rally in Washington, D.C., demonstrated the increasing solidarity of secular and religious environmental activists. Catherine Woodiwiss of the Center for American Progress noted that the protests were "couched in sweeping moral language—an example of the increasingly values-based lens being applied to public discourse about climate change and green energy technology."[14]

Faith-based Action. A growing number of religious denominations and leaders continue to move into the world of environmental activism, driven by a sense of moral responsibility to address human injustice, to relieve human suffering, and to serve their Creator as stewards of divine creation. In the past year, religion-based campaigns included a Global Day of Prayer for Creation Care organized by the Evangelical Environmental Network, with presentations by evangelical leaders from the United States, Europe, Latin America, and Africa. Interfaith Moral Action on Climate, a newly formed collaborative endorsed by 45 groups and scores of religious leaders, sponsored a Cultural Implications of Climate Change program with talks by leaders from Christian, Islamic, Jewish, Baha'i, Hindu, and Native American faith traditions. To traditional religious concerns of social justice and compassion, these initiatives bring a powerful commitment to "creation care," the obligation to protect divine creation and to honor Nature—a spiritual imperative especially strong in indigenous religions, Taoism, Confucianism, and Buddhism.[15]

Activists deliver petitions with 160,000 signatures to ban fracking to New York Governor Cuomo's office in October 2012.

Reimagining Ethics. Evolutionary science, ecological science, and almost all the religious and spiritual traditions of the world tell us that human/nature dualism and human exceptionalism are fundamentally mistaken; rather, humans are deeply of the earth, embedded in emergent systems that are interconnected, interdependent, finite, and beautiful. Recognizing that a truly adaptive civilization will align its ethics with the ways of the earth, a number of organizations are articulating or calling for an earth-based ethic to replace anthropocentric utilitarianism, which measures acts by their usefulness to human ends. An example of such an ethic is the Blue River Declaration, written by an interdisciplinary seminar convened by the Spring Creek Project in Oregon's Cascade Mountains in 2011. The authors concluded: "Humanity is called to imagine an ethic that not only acknowledges, but emulates, the ways by which life thrives on Earth. How do we act, when we truly understand that we live in complete dependence on an Earth that is interconnected, interdependent, finite, and resilient?"[16]

Reimagining Institutions. An ethic of care for the earth calls into question many of the institutions of "business-as-usual," including the corporation. Traditional corporations maximize for one and only one value: shareholder profits. So far, 12 states have passed legislation to create a new kind of corporation, called the B-corporation—the "B" standing for benefit. B-corporations integrate social benefit directly into the missions and charters of their businesses, offering if not a moral shift, at least a moral promise. By November 2012 there were 650 B-corporations in 60 industries in 18 countries, with a combined worth of $4.2 billion.[17]

A Paradigm Shift in Worldviews

Along with these moral responses to climate change comes the call for a Great Turning, as Joanna Macy puts it, toward a paradigm shift in worldview, away from the conviction that humans are separate from and superior to the rest of creation. Humans are part of this world, fully and deeply nested into intricate, delicately balanced systems of living and dying that have created a richness of life greater than the planet has ever seen. In our common origins and in our common fates, in the interdependence of our functioning, we and the rest of the natural world are kin. Because we are part of the earth's systems, humans are utterly dependent on their resilience and thriving. How soon we grasp that reality will determine not only our ecological and social future but our moral future as well.[18]

Pathways to Sustainability: Building Political Strategies

Melissa Leach

In Rio de Janeiro in June 2012, two vast political gatherings deliberated about the future of sustainability. On the campus of RioCentro, heads of state, ministerial representatives, and other national delegates sat in conference chambers and roundtable rooms at the United Nations Conference on Sustainable Development, attempting to negotiate formal agreements on sustainable development. Across the city, in Flamengo Park, civil society and citizens' groups struck a sharp contrast at the People's Summit—with an impassioned festival atmosphere of tent talks, demonstrations, and participatory events. Agendas ranged from agroecological farming to alternative currencies, renewable energy to recycling, and the rights to land, water, reproductive choice, and alternative ways of living with nature.[1]

The political strategies and styles on display could not have been more different. They exemplified contrasting approaches to the fractured politics of sustainability: global versus grassroots, top-down versus bottom-up, state-led versus citizen-led, formal versus informal. Cross-cutting these were distinctions between dominant "reformist" approaches, seeking sustainability through tweaks to existing social and economic systems under the current rubric of "green economies," and more-marginal "radical" arguments that sustainability requires more-fundamental overhauls of social and economic systems, whether based on anti-capitalist or socialist principles or on alternative eco-philosophies.[2]

Both gatherings made it clear that sustainability is not primarily a technical challenge. It is fundamentally a matter of politics. What political strategies are needed to break the political logjam? Sustainability is not just one thing, and there is a need to recognize the multiple sustainability goals and possible futures given priority by different people and groups and across scales, as well as the disputes and trade-offs among them. The challenge is thus to open up the politics of sustainability to recognize and enable negotiation among different possible pathways.

Melissa Leach is a social anthropologist and Professorial Fellow at the Institute of Development Studies, University of Sussex, United Kingdom. She directs the ESRC STEPS (Social, Technological and Environmental Pathways to Sustainability) Centre, an interdisciplinary research and policy engagement organization with partners in Africa, Asia, and Latin America.

www.sustainabilitypossible.org

Pathways and Politics

Pathways of change toward sustainability must steer us toward a safe ecological and economic operating space for humanity, as well as toward a social space that respects basic standards of human dignity, well-being, and rights. This challenge is inherently political, requiring the recognition and realignment of the political-economic interests, institutions, and power relations that constrain us to well-worn pathways. Examples of such pathways include fossil-fueled energy regimes that have developed along with incumbent political interests, patterns of economic activity, and established technologies and infrastructures in both older and newly industrializing countries and the heavily industrialized agriculture and high meat consumption that threaten biodiversity, land, and freshwater use and that are interlocked with the political-economic interests of the food industry and the lifestyles and preferences of many consumers.[3]

Yet the challenges do not stop there. Even agreeing on the general need to move toward sustainability leaves us facing a multiplicity of diverse possible goals and related pathways. In global, national, and local settings, there are inevitably contested versions of sustainability and "sustainable development," implying different winners and losers. These specificities were glossed over in the 1987 definition of sustainable development by the World Commission on Environment and Development, and they are equally downplayed in current debates around "the future we want." Seeking "true sustainability" requires addressing far more precisely who exactly "we" are in different contexts and whose needs and goals are at stake.[4]

To consider just one example, take the challenge of combating hunger in various rural settings across the world. Does sustainable development mean improving food security through boosting agricultural productivity, using modern plant breeding and genetic engineering to roll out technical solutions at scale? Or does it mean tackling diverse local food insecurities shaped by ecological, market, social, and institutional contexts through farmer-participatory approaches? Or some other approach not yet developed?

The same abundance of choices arises with respect to energy, water, and many other sustainability challenges. Of course, these are not clear-cut either-ors. What might work, or be desirable, will vary from place to place and for different groups of people. And keeping open a diversity of policy, technology, and economic options and approaches is itself desirable. Given the complexities and uncertainties surrounding so many environmental and economic processes, it makes sense to avoid putting all our eggs in one basket. Diversity of possible pathways also allows for decisionmakers and users to select, adapt, and innovate creatively to suit what inevitably are highly diverse contexts and values. The point, though, is that not all pathways can be pursued; there are always going to be trade-offs between and controversies about alternatives. Politics and power are thus critical at this level, too, in shaping which possible versions of sustainable development are recognized and how these disputes play out in global, national, and local settings.

This means that the challenge for sustainability politics is not just to attempt a shift or a reorientation from unsustainable pathways to sustainable ones, as if this were about redirecting a super-highway. And it is not just about building support for top-down, singular policy, technological, and economic approaches to sustainable development of the kind that have dominated so much debate and attempted action. The challenge is also to open up understanding and action around sustainability to reveal and empower alternative pathways that might currently be hidden, including those that emerge from the experiences, knowledge, and creativity of poorer women and men, rural and urban dwellers, and citizens and small businesses in particular places.

How might this be done? There are no simple answers. Four practical ways forward are offered here: deliberating goals, mobilizing citizens, building networks, and exploiting openings in political and policy structures. Political strategies and actions along these lines are already unfolding around the world and offering valuable lessons, guidelines, and clues for those seeking transformative change. Taken together, these four strategies offer ways of bridging and connecting top-down and bottom-up as well as reformist and radical approaches.

Deliberating Goals. Strategies for deliberative governance aim to bring diverse people and perspectives together into forums for debate, dialogue, negotiation, and engagement around particular problems. These in turn draw on ideas of direct and participatory democratic politics, in which people with a stake in an issue engage directly in forums where it will be debated or decided rather than just through voting for political candidates to represent them. Giving voice to alternative perspectives that may point in sustainable directions is, in itself, a way to counter lock-in to singular, dominant pathways.[5]

There are many examples of such deliberative approaches convened by governments, nongovernmental organizations (NGOs), or researchers and linked with an array of practical tools and methods. Many have a local focus. Community trade-off assessments have been pioneered in Guyana, for example, in which local community members assess different sustainable development options in terms of their own worldviews and aspirations. In India, citizens' juries have been used to open up discussion of genetically modified crops and sustainability among farmers, businesses, and political leaders. Other examples aim to link local perspectives with national actors and policies. Thus, for instance, multicriteria mapping (MCM) methods have been used effectively to generate debate about different goals and pathways for agricultural development in dryland Kenya in the context of climate change. (See Box 22–1.) "The pyramid" is a deliberative framework and approach that has been used to promote participatory dialogue and target setting in forestry policy at the national level in Brazil and elsewhere.[6]

Deliberative dialogues have also been attempted at the global scale. For several years starting in 2003, for instance, the International Assessment of Agricultural Knowledge, Science and Technology for Development had over 900 contributors from across the world discussing possible futures for agricultural development. The process had some success in opening up what had been a rather black-and-white debate about the merits of high-tech modern biotechnology and market-based solutions, highlighting the need for varied social and technical approaches suited to different socioeconomic and agroecological conditions.[7]

And in 2012 an innovative attempt was made to enrich the Rio+20 Conference through a process to include civil society perspectives and priorities. The Rio+20 Dialogues for Sustainable Development, initiated by the government of Brazil and supported by the United Nations, involved a multistage process of online discussion; selection and open online voting on 10

Box 22–1. Multicriteria Mapping of Agricultural Pathways in Dryland Kenya

Four out of five people in Kenya rely on agriculture. There is a virtual "lock-in" to maize—the region's culturally and politically valued staple crop—as the dominant pathway to food security. Amid growing concern with climate change in Sakai, a semiarid and risk-prone area, a Kenyan and British research team facilitated a deliberative process using multicriteria mapping to identify and explore how farmers might better deal with the challenges posed by frequent droughts. Farmers identified nine possible pathways, differentiated according to whether they depended on high or low levels of external inputs, such as commercially bought seeds, fertilizers, and irrigation, and the farmers' respective focus on maize or on other crops such as sorghum, cassava, vegetables, or tree fruits. Using the MCM tool, different groups—including richer and poorer farmers, crop researchers, policymakers, extension workers, and executives in commercial seed companies—appraised these different pathways. The MCM software package helped stakeholders to identify criteria of their own

choosing; to score each pathway numerically against all criteria, providing both "optimistic" and "pessimistic" scores; and to weight the relative importance they attached to each criterion. The MCM tool then provided the stakeholders with a graphic representation of their comparative assessments of all the pathways. This provided a powerful basis for debate and discussion about the ways they had scored each pathway and their underlying reasoning.

The MCM exercise revealed the interests of many poorer and women farmers, especially, in diversification into non-maize crops. But it also revealed farmers' concerns and uncertainties about their ability to sell different produce, as well as the strong political-economic interests of agricultural researchers and seed companies in a continued focus on maize. By making these interests and ambiguities explicit, the MCM-assisted deliberation paved the way for better-informed and more-inclusive dialogue about policy options.

Source: See endnote 6.

recommendations; a live discussion at Rio Centro that involved further recommendations from expert panels, public discussion, and a vote; and presentation of the recommendations to a roundtable of leaders gathered for the high-level segment of the Rio+20 Conference. Unfortunately, although more than 63,000 people from 193 countries cast nearly 1.4 million votes, the ballot was on recommendations that had been watered down through the Internet-mediated process to an almost meaningless level of generality—and with no compulsion for those leading the intergovernmental dialogue to respond.[8]

Whatever the setting or scale, experience with such approaches to deliberating goals suggests a range of lessons and challenges. Politics and power relations often pervade deliberative processes themselves, making it vital to attend carefully to who has framed the agenda. Which issues and angles are included and which are off-limits? Who is represented and who is not? Which voices dominate the dialogue and which remain marginal? Facilitating deliberative dialogues involves negotiating such relations, balancing the needs of different participants, remaining as open and inclusive as possible, recognizing conflict and dissent as valid contributions, and encouraging learning.[9]

There is value in recognizing diversity and making conflicts and trade-offs explicit rather than acceding to an apparent consensus view that in some cases might merely represent the interests of the contextually powerful and in others may be a lowest common denominator that loses the richness and sharpness of participants' views. In the Rio Dialogues, for instance, the knowledge and ideas captured through the online process were both more radical and more detailed and specific than the handful of final recommendations.

A related challenge concerns whether such deliberation over goals is actually allowed to shape wider political or policy processes. Despite the innovative opening up of the Rio Dialogues, for example, the intergovernmental process was not geared up to receive the resulting recommendations. In some cases governments have convened public participation processes only to ignore inconvenient outcomes that challenged established policy directions. Policy processes must be opened up in order to profit from the plurality of views. Involving decisionmakers themselves in deliberative approaches can help by getting them to engage with other stakeholders.

Mobilizing Citizens. Deliberating goals may play a role in directing and opening up alternative pathways to sustainability. But especially where political and economic positions are entrenched and power relations are deeply unequal, this will not be enough. There are many examples of citizens expressing themselves around sustainability more spontaneously, linked with action and activism of various kinds. Such active citizen mobilization suggests further crucial political strategies in directing and opening up pathways to sustainability.

As many of the Rio People's Summit events showed, citizen mobilization is not always geared to building consensus. It can also involve dissent, protest, and resistance against state, global, or business interests. Such antagonistic counterpolitics is an important complement to argumentation, deliberation, and reasoning, and it can be crucial both in getting new issues and directions onto political agendas and in seeing them through.

For example, water issues in India have generated many examples of activism and mobilization. Large dams and river-linking systems have often been undertaken there by government and industry, with international backing, as large-scale technological "solutions" to assumed problems of water scarcity (and now in response to the need for low-carbon hydroelectric energy systems). These have long been a focus of mobilization and protest. Anti-dam movements such as the Save the Narmada Movement globally projected citizens' concerns about the loss of forest-based livelihoods

The Sardar Sarovar Dam on the Narmada River in India.

and cultural values threatened by upstream flooding, about whether India's Sardar Sarovar Dam would really resolve the downstream water problems of local farmers and pastoralists, and about the elite industrial and political interests perceived to drive large dam approaches.[10]

Linking up with similar movements across the world, the Narmada mobilization helped to provoke a wave of questioning (for instance, in the report and guidelines issued by the World Commission on Dams) around the appropriateness of large-scale engineering technologies compared with approaches that are better attuned to local ecological and social conditions. More recently, while the life-and-death struggle for villagers faced with submergence by the Sardar Sarovar Dam continues, mobilization and protests around water in India, as elsewhere, have come to focus more on the problems of large-scale privatization of water resources and "water grabbing"— another blanket solution to so-called problems of scarcity that threatens to ride roughshod over the rights and concerns of marginalized people.[11]

Activism relevant to sustainability can be motivated and held together by quite diverse concerns that are not always labeled "environmental." It may reflect shared struggles for livelihoods and justice, as in the dams example, or struggles for sociocultural autonomy and identity, as in many indigenous peoples' movements around the world. Or it may reflect frustration with the perversities and injustices of dominant political-economic systems, in which their (un)sustainability is only one concern. The Occupy movement in many countries following the financial crisis of 2008–09, protesting the inequity of global and national economic orders, is an example.[12]

Movements often draw together people of diverse backgrounds and positions who coalesce around a particular issue and moment. Contemporary forms of sustainability activism are not directed just at governments and corporations but also at regional and global arenas and agencies such as the World Bank and International Monetary Fund and, as the Occupy movement shows, the networks of powerful actors who steer dominant political, economic, and environmental pathways. Citizen mobilization also involves a wide range of political styles and tactics—from face-to-face demonstrations, marches, and sit-ins to media campaigns, claims through legal channels, and the use of online forums and social media. The most successful mobilizations have often combined tactics in shifting combinations, gearing them to unfolding political processes.[13]

While mobilization often starts locally and retains local roots, in this Age of the Internet it increasingly also links participants in many local sites into global movements. Some become formalized, such as the international peasant movement La Via Campesina, which links land rights activist groups across the world and has campaigned successfully for the introduction of voluntary guidelines to regulate global land deals. Events such as the World Social Forum or the People's Summit in Rio offer venues in which local movements can build their connections and find common ground. Such "globalization from below" is particularly significant for sustainability issues, which have both global and local manifestations.[14]

Building Networks. Multiple actors and institutions—governments, businesses, civil society groupings, and international agencies—have long been involved in making and implementing sustainability-related policy and political decisions. Increasingly, state power has diminished and altered with the rise of public-private partnerships, market actors, and new mechanisms—from financial instruments to green corporate accounting and ecosystem service payments. The disappointing outcomes of Rio's multilateral negotiations are intimately linked with these developments. They might be lamented as a political crisis for sustainability insofar as governments, which are at least formally accountable downward to their citizens and upward to agreed global regulations, are losing their power—to be replaced by an unaccountable world of green wheeling and dealing. But the move to networked governance also opens up new opportunities for political strategies in resteering and building pathways to sustainability. If it is networks that now steer politics and policy, then sustainability strategies need to first understand how they operate and then identify and build alternative networks to influence or counter them.[15]

For example, interactions among ministries of agriculture, seed companies, agro-dealers, and NGOs have emerged as central to the shaping of agricultural policies in many African settings. Equally, new networks linking electricity supply companies with government agencies and consumer groups have helped steer policy in the energy sectors of many countries. Such networks often operate across national borders and across spatial scales; indeed, multilevel approaches to politics and governance are particularly significant for environmental problems whose causes and manifestations so often cut across local and global levels. Multiscale networks have emerged particularly fast in the climate and energy realm. Climate policy and politics now involve international institutions; carbon-market arrangements; nongovernmental, civic, and business groups; national ministries; technical agencies and supply firms; and formal and informal consumer institutions.[16]

Where powerful networks are supporting unsustainable pathways, political strategies may be geared toward undermining them or influencing them to bring about change. Likewise, alternative networks may be built up to counteract dominant ones or support alternative political or policy ideas. Understanding where the power lies—knowing which actors and institutions are important, understanding the jostling of positions and interests at global, national, and local levels, and tracing the connections between them—helps to identify who to target, where, and with what sorts of message. Experience points to the importance of informal "shadow networks" (such as the networks of scientists, activists, and local people who have made the case for adaptive river basin management in Southeast Asia) and their coordinated efforts to develop alternatives, build the case for them, and identify and exploit political opportunities.[17]

Exploiting Openings. Can alternative ideas and options for pathways to sustainability, and for generating support and momentum for these through citizen mobilization and network-building, trigger the required shifts in political-economic and policy direction? Sometimes current structures and regimes are too deeply entrenched, too powerful and resilient, for change to happen just in response to a push from outside. In these circumstances, crisis can create opportunity. Breakages or openings in existing structures can provide political windows for new ideas and network positions.

Effective leveraging of policy or political change demands an aptitude for seizing particular policy opportunities as they arise. Such opportunities may be triggered by acknowledged crises in the management of a particular issue. To take one example, the Florida Everglades in the United States underwent four transformations in management during the twentieth century as changing conditions triggered successive crises and new management needs to control unwanted floodwater, sustain the water supply for a growing population, control the nutrients associated with land-use interactions, and then begin restoring the ecosystem.[18]

Opportunities may also be triggered by wider political transitions and changes, for instance by an election or civil conflict that brings in a new government. In a number of countries the financial crises since 2008 have been seen as an opportunity for fundamental challenges to economic orders. Movements and coalitions advocating new approaches to green, service, and employment-oriented economies have actively sought to insert their arguments into this political window. But the opening has been constrained by the ability of dominant banking and financial infrastructures and interests to bounce back and reassert their power. Nor is there any guarantee that policy reforms and transformations enacted in moments of opening will necessarily stick. Even legislation can be undone. Attention therefore also needs to be given to the conditions that make shifts politically durable. This in turn requires strategies and approaches that build up networks and critical masses of public support once a change has happened, to ensure that newly established pathways to sustainability continue to build strength and momentum.

Toward Transformative Change

The political challenge of building pathways to sustainability is urgent. It involves both realigning current pathways toward a safe and socially just operating space and opening up sustainability politics to facilitate debate and negotiation. Without such an opening up, sustainability politics and policies risk imposing blanket targets and "solutions" that do not fit real, diverse ecological and social contexts, and over time they will simply fail or provoke resistance.

State-based and multilateral politics still have key roles to play in negotiating pathways to sustainability, but they need to be reinforced and complemented by the political strategies just described of deliberation, citizen mobilization, network-building, and exploitation of political openings. Each of these clusters of strategies transcends distinctions between reformist and radical approaches. Identifying and pursuing alternative pathways to sustainability will involve both approaches in different measures and combinations, depending on the issue and context.

These strategies also connect people and places across local, national, and global scales, blurring distinctions between global and grassroots action. Increasingly, sustainability politics must connect bottom-up with top-down and be concerned not just with the allocation of material resources, ecological space, status, and authority but also with who defines the future and what perspectives and experiences matter. Opening up sustainability is about cultivating a wider breadth of knowledge and experience to define goals and appropriate ways of reaching them, enabling the diversity that is required to respect different ecological and social contexts and to keep options open in the face of the unexpected.

Political contexts also matter. Political histories, cultures, and styles of decisionmaking vary between nations, regions, and localities and around particular issues—shaping which political strategies and combinations are feasible and desirable. A diversity of strategies and styles will therefore be needed, adapted to issues and settings, from within the repertoire laid out here of deliberating goals, mobilizing citizens, building networks, and exploiting openings. With these strategic options, we will be better equipped to meet the major political challenge of building a future we can all want, a future that keeps humanity within a safe and just operating space while striving for inclusive processes that recognize the diverse sustainable futures that people do not just want but need.

CHAPTER 23

Moving from Individual Change to Societal Change

Annie Leonard

In one of the most iconic ads of the twentieth century, a Native American (actually, it was an Italian dressed up as a Native American) canoes through a river strewn with trash. He disembarks and walks along the shore as the passenger in a car driving past throws a bag of litter out the window. As the camera zooms in to a single tear rolling down his cheek, the narrator announces, "People start pollution. People can stop it."[1]

This 1971 ad, just a year after the first national Earth Day celebration, had a huge impact on a generation awakening to environmental concerns. Children and young adults watched it over and over, shared the faux-Indian's grief, and vowed to make changes in their individual lives to stop pollution. That response was exactly what the ad's creators hoped for: individual action. For the ad was produced not by a campaign to protect the environment but by a campaign to protect the garbage-makers themselves.

In 1953, a number of companies involved in making and selling disposable beverage containers created a front group that they maintain to this day, called Keep America Beautiful (KAB). Since the beginning, KAB has worked diligently to ensure that waste was seen as a problem solved by improved individual responsibility, not stricter regulations or bottle bills. It even coined the term "litterbug" to identify the culprit—individuals. By spreading slogans like "people start pollution, people can stop it," KAB effectively shifted attention away from those who design, produce, market, and profit from all those single-use disposable bottles and cans that were ending up in rivers and on roadsides. As part of this effort, KAB created the infamous "crying Indian" ad against litter.[2]

It worked. Over the last few decades, the theme of the individual's role in wrecking the environment, and the individual's responsibility in fixing it, has only grown stronger—driven not just by KAB but by hundreds of businesses, by the government, even by well-meaning individuals and organizations. Today, lists of "10 simple things you can do to save the envi-

Annie Leonard is the host and author of the Internet film and book, *The Story of Stuff*, and codirector of The Story of Stuff Project.

www.sustainabilitypossible.org

ronment" abound. The *Lazy Environmentalist* website will send you regular emails with tips on greening your shopping and household maintenance, implying that we really can save the environment without even breaking a sweat. Recyclebank, which is sponsored by Coca-Cola, rewards individuals for increasing their use and recycling of single-use beverage containers and other packaging. Participants who throw more single-use containers into the recycling bin are rewarded with more points—points that can be used to go shopping.[3]

Picking up litter, carrying reusable bags to the store, biking instead of driving—all these are good things to do and there are many reasons to do them. They demonstrate our concern to those around us, hopefully providing inspiration and social proof for friends and neighbors to follow our lead. Greening our small daily acts brings into alignment our values and our actions, which feels good. As political science professor Michael Maniates says, "Small, everyday acts of green consumption are important moments of 'mindful living': they serve as daily reminders of our values, and of the larger struggles before us. But these individual actions are puny when compared to the challenges before us, and can't achieve the kind of change we desperately need today." As explained in *The Story of Change*, the latest Internet film by The Story of Stuff Project, these small actions are a fine place to start. But they are a terrible place to stop.[4]

The Behavior-Impact Gap

Even if we could convince everyone to make all the adjustments advocated by the *Lazy Environmentalist* or the "10 simple things" lists, it simply would not significantly change our environmental trajectory—which is headed toward an ecological cliff. Maria Csutora of Corvins University in Budapest has studied the gap between pro-environment attitudes and behaviors and actual environmental impacts, a problem she calls the Behavior-Impact Gap, or BIG, problem. (See Figure 23–1.) The BIG problem occurs when green-oriented behavior change is adopted with the expectation of making change, but little or no positive environmental impact follows.[5]

Csutora explains that the "BIG problem means that even when consumers act in an environmentally aware manner, their carbon footprint or ecological footprint may improve only slightly, if at all. Wishful thinking about prospective gains from pro-environmental behavior is common, which is actually more a policy-making problem than a consumer behavior problem." The result, in Csutora's words, is that "environmental actions may serve as green means for relieving our guilty ecological consciences without actually or genuinely reducing impacts."[6]

There are many theories as to why the BIG problem exists. Some scientists attribute the lack of meaningful impact of all these green activities to

Figure 23–1. The Behavior-Impact Gap (BIG) Problem

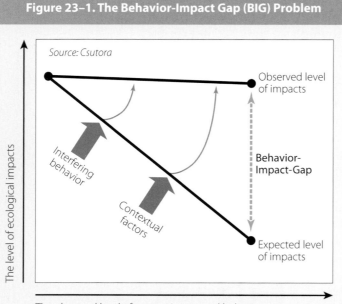

Source: Csutora

The level of ecological impacts

Observed level
of impacts

Interfering
behavior

Contextual
factors

Behavior-
Impact-Gap

Expected level
of impacts

The observed level of pro-environmental behavior

the rebound effect: our tendency to increase our use of more-efficient appliances. The most common example of this is the driver who gets a new hybrid car, doubling his gas mileage, but then ends up doubling the miles driven in part because driving is relatively cheaper, cancelling out the benefit. Or the urban dweller who, able to live a car-free lifestyle, uses the thousands of dollars she saves each year from not owning a car to take an exotic far-off vacation, burning more carbon in one week than she would have in an entire year of driving.

Others point out that individuals may think they are engaging in pro-environment behavior, such as buying shampoo with the terms "natural" or "organic" on the label, when in reality the products they buy do not differ in environmental impacts from conventional products. Or people may decrease one environmentally destructive behavior with good intentions, only to offset the gains by increasing a different and more destructive activity. An example of this is the individual who decreases meat consumption out of environmental concern, only to then increase consumption of imported nuts that may have a greater carbon footprint than local meat.

Unfortunately, even if we overcome the rebound effect, if we really do decrease our driving, stop littering, and refuse plastic carry bags—which are all good things to do—the broader impacts are still negligible, since day-to-day individual actions do not contribute the bulk of today's environmental harm.

Take garbage. Many conscientious householders are going to extremes to reduce their household garbage generation. A number of "Zero Waste" families have been profiled in the popular press after reducing their annual household garbage production to a single bag.[7]

Reducing waste in our daily lives is surely a good thing to do. Recycling reduces household waste sent to landfills and incinerators and creates jobs. The catch is that the garbage coming out of U.S. households accounts for less than 3 percent of the country's total waste. (See Figure 23–2.) If we focus the bulk of our attention on reducing waste in our kitchens, we miss the much larger potential to promote reducing waste in our industries and

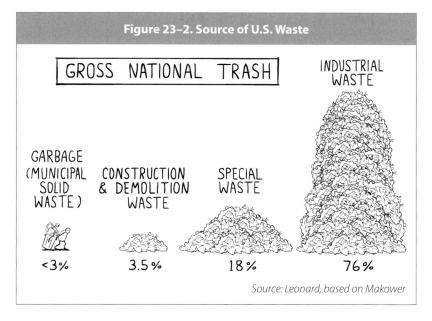

Figure 23–2. Source of U.S. Waste

GROSS NATIONAL TRASH

INDUSTRIAL WASTE

GARBAGE (MUNICIPAL SOLID WASTE)

CONSTRUCTION & DEMOLITION WASTE

SPECIAL WASTE

<3%   3.5%   18%   76%

Source: Leonard, based on Makower

businesses—where it is truly needed. And if someone really wants to work on reducing household waste, civic organizing to get a mandatory curbside recycling and composting program is a far more effective way to increase recycling and reduce waste than trying to maintain an eco-perfect household. But this focus on individual behavior is exactly where the companies behind Keep America Beautiful hoped to channel public concern about waste.[8]

Framing environmental deterioration as the result of poor individual choices—littering, leaving the lights on when we leave a room, failing to carpool—not only distracts us from identifying and demanding change from the real drivers of environmental decline. It also removes these issues from the political realm to the personal, implying that the solution is in our personal choices rather than in better policies, business practices, and structural context. Environmental decline is framed as the result of an epidemic of bad individual choices rather than of an economic, regulatory, and physical infrastructure that facilitates environmentally destructive activities over environmentally restorative ones. And the solution, then, is to perfect our own day-to-day choices rather than build political power to change the context, making environmentally beneficial actions the new default.

Describing today's environmental problems and solutions as individual issues also has a disempowering effect, leaving people to feel that their greatest power lies in perfecting their daily choices. Traditionally, the main strategies used to influence individual choice on environmental issues have focused on providing information and persuasion rather than working together to change the context in which the choices are made. As University of California at Santa Cruz sociology professor Andrew Szasz explains, this focus on changing individual behavior in response to environmental concerns is

> a strange, new, mutant form of environmentalism. There is awareness of hazard, a feeling of vulnerability, of being at risk. That feeling, however, does not lead to political action aimed at reducing the amounts or the variety of toxics present in the environment. It leads, instead to individualized acts of self-protection, to just trying to keep those contaminants out of one's body. And that is not irrational if one feels that there is nothing to be done, that conditions will not change, cannot be changed. I think, therefore, that we can describe this as a resigned or fatalistic expression of environmental consciousness.[9]

Making Change—Past, Present, and Future

If perfecting our everyday individual choices is not the answer to creating a sustainable society, what is? Clearly, much needs to change beyond the level of our individual actions. Society-wide, we need to implement new technologies, cultural norms, infrastructure, policies, and laws. Many of these already exist, so the problem is less about inventing new ways to do things than about building the political power to demand them.

Consider some previous movements for major social change: in the United States, the civil rights and United Farm Workers of America movements, as well as national-level environmental victories of the 1970s, and internationally the South African anti-apartheid movement and the Indian Independence Movement. In each case organizers did appeal to the public to change their daily actions. Throughout the civil rights movement, supporters were asked to patronize black-owned businesses and avoid shopping at segregated ones. Millions heeded Cesar Chavez's call to boycott California grapes in protest of farmworker conditions. During the 1970s, in the wake of *Silent Spring* and the first Earth Day, people were asked to choose pesticide-free produce and to save newspapers for recycling. Around the world, opponents of South Africa's apartheid system boycotted companies invested in that racist regime. And most people have heard of Mahatma Gandhi's famous pleas to buy Indian-made *swadeshi* goods rather than imported British ones.

But the organizers in each of these movements did not stop with pleas for individuals to make different shopping choices. They did not argue that individual people cause segregation or British colonialism and that different individual behaviors can stop these wrongs. They shared a compelling vision of how things could be better, they worked together as engaged citizens, and they changed the rules of the game. The calls for changes in individual behavior were tactical elements within broader political campaigns—campaigns that engaged people as citizens working together, using the range of tools available to them, including protesting, lobbying, legal action, economic sanctions, creating alternatives, and civil disobedience.

Integrated into broader political campaigns, calls to alter a person's individual choices can be used to educate and recruit supporters and to demonstrate commitment—all good tactical steps toward real victories. But too many of today's "green living" advocates are missing the broader political strategies that would enable the small acts to be more than just symbolic, feel-good activities.

A vigorous debate is currently under way about whether greening our daily individual acts leads people to the kind of deeper civic engagement that makes meaningful change or instead lulls them into a false sense of security and accomplishment. In other words, are these individual acts "on-ramps" to greater engagement, or are they "dead ends"?[10]

This debate has existed as long as campaigners have been extolling individuals to get involved in working for change. In the early nineteenth-century abolitionist movement, for example, "Free Produce" activists called on people to go out of their way to avoid purchasing goods made with slave labor. While the Free Produce approach was initially welcome in the broader campaign to end slavery, a growing number of abolitionists began questioning it as ineffective and distracting from the political work, which promised greater results. Abolitionist William Lloyd Garrison argued that Free Produce advocates were "so occupied by abstinence as to neglect THE GREAT MEANS of abolishing slavery."[11]

In his history of consumer activism in America, *Buying Power*, Lawrence Glickman explains that Garrison felt the Free Produce movement was a dead end because shoppers had "'a pretext to do nothing more for the slave because they do so much' in the exhausting efforts to find non-slave-made goods and the uncomfortable job of wearing and eating them. In other words, even if it were possible to divest oneself of all slave-made goods, the quest for what one free produce advocate called 'clean hands' diverted energy from the antislavery struggle by shifting the focus to what amounted to a selfish obsession with personal morality."[12]

Academics and activists on both sides of this debate have amassed studies documenting that small acts hasten or distract from greater engagement. It seems that the most honest answer is that it depends. Some people start with separating waste for recycling and move on to campaign for their local government to implement curbside recycling programs and to pressure companies to make products more recyclable. Others start recycling, and then stop worrying about waste—even increase the waste they produce—comforted by the fact that they can now put more in the recycling bin and are even rewarded for doing so if they live in a community partnering with Recyclebank. Rather than get stuck in this on-ramp versus dead-end debate, people concerned about transitioning to a sustainable society need to clearly and consistently link calls for individual action to bigger visions and bolder campaigns to ensure the individual first steps become on-ramps to making meaningful change.[13]

Making Broader Change

While making change in our kitchens may be easy, figuring out how to make change in larger communities and in broader societies is less so. The question ultimately revolves around what it takes to bring about change. Looking back over case studies where change has happened, it seems that change almost always involves at least three things.

First, there is a big idea of how things could be better. To move people beyond the easy green actions, we need to put forward an inspiring, morally compelling, powerful, and inviting vision comparable to that in transformative social movements of the past—compelling enough that people are eager to work long and hard to achieve it, because that is what it is going to take. Fortunately, we have that: Let's build a new economy that puts people and the planet first. Let's aim for nothing less than healthy, happy communities and a clean and thriving environment. Let's ensure that economic activity serves the goals of public health and well-being, environmental sustainability, and social justice rather than undermining them in the name of growth and profit.

Second, there needs to be a commitment to move beyond individual actions. Once we have a compelling vision, we need to join with others to build the power necessary to make it real. Building a mass movement strong enough to achieve the level of change needed is an inherently collective endeavor. To do this, we've got to reach beyond the traditional environmental community to create what Vermont Law School professor Gus Speth calls a "Progressive Fusion":

> Coming together is imperative because all progressive causes face the same reality. We live and work in a system of political economy that cares profoundly about profit and growth and about international power and prestige. It cares about society and the natural world in which it operates primarily to the extent the law requires. So the progressive mandate is to inject values of justice, democracy, sustainability, and peace into this system. And our best hope for doing this is a fusion of those concerned about environment, social justice, true democracy, and peace into one powerful progressive force. We have to recognize that we are all communities of a shared fate. We will rise or fall together, so we'd better get together.[14]

Good old-fashioned organizing basics, combined with new social media and networking tools, make it easier than ever to connect with others in our own neighborhoods or around the world to build that powerful unified force for change.

And third, action must follow. Right now, high percentages of people—in most cases a significant majority—support a cleaner environment, safer products, and a better functioning democracy, but these people are not yet actively working for change. The missing ingredient is not more information or more individual eco-perfectionists, it is collective engagement for political and structural change. Once we have a vision and a commitment to work together, there are an almost infinite number of ways to take action beyond the individual level: join or form an organization, draft legislation, gather signatures, litigate to stop a problem and advance a solution, launch campaigns to get companies to change their practices, run for office, write articles and educational material, invite others to join, organize protests and parades to make your opinion visible, engage in nonviolent civil disobedience, and much, much more.

There are already stellar examples of coalitions of groups doing just this—tackling a variety of environmental and social issues, from chemical pollution to climate change. The Safer Chemicals, Healthy Families Coalition in the

GAIA members and allies conduct a waste audit at Manila Bay to support their campaign for better enforcement of Philippine waste policies.

United States, for example, includes 440 organizations representing more than 11 million individuals concerned about toxic chemicals in their homes, workplaces, and products. Members include parents, health professionals, advocates for people with learning and developmental disabilities, reproductive health advocates, environmentalists, community-based organizations, and businesses from across the nation. Yes, they offer advice on identifying and avoiding toxin-containing products, but their work focuses on advocacy campaigns for stronger policies and laws, along with market campaigns to affect broader shifts in the industry. Campaign director Andy Igrejas explains: "You can't shop your way around the problem and you shouldn't have to. There is no app for the kind of change we need. The problem is large and pervasive enough that we need broad changes in policy and by companies themselves. Consumer action can be a tool in that process—to send a message to a particular company for example—but it is not a substitute."[15]

Another example, the international climate change campaign 350 .org, was founded around the idea that individual action is not going to be enough to solve the climate crisis. It is going to take a movement. The group's first day of action in 2009 brought together over 5,200 events in 181 countries, what CNN called "the most widespread day of political action in the planet's history." Instead of changing lightbulbs, people dove underwater with banners carrying climate change messages, hung signs off mountains, biked by the hundreds through their capitols, and found other creative ways to take action together and make their voices heard. Since then, 350.org has continued to push the boundaries of traditional environmentalists, from organizing the world's largest climate art exhibit to getting more than 1,200 people arrested in front of the White House over several weeks to protest the Keystone XL pipeline—a 4,300 kilometer (1,700-mile) fuse to the largest carbon bomb on the planet, the Canadian tar sands. As 350.org founder Bill McKibben says, "First change your politicians, then worry about your lightbulbs."[16]

The Global Alliance for Incinerator Alternatives (GAIA) is a leading catalyst for change in an area where historically most effort has been directed toward changing individual actions: waste. This global network promotes Zero Waste by providing its members with advice on setting up composting and local recycling programs while it simultaneously lobbies governments around the world to end subsidies for polluting waste incineration and to adopt ambitious policies to reduce all kinds of waste. According to GAIA U.S. coordinator Monica Wilson, "Providing tips for reducing waste at the individual level is important since many of our members come to us eager to get started right away in their own lives, but we know that real solutions to waste can't be achieved at the individual level alone. Ultimately we need stronger standards and laws, as well as shifts in societal and cultural norms, to achieve the solutions we know are possible."[17]

The good news is that we have everything we need to make big change in the years ahead. We have model policies and laws. We have innovative green technologies to help with the transition. We have an informed and concerned public; millions and millions of people know there is a problem and want a better future. The only thing we are missing is widespread citizen action on the issues we already care about. As American author and activist Alice Walker says, "The most common way people give up their power is by thinking they don't have any." Our real power lies not in perfecting our ability to choose from items on a limited menu but in deciding what gets on that menu. Let's ensure that all the options offered move us closer to sustainability and justice. That is the kind of change we need. And we can only get it by working together.[18]

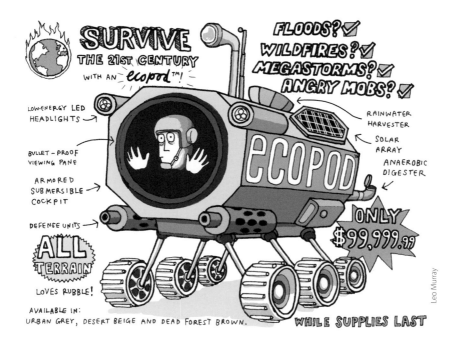

SURVIVE THE 21ST CENTURY WITH AN ecopod™!

FLOODS? ✓
WILDFIRES? ✓
MEGASTORMS? ✓
ANGRY MOBS? ✓

LOW-ENERGY LED HEADLIGHTS

BULLET-PROOF VIEWING PANE

ARMORED SUBMERSIBLE COCKPIT

DEFENSE UNITS

ALL TERRAIN

LOVES RUBBLE!

AVAILABLE IN:
URBAN GREY, DESERT BEIGE AND DEAD FOREST BROWN.

RAINWATER HARVESTER

SOLAR ARRAY

ANAEROBIC DIGESTER

ECOPOD

ONLY $99,999.99

WHILE SUPPLIES LAST

Leo Murray

Open in Case of Emergency

In November 2012, the Big Four accounting firm PricewaterhouseCoopers released a report that concluded it was too late to hold the future increase in global average temperatures to just 2 degrees Celsius. "It's time," the report announced, "to prepare for a warmer world."

The same month, the World Bank released *Turn Down the Heat*, which soberly set forth why a 4-degree warmer world must be avoided. Meanwhile, accounts of myriad emergent calamities were easy to find in the press: the failure of the Rio+20 talks, "zombie" coral reefs, calls for higher birth rates, declining Arctic sea ice, an approaching "state shift" in Earth's biosphere, and other evidence of strain in natural systems and of human blindness, ignorance, or denial.

Time to buy an Ecopod?

Clearly, trouble is coming—but there are better responses to it than stockpiling canned goods and weaponry. In view of humanity's failures of foresight and political will to address the array of sustainability problems ahead, we asked some notable thinkers to ponder what we might do to make the best of it.

A central theme of their answers is "build resilience." That requires, according to Laurie Mazur, diversity, redundancy, modularity, social capital, agency, inclusiveness, tight feedbacks, and the capacity for innovation. To begin strengthening our resilience, Erik Assadourian urges the construction of an enduring environmental movement that can engage people and ground their ethics and behavior in ecological reality. Michael Maniates echoes the grounding theme in his call for environmental education to stop misleading and underpreparing students for the challenges ahead: that the coming crises will galvanize action rather than generate anger, fear, and conflict. Paula Green stresses the value of community roots and strong social capital, including intergroup networks to bridge different communities. Bron Taylor argues, carefully, for an ecological resistance movement. "Given the urgency of the situation," he writes, "extralegal tactics should be on the table, as they were in earlier causes where great moral urgency was properly felt."

If the crises do threaten conflict, that risk will be aggravated by a rising tide of environmental refugees. Michael Renner writes that tens or even hundreds of millions of people are likely to be displaced by 2050, yet money spent on adaptation measures in developing countries is already inadequate—a shortfall that must be remedied. Failing that, such migrations will join other pressures driving us to deploy geoengineering techniques—giant space mirrors, carbon-capturing cement—as quick fixes for a disrupted climate. In reviewing these schemes, Simon Nicholson urges research to continue but notes that the least of their problems are the technical uncertainties and unpredictable effects; many are fraught with grave geopolitical risks too.

Governance will figure crucially in our response to the coming "long emergency," as David Orr terms it (following James Howard Kunstler). Brian Martin argues that governance should be flexible, not stiff. That requires participation, high skill levels, robust debate, and mutual respect. If this sounds like a deepened democracy, Orr agrees: he calls for "a second democratic revolution" in which we "master the art and science of governance for a new era."

If circumstances overtake our best efforts, there may be some comfort in Pat Murphy and Faith Morgan's telling of Cuba's story. Forced to the brink by the Soviet Union's collapse, Cuba suffered a period of harsh adjustment but has scavenged a culture with a small environmental footprint and remarkably high levels of nonmaterial well-being.

Is it too late? In the concluding essay, science fiction writer Kim Stanley Robinson says the real question is, How much will we save? "We can see our present danger, and we can also see our future potential. . . . This is not just a dream but a responsibility, a project. And things we can do now to start on this project are all around us, waiting to be taken up and lived."

—*Tom Prugh*

CHAPTER 24

Teaching for Turbulence

Michael Maniates

In late 2010, a respected research team led by Yale University professor Susan Clark released a two-part assessment of college and university programs in environmental studies and science (ESS). The team's conclusions were hard-hitting and pointed. Too many ESS programs, they wrote, do too much too quickly with insufficient clarity of purpose and method. They "suffer from muddled goals, disciplinary hodge-podge, and an educational smorgasbord of course offerings." At a time when the need for dynamic college and university programs in environmental science and studies has never been greater, those who plan and deliver these programs appear to be selling their students and the planet short.[1]

Clark's assessment is the latest in a series of warnings about the incoherence of environment and sustainability programs in higher education. In a seminal 1998 essay, for example, University of California at Santa Cruz professor of environmental studies Michael Soulé and his colleague Daniel Press lamented a persistent and structural "multidisciplinary illiteracy" among ESS undergraduates. Even critics of their argument had to admit that at least 30 percent of ESS programs were fragmented and poorly conceived.[2]

As the planet's health declines and undergraduate interest in environmental issues soars, concern about the effectiveness of ESS programs will surely intensify. At first glance, this is welcome. Who, after all, could be in favor of diffuse goals and multidisciplinary illiteracy around educational programs so critical to the transition to sustainability? Architects of ESS programs and professors who work within these programs must redouble their efforts to clarify the field's core competencies while implementing curricular mechanisms that enforce focus and integration. And students should ask tougher questions about curricular form and focus. Flashy websites, green buildings, and environmentally responsible campus practices do not necessarily translate into strong ESS programs, regardless of first impressions.

But aspiring students and program architects must also remember that

Michael Maniates is a professor of environmental science and political science at Allegheny College and a visiting professor of environmental studies at Oberlin College.

www.sustainabilitypossible.org

the college student of today will graduate into a world that will be singularly defined by turbulence—a white-water turbulence of climate instability, eco-logic decline, and attendant economic and political dislocation, with winners, losers, and persistent inequality. Merely sharpening the focus of programs built for placid waters will not be enough. Now is the time to explore how current ESS programs undermine student capacity to navigate a turbulent world—and to entertain new curricular features that foster nimbleness and wisdom in times of crisis.

Patterns of Teaching and Learning

Not long ago, the notion that ESS programs could play a pivotal role in the transition to sustainability was a distant thought. They were often viewed on campus as marginal programs, a place where students who could not succeed in traditional natural-science fields (biology, chemistry, or geology, for instance) could complete their studies and graduate. On more than a few campuses, "ES" came to stand for "easy science."

For a time it looked as if a multidisciplinary assessment of environmental problems that integrated the social and policy sciences could only occur outside of ESS. North Carolina State professor Marvin Soroos, a prominent scholar of environmental politics, spoke for many when he argued, in 1991, that professors of political science and international relations had best begin teaching about sustainability if academia harbored any hope of "preparing students for the historically unprecedented challenges that their generation will face." Soroos had no quarrel with the investigatory power of the natural sciences but, like others, saw a natural-science focus as insufficient to the demands of sustainability. If ESS would not change, then it would be marginalized, in part by political science and international relations programs with their own programs in environmental studies.[3]

Those days of doubt about ESS programs are long gone, at least in the United States, which boasts the greatest concentration of such programs. According to Shirley Vincent, perhaps the nation's premier authority on the focus and trajectory of ESS programs, there were some 500 such programs in the United States in 1990. By 2010, there were 1,200, with 90 percent of them at the undergraduate level. By 2015 that number could easily expand to 1,400 or more, making ESS one of the fastest-growing fields of undergraduate study in the country. This explosion in programs has been matched by an expansion in disciplinary diversity and intellectual focus. Some ESS programs, notes Vincent, prepare natural scientists capable of analyzing environmental science problems, while others strive to foster a deeper understanding of the policy process and environmental citizenship. Still others focus on training managers in collaborative processes of environmental problem solving. Almost all programs strive to imbue their students with critical thinking and problem-solving skills appropriate to the challenges ahead.[4]

Three patterns of teaching and learning emerge from today's mélange of programs. The first is a general trend toward urgency and alarm, coupled with a focus on the inability of prevailing systems of economic accounting and political decisionmaking to address looming environmental ills. ESS courses, and especially introductory courses that summarize the extent of the human assault on nature, can be jarring. Students quickly learn that the planet's health is declining more rapidly and systematically then they might have imagined. They discover that the damage often flows from the very institutions—the market, pluralist democracy, education— that we often look to for solutions. Left unchallenged, this "urgency + inability" equation can overwhelm students with a sense of hopelessness and despair and can foster the expectation that system-jarring crises are just around the corner.[5]

NOAA Photo Library

North Carolina State University students are involved in a joint EPA/NOAA Air Resources Laboratory project to measure and model ammonia fluxes in forest and agricultural landscapes.

To battle this despair and to create opportunities for interdisciplinary integration of course material, ESS programs turn to applied research and hands-on problem solving. This second pattern of teaching and learning is perhaps the most essential feature of ESS. It is not enough in most programs to simply understand the major environmental problems. Students must critically assess them and carefully evaluate competing solutions. To this end, program websites and brochures emphasize the acquisition of problem-solving approaches and research skills.

Required courses focus on environmental problems on campus or in the community and engage students in community projects and applied research. Campus administrators, sympathetic community groups, and local political actors are frequently part of the mix so that students can practice communicating environmental information to disparate groups. Sustainability coordinators responsible for college- or university-wide environmental initiatives chip in by coordinating campus-wide recycling and energy conservation challenges. The problem-solving focus is typically local, with the hope that these small-scale interventions will scale up to match regional, national, and even international challenges.

Indeed, perhaps more than any other higher-education field of study, ESS understands and justifies itself as a problem-solving discipline. Writing in 2005 to the Andrew W. Mellon Foundation, for example, professors Sharon Hall, Tom Tietenberg, and Stephanie Pfirman, representing Colorado College, Colby College, and Barnard College, observed that "service learning and community-based learning (CBL) courses or experiences are among the most successful and empowering experiences for ES students during their time in college." They noted that these experiences, together with courses that focus on the local environment, provide "a productive source of inspiration for 'hands on' student research" while fostering engagement with interdisciplinary approaches to real-world problem solving. In 2005, this assessment illuminated the best practices of the top ESS programs. Today it describes the curricular norm in the field.[6]

It comes as no surprise that ESS students and their mentors are unusually active—in the lab, the classroom, the library, on campus, and in the community. Their work is both positive and normative: they seek to understand the causes of environmental ills, and they strive to implement solutions. By and large, though, this work occurs without any systematic assessment of how it fits into a larger mosaic of political power, cultural transformation, and social change. As Shirley Vincent notes, few programs ask their students to study competing theories of social change or to critically assess how their research or work on local projects fit into larger models or ideas about cultural transformation. This is an odd oversight, since ESS students are almost always asked to think critically about how change happens in natural systems. But such "systems analysis" rarely spills over into the social sciences, at least not in any concerted or focused way.[7]

Why the omission? For Richard Wallace, an environmental studies professor at Ursinus College who studies the dynamics of interdisciplinary education, the "big tent" approach in ESS is largely at fault. As a field of study and a guide to problem solving, ESS invites and includes a diversity of disciplinary approaches to environmental problem solving. Under this sprawling canopy, no single notion about how or why social change occurs is privileged. Students are to glean theories of social change from their courses outside of ESS and then integrate them during their research and project work. It is decidedly a do-it-yourself affair. Wallace's diagnosis enjoys support from other scholars, including Indiana University professor Matt Auer, whose analysis of graduate-level ESS programs paints a similar "big tent" picture of teaching and learning.[8]

Another explanation, according to analysts like journalist Mark Dowie and scholars like Wallace, is the strong influence of the natural sciences on the evolution of ESS. This influence privileges the notion that societies change naturally and rationally in response to new scientific information.

Social change becomes an exercise in finding the facts and electing policymakers who will act on the data. It is a straightforward process in need of no serious interrogation other than reflecting on how natural scientists can more effectively communicate their findings to policymakers.[9]

Finally, faculty in ESS programs may shy away from developing courses that focus on social activism and political change for fear of looking as if they are training environmental activists rather than environmental scientists and analysts. U.S. environmentalism, notes Dowie, has historically been a "polite movement," where offering additional research and compelling facts has been a more comfortable way of promoting change than noisy activism or social protest. Vermont Law School professor and author Gus Speth, a pivotal figure in the U.S. environmental movement, makes the same point in his clarion call to the environmental community to abandon its safe but largely ineffective reliance on facts, studies, and data to drive political change and social transformation.[10]

Disabling Assumptions

Too often, students are left to cobble together their own theories of social and cultural change amid a backdrop of troubling urgency, looming crisis, and a focus on research and project-implementation skills. What do they conclude? This question weighed heavily on Sam Rigotti, an environmental studies student and researcher at Allegheny College until his graduation in 2010. In a path-breaking study, Rigotti began by observing how "10 Easy Ways to Save the Planet" lists and similar publications have inundated his generation. He hypothesized that the lack of sustained analysis of processes of social change within ESS programs creates a vacuum that the "easy ways to save the planet" narrative quickly fills: buy green, initiate a few lifestyle changes, spread the word to others, and wait for the totality of these small changes to sum into fundamental social change. Rigotti feared that students who assimilated this "small and easy" view would later come to grips with its limitations and in frustration fall back on notions from their introductory classes about the inevitability of crisis.[11]

Working with faculty and other students at Allegheny, Rigotti conducted the first national survey that explores these issues. His results, from 437 randomly selected ESS students at 15 colleges and universities, are provocative. Some three quarters of students surveyed, for example, identified green consumption and "voting with your dollar" as among the very best strategies for promoting environmentally conscious social change. By contrast, students thought that supporting or joining environmental interest groups, pressuring legislators, engaging in electoral politics, and other forms of civic engagement were too diffuse or decidedly utopian. For these students, the small and easy theory of social change seemed natural and obvious—and empowering too. Being a meaningful part of social change is as straightforward and accessible as driving less, recycling more, eating less meat, buying vegetables at a farmers market, or making a point of purchasing environmentally oriented products.[12]

The most startling insight from Rigotti's analysis, however, may be around the notion of crisis. Seventy percent of students surveyed blamed "poor environmental values" for our current predicament and pointed to the need for more education and a compelling crisis to drive a meaningful transition to sustainability. According to these students, the average American does not know about environmental problems or knows but does not deeply care. For more than half the students in the sample, a crisis that will make Americans care—that will compel them to heed well-trained experts in environmental problem solving—is something to anticipate and welcome.[13]

This naive faith in crisis and the dim view of human nature upon which it rests reflects the literatures to which ESS students are commonly exposed. Mainstay introductory textbooks, like *Environmental Science* by G. Tyler Miller and Scott Spoolman, as well as Daniel Chiras's text of the same name, underscore the power of crisis in driving needed change. When explaining policy change, for instance, Chiras shares former secretary of state Henry Kissinger's observation that "in government, the urgent often displaces the important" to make the case that change occurs only in the face of compelling crisis. Miller and Spoolman are more direct; they simply state that "U.S. political and cultural systems are slow moving" and that "change happens slowly" in the absence of crisis.[14]

In the same vein, the core environmental policy texts in the field, including those by Walter Rosenbaum and by Norman Vig and Michael Kraft, attribute the spate of environmental regulation in the 1970s to crisis events like air pollution alerts in Los Angeles and burning rivers in Ohio. Key intellectual frameworks, finally, underscore the shortsightedness of human behavior and the inevitability of crisis. ESS students need not go much further in their early studies than Garrett Hardin's famous "The Tragedy of the Commons" essay to learn that an environmental crisis, driven by human failing, is both necessary and inevitable.[15]

The small and easy theory of social change, which promises big change when large masses of people commit themselves to small acts of personal sustainability, only amplifies this kind of crisis thinking. This is because social change does not happen through mass, uncoordinated shifts in lifestyles or consumption choices: small and easy is attractive, plausible, and dead wrong. It is the rare social movement that crystalizes and advances because of the initial mobilization of large majorities of the population, and the environmental movement is no exception. After all, some people will always refuse to adopt any lifestyle or consumption change.

And in the realm of environmental action, the proportion of the reluctant remains consistently large, despite decades of aggressive environmental education and untold millions spent by marketers of green products. More than 80 percent of Americans fail to consistently practice a small suite of environmentally sound behaviors, like reducing their energy use, driving smaller cars, and buying green products. Almost 25 percent of Americans do not recycle, often because they cannot be bothered or believe that doing so makes little difference. More generally, consumer commitment to environmental practices appears to be waning. Harris Interactive, which regularly polls Americans on their environmental behaviors and attitudes, reports a decline in overall "green" activities and concerns in 2012.[16]

Bill Owen

Allegheny College students and faculty work with a local farmer on an aquaponics project that raises tilapia and grows lettuce in the same facility.

These data and the behaviors they document generate a predictable set of responses among adherents to small and easy. Confronted by low rates of green consumerism in the general population, well-meaning environmentalists rabidly promote green lifestyles with a heavy dose of guilt and almost missionary zeal. They offer pronouncements meant to underscore the importance of unified commitment to environmental aims, like "If everyone in America used energy-efficient lighting, we could retire 90 average-sized power plants, reducing CO_2 emissions, sulfur oxide, and high-level nuclear waste." They offer more and more information on the virtues of environmental living. And they often heap disdain on those who do not, for instance, recycle or drive small cars or otherwise live sustainability. When all this fails, what remains is a natural, logical, altogether understandable tendency to conclude that people themselves are at fault—they are too selfish, too ignorant, too irresponsible—and that, ultimately, only a crisis will move them.[17]

Of course, all this is both unproductive and misdirected. A politics of guilt can never mobilize and inspire. And even if most Americans did suddenly "green" their lifestyles, underlying processes of production and disposal that are largely insulated from personal consumption decisions would still drive the planetary ecosystem toward collapse, albeit just a bit more slowly. This point is vividly illustrated by the "personal footprint calculator" offered by the highly respected Global Footprint Network. (See Chapter 4.) As the calculator consistently demonstrates, large changes in lifestyle translate into disappointingly small effects on anyone's environmental footprint.[18]

Sam Rigotti's study was the first of its kind and thus awaits further verification and refinement. On its face, though, it is both plausible and compelling. It resonates deeply with the experience of many ESS educators who find their students to be overly enamored with the power of crisis and too often dismissive of the capacity of Americans to sacrifice for the common good. The risk here is not that students see crisis on the horizon, for crisis is surely coming. The danger instead is that ESS graduates increasingly view crisis as a benevolent force that will rally the public and enhance the power of environmental problem solvers like themselves. This idea of crisis as a welcome lubricant in the transition to a sustainable world is a lovely, if unpromising, notion. Preparing students for turbulence involves making them aware of less-benign species of crisis and enabling them to react in kind.

The Real Face of Crisis

Early in President Obama's first term, in the midst of a financial meltdown in the United States, chief of staff Rahm Emanuel was quoted as saying, "Emanuel's Rule One: Never allow a crisis to go to waste. They are opportunities to do big things." Emanuel's theory of crisis reaches back to the sixteenth century, when Niccolo Machiavelli wrote, in *Il Principe*, "Never waste the opportunities offered by a good crisis."[19]

Students of environmental issues would undoubtedly agree with Emanuel, but in doing so they may have in mind a kind of crisis that author Rebecca Solnit writes about so eloquently in *A Paradise Built in Hell*. Disasters, Solnit says, demonstrate "the resilience and generosity of those around us and their ability to improvise another kind of society. . . .They demonstrate how deeply most of us desire connection, participation, altruism, and purposefulness."[20]

In ways both compelling and persuasive, Solnit profiles five disasters, ranging from the San Francisco earthquake in 1906 to Hurricane Katrina's assault on New Orleans in late summer 2005. She documents striking heroism, ingenuity, and compassion among ordinary people, and she shows how communities traumatized by crisis self-organize in effective and humane ways. For Solnit, sudden disaster reveals a generosity, resourcefulness, and bravery latent within us, ready to be called forth in service of a "new paradise." Look closely at disaster-driven crises, she says, and you can see how a new world might be possible, with all that is necessary already within each of us.[21]

A Paradise Built in Hell should be required reading for ESS students, regardless of their disciplinary orientation. Solnit complicates the dim view of human nature to which many ESS students subscribe and offers hope of a better world rooted in existing abilities and widely felt yearnings. Read closely, her work suggests that ESS students might best think of themselves as midwives working to deliver something already present within society rather than as experts trained to educate the uninformed and motivate the uninspired.

The difficulty with Solnit's work, and its notion of "crisis as deliverance," lies with the type of crises she documents and that ESS students so commonly imagine. They are sudden, cataclysmic events with jarring psychological and political impact. They bring to the forefront underappreciated or nascent networks of human connection while, for a time, throwing existing power structures back on their heels. These sudden disasters, moreover, expose stark divisions in wealth and power that, so brightly illuminated, are questioned or rejected, at least for a time.

By contrast, the disasters that ESS graduates will confront are likely to be slow-motion affairs: gradual and persistent, with moments of upheaval punctuating slow decline. Water will grow scarcer, food prices will rise, coastal cities will periodically flood as increasingly intense storms lash their shores, droughts will become more commonplace, livelihoods will be disrupted, economies may falter, and inequality will deepen. The threat of these crises is not so much that they generate catastrophes of unthinkable proportion but rather that they will become the norm, freighted with a deepening sense of inevitability.

These slow-motion crises risk evoking three dynamics that ESS graduates are poorly prepared for. One is what environmental analysts Michael Shellenberger and Ted Nordhaus call "insecure affluence": the growing sense among a large slice of Americans that their economic position in life is unstable at best and more likely at imminent risk. As insecure affluence deepens, Americans may be especially reluctant to accept even the smallest of material sacrifices, especially if these sacrifices are imposed on them by elites, a point emphasized by political theorist John Meyer, who observes that "an environmentalist call to sacrifice" will be resisted "not just for its paternalistic attitude, but also for its blindness to the lived experience of sacrifice central to the lives of many." Alas, too many ESS students are trained to play the very role of elites who, in one way or another, will make arguments supporting present sacrifice for future gain. Crisis will be no friend to these graduates.[22]

Nor will a politics of anger, which is another likely result of economic and ecologic upheaval. As former U.S. secretary of labor Robert Reich notes, prolonged periods of stress and insecurity lead to "an increasing bitterness and virulence of the nation's politics" and can quickly morph into "an underlying readiness among average voters to see conspiracies among powerful elites supposedly plotting against them." If it is true, as Shellenberger and Nordhaus assert, that environmentalists naively "hoped that the environmental crisis would bring us together and make us happier," then Reich and scholars like historian Richard Hofstadter, who studied paranoia in politics, or Thomas Edsall, who reflects on American politics under conditions of scarcity, offer a rude awakening. It is more likely that crisis will generate widespread anger, fear, conflict, and a deepening paranoia than a spiritual awakening and ecological reckoning. ESS graduates expecting the latter and ill-prepared for the former may wonder why their false expectations were not more thoroughly challenged by their professors.[23]

Finally, while crisis may provoke suspicion and fear of elites among some citizens, it is likely to fuel a desire among others for greater government power and control. In this way, as observed by economic philosopher Robert Heilbroner in the late 1960s, ecological crisis can bring about a slow slide to authoritarianism, as people become more willing to trade their freedom away for the promises of strong leaders who will fix pressing problems. The danger that Heilbroner highlighted is familiar to Americans worried about the erosion of civil liberties after 9/11. And Heilbroner's warnings are not without empirical support. In her classic study of crisis and dictatorship, which spurred an entire line of scholarship, sociologist J. O. Hertzler showed how crisis—often but not always economic—erodes democratic impulses and structures and produces a consolidation of power friendly to dictatorial regimes. Studies like these suggest that crisis is inimical to progressive social causes, environmentalism included.[24]

Despite these tendencies and dangers, it may yet be possible to follow Rahm Emanuel's Rule One. But using the crises to do big things means seeing them for what they are and training a new generation of college students to think strategically, rather than wishfully, about the possibilities that crises present.

A Curriculum for Turbulence

White-water rafting is a growing tourist activity, and young people willing to serve as raft guides are in high demand. New employees who would steer rafts down turbulent rivers are educated in the art of "reading" rivers, navigating boats, and coaxing effective and timely paddling from their guests, who help propel and steer their craft through bumpy waters. White-water guides-to-be are also trained to anticipate worst-case scenarios: an overturned raft, a guest dumped into the water, broken bones, or equipment failure. It is impossible, of course, to prepare raft guides to handle unanticipated risks and problems—but they can be and are primed to expect the unknown and to approach it with humility and equanimity.

What might a course of instruction look like for students in ESS programs who will be asked to negotiate a similar kind of turbulence? Five characteristics loom large, especially in light of patterns of existing curricular deficiencies. (See Box 24–1.)[25]

First, ESS programs must stay true to their founding passions and intent, even as they seek to address curricular aimlessness and incoherence. Not every student must become an expert in processes of social change or prove capable of thinking creatively about political behavior during prolonged crisis. Nor must every program undergo radical change to effectively teach for turbulence. The best curricular reforms will be those that achieve the greatest effect with the least intrusion and that anticipate and prevent student misperceptions about social change and crisis before they take deep root.

Box 24–1. Gaps and Opportunities in Environmental Studies

Internationally known climatologist Richard Alley, a professor at Penn State, writes and performs rock songs on climate change and does a spirited dance illustrating how Earth's orbital variations influence climate. Humboldt State University in Arcata, California, recently launched an environmental studies program that aspires to train students to think creatively about power, privilege, and social change. And faculty at Wheaton College in Norton, Massachusetts, pioneered, for a time, an undergraduate course on the theory and practice of environmental conflict resolution that used case studies, community engagement, and scenario building to prepare students for an increasingly contentious world.

These examples of innovative ESS pedagogy and curriculum stand out because they remain the exception rather than the rule. A review of the most prominent ESS programs in the United States reveals that few programs expose their students in systematic ways to a range of ideas about how change occurs in political and cultural systems. Even fewer still put students in the way of experiences that will help them rigorously analyze and initiate social change and reflect on how locally successful initiatives might "scale up" or "network out." That is why new programs like that at Humboldt State University are so exciting.

Likewise, although many ESS programs ask their students to engage in community projects, almost all such work occurs in no- or low-conflict situations. These courses emphasize research skills, data collection, and

communication across disciplinary boundaries—important goals, to be sure, but insufficient in the face of growing social turbulence. Wheaton College's openness to courses that bring political conflict and cultural discord into the mix is laudable and worthy of emulation.

Finally, despite the centrality of the natural sciences to most environmental programs, there are surprisingly few places in the ESS curriculum where students explore the changing role of science and scientists in the struggle for sustainability. Such exploration might begin with how scientists better communicate their ideas in politically charged environments and then extend to deeper questions about the politics of expertise around contentious environmental issues. During turbulent times, natural scientists and the insight they generate will be greeted with increasing skepticism and hostility. The best-trained ESS students, and especially those with strong natural-science interests, will be those who have given careful thought to these dynamics, beginning but certainly not ending with Richard Alley's playful approach to scientific communication.

Most ESS programs fail to acclimate students to contentious environments, neglect to analyze the changing nature of natural-science expertise, and gloss over processes of social and cultural change. But this is changing, slowly. ESS programs that consciously train students for turbulence by filling these gaps are the promise of the future.

Source: See endnote 25.

Second, early courses in ESS programs might ask students to think critically and imaginatively about human nature and the nature of crisis, separately and together. Instructors could take a page from Rebecca Solnit's work and push students to explore the often latent capacity of humans to connect with and care for one another, to take the long view, and to work in common for the common good. Even as these introductory courses document growing environmental threats to human well-being, they might also explore the conditions under which humans regularly sacrifice for their family, faith, and community. Ideally, students would leave this course work preoccupied with how sustainability initiatives could more consistently bring these latent and noble human capacities to the surface rather than reflecting on how looming crises will nicely teach selfish and narrow-minded people an important lesson or two.

An important curricular pivot, of course, is a rigorous course or courses that interrogate overlapping and competing theories of political and cultural change. The successful integration of this third curricular element will produce students whose thinking about social change will transcend the "small and easy" frame that is so unproductive to enlightened and empowering action. ESS programs that focus on feedback, thresholds, and dynamics of change in their natural-science courses must now bring the same level of rigorous analysis to their discussion of social and cultural change. To continue hoping that other departments or students' own initiative will fill the "theory of social change" hole in the ESS curriculum is at best wishful thinking. Some of the most exciting work in ESS over the next few years will revolve around the design and delivery of such courses.

In their applied and experiential courses, most ESS students engage with campus and community partners who are broadly sympathetic to their work. During times of crisis, however, such natural sympathy will be the exception rather than the norm. To teach for turbulence, ESS programs could expose students to more-contentious environments and create classroom moments that foster strategic thinking about managing—and even taking advantage of—a politics of anger or the anxiety that comes with insecure affluence. In advancing this fourth curricular element for turbulence, ESS programs might also consider how to draw on campus resources around conflict management and resolution.

Finally, teaching for turbulence means providing students with the theoretical background and classroom practice to explore how they can best pursue their passions in rough water. Natural scientists might focus on the increased politicization of science in a turbulent world and what that may imply for their own work. Students with a talent for project-based community work might be engaged in thinking critically about how local-level initiatives can scale up in ways that address or capitalize on insecure af-

fluence or a politics of anger. And ESS majors who see themselves working as managers or practitioners in organizations of environmental governance or stewardship could be similarly challenged to analyze the shifting
role and power of organizations during times of political paranoia and a tilt
toward authoritarianism. After all, these three groups of students imagine
themselves as "boundary spanners" who will work at the intersection of the
multiple disciplines and disparate concerns. Their training will be complete
when they can anticipate greater discord at these intersections and react
with strategic balance.

A New Coherence

Ocean Conservation Society executive director Charles Saylan and professor Daniel Blumstein of the University of California at Los Angeles paint a
dim picture of environmental education in the United States in their recent
book, *The Failure of Environmental Education*. Despite decades of environmental education, they say, significant change in human behaviors that matter most are scarce. Indeed, based on behavior, it is difficult to distinguish
students who have participated in environmental education from those who
have not. It is time for a better curriculum, one that moves students to new
ways of thinking and acting. That curriculum, they say, would focus on consumption and overconsumption, underscore the necessity of sacrifice, and
tease apart the dynamics of policy change.[26]

While their work has generated controversy, in the end Saylan and Blumstein probably do not go far enough. The real danger, at least when it comes
to ESS education within colleges and universities, is not the puny effect of
environmental education on behavior. The danger is the impact of this education on students' sense of the possible and of their own role and power
in transforming the world around them. Educational programs that leave
students with an emaciated theory of social change and that fuel a politics of
guilt and crisis do little to foster the creativity and compassion that sustains
personal and collective transformation.

It is time for a new coherence in undergraduate ESS programs—not just
among the hodgepodge of courses that produce multidisciplinary illiteracy
but also within the story that students hear as they move through the curriculum. These students come to understand, with great clarity, that industrial civilization as we know it stands at a precipice of change, where existing political, economic, cultural, and technological patterns must quickly be
supplanted by new arrangements and habits. But they are rarely presented
with a coherent picture of how to bring about these arrangements or of how
exploring competing processes of social, scientific, and technological change
can illuminate pressure points for change. Instead they are offered, in sometimes intricate detail, the blueprints of a sustainable future—renewable

energy, sustainable agriculture, reconfigured cities, and a plentitude economy—but with little integrated, systematic sense of how to get from here
to there.

In defense of those who teach in and design ESS programs, the path from
here to there is profoundly unclear. But this uncomfortable fact only under-
scores the importance of preparing students for times of turbulence in the
hope that when white water hits, they have both the tools and the vision to
see the route down the river and coax effective and timely paddling from
their fellow rafters. The future, as most people who work or study within
ESS programs know, will not be like the present. Now is the time to carefully
consider how students are best prepared to be thoughtful and anticipatory
agents of change in the tumult to come.

CHAPTER 25

Effective Crisis Governance

Brian Martin

When a crisis develops, what sort of governance—what sort of system for running society—is most resilient? Does centralized control give the best prospect of survival? Or is something more decentralized needed?[1]

Possible political sources of crisis include military invasion, internal coups, political paralysis, major corruption, and revolutionary change. Wars in the past century triggered changes in governance in countries such as Germany, Japan, and Cambodia. Coups affected dozens of countries, from Chile to Greece. Revolutions transformed Russia, China, and Iran.

At least as significant are changes enabled by belief systems. The spread of neoliberalism—based on belief in unfettered markets—has transformed political systems, especially in the United States, the United Kingdom, and other English-speaking countries. Belief in political freedoms and fair elections has underpinned challenges to repressive regimes in Serbia, Georgia, Ukraine, and elsewhere. Belief in racial equality was behind the successful struggle against apartheid in South Africa.

Environmental impacts intersect with political and economic systems and crises in various ways. Disasters with environmental impacts can affect politics, as when the devastation from the 2004 Indian Ocean tsunami encouraged the signing of a peace agreement in Indonesia's war-torn province of Aceh. Governments can influence responses to crises with environmental impacts, as when the Burmese government hindered international relief efforts following the devastating 2008 cyclone Nargis. Some types of political and economic systems are more prone to contributing to environmental problems, and some systems are better at responding to emerging or full-blown environmental crises.[2]

War, which can be considered a type of political crisis, is devastating to humans and the environment and in fact can be a source of environmental crisis. Massive refugee movements—themselves a source of political crisis—can be triggered by war and political repression but also by environmental

Brian Martin is a professor of social sciences at the University of Wollongong, Australia.

www.sustainabilitypossible.org

disasters. Global warming has the potential of creating huge numbers of "environmental migrants."[3]

Resilience is the capacity of a system to respond effectively to assaults like these on its functioning or very existence. Resilience in the case of communication technology includes the capacity to keep functioning despite breakdowns or attack: the Internet was originally designed, remember, to maintain communication in the face of nuclear attack. The resilience of political systems includes both the survival and the maintenance of formal decisionmaking processes and of associated systems—such as transport, food, and communication—for maintaining the survival and social functioning of the population.[4]

When considering responses to crises, it is useful to distinguish two contrasting sorts of governance: stiff and flexible. Stiff governance can be well suited for a particular task, often for a particular threat. The classic example is a dictatorship with a command economy, ideally designed for warfare: central direction can be used to mobilize resources for defense or attack. Such a system can have great difficulty dealing with other sorts of threats, however. A command economy cannot innovate easily because the initiative of the populace is suppressed, which means that retooling for a different sort of threat—economic competition, for instance, or a shortage of liquid fuels—is more difficult.

Flexible governance, in contrast, is based on the capacity to adapt, improvise, and change directions. It may not be ideally designed for any specific threat, but it is able to deal credibly with a variety of threats. In general, systems based on participation, high skill levels, robust debate, and mutual respect are more likely to be flexible.

Command systems might seem to have a greater capacity to respond to a new type of threat because the people in command can simply direct people and resources to deal with it. But these systems have several inherent difficulties in actually doing this. Because relatively few people have an input into decisionmaking, there is lower capacity to recognize novel threats and to innovate against them. Subjects—those who are expected to follow orders—are typically less than enthusiastic in obeying. Finally, change can be threatening to those with power and privilege, so maintaining the relations of power can become more important than making sure the system survives.

An example of stiff governance is China in the 1950s, with a command economy driven by political ideology. The Great Leap Forward, launched in 1957, was an attempt to accelerate economic development. But the result was a vast famine that killed tens of millions of people and caused massive destruction of property and damage to the environment. The political system was incapable of responding to the catastrophe it created. Had there been a more flexible, open system in China, with independent media, things might have turned out differently. Countries with a flexible governance system are far less susceptible to famine because leaders are under greater pressure to respond to emerging crises. In essence, there is a feedback mechanism to stimulate political responses to a crisis, preventing cover-up and making inaction untenable.[5]

Centralized rule thus can be a threat in itself as well as an obstacle to responding to other sorts of threats. Fiji was a thriving multicultural democracy when, in 1987, there were two military coups. The result was mobilization of racism, emigration of skilled professionals, decline in the economy, general cultural stagnation, and ongoing political instability.[6]

Lessons from Civil Resistance

The history of civil resistance against repressive regimes reveals features that raise the odds of governance systems responding effectively to technological or political threats. The power of a mobilized citizenry is dramatically revealed in popular challenges to autocratic governments through demonstrations, strikes, boycotts, sit-ins, and other forms of protest, but without physical violence. This method of struggle is called nonviolent action, civil resistance, or "people power." In country after country, repressive rulers have succumbed to people power, for example in the Philippines in 1986, Eastern Europe in 1989, and Egypt in 2011. In these dramatic episodes, large numbers of people protested by using rallies, strikes, boycotts, and a host of other techniques, usually with little or no violence by the protesters.[7]

Erica Chenoweth and Maria J. Stephan, in a path-breaking study of people-power movements between 1900 and 2006, showed that regime change and anti-occupation nonviolent movements are more likely to be successful than armed movements in achieving their goals when facing similarly repressive opponents. (See Table 25–1.) They also found that success is more likely when large numbers of people are mobilized and when protesters are tactically and strategically innovative. When more people are actively involved, there is a greater capacity to try out creative ideas for resistance, which are needed to counter new repressive moves by the government. Greater participation needs to be accompanied by an ethos of inclusiveness, so that diverse groups can support the common cause. Groups with skills in many areas—including communication, organization, finance, languages, persuasion, and psychology—are valuable to help the movement operate effectively and survive attacks. If, for example, the movement depends on a single sector, such as students, it is easier for the government to repress or co-opt it. Wider participation provides a greater capacity for learning. This also provides a better basis for a stable, free society if the movement is successful in toppling a ruler.[8]

People power can be used to resist coups, as happened in Germany in

Table 25–1. Outcomes of Violent and Nonviolent Campaigns Aimed at Regime Change, Anti-occupation, or Secession, 1900–2006

	Regime Change		Anti-occupation		Secession	
Outcome	Violent (111 campaigns)	Nonviolent (81 campaigns)	Violent (59 campaigns)	Nonviolent (17 campaigns)	Violent (41 campaigns)	Nonviolent (4 campaigns)
			(percent)			
Success	27	59	36	35	10	0
Limited success	12	24	10	41	22	0
Failure	61	17	54	24	68	100

Source: See endnote 8.

1920, Algeria in 1961, and the Soviet Union in 1991. In each case, the key was the willingness of large numbers of people to take action—without using violence. In contrast, armed resistance to coups easily degenerates into civil war, which is a different sort of crisis, and a highly damaging one.[9]

Flexible Governance

Flexible governance means that there are methods for making and implementing decisions affecting entire communities in ways that enable rapid adaptation to new situations. This form of governance virtually requires flexible technological systems, which typically are modular, adaptable, and low cost.

In the energy sector, the best example of a rigid, inflexible technology is nuclear power, with its high capital cost, long lead times for construction, large unit sizes, and potential for causing environmental catastrophe through reactor accidents, terrorist attacks, or the proliferation of nuclear weapons. Because of its scale and potential risk, nuclear power requires special security measures, which in turn limit the possibility for citizen participation. Introduction of a "plutonium economy" based on the nuclear fuel cycle would drastically limit flexibility in both energy systems and governance.[10]

Small-scale renewable energy systems are better matched to flexible governance. Community-level solar and wind systems are relatively low cost, quick to construct, and small in scale, with only a small potential for environmental risk: for example, terrorists are unlikely to attack them. These features mean that communities are less locked-in to the technology and, just as important, that corporate and government commitments to the system are less entrenched.[11]

Most technologies are intermediate in scale between nuclear power and a solar hot water heater, but the same sort of analysis applies: technologies with lower unit costs and lower potential risks to health and the environment are usually also more amenable to citizen control. In short, flexible technological systems are well suited to flexible governance.

The experience of people power against repression provides a template for the sort of governance most likely to be effective in crises. There are four key features:

- *Participation of significant numbers of people.* Significant participation is essential for rapidly responding to crises. People's commitment comes from being involved in decisionmaking and feeling part of the solution. Genuine participation is greatest when power is shared. Governance with extensive participation goes under various names, including participatory democracy, self-management, workers' control, and neighborhood power.[12]
- *Resources, including food, transport, and especially communication.* Resources, including material and technological resources, need to be available and ready. A society needs to have the capacity to deal with future contingencies rather than putting all its resources into one development path.
- *Openness, tolerance, and inclusion, with involvement of many different sectors of the population.* Openness, tolerance, and inclusion are necessary to be able to mobilize the entire society to meet the challenge. When significant groups are opposed to action, this can paralyze efforts. The governance form most suited to inclusion is consensus, sometimes called *unitary democracy*, in contrast to representative government, which can be called *adversary democracy*. But just as electoral systems require innovation and modification to address problems such as voting fraud, consensus systems require experience, testing, and innovation to address problems such as entrenched resistance to a near-unanimous agreement. There is now considerable experience with consensus-building processes.[13]
- *Learning skills for struggle and developing strategic acumen.* Skills and strategic acumen are needed to be effective in responding to threats intelligently rather than in an instinctive, unreflective way. Strategic insight is most likely to flourish in a form of governance that gives considerable autonomy to smaller units, while enabling communication between them so that insights can be shared, tested, and applied.

These four features are mutually supportive. Widespread participation is necessary for collective change or response, but it needs to be coordinated, hence the need for communication infrastructure and skills. Strategy can be more adaptable when there is openness to participation by a wide diversity of individuals with different perspectives and recognition that their perspectives and ideas may be worthwhile.

Openness, tolerance, and inclusion include forging links with sectors of the population often seen to be part of the problem. In a military coup, soldiers are at the heart of the threat. People-power resistance requires winning over some of the soldiers, weakening their resolve or convincing them to join the opposition. Armed resistance is counterproductive for this purpose when it stimulates unity within the regime, as often occurs. By analogy, in dealing with other sorts of threats, tactics need to be chosen that win over some people normally seen as being on the "other side," whether corporate elites, government personnel, or security forces.[14]

Adding these elements together, the form of governance most promising for responding to threats will have significant citizen participation in decisionmaking, will allocate ample resources for communication and contingencies, will include diverse groups in the population, and will allow decentralized yet coordinated action.

Transforming Governance

Rather than try to describe this flexible form of governance—which can quickly degenerate into arguments about preferred models—it is useful to look at methods for moving toward these four elements. In other words, rather than fixating on the desirable end state, which might not be knowable anyway, it is worth turning each of the elements of flexible governance into methods for transforming governance.

Significant Participation. Initiatives to foster participation can be taken at all levels. Within local groups—including formal associations from sports clubs to churches, and informal groups—leaders and members can foster greater participation. Local governments can introduce various forms of citizen participation. Companies can promote worker participation.

One of the most promising initiatives is the movement for "deliberative democracy," which includes experiments in direct decisionmaking by citizens on important policy issues. An example of this is inviting a randomly selected group of 12–25 citizens—called *planning cells* or *citizens' juries*—to address a policy issue over several days by reading materials, hearing from experts and partisans, and developing recommendations, all under the guidance of neutral facilitators. Hundreds of such exercises have been held in various countries, including Australia, the United Kingdom, Germany, and the United States. Many of these deliberative democracy initiatives are taking place below the radar of mainstream politics and the mass media, so few people realize how much of this activity is occurring.[15]

In crises, opportunities can exist for dramatically increased participation. Historically, there are numerous examples of popular participation in crisis situations, such as in Paris in 1871, Russia in 1917, Spain in the late 1930s, and France in 1968. These revolutions of popular control were all suppressed by the state, but they do show the possibility for citizens to reorganize decisionmaking at short notice.[16]

In contrast, after the breakup of the Soviet Union in 1991 there was a rapid transition to predatory capitalism involving massive corruption: popular mobilization was restricted to resisting a coup rather than creating a participatory alternative. This suggests the importance of local initiatives that build the foundation for a genuinely participatory alternative.[17]

In Argentina, following the 1999 economic collapse and the freezing of bank assets in December 2001, in a surge of local initiatives workers took over failed businesses, and communities made decisions in neighborhood assemblies. The Argentine initiatives have succeeded more than some previous ones, perhaps because there was less of an attempt to take over the state and more emphasis on creating a living alternative.[18]

Environmental movements can contribute to transforming governance through the way they operate. When movements are made up of many local groups that foster participation—for example, through consensus decision-making—and are not dominated by central offices and paid staff, they are ideally poised to react quickly and creatively to existing and new crises. They also provide a model of flexible governance.

Resources for Struggle. Promoting the development of resources for any struggle is an ongoing process in which many groups are involved. The movement for appropriate technology—typically small-scale, low-cost, locally produced, and locally managed technology for energy, agriculture, transport, and other sectors—is a model for building resources that support resilient governance. Communities using appropriate technology are better able to survive in the face of economic or physical-system collapse: they can rely on their own resources without excessive dependence on imports or specialist skills.[19]

A demonstration garden in the Transition Town Linlithgow, Scotland.

yellow book

The Transition Towns movement, motivated by preparation for a looming shortage of cheap liquid fuels and the impacts of climate change, combines local participation in planning with the promotion of community resilience, including local production of food, energy, and housing. In this model, resources for struggle are developed as part of the struggle itself.[20]

In the communication sector, the key is the ability to maintain communication in a crisis. The technology for network communication is becoming ever-more developed with the Internet, Web 2.0, and social media. These provide powerful tools for rapidly and flexibly responding to emergencies, and when people gain practice in coordinating responses, this has relevance for both political and environmental crises.

Working against this ability are governments and corporations that seek to limit communication freedom, for example through censorship, surveillance, and controls over innovation in the guise of intellectual property. If governments can shut down or restrict the Internet for political purposes—as happened in Egypt in 2011, among other places—and use digital surveillance techniques to track dissidents, the ability and willingness of citizens to coordinate against threats, whether political or environmental, will be reduced. The struggle for free communication can be considered an essential part of the struggle for more flexible governance.[21]

Openness, Tolerance, and Inclusion. Movements that polarize society, turning some groups into enemies, contribute to stiff governance. U.S. foreign and domestic policies have done this. Foreign military interventions such as in Afghanistan and Iraq, with civilian deaths as "collateral damage," create enmity and enemies and then, when foreign groups retaliate, become justifications for further interventions. The domestic response to 9/11, which involved labeling terrorists as enemies to be destroyed, did little to include a range of groups in a struggle against the roots of terrorism. In this context, efforts to promote tolerance and inclusion—nationally and internationally—are important in moving toward flexible governance.[22]

One of the biggest challenges ahead is growing economic inequality, leading to disenfranchisement of all but the wealthy. Responding to economic, resource, or political crises will be much more difficult in societies divided into "haves" and "have-nots." This suggests that movements for greater economic equality can, as a side effect, help build resilience. The Occupy Movement has put the issue of inequality on the popular and political agenda, but it remains to be seen if this can slow or reverse the continuing increase in inequality stimulated by corporate globalization.

Pervasive corruption is a major obstacle to good governance. One of the most powerful tools against corruption is nonviolent action; some popular challenges to repressive regimes, such as in Egypt in 2011, have been stimulated by opposition to high-level corruption. Political and economic systems that permit fair participation by a wide range of groups rather than siphoning spoils to the ruling elite are more likely to lead to prosperity. Inclusion thus is a key to greater commitment in addressing social problems.[23]

Learning Skills for Struggle and Developing Strategic Acumen. Numerous initiatives and movements around the world foster greater skills for satisfying human needs, from agriculture to software development. A prime example is the open source movement, building software and other products that draw on contributions from numerous volunteers. Another example is the ever-increasing information and tools for learning available on the Internet, enabling learning outside of institutions. Community renewable-energy projects foster learning of practical skills; the Danish community wind-power movement in the 1970s did this while sparking development of what is now a major industry. Also relevant are self-help groups—for example, addressing particular diseases or experiences ranging from breast cancer to having a family member in prison. There are a growing number of activist handbooks and activist training programs.[24]

As more and more people increase their education (formal and informal) and engage in civic initiatives (face-to-face or online), skills and strategic flexibility increase. Especially relevant for this are initiatives to provide experience in governance, such as the participatory budgeting pioneered in cities in Brazil. In a typical process of participatory budgeting, multiple citizen assemblies discuss priorities, and then a participatory budget council, with representatives from the assemblies, deliberates on priorities, negotiating between the assemblies and the city administration.[25]

In a Crisis

International governance is particularly unsuited for dealing with crises. The United Nations might give the appearance of having a centralized response capability, but in reality it is the tool of powerful governments that have their own agendas. There is little citizen participation and little capacity for skill development. The result is a form of symbolic politics that gives only the illusion of authority and response capacity.[26]

In Rwanda in 1994, for example, when mass killings commenced, western governments pulled out their citizens, thereby removing sources of information on and witnesses to human rights violations. The United Nations Security Council dithered and then withdrew most U.N. peacekeepers. In this case, international governments utterly failed to avert or confront a genocide in which over half a million people were killed.[27]

Rapidly developing crises are obvious and hence are more likely to stimulate responses. Far more challenging are slow-moving crises, which escape attention but can cause just as much damage. An example is the oil spill in Guadalupe Dunes on the central Californian coast, which released as much oil as in the famous 1989 *Exxon Valdez* spill but which is virtually unknown. Because it happened more slowly, over decades, people in the region accommodated the oil releases, psychologically and socially.[28]

Climate change is the most prominent slow-moving crisis. As in the case of war and genocide, many governments and international bodies have provided only symbolic gestures. By far the most effective response arguably has come from grassroots groups and local governments, indicating the importance of participation in dealing with environmental crises.

Moving toward Flexible Governance

Governance is often seen as a comprehensive package: an entire system, operating according to a consistent set of principles, whether it be dictatorship, representative government, or a modern-day plutocracy in which the rich rule via captive politicians. Any such pure system of governance would be suited for one set of conditions but be vulnerable to sudden changes. However, actual systems in the world today are mixed. The United States, for example, could be considered a combination of representative government, plutocracy, a security state, and pockets of participatory democracy ranging from cooperatives to the free-software movement. Governance in practice contains rigidities, capacities, and possibilities.

In the face of threats and crises—political, economic, and resource-based—the most promising sort of governance is flexible, able to draw on widespread participation and an abundance of human and material resources. The inclusion of different groups provides a greater diversity of knowledge and experience for meeting challenges. Whether or not there is an ideal system with all these characteristics, it is possible to move in the direction of flexible governance by taking initiatives that support participation, resources for struggle, inclusion, and skills development.

In responding to environmental and resource crises, activists usually focus primarily on the immediate issues—trying to stop logging, for example, or the burning of fossil fuels and other damaging activities. To maximize long-term effectiveness, it is valuable to complement these actions with efforts to transform governance, as otherwise the same problems will recur. Ideally, responses to environmental problems should themselves incorporate the elements of flexible governance, so that current actions can help create the sort of institutions that are more capable of dealing with problems and preventing them in the first place.

CHAPTER 26

Governance in the Long Emergency

David W. Orr

The first evidence linking climate change and human emissions of carbon dioxide was painstakingly assembled in 1897 by Swedish scientist Svante Arrhenius. What began as an interesting but seemingly unimportant conjecture about the effect of rising carbon dioxide on temperature has turned into a flood of increasingly urgent and rigorous warnings about the rapid warming of Earth and the dire consequences of inaction. Nonetheless, the global dialogue on climate is floundering while the scientific and anecdotal evidence of rapid climate destabilization grows by the day.[1]

We have entered a "long emergency" in which a myriad of worsening ecological, social, and economic problems and dilemmas at different geographic and temporal scales are converging as a crisis of crises. It is a collision of two non-linear systems—the biosphere and biogeochemical cycles on one side and human institutions, organizations, and governments on the other. But the response at the national and international levels has so far been indifferent to inconsistent, and nowhere more flagrantly so than in the United States, which is responsible for about 28 percent of the fossil-fuel carbon that humanity added to the atmosphere between 1850 and 2002.[2]

The "perfect storm" that lies ahead is caused by the collision of changing climate; spreading ecological disorder (including deforestation, soil loss, water shortages, species loss, ocean acidification); population growth; unfair distribution of the costs, risks, and benefits of economic growth; national, ethnic, and religious tensions; and the proliferation of nuclear weapons—all compounded by systemic failures of foresight and policy. As a consequence, in political theorist Brian Barry's words, "it is quite possible that by the year 2100 human life will have become extinct or will be confined to a few residential areas that have escaped the devastating effects of nuclear holocaust or global warming."[3]

Part of the reason for paralysis is the sheer difficulty of the issue. Climate change is scientifically complex, politically divisive, economically costly,

David W. Orr is the Paul Sears Distinguished Professor of Environmental Studies and Politics at Oberlin College in Ohio.

www.sustainabilitypossible.org

morally contentious, and ever so easy to deny or defer to others at some later time. But the continuing failure to anticipate and forestall the worst effects of climate destabilization in the face of overwhelming scientific evidence is the largest political and moral failure in history. Indeed, it is a crime across generations for which we have, as yet, no name.

Barring a technological miracle, we have condemned ourselves and posterity to live with growing climate instability for hundreds or even thousands of years. No government has yet shown the foresight, will, creativity, or capacity to deal with problems at this scale, complexity, or duration. No government is prepared to make the "tragic choices" ahead humanely and rationally. And no government has yet demonstrated the willingness to rethink its own mission at the intersection of climate instability and conventional economic wisdom. The same is true in the realm of international governance. In the words of historian Mark Mazower: "The real world challenges mount around us in the shape of climate change, financial instability . . . [but there is] no single agency able to coordinate the response to global warming."[4]

The Problem of Governance

In *An Inquiry into the Human Prospect*, in 1974, economist Robert Heilbroner wrote: "I not only predict but I prescribe a centralization of power as the only means by which our threatened and dangerous civilization will make way for its successor." Heilbroner's description of the human prospect included global warming but also other threats to industrial civilization, including the possibility that finally we would not care enough to do the things necessary to protect posterity. The extent to which power must be centralized, he said, depends on the capacity of populations, accustomed to affluence, for self-discipline. But he did not find "much evidence in history—especially in the history of nations organized under the materialistic and individualistic promptings of an industrial civilization—to encourage expectations of an easy subordination of the private interest to the public weal."[5]

Heilbroner's conclusions are broadly similar to those of others, including British sociologist Anthony Giddens, who somewhat less apocalyptically proposes "a return to greater state interventionism"—but as a catalyst, facilitator, and enforcer of guarantees. Giddens believes the climate crisis will motivate governments to create new partnerships with corporations and civil society, which is to say more of the same, only bigger and better. David Rothkopf of the Carnegie Endowment for International Peace likewise argues that the role of the state must evolve toward larger, more innovative governments and "stronger international institutions [as] the only possible way to preserve national interests."[6]

The performance of highly centralized governments, however, is not encouraging—especially relative to the conditions of the long emergency. Governments have been effective at waging war and sometimes in solving— or appearing to solve—economic problems. But even then they are cumbersome, slow, and excessively bureaucratic. They tend to fragment agencies by problem, rather like mailbox pigeonholes, but the long emergency will require managing complex systems over long time periods. Might there be more agile, dependable, and less awkward ways to conduct the public business in the long emergency that do not require authoritarian governments, the compromises and irrational messiness of politics, or even reliance on personal sacrifice? Can these be made to work over the long time spans necessary to stabilize the climate? If not, how else might we conduct the public business? Broadly, there are three other possibilities.[7]

First, champions of markets and advanced technology propose to solve the climate crisis by harnessing the power of markets and technological innovation to avoid what they regard as the quagmire of government. Rational corporate behavior responding to markets and prices, they believe, can stabilize climate faster at lower costs and without hair-shirt sacrifice, moral posturing, and slow, clumsy, overbearing bureaucracies. The reason is said to be the power of informed self-interest plus the ongoing revolution in energy technology that has made efficiency and renewable energy cheaper, faster, less risky, and more profitable than fossil fuels. In their 2011 book, *Reinventing Fire*, Amory Lovins and his coauthors, for example, ask whether "the United States could realistically stop using oil and coal by 2050? And could such a vast transition toward efficient use and renewable energy be led by business for durable advantage?" The answer, they say, is yes, and the reasoning and data they marshal are formidable.[8]

But why would corporations, particularly those in highly subsidized extractive industries, agree to change as long as they can pass on the costs of climate change to someone else? Who would pay for the "stranded" oil and coal reserves (with an estimated value in excess of $20 trillion) that cannot be burned if we are to stay below a 2 degree Celsius warming—often thought to be the threshold of catastrophe? Would corporations continue to use their financial power to manipulate public opinion, undermine regulations, and oppose an equitable sharing of costs, risks, and benefits? How does corporate responsibility fit with the capitalist drive to expand market share? Economist Robert Reich concludes that given the existing rules of the market, corporations "*cannot* be socially responsible, at least not to any significant extent. . . . Supercapitalism does not permit acts of corporate virtue that erode the bottom line. No corporation can 'voluntarily' take on an extra cost that its competitors don't also take on." He further argues that the alleged convergence of social responsibility and profitability is unsupported by any factual evidence.[9]

There are still larger questions about how large corporations fit in democratic societies. One of the most insightful students of politics and economics, Yale political scientist Charles Lindblom, concluded his magisterial *Politics and Markets* in 1977 with the observation that "the large private corporation fits oddly into democratic theory and vision. *Indeed, it does not fit*" (emphasis added). Until democratized internally, stripped of legal "personhood," and rendered publicly accountable, large corporations will remain autocratic fiefdoms, for the most part beyond public control.[10]

These issues require us to ask what kind of societies and what kind of global community do we intend to build? It is certainly possible to imagine a corporate-dominated, hyper-efficient, solar-powered, sustainable world that is also grossly unfair, violent, and fascist. To organize society mostly by market transactions would be to create a kind of Ayn Randian hell that would demolish society, as economist Karl Polanyi once said. Some things should never be sold—because the selling undermines human rights; because it would violate the law and procedural requirements for openness and fairness; because it would have a coarsening effect on society; because the sale would steal from the poor and vulnerable, including future generations; because the thing to be sold is part of the common heritage of humankind and so can have no rightful owner; and because the thing to be sold—including government itself—should simply not be for sale.[11]

A second alternative to authoritarian governments may lie in the emergence of national and global networks abetted by the Internet and advancing communications technology. They are decentralized, self-replicating, and sometimes self-correcting. In time, they might grow into a global system doing what traditional governments and international agencies once did—but better, faster, and cheaper. Some analysts believe that the old model of the nation-state is inadequate to meet many of the challenges of the long emergency and is losing power to a variety of novel organizations. Anne-Marie Slaughter of Princeton University, for one, envisions networks of "disaggregated states in which national government officials interact intensively with one another and adopt codes of best practices and agree on coordinated solutions to common problems," thereby sidestepping conventional intergovernmental practices and international politics.[12]

Below the level of governments there is, in fact, an explosion of nongovernmental organizations, citizens' groups, and professional networks that are already assuming many of the functions and responsibilities once left to governments. Writer and entrepreneur Paul Hawken believes that the world is already being reshaped by a global upwelling of grassroots organizations promoting sustainable economies, renewable energy, justice, transparency, and community mobilization. Many of the thousands of groups Hawken describes are linked in "global action networks," organized around specific issues to provide "communication platforms for sub-groups to organize in ever-more-specialized geographic and sub-issue networks." Early examples include the International Red Cross and the International Labour Organization.[13]

Recently clusters of nongovernmental groups have organized around issues such as common property resources, global financing for local projects, water, climate, political campaigns, and access to information. They are fast, agile, and participatory. Relative to other citizens' efforts, they require little funding. But like other grassroots organizations, they have no power to legislate, tax, or enforce rules. In Mark Mazower's words, "Many are too opaque and unrepresentative to any collective body." Much of the same, he believes, can be said of foundations and philanthropists. By applying business methods to social problems, Mazower writes, "Philanthrocapitalists exaggerate what technology can do, ignore the complexities of social and institutional constraints, often waste sums that would have been better spent more carefully and wreak havoc with the existing fabric of society in places they know very little about." Moreover, they are not immune to fashion, delusion, corruption, and arrogance. Nor are they often held accountable to the public.[14]

School children joined with local organizations in Nagpur, India, to form a giant 350 on the International Day of Climate Action, 2008.

So what is to be done? Robert Heilbroner proposed enlarging the powers of the state. Green economy advocates believe that corporations can lead the transition through the long emergency. Others argue that an effective planetary immune system is already emerging in the form of networks. Each offers a piece in a larger puzzle. But there is a fourth possibility. Canadian writer and activist Naomi Klein proposes that we strengthen and deepen the practice of democracy even as we enlarge the power of the state. "Responding to climate change," she writes:

> requires that we break every rule in the free-market playbook and that we do so with great urgency. We will need to rebuild the public sphere, reverse privatizations, relocalize large parts of economies, scale back overconsumption, bring back long-term planning, heavily regulate and tax corporations, maybe even nationalize some of them, cut military spending and recognize our debts to the global South. Of course,

none of this has a hope in hell of happening unless it is accompanied by a massive, broad-based effort to radically reduce the influence that corporations have over the political process. That means, at a minimum, publicly funded elections and stripping corporations of their status as "people" under the law.[15]

Democracy, Winston Churchill once famously said, is the worst form of government except for all the others ever tried. But has it ever been tried? In columnist Harold Myerson's words, "the problem isn't that we're too democratic. It's that we're not democratic enough." The authors of the U.S. Constitution, for example, grounded ultimate power in "we the people" while denying them any such power or even much access to it.[16]

Political theorist Benjamin Barber proposes that we take some of the power back by revitalizing society as a "strong democracy," by which he means a "self-governing community of citizens who are united less by homogeneous interests than by civic education and who are made capable of common purpose and mutual action by virtue of their civic attitudes and participatory institutions rather than their altruism or their good nature." Strong democracy requires engaged, thoughtful citizens, as once proposed by Thomas Jefferson and John Dewey. The primary obstacle, Barber concedes, is the lack of a "nationwide system of local civic participation." To fill that void he proposes, among other things, a national system of neighborhood assemblies rebuilding democracy from the bottom up.[17]

Political theorists Amy Gutmann and Dennis Thompson similarly propose the creation of deliberative institutions in which "free and equal citizens (and their representatives), justify decisions in a process in which they give one another reasons that are mutually acceptable and generally accessible, with the aim of reaching conclusions that are binding in the present to all citizens but open to challenge in the future." Reminiscent of classical Greek democracy, they intend to get people talking about large issues in public settings in order to raise the legitimacy of policy choices, improve public knowledge, and increase civil discourse. (See Box 26–1.) A great deal depends, they concede, on the durability and vitality of practices and institutions that enable deliberation to work well.[18]

Political scientists Bruce Ackerman and James Fishkin propose a new national holiday, Deliberation Day, on which citizens would meet in structured dialogues about issues and candidates. They believe that "ordinary citizens are willing and able to take on the challenge of civic deliberation during ordinary times" in a properly structured setting that "facilitates genuine learning about the choices confronting the political community."[19]

Legal scholar Sanford Levinson believes, however, that reforms will be ineffective without first repairing the structural flaws in the U.S. Constitution, which is less democratic than any of the 50 state constitutions in the

Box 26–1. A More Sustainable Democracy

Philosophers have argued through the ages that democracy is the best form of government, and some have claimed that the deeper it is, the better. By "deeper" they mean a structure that spreads power widely, engages more people, and invites them to take a more direct role in the shaping of policy.

Most liberal (current) democracies do not meet that definition, being republican in form and thus giving most power and decisionmaking responsibility to elected representatives. In some of these republics, democracy is even further degraded. In the United States, for instance, Supreme Court decisions over the years have established that there is essentially no difference in civic standing between individual citizens and corporations or other private interests that can and do spend billions of dollars on political advertising, lobbying, and propaganda (over $8 billion in the 2010 election cycle).

But it is not simply such distortions of democracy that compel a closer look at the benefits of deepening it. The democracies that most of the industrial world lives in have been derided by political theorist Benjamin Barber as "politics as zookeeping"—systems designed "to keep men safely apart rather than bring them fruitfully together." In fact there are major potential advantages in bringing people fruitfully together in the political arena, not least with respect to the environmental crises that beset humanity now. Paradoxically, one of the weaknesses of liberal democracy may be not that it asks too much of its citizens but that it asks too little. Having mostly handed off all responsibility for assessing issues and setting policy to elected politicians, voters are free to indulge themselves in narrow and virulently asserted positions rather than having to come together, work to perceive the common good, and plot a course toward it.

One antidote to this is deliberation. Deliberative democracy can take many forms, but its essence, according to social scientist Adolf Gundersen, is "the process by which individuals actively confront challenges to their beliefs." It can happen when someone reads a book and thinks about what it says, but in the public sphere more generally it means engaging in pairs or larger groups to discuss issues, compare notes, probe (not attack) one another's assertions, and take the opportunity to evolve a personal position in the interests of forging a collective one. Deliberative democracy, in Gundersen's words, "challenges citizens to move beyond their present beliefs, develop their ideas, and examine their values. It calls upon them to make connections, to connect more firmly and fully with the people and the world around them." When arranged to address environmental aims, deliberative democracy "connects the people, first with each other and then with the environment they wish not simply to visit, but also to inhabit."

Given the uneven record of democracies in educating their people into citizenship, true deliberation might be difficult to learn, especially in countries where the politics are strongly adversarial. Deliberative democracy is a "conversation," Gundersen says, "not a series of speeches." Conversations involve respectful listening—not just waiting to talk—as well as speaking. Yet there is an untapped hunger for it that can be released when the circumstances are conducive. And Gundersen has established through 240 hours of interviews with 46 Americans that deliberation about environmental matters "leads citizens to think of our collective pursuit of environmental ends in a more *collective, long-term, holistic, and self-reflective* way." Such thinking might be the indispensable foundation for achieving anything like sustainability.

—Tom Prugh
Codirector, State of the World 2013
Source: See endnote 18.

United States. He proposes a Constitutional Convention of citizens selected by lottery proportional to state populations to remodel the basic structure of governance. Whether this is feasible or not, the U.S. Constitution has other flaws that will limit effective responses to problems of governance in the long emergency.[20]

In this regard the U.S. Constitution is typical of others in giving no "clear, unambiguous textual foundation for federal environmental protection law," notes legal scholar Richard Lazarus. It privileges "decentralized, fragmented, and incremental lawmaking . . . which makes it difficult to address issues in a comprehensive, holistic fashion." Congressional committee jurisdiction based on the Constitution further fragments responsibility and legislative results. The Constitution gives too much power to private rights as opposed to public goods. It does not mention the environment or the need to protect soils, air, water, wildlife, and climate and so it offers no unambiguous basis for environmental protection. The commerce clause, the source for major environmental statutes, is a cumbersome and awkward legal basis for environmental protection. The result, Lazarus notes, is that "our lawmaking institutions are particularly inapt for the task of considering problems and crafting legal solutions of the spatial and temporal dimensions necessary for environmental law."[21]

The U.S. Constitution is deficient in other ways as well. Posterity is mentioned only in the Preamble, but not thereafter. The omission, understandable when the Constitution was written, now poses an egregious wrong. In 1787, the framers could have had no premonition that far into the future one generation could deprive all others of life, liberty, and property without due process of law or even good cause. And so, in theologian Thomas Berry's words: "It is already determined that our children and grandchildren will live amid the ruined infrastructures of the industrial world and amid the ruins of the natural world itself." The U.S. Constitution gives them no protection whatsoever.[22]

Further, with a few notable exceptions—such as in Ecuador—most constitutions pertain only to humans and their affairs and property. We privilege humans, while excluding other members of the biotic community. A more expansive system of governance would extend rights of sorts and in some fashion to species, rivers, landscapes, ecologies, and trees, as legal scholar Christopher Stone once proposed. In Thomas Berry's words: "We have established our human governance with little regard for the need to integrate it with the functional order of the planet itself." In fact, from our bodies to our global civilization we are part of a worldwide parliament of beings, systems, and forces far beyond our understanding. We are kin to all that ever was and all that ever will be and must learn what that fact means for governance.[23]

Building the Foundations of Robust Democracies

The history of democracy is complex and often troubled. In classical Athens it lasted only 200 years. Political philosopher John Plamenatz once wrote that "democracy is the best form of government only when certain conditions hold." But those conditions may not hold in established democracies in the long emergency ahead and may be impossible in less stable societies and failed states with no history of it. The reasons are many.[24]

For one, citizens in most democratic societies have become accustomed to comfort and affluence, but democracy "requires citizens who are willing to sacrifice for the common good and [restrain] their passions," notes political theorist Wilson Carey McWilliams. How people shaped by consumption will respond politically in what will certainly be more straitened times is unknown. Political analyst Peter Burnell cautions that "democratization does not necessarily make it easier and can make it more difficult for countries to engage with climate mitigation."[25]

Even in the best of times, however, representative democracies are vulnerable to neglect, changing circumstances, corruption, the frailties of human judgment, and the political uses of fear—whether of terrorism or subversion. They tend to become ineffective, sclerotic, and easily co-opted by the powerful and wealthy. They are vulnerable to militarization, as James Madison noted long ago. They are susceptible to ideologically driven factions that refuse to play by the rules of compromise, tolerance, and fair play. They work differently at different scales. And they cannot long endure the many economic and social forces that corrode political intelligence and democratic competence.[26]

Democracies are also vulnerable to what conservative philosopher Richard Weaver once described as the spoiled-child psychology, "a kind of irresponsibility of the mental process . . . because [people] do not have to think to survive . . . typical thinking of such people [exhibits] a sort of contempt for realities." Psychologists Jean Twenge and Keith Campbell believe that the behavior Weaver noted in the 1940s has now exploded into a full-blown "epidemic of narcissism." Such failures of personality, judgment, and character could multiply under the stresses likely in the long emergency.[27]

We are between the proverbial rock and a hard place. There is no good case to be made for smaller governments in the long emergency unless we wish to sharply reduce our security and lower our standards for the public downward to a libertarian, gun-toting, free-for-all—Thomas Hobbes's nightmare on steroids. On the contrary, it will be necessary to enlarge governments domestically and internationally to deal with the nastier aspects of the long emergency, including relocating people from rising oceans and spreading deserts, restoring order in the wake of large storms, managing conflicts over diminishing water, food, and resources, dealing with the spread of diseases, and managing the difficult transition to a post-growth economy. On the other hand, we have good reason to fear an enlargement of government powers as both ineffective and potentially oppressive.[28]

Given those choices, there is no good outcome that does not require something like a second democratic revolution in which we must master the art and science of governance for a new era—creating and maintaining governments that are ecologically competent, effective at managing complex systems, agile, capable of foresight, and sturdy over an extraordinary time span. If we intend for such governments to also be democratic, we will have to summon an extraordinary level of political creativity and courage. To meet the challenges of the late eighteenth century, James Madison argued that democracy required a free press that served a well-informed and engaged citizenry, fair and open elections, and reliable ways to counterbalance competing interests. But he feared that even the best government with indifferent and incompetent citizens and leaders would sooner or later come to ruin.

In our time, strong democracy may be our best hope for governance in the long emergency, but it will not develop, persist, and flourish without significant changes. The most difficult of these will require that we confront the age-old nemesis of democracy: economic oligarchy. Today the majority of concentrated wealth is tied, directly or indirectly, to the extraction, processing, and sale of fossil fuels, which is also the major driver of the long emergency. Decades of rising global inequality have entrenched control in a small group of super-wealthy individuals, financiers, corporations, media tycoons, drug lords, and celebrities in positions of unaccountable authority.[29]

Angelo Lopez/Cartoon Movement

In the United States, for example, the wealthiest 400 individuals have more net wealth than the bottom 185,000,000 people. Six Walmart heirs alone control as much wealth as the bottom 42 percent of the U.S. population. Rising inequality in the United States and elsewhere reflects neither efficiency nor merit. And beyond some threshold it divides society by class, erodes empathy, hardens hearts, undermines public trust, incites violence, saps our collective imagination, and destroys the public spirit that upholds democracy and community alike. Nonetheless, the rich do not give up easily. According to political economist Jeffrey Winters, the redistribution of wealth has always occurred as a result of war, conquest, or revolution, not as a democratic decision or from the benevolence of plutocrats.[30]

Toward the end of his life, historian Lewis Mumford concluded that the only way out of this conundrum is "a steady withdrawal" from the "megamachine" of technocratic and corporate control. He did not mean community-scale isolation and autarky, but rather more equitable, decentralized, and self-reliant communities that met a significant portion of their needs for food, energy, shelter, waste cycling, and economic support. He did not propose secession from the national and global community but rather withdrawal from dependence on the forces of oligarchy, technological domination, and zombie-like consumption. Half a century later, that remains the most likely strategy for building the foundations of democracies robust enough to see us through the tribulations ahead.[31]

In other words, the alternative to a futile and probably bloody attempt to forcibly redistribute wealth is to spread the ownership of economic assets throughout society. From the pioneering work of progressive economists, scholars, and activists such as Scott Bernstein, Michael Shuman, Gar Alperovitz, Ted Howard, and Jeff Gates we know that revitalization of local economies through worker-owned businesses, local investment, and greater local self-reliance is smart economics, wise social policy, smart environmental management, and a solid foundation for both democracy and national resilience.[32]

Simultaneously, and without much public notice, there have been dramatic advances in ecological design, biomimicry, distributed renewable energy, efficiency, ecological engineering, transportation infrastructure, permaculture, and natural systems agriculture. Applied systematically at community, city, and regional scales, ecological design opens genuine possibilities for greater local control over energy, food, shelter, money, water, transportation, and waste cycling. (See Box 26–2.) It is the most likely basis for revitalizing local economies powered by home-grown efficiency and locally accessible renewable energy while eliminating pollution, improving resilience, and spreading wealth. The upshot at a national level is to reduce the need for government regulation, which pleases conservatives, while improving quality of life, which appeals to liberals. Fifty years ago, Mumford's suggestion seemed unlikely. But in the years since, local self-reliance, Transition Towns, and regional policy initiatives are leading progressive changes throughout Europe and the United States while central governments have been rendered ineffective.[33]

A second change is in order. Democracies from classical Athens to the present are only as vibrant as the quality and moral power of the ideas they can muster, mull over, and act upon. Debate, argument, and civil conversation are the lifeblood of the democratic process. In our time, said to be an

Box 26–2. Resilience from the Bottom Up

At the dawn of the modern environmental era, in 1970, the National Environmental Policy Act required all federal agencies to "utilize a systematic, interdisciplinary approach which will insure the integrated use of the natural and social sciences and the environmental design arts in planning and in decision-making." Nonetheless, the government and corporations, foundations, and nonprofit organizations still work mostly by breaking issues and problems into their parts and dealing with each in isolation. Separate agencies, departments, and organizations specialize in energy, land, food, air, water, wildlife, economy, finance, building regulations, urban policy, technology, health, and transportation as if each were unrelated to the others.

Reducing wholes to parts is the core of the modern worldview we inherited from Galileo, Bacon, and Descartes. And for a time it worked economic, scientific, and technological miracles. But the price we pay is considerable and growing fast. For one, we seldom anticipate or account for collateral costs of fragmentation or count the benefits of systems integration. We mostly focus on short-term benefits while ignoring long-term risks and vulnerabilities. Imponderables and non-priced benefits are excluded altogether. The results corrupt our politics, economics, and values, and they undermine our prospects.

Nonetheless, we administer, organize, and analyze in parts, not wholes. But in the real world there are tipping points, surprises, step-level changes, time delays, and unpredictable, high-impact events. To fathom such things requires a mind-set capable of seeing connections, systems, and patterns as well as a perspective far longer than next year's election or an annual balance sheet. Awareness that we live in systems we can never fully comprehend and control and humility in the face of the unknown gives rise to precaution and resilient design.

One example of this approach comes from Oberlin, a small city of about 10,000 people with a poverty level of 25 percent in the center of the U.S. "Rust Belt." It is situated in a once-prosperous indus-

trial region sacrificed to political expediency and bad economic policy, not too far from Cleveland and Detroit. But things here are beginning to change. In 2009, Oberlin College and the city launched the Oberlin Project. It has five goals: build a sustainable economy, become climate-positive, restore a robust local farm economy supplying up to 70 percent of the city's food, educate at all levels for sustainability, and help catalyze similar efforts across the United States at larger scales. The community is organized into seven teams, focused on economic development, education, law and policy, energy, community engagement, food and agriculture, and data analysis. The project aims for "full-spectrum sustainability," in which each of the parts supports the resilience and prosperity of the whole community in a way that is catalytic—shifting the default setting of the city, the community, and the college to a collaborative post-cheap-fossil-fuel model of resilient sustainability.

The Oberlin Project is one of a growing number of examples of integrated or full-spectrum sustainability worldwide, including the Mondragón Cooperative in Spain, the Transition Towns movement, and the Evergreen Project in Cleveland. In different ways, each is aiming to transform complex systems called cities and city-regions into sustainable, locally generated centers of prosperity, powered by efficiency and renewable energy. Each is aiming to create opportunities for good work and higher levels of worker ownership of renewably powered enterprises organized around necessities. The upshot is a global movement toward communities with the capacity to withstand outside disturbances while preserving core values and functions. In practical terms, resilience means redundancy of major functions, appropriate scale, firebreaks between critical systems, fairness, and societies that are "robust to error," technological accidents, malice, and climate destabilization. In short, it is human systems designed in much the way that nature designs ecologies: from the bottom up.

Source: See endnote 33.

age of information, one of the most striking characteristics is the triviality, narrowness, and often factual inaccuracy of our political conversations. Much of what passes for public dialogue has to do with jobs and economic growth, but it is based on economic theories that fit neither biophysical reality nor the highest aspirations of humankind. The rules of market economies are said to date from Adam Smith 237 years ago, but those of natural systems are 3.8 billion years old. Allowed to run on much longer, the mismatch will destroy us.

It is time to talk about important things. Why have we come so close to the brink of extinction so carelessly and casually? Why do we still have thousands of nuclear weapons on hair-trigger alert? How can humankind reclaim the commons of atmosphere, seas, biological diversity, mineral resources, and lands as the heritage of all, not the private possessions of a few? How much can we fairly and sustainably take from Earth, and for what purposes? Why is wealth so concentrated and poverty so pervasive? Are there better ways to earn our livelihoods than by maximizing consumption, a word that once signified a fatal disease? Can we organize governance at all levels around the doctrine of public trust rather than through fear and competition? And, finally, how might *Homo sapiens*, with a violent and bloody past, be redeemed in the long arc of time?[34]

Outside of Hollywood movies, stories do not always have happy endings. Human history, to the contrary, is "one damn thing after another" as an undergraduate history major once famously noted. And one of those damn things is the collapse of entire civilizations when leaders do not summon the wit and commitment to solve problems while they can. Whatever the particulars, the downward spiral has a large dose of elite incompetence and irresponsibility, often with the strong aroma of wishful thinking, denial, and groupthink abetted by rules that reward selfishness, not group success.[35]

In the long emergency ahead, the challenges to be overcome are first and foremost political, not technological or economic. They are in the domain of governance where the operative words are "we" and "us," not those of markets where the pronouns are "I," "me," and "mine." At issue is whether we have the wherewithal, wisdom, and foresight to preserve and improve the human enterprise in the midst of a profound human crisis. Any chance for us to come through the trials of climate destabilization in a nuclear-armed world with 10 billion people by 2100 will require that we soon reckon with the thorny issues of politics, political theory, and governance with wisdom, boldness, and creativity.

Look for "State of the World 2014: Governing for Sustainability" coming in April 2014. Follow the development of the book here: http://islandpress.org/sotw2014

CHAPTER 27

Building an Enduring Environmental Movement

Erik Assadourian

In the early 1980s—not long after monumental victories in improving air and water quality—some within the environmental movement questioned the true value of these successes. Environmentalist Peter Berg pointed out that "rescuing the environment has become like running a battlefield aid station in a war against a killing machine that operates just beyond reach, and that shifts its ground after each seeming defeat. No one can doubt the moral basis of environmentalism, but the essentially defensive terms of its endless struggle mitigate against ever stopping the slaughter."[1]

Decades later, the moral basis of environmentalism is still undoubted, though the design and execution of many environmental campaigns have received increased scrutiny. And the deeper critique has yet to be answered. Environmentalism, first and foremost, continues to be a game of defense—working to reduce overall carbon emissions, chemical releases, forest loss—rather than a battle to transform the dominant growth-centric economic and cultural paradigm into an ecocentric one that respects planetary boundaries. And today, more than ever, environmentalists are outmaneuvered by better funded, better organized, and better connected adversaries, which keeps victory well beyond reach.

The current focus of environmentalism leaves little hope of successfully defeating the ecologically destructive political, economic, and cultural forces that undermine the very foundations of life. It will require a dramatic reboot if the movement is going to reverse Earth's rapid transformation and help create a truly sustainable future—or at least help humanity get through the ugly ecological transition that most likely lies ahead.

Are Today's Environmental Organizations Succeeding?

There have been plenty of internal critiques of the environmental movement since it appeared on the scene in the 1960s—from deep ecology and bioregionalism in the 1970s to the recent reports *The Death of Environ-*

Erik Assadourian is a senior fellow at Worldwatch Institute and director of the Transforming Cultures Project. He is codirector of *State of the World 2013*.

www.sustainabilitypossible.org

mentalism and *Weathercocks and Signposts: The Environment Movement at a Crossroads.*[2]

In 2004, in *The Death of Environmentalism*, Michael Shellenberger and Ted Nordhaus made two important criticisms of modern environmental advocacy: that it fails to provide any bold vision of a sustainable future and that it is essentially "just another special interest," unable to capture "the popular inspiration nor the political alliances the community needs to deal with the problem."[3]

In the 2008 WWF-UK report *Weathercocks and Signposts*, Tom Crompton noted that in environmentalists' urgent efforts to change people's behavior, they have often reinforced dominant consumeristic values rather than tapping more-sustainable values, like altruism. This, he noted, has proved to be a strategy that offers some short-term success but undermines itself in the long run, for example, as people who were encouraged to save money by buying energy-efficient lightbulbs then spend their savings on new consumer products.[4]

And recently the Smart CSOs Lab noted that environmental organizations are typically focused on a single issue—climate change, biodiversity, deforestation, toxic chemicals, conservation—and thus fail to think holistically about solutions, focusing on short-term fixes rather than addressing root causes.[5]

There is validity in all of these critiques. Many campaigns focus on treating environmental problems rather than addressing their roots, and they typically do so in ways that fail to build an alternative vision for a species not in a permanent state of conflict with the planet.

Worse still is that the movement is not even battling immediate threats all that well. Along with often being a marginalized special interest—failing to build strong-enough alliances to pass Earth-saving legislation—many conservation and environmental groups have also fallen prey to the same conflicts of interest observed in other philanthropy-dependent sectors. Just as more medical researchers have accepted funding from pharmaceutical companies, and breast cancer advocacy groups from companies that produce cancer-causing products, some environmental groups—seeking to have as large an impact as possible—are taking more funds from corporations with questionable environmental track records.[6]

As journalist and former Conservation International employee Christine MacDonald describes in *Green, Inc.*, accepting funding from corporations—which have a lot to spread around and are willing to do so to "greenwash" their image—has misdirected organizations from the true challenges facing them. Moreover, it has led some groups to soften their criticism of supportive companies and in some cases has even led to questionable endorsements of polluting companies or their products.[7]

This cozy relationship has also provided some of the most unsustainable corporations a way to mitigate the public relations challenges of being major polluters. MacDonald found that 29 of "The Toxic 100"—the worst corporate air polluters in the United States according to the Political Economy Research Institute—are major contributors to conservation organizations. Whether these and other corporations have just used environmental groups as greenwashing vehicles or have also influenced the agendas of the organizations that they donate to is harder to measure. But considering the size of some donations and the presence of corporate representatives on many organizations' boards, it is hard to imagine that these relationships have no influence at all. David Morine, a former vice president in charge of land acquisition at The Nature Conservancy, said after leaving the organization that his pioneering effort to bring in corporate funders "was the biggest mistake in my life," as he told the *Washington Post*. "These corporate executives are carnivorous. You bring them in, and they just take over."[8]

Toyota

The environmental group Audubon displays the new car it won from Toyota in a social media popularity contest.

What is more, most environmental organizations, including Worldwatch Institute, receive funding from affluent donors, foundations, and corporations that depend on a growing economy to keep their endowments robust enough to continue their philanthropy. Ironically, if environmental groups actually succeed in building a sustainable, equitable, steady-state economy, there is a good chance that their donors' philanthropic giving would shrink as wealth is better distributed and as stock markets stop growing. And if environmentalists fail in their mission, there's also a good chance the economy will contract: a 2012 report by DARA International projects that gross domestic product worldwide will shrink 3.2 percent a year by 2030 if climate change and air pollution are not dealt with. A shrinking economy is rarely a boon to philanthropy.[9]

Even if most environmental groups had secure forms of funding that did not lead to conflicts of interest, the broader critique remains. The movement is trying to stem the tide of global ecocide with strategies that fall far short of what is necessary to create a truly sustainable civilization—whether that is due to short-term thinking, overspecialization, lack of vision, or the realities of making political compromises, especially when at the table with much more powerful actors.

Thus it is time for the environmental movement to evolve. It needs to accelerate the shift to a sustainable society and to become more independent and resilient, even in the worst-case scenario of a rapid ecological transition. The only question is, How?

A Deeper Environmentalism

In 2007, a group of prominent environmentalists gathered in Aspen, Colorado, to discuss how to redesign the environmental movement to combat the linked environmental, social, and spiritual crises facing humanity. The group concluded that humanity needs a "new consciousness," new stories, new values—including an "ethics of reverence for the Earth" and a sense of intergenerational responsibility. And that to spread these, the movement will need to redevelop its grassroots potential, diversify its sources of funding, and use a variety of innovative strategies like embedding environmental education into schools' core curricula, doing a better job using media programming to spark environmental awareness, and establishing a Peace Corps–like effort that could help restore ecosystems and tackle global environmental challenges.[10]

The idea of deepening humanity's environmental consciousness and redesigning the movement to help do this is certainly not new. In 1973 Norwegian philosopher Arne Naess coined the term *deep ecology*, criticizing the "shallow" anthropocentric approach to environmentalism and instead advocating an ecocentric ecological philosophy to guide individuals and the movement. One of his main conclusions was that we need a set of principles to guide our behavior and to reinforce our commitment to help our planet flourish. His hope was that each of us would make a personal "ecosophy" (ecological philosophy) stemming from these principles that would shape our broader values and lives—from what we buy and eat and how many children we have to how we spend our time. Naess, with deep ecology, was perhaps the first to propose making environmentalism a fully lived philosophy.[11]

But deep ecology and its critique have remained marginal ideas in the broader movement, with environmentalists continuing to focus instead on short-term or shallow campaign goals. So it is not surprising, then, that environmental groups continue to engage their members in shallow ways—asking for donations, signatures on petitions, support of a specific political candidate, perhaps participation in a local protest. Yet within the movement, rare are the deeper opportunities to engage—community potlucks, for instance, or weekly meetings filled with stories of celebration or hope.

Defensive advocacy remains the environmental movement's primary role. As theologian and environmentalist Martin Palmer notes, "Environmentalists have stolen fear, guilt and sin from religion, but they have left behind celebration, hope and redemption." The problem is that fear without hope, guilt without celebration, and sin without redemption is a model that fails to inspire or motivate.[12]

Environmentalists must create a more comprehensive philosophy—complete with an ethics, cosmology, even stories of redemption—that could deeply affect people and change the way they live. Vaclav Havel, the Czech writer and political leader, once asked, "What could change the direction of today's civilization?" He answered that "we must develop a new understanding of the true purpose of our existence on this Earth. Only by making such a fundamental shift will we be able to create new models of behavior and a new set of values for the planet."[13]

This, naturally, should be the starting point of any philosophy, ecological or otherwise. Why are we here? and What is our purpose? are questions as old as human beings. And while religions have offered one set of explanations, and science another, neither have proved up to the task of answering in a way that enables humanity to live within the bounds of Earth.

The first principle of deep ecology points out that "the flourishing of human and nonhuman life on Earth has inherent value. The value of nonhuman life-forms is independent of the usefulness of the nonhuman world for human purposes." This ecocentric view of the planet offers a possible answer. Humanity's purpose may be as straightforward as helping the earth to flourish—and certainly not impeding its ability to do so.[14]

The ethics of an effective eco-philosophy must be grounded, completely and fully, in Earth's ecological realities and should facilitate humanity's Earth-nurturing purpose. As conservationist Aldo Leopold noted over 60 years ago, "A thing is right when it tends to preserve the integrity, stability, and beauty of the biotic community. It is wrong when it tends otherwise." This simple rule could serve as a foundation for a broader ecological ethics.[15]

Granted, this will not be an easy ethical code to follow. As the fourth principle of deep ecology notes, "the flourishing of human life and cultures is compatible with a substantial decrease of the human population. The flourishing of nonhuman life requires such a decrease." Decreases in both human population size and its impact (as much an outcome of how we consume as our total numbers) may raise some uncomfortable questions, such as, Can we have a sustainable civilization while fully respecting people's freedom to reproduce or consume without limits? However, not wrestling with these limits may prove much more perilous. And perhaps over time, norms around optimal family size and consumption levels will evolve, facilitating the transition to cultures in balance with a flourishing Earth.[16]

In order for this philosophy to attract people, it will also need to answer broader philosophical questions like Where did we come from? (cosmology) and Why do we suffer? (theodicy)—an essential component of any comprehensive philosophy, and one that will be especially necessary in getting through the difficult centuries to come.

Of course, other elements will have to emerge as well. Stories, exemplars, ways to cultivate fellowship among adherents, and ways to celebrate life's rites of passage—birth, coming of age, marriage, and death—and other cycles of life like the advent of a new year. Together, these elements could add up to a robust, holistic ecological philosophy that could inspire people across cultures to follow a new ecocentric way of life and encourage others to join them.

For that to happen, however, environmentalists must build the mechanisms to cultivate community among members and to spread this philosophy to new populations. In other words, for the environmental movement to succeed it will have to learn from something it often ignores or even keeps its distance from—religion, and specifically missionary religions, which have proved incredibly successful in orienting how people interpret the world for millennia, effectively navigating across radically different eras and geographies.

Missionary Movements and Their Potential

Let's start with a basic question. How have missionary religious philosophies spread so completely around the world? (Religions, while they are understandably more than this to adherents, are essentially orienting philosophies.) Yes, swords and guns were part of the success equation, as was the adoption of these philosophies by governments. But a larger part of these philosophies' success was a powerful, timeless vision, beautiful stories, inspiring exemplars, committed adherents, and the promise of immediate assistance— the offering of food, clothing, education, livelihoods, medical care, even a community.

The advent of Christian Socialism in the mid-nineteenth century offers a powerful and relevant case study on the spread of Christianity in a disrupted, rapidly industrializing, and urbanizing Europe and United States. Recognizing the corrosive effects of cities and urban poverty, many Christian reformers worked to spread the Gospel through the creation of social programs—including providing job trainings, food, safe shelters for people migrating to the cities, and so on.[17]

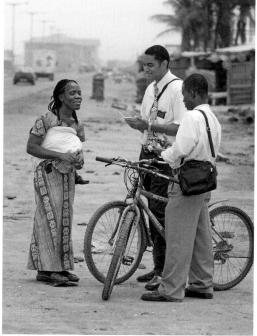

Two Mormon missionaries speaking to an African woman with a baby.

Both the Salvation Army and the YMCA were founded in the United

Kingdom in this era, spreading Christian values and the faith through the provision of social services. Today, both organizations continue to have a global reach, and combined they have several million volunteers reaching out to tens of millions of people in more than 110 countries. In 2011, the Salvation Army alone provided $3 billion worth of basic social service assistance to nearly 30 million people.[18]

The Catholic Knights of Columbus—founded in Connecticut in 1882 and now boasting 1.8 million members worldwide—also used a powerful communitarian model, offering support for recent Catholic immigrants to the United States (who often worked dangerous jobs and were excluded from labor unions). The Knights provided life insurance to care for widows and orphans if members were killed. Today it underwrites more than $80 billion in life insurance policies and continues to be active in charitable and political activities.[19]

Providing social services is not only a worthy goal in itself but also a means to build broader influence—growing the ranks of adherents and changing how people view the world and live their lives, and then using that influence to shape broader social, cultural, economic, and political norms. The Shakers, a Christian sect founded in England in 1771, offer a valuable lesson in how to grow influence and even in how to prepare for the coming economic and ecologic transition. (See Box 27–1.)[20]

Another Christian offshoot, the Church of Jesus Christ of Latter-day Saints (the Mormons), offers one more successful strategy to spread a philosophy—going door to door. Each year 55,000 full-time Mormon missionaries fan out around the world (with more than 1 million missionaries having served since the Church's founding), going on two-year missions to convert people to their philosophy slowly and methodically—a leading reason that a religion that is less than 200 years old has more than 14.4 million adherents worldwide. For these missionaries—typically young adults supported by family and friends or by their own childhood savings—this rite of passage is often life-changing. It deepens their own commitment to their beliefs while also spreading the ideas of this religion and drawing new members to the Mormon faith.[21]

Box 27–1. The Shakers' Relevance in a Post-Consumer Era

While often dismissed as a failed experiment—as their community no longer exists today—at their peak the Shakers were a powerful religious, economic, and social force, growing to 6,000 members in 1840 even while practicing celibacy. At the time, the group was a leading producer of herbal medicines. And its members were celebrated architects and craftspeople as well as renowned inventors: they invented the circular saw, clothespins, and ironing-free cloth. Believing that God dwelt in the quality of their craftsmanship, the Shakers strove for perfection in crafting their simple but beautiful products. And this success drew many new adherents to their faith.

But the Industrial Revolution and the mass-produced goods it led to were the Shakers' undoing. As markets for their high-quality, higher-cost products collapsed in the mid-1800s, so did their economic niche and their total number of adherents. The Shakers offer an important lesson, however: strong community and a relevant economic niche can attract people and provide the foundation for broader influence, even when certain elements of the philosophy are hard to stomach.

As access to cheap energy sources wanes, and with it mass-produced goods and globalized trade, many aspects of this model could once again flourish, providing one possible way to spread an ecocentric philosophy.

Source: See endnote 20.

Compare this to modern environmental canvassers who also go door-to-door asking for campaign donations. They are typically told by their managers to get a donation and move to the next door as quickly as possible, forgoing true engagement with the people they meet. Rather than growing supporters and political power, most of today's environmental door-knockers are merely neighborhood money-miners.[22]

Other missionary religious philosophies, such as Buddhism and Islam, also use a variety of social service provisions to spread their philosophies. Islamic madrassas are a leading provider of education in many countries. Today, madrassas educate millions of students around the world, providing literary, math, and science education in addition to knowledge of the Koran and Islam.[23]

As the provision of basic services led to new members being integrated into these various communities, social modeling played an important role in shaping their behaviors, and the routine professing of values and myths helped reinforce a new way of living. As numbers grew, so did their political, economic, and cultural influence—both at the aggregate and through the spread of smaller sects of broader philosophical persuasions. Quakers, Jesuits, Jehovah's Witnesses, Shriners (with their network of children's hospitals), and Scientologists have effectively spread their orienting philosophies—no matter how controversial they might have been—through the concerted proliferation of social services, designed in ways that help people in their moment of need and, as important, fold them into a broader philosophical community. Unfortunately, there have been few equivalent efforts by the environmental community.

The Rise of a Missionary Eco-Philosophy?

An informal survey of Kibera, one of the largest slums in Africa, found that nearly half of the roughly 250 schools serving the 200,000–250,000 Kenyans living there are religious in nature. The goal of these Pentecostal, Catholic, Protestant, Jehovah's Witness, YMCA, Salvation Army, Quaker, and other religious schools is to charitably provide the basic service of education—a service the Kenyan government cannot provide enough of. But these schools are also there to save souls and to add members to their philosophical communities.[24]

At the same time, there appear to be no schools in Kibera teaching an ecological philosophy. But imagine if there were. Imagine a school that, at every turn reinforced the idea that humanity depends completely and utterly on Earth and its complex systems for our well-being. That it is unjust to consume more than your fair share and to have a lifestyle that depends on the exploitation of ecosystems, workers, and communities polluted by factories, mines, and dumps. That the best life to live is one committed to changing this untenable, inhumane, and unsustainable system in ways that improve the well-being of your local community, your broader philosophical community, and above all the planetary community.[25]

This is a philosophy that could be reinforced in every aspect of the school—from what is taught in the classroom (ecology, ethics, activism, and permaculture along with basic math and literacy) to what is served in the lunchroom and everything in between. Some students would walk away just with knowledge, including a better understanding of our dependence on Earth and perhaps a basic livelihood and trade skills—skills that will only grow in value in a post-consumer future. But others would walk away with a deep commitment to this way of thinking, and perhaps even become missionaries of that ecological philosophy, starting new schools or other social services that could improve people's lives while spreading a way of life that could compete with the seductive consumerist philosophy so dominant today.[26]

And this model could be applied to a variety of needs. Eco-clinics could provide basic medicine but also focus on prevention that will help both people and the planet. For example, people with adult-onset diabetes might be asked to spend time tending the eco-clinic garden in partial payment for treatment, growing healthy food to replace the toxic, processed fare that contributed to their diabetes and so many other modern ailments. The clinic could also provide cooking and lifestyle courses as well as engaging with the larger community to help patients eat well and regain their health. In the process, their ecological impact would shrink along with their waistlines as they reduced their consumption of meat and processed food, both of which have higher ecological impacts than locally grown vegetables.[27]

Of course, religious social service providers are embedded in a broader community with a somewhat unified belief system—something environmentalists currently lack. But as ecosystems decline further, as the consumerist philosophy is revealed as no longer workable, the philosophies with alternative visions that also offer help and community solidarity will flourish—whether they are ancient religions, new religions, or perhaps even philosophies like environmentalism.

Ideally, social services should not be provided piecemeal by civil society organizations of any type. They should be the responsibility of a functioning government. But in reality, even at the peak of our unsustainable levels of wealth today, many governments fail in their duty to provide basic services for their citizens. As ecosystems unravel, as economies falter, and as local and national governments go bankrupt or adopt austerity measures to appease lenders, there is a good chance that social services will be cut. In that case, the need for nongovernmental actors to provide these services will only increase.

Just like advocacy campaigns, these efforts cost money, of course. Some of the funding could come from foundations perhaps. But groups could also use strategies more typical of religious organizations, generating money directly from adherent communities. Of the $298 billion donated to charity in 2011 in the United States, 32 percent went to religious groups, while just 2.6 percent was given to environmental groups. People are more likely to give to their own communities—those who are there for them through thick and thin—as well as to those who share deeply in their beliefs and understanding of the world.[28]

Funding could also come from social enterprises. Just as the Salvation Army earns hundreds of millions of dollars a year from the sale of used household goods and clothing (while also providing a valuable service), the environmental movement could take a more active role in setting up profitable social enterprises that generate revenue for its social service provision arm, as well as for efforts focused on advocacy and shifting broader cultural norms.[29]

These social-service providers and social enterprises—from cafes, bookstores, and used item stores to renewable energy utilities, energy retrofit providers, and permaculture training programs—would not only generate revenue but also offer a key mechanism to spread the eco-philosophy and recruit new members.

As eco-philosophies spread, and their followers grow in number, new opportunities would grow too. The Quakers, a small Christian sect, became a dominant economic and political force of Pennsylvania in the 1700s as well as a major force in the abolition movement. Even today Quakers remain a powerful voice in international peace and governance processes—far beyond what their total membership of 340,000 would seem to warrant. Eco-philosophical adherents could also play an outsized role in driving cultural change, particularly working to shift the consumer culture to be more sustainable by taking leadership roles in government, the media, business, and education. (See Chapter 10.)[30]

As the need for resistance to the modern industrial socioeconomic model grows (see Chapter 28), a committed community of environmentalists could be a powerful force, helping to use these tactics—whether as a distinct philosophical group or embedded in other philosophical traditions. (See Box 27–2.)[31]

Getting from Vision to Reality

The odds are that the state of the world is going to get really bad—and much sooner than we think. Reports about the fallout from climate change alone make it clear that the twenty-first century is unlikely to follow a linear path of more growth, more progress, more "development." There are probably going to be major political, social, and economic disruptions, a flood of fail-

Box 27–2. The Relationship Between Ecological and Religious Philosophies

Are ecological and religious philosophies incompatible? Not at all. Effective missionary philosophies can live beside other philosophies or incorporate those traditions into their practices: witness the syncretic relationship between Shintoism and Buddhism in Japan and the way Christianity incorporated folk religions as it spread.

An ecological philosophy may grow up alongside the dominant religious philosophies of today or even be absorbed by religious reformers, which could prevent the latter from losing their followers as ecological philosophies grow in attractiveness.

Indeed, the greening of religious traditions has already started at the margins, with more Christian sects drawing attention to green teachings from the Bible and designing programs to appeal to environmentally minded adherents. Buddhist monks are establishing sacred forests, Muslims are developing ways to celebrate Ramadan sustainably, and Hindus are finding ways to make ritual sacrifices greener.

In Sri Lanka, the Buddhist movement Sarvodaya Shramadana has created a comprehensive path to both material and spiritual development—emphasizing community, basic economic security, and sustainability at the heart of their model. The movement, which literally means "awakening through sharing," has focused on small community projects—building latrines, schools, and cultural centers—that improve village well-being and has simultaneously discouraged adoption of consumerism (or in Buddhist terms, attachment and desire). Today this sustainable Buddhist movement has a presence in more than half of Sri Lanka's 24,000 villages.

As these ideas incubate in coming centuries and the world undergoes dramatic changes, ecological philosophies may form independently and stay independent, they may be absorbed by today's dominant philosophies (or come into conflict with those philosophies as they compete for members), or they may even absorb or replace older philosophies.

Source: See endnote 31.

ing states, the dislocation of millions of people. Will people in environmental organizations simply close their doors as things unravel, as their funding dries up, and turn instead to simply surviving—taking any job still available in order to feed their families? Who will serve as a voice for Earth? Who will help steer us through this historically unique global ecological transition? Will it be fundamentalist religious institutions that read the unraveling ecosystems as signs of the end times? Or authoritarian governments that offer security in exchange for the last remnants of freedom?[32]

The future increasingly looks like it could take a page from a dystopian science fiction novel. Perhaps from *A Canticle for Leibowitz*—the story of a post-collapse civilization where one occupation is harvesting iron rebar out of concrete rubble, with the workers musing on how their ancestors got iron bars into stone in the first place. Over the course of the novel, modern knowledge is rediscovered, and once again people invent electricity, engines, even nuclear power. And how does it end? With humanity once again pursuing growth and empire, and once again destroying itself in the process.[33]

The hope is that we prevent collapse by following a new set of philosophical, ethical, and cultural norms that bring about a life-sustaining civilization, or what eco-philosopher Joanna Macy has called "the Great

Turning." The second hope is that, failing this—and failing to prevent "the Great Unraveling"—we preserve enough knowledge and wisdom so that as the dust settles in a few centuries, with the population stabilized at a lower number that a changed planetary system can sustain, our great-great-great-great-great grandchildren do not reinvent our mistakes. That they do not once again start worshipping growth and consumption but instead stay true to a philosophy that allows them to sustain the planet that sustains them. As Macy notes, "The awesome thing about the moment that you and I share is that we don't know which is going to win out, how

Tree seedlings being distributed in Uganda as part of The Alliance of Religions and Conservation's long-term environmental action plan for sub-Saharan Africa.

ARC-The Alliance of Religions and Conservation

the story is going to end. That almost seems orchestrated to bring forth from us the biggest moral strength, courage, and creativity. When things are this unstable, a person's determination—how they choose to invest their energy and heart-mind—can have much more effect on the larger picture than we are accustomed to think."[34]

Let us hope that this proves to be the case. And that centuries from now an ecocentric civilization—celebrating its nurturing niche on a once-again flourishing planet—tells stories of the bold individuals and communities that changed humanity's path in such a glorious way.

CHAPTER 28

Resistance:
Do the Ends Justify the Means?

Bron Taylor

Has the time come for a massive wave of direct action resistance to accelerating rates of environmental degradation around the world—degradation that is only getting worse due to climate change? Is a new wave of direct action resistance emerging, one similar but more widespread than that sparked by Earth First!, the first avowedly "radical" environmental group?

The radical environmental movement, which was formed in the United States in 1980, controversially transformed environmental politics by engaging in and promoting civil disobedience and sabotage as environmentalist tactics. By the late 1980s and into the 1990s, when the most militant radical environmentalists adopted the Earth Liberation Front name, arson was increasingly deployed. The targets included gas-guzzling sport utility vehicles, U.S. Forest Service and timber company offices, resorts and commercial developments expanding into wildlife habitat, and universities and corporations engaged in research creating genetically modified organisms. Examples of such militant environmentalism can be found throughout the world, and they are increasingly fused with anarchist ideologies. Given this history, the question arises as to whether direct action resistance is becoming unambiguously revolutionary, or perhaps even purposefully violent.[1]

People attending the Earth at Risk: Building a Resistance Movement to Save the Planet conference in Berkeley, California, in November 2011 might well have thought so. Some 500 people joined this conference, which called for a new "deep green resistance" movement in response to intensifying environmental decline and increasing social inequality. The format of the conference was a scripted dialogue, or what might be called political performance art, with the writer and activist Derrick Jensen posing questions to a series of environmental activists and writers, including, most prominently, the Man Booker Prize winner from India, Arundhati Roy.[2]

The tone of the meeting was sober and its messages radical. Succinctly put, the speakers issued the following diagnoses: Electoral politics and lob-

Bron Taylor is a professor of religion and nature, environmental ethics, and environmental studies at the University of Florida, and a fellow of the Rachel Carson Center in Munich, Germany.

www.sustainabilitypossible.org

bying, as well as educational and other reformist conversion strategies that give priority to increasing awareness and changing consciousness, have been ineffective. Such strategies do not work because for 10,000 years agricultures have been established and maintained by violence. This violence has foremost targeted foraging societies (and later indigenous and poor people), nonhuman organisms, and nature itself. Fossil-fueled industrial-agricultural civilizations are especially destructive and unsustainable. Popular and democratic movements have been overwhelmed by the increasingly sophisticated ways that elites justify and enforce their rule and promote materialism and the domination of nature.

In concert, the conference speakers offered radical prescriptions. They called for direct and aggressive resistance to plutocracy and environmental destruction. The immediate objective, several of them contended, should be to bring down industrial civilization—which, they claimed, has structural vulnerabilities. Specifically, they urged those gathered to form or support secret cells that would, as their first priority, sabotage the energy infrastructure of today's dominant and destructive social and economic systems. It is also critical, they contended, that activists avoid pacifist ideologies and even carefully consider whether, and when, the time might be ripe to take up arms to overturn the system. After the most inflammatory of these statements, at least a third of the crowd rose in standing ovation.[3]

It is not necessary to hold an anarchist or anti-civilization ideology to wonder if electoral politics, lobbying and educational efforts, or litigation-based strategies are enough. Indeed, one reason that many people in mainstream environmental organizations sympathize with these radicals is that they often share a despairing view of the current destructive trends and recognize that, despite their best efforts, they have been unable to slow or reverse them.

It is also not necessary to be willing to contemplate violent tactics when considering or engaging in resistance. Although definitions of resistance typically include underground organizations opposing an occupying or authoritarian power or regime, often with acts of sabotage or guerilla warfare, the term can also refer to nonviolent, extralegal opposition to a regime or its practices—even a regime that is considered politically legitimate, such as in democratic countries. Examples of such resistance include disruptive protest, civil disobedience, and noncompliance with government laws or with the dictates or operations of public or private institutions considered to be engaged in wrongdoing.

Anyone paying attention can easily identify both actions and negligent inaction on the part of public and private actors that are exacerbating exceptionally harmful environmental and social trends. Is it time, then, for resistance? Has it been effective or counterproductive? If effective or potentially so, which kinds are, under what circumstances, and by whom? What should the posture of mainstream environmental organizations be toward those who engage in resistance?

It is time to break the taboo against talking about this and to consider what lessons can be drawn from decades of experimentation with direct action resistance.

Premises

This is ethically fraught terrain. To be as clear as possible, let's begin with a forthright statement of the premises underlying the analysis in this chapter.

First, sometimes it is permissible or even obligatory to resist legally constituted laws and policies. This statement is uncontroversial when it comes to long-settled social conflicts. In hindsight, at least, nearly everyone would agree that the Confessing Church's resistance to the duly elected Nazi regime and its laws was not only morally permissible but obligatory. To this a host of additional examples could easily be added: Mahatma Gandhi leading the resistance to British imperial rule, Martin Luther King, Jr. in his often illegal pursuit of full citizenship for African-Americans, and even Nelson Mandela and the African National Congress' insurrectionary strategy to topple South Africa's apartheid regime.

Once it is acknowledged that laws and policies have been and can be unjust, whether to resist becomes a muddier moral terrain. When laws are enacted through democratic processes, of course, they are generally considered on first appearance to be legitimate, so any decision to break them ought not be taken lightly. Such a decision often requires someone to choose between competing goods, between moral principles that ordinarily would not be in conflict but that can be in specific cases. The best laws try to anticipate exceptions and complexity, including by fashioning penalties that recognize moral ambiguity and unusual circumstances. Breaking into someone's home, for example, is normally and properly judged illegal, but in the case of a fire, it becomes permissible so that lives can be saved.

Criminal codes at their best carefully consider the intent of the accused, and penalties increase according to a crime's maliciousness. But exigent circumstances are not usually factored into criminal statutes. Nor do lawmakers always anticipate and incorporate into law, as they should, new circumstances or understandings. It is not uncommon, therefore, that deeply ethical and well-informed people will decide that some laws are inadequate, outdated, or just plain wrong, that the processes for changing them are too corrupt or the time too short, and that the stakes just too high to justify obeying such laws.

Second, it is wrong for one species to dramatically reduce Earth's biological diversity, and preventing anthropogenic species extinctions should be a high moral priority. This ethical premise has been defended on many grounds, a survey of which is not possible here, but they include prudential and anthropocentric concern for human welfare, biocentric philosophy or spirituality, and diverse religious grounds in which protecting species is a religious duty.[4]

Third, the best available consensus science indicates that our species is precipitating a rapid decline of biological diversity, and this process is accelerating due to anthropogenic climate change. It is also clear that political systems have not halted these processes.

Fourth, and finally, since species that go extinct are lost forever, the stakes are high and an exigent response is urgently needed. Political systems have utterly failed to arrest biodiversity decline, nor are they poised to respond quickly and effectively.

Given these ethical and factual premises, individuals and organizations should consider the reasons for this decline and how to overcome it. Since current laws and political activities have failed to redress the situation and appear unlikely to do so, it is incumbent to ask what strategies and tactics might be successful. Such an assessment should include determining whether strategies and tactics must be constrained by existing laws and prevailing assumptions about what constitutes acceptable political action.

Put more simply: anthropogenic environmental decline in the light of life-affirming values and political inaction demands analysis of the obstacles to effective action, including laws and mores that might constrain it. Given the urgency of the situation, extralegal tactics should be on the table, as they were in earlier causes where great moral urgency was properly felt.

This does not, however, answer the question of whether the time for resistance has come. For this, we would need to diagnose the reasons for the present predicaments, determine what resources can be acquired, the sort of resistance needed, and whether a given action or campaign would be morally permissible, likely to be effective, and unlikely to be counterproductive. Venturing answers is perilous, in part, because there is so much complexity and uncertainty in the deeply entwined environmental and human socioeconomic systems we seek to understand and affect. Yet the urgency of the situation requires nothing less.

Types of Resistance

Recognizing that social reality never perfectly reflects our maps of it, it is nevertheless useful to proceed with a review of the main types of resistance.

First, but not least, there are many ways that people of conscience resist current trends, including by battling ideas that consider the world to be a smorgasbord for ever-swelling human numbers and appetites and that view human beings as somehow exempt from nature's laws. More important, there is a revolution going on with regard to understanding the human place in and responsibilities to nature. These are unfolding rapidly and globally, and while the trends have diverse tributaries and expressions, they also have common emotional and spiritual dimensions, including deep feelings of belonging and connection to nature, as well as convictions about the value of all living things. There are, put simply, many forms of cultural resistance to beliefs and practices that do not cohere with science or progressive environmental ethics. These trends are important to note if we are to avoid the disempowering influence of cynicism.[5]

March against the proposed Keystone XL pipeline, Minneapolis.

While contemplating the possibility and promise of resistance, it is also important to note that not everyone has the ability to participate in its more radical forms. Economically vulnerable populations, for example, might have few resources or opportunities to directly confront forces they understandably fear or upon whom they directly or indirectly depend. People in such situations, who have much to lose from direct confrontation with workplace authorities or rulers, sometimes engage in what might be labeled *passive resistance*. This generally involves noncooperation and noncompliance, such as through work slowdowns, theft, feigned ignorance, and sometimes difficult-to-detect forms of sabotage. Such tactics are designed to avoid attention or detection. The focus here, however, is on whether more direct and aggressive forms of resistance are warranted.[6]

For radical environmentalists, the answer is a resounding yes, because they agree that the agricultural-capitalist-industrial system is fundamentally destructive and inherently unsustainable. The earliest Earth First! activists, for example, hoped that a combination of public protest—including civil disobedience and sabotage to thwart and deter the greatest assaults on biodiversity—would increase public sympathy and demands for environmental protection. Often, but not always, a connection was made between the erosion of biological diversity and cultural diversity (especially as represented in indigenous and peasant cultures). And concern for both animated the efforts.

Some also supported the political theory that creating an environmental extreme would serve as a counterweight to the extreme right in political battles, pulling the political center more toward the environmentalist pole of the right/left continuum, which is where laws and policies tend to end up. Yet others, such as the radical environmental activists who, after a number of their comrades were arrested, concluded that they could save nothing from prison, established the Greater Gila Biodiversity Project in 1989, which eventually became the Center for Biological Diversity. These activists were among the ones who pioneered tenacious litigation strategies, using existing laws and rules written by resource agencies to challenge, with great success, practices they considered destructive. This is another form of resistance, although it is seldom recognized as such.[7]

While these early radical environmental activists maintained an apocalyptic view that modern society would collapse of its own unsustainable weight, their priority was to save what they could of the genetic and species variety of the planet before that inevitable collapse. They welcomed the envisioned collapse, believing it would halt the destruction and give the planet a chance to heal.[8]

This stream of thought thus had both radical and reformist dimensions. The more optimistic activists thought that direct action resistance might help precipitate widespread consciousness change, preventing humans from overshooting their carrying capacity and precipitating the collapse of environmental and thus social systems. The more reformist participants resembled those from more mainstream environmental movements, who consider mass protests, accompanied by nonviolent civil disobedience and sometimes spectacular acts of protest and resistance (such as by Greenpeace), as a way to educate and transform public opinion and thus to change behaviors, laws, and policies.

The revolutionary stream of these activists find hope only in actions that would accelerate the collapse of the societies they do not believe can be reformed voluntarily. These activists believe that, given the propaganda power of the elites who are most responsible for the destruction and who control political systems, more egalitarian, democratic, and environmentally sustainable systems have no chance of being established until this system is demolished or falls of its own unsustainable weight.[9]

In sum, when it comes to ecological resistance movements, there is a continuum of types, with varying diagnoses, strategies, and tactics. One extreme of the continuum of activists, who grew in number soon after the founding of Earth First!, is represented by the Earth Liberation Front, green anarchism, and Deep Green Resistance. These forms can be labeled revolutionary resistance, and they boldly proclaim an intention to bring down, "by any means necessary," an industrial system considered inherently destructive.

More-moderate sectors of radical environmentalism represent a kind of revolutionary/reformist hybrid, which shares many of the critical perspectives about the roots and current drivers of environmental degradation but which draws more eclectically and pragmatically on revolutionary and reformist ideas, strategies, and tactics. These activists do not absolutely dismiss the possibility that, with the right combination of resistance and reform strategies, there could be an upwelling of public support for environmental health and social equity and therefore that a less catastrophic transition toward sustainability might yet be possible.

On the other end of this spectrum is reformist resistance, which endorses demonstrations, including extralegal tactics such as civil disobedience (which can be highly disruptive, as for example when logging roads or highways are blockaded) as well as diverse pedagogical efforts, hoping to sway public opinion and pressure public officials into changing laws and policies while also affecting whether they honestly and successfully enforce current laws and policies. More so than the previous two types, here the goal is to force a democratic revolution or restore it where it has been subverted. And the hope is that this could create the conditions needed for dramatic action to address the most trenchant environmental and social problems.

Activists taking this approach may share the critical perspective of the more radical advocates of resistance about agriculture and industrialism, but they nevertheless take a more pragmatic approach, sometimes acknowledging that the current systems are powerful, resilient, and difficult to bring down. Or they may conclude that the threat to human beings, to other species, and to environmental systems would be too great should the current systems precipitously collapse and that therefore such an outcome should not be pursued.

Assessing Resistance

With premises about and types of resistance established, and with humility given the diverse variables in play and the difficulty in predicting the effects of different courses of action, it is possible to venture a broad assessment of resistance strategies. These views are quite properly subject to change, given changed circumstances and understandings.

The radical critique of agricultural, industrial civilization cannot be easily dismissed. It is true that as agricultural societies spread around the world, cultural diversity has dramatically eroded. Agricultures have displaced, murdered, or assimilated foraging peoples, whether through superior numbers and force, through the diseases their lifestyles brought with them, or through processes of settler colonialism. The erosion of biological diversity has gone hand-in-hand with these processes, all of which intensified with the power of the fossil-fuel-driven industrial age.[10]

Modern societies are unduly celebratory of their achievements when they have amnesia about what has been lost and by whom. With an understanding of the tragic aspects of this history and recognition that these very processes are ongoing, it is clear that dramatic actions to halt these processes and engage in restorative justice and healing where possible are morally obligatory.

This does not mean, however, that the revolutionary prescription of the Deep Green Resistance activists—attacking the energetic infrastructure of industrial civilization—is warranted. Indeed, the claim that this could cause the collapse of industrial civilization is fanciful. Natural disasters (including those intensified or worsened by human activities) demonstrate that as long as energy is available, large-scale societies will rebuild. Even if resisters were to disrupt the system significantly, not only would the system's rulers rebuild, recent history has shown that they would increase their power to suppress resisting sectors.

Moreover, as many radical activists have acknowledged in interviews—even those who have supported sabotage—the more an action risks or intends to hurt people, the more the media and public focus on the tactics rather than the concerns that gave rise to the actions. This means that the most radical tactics tend to be counterproductive to the goal of increasing awareness and concern in the general public.

When accessing the effectiveness of resistance, it is also important to address how effective authorities will be at preventing and repressing it. The record so far does not lead easily to enthusiasm for the most radical of the tactics deployed thus far. Authorities use tactics that are violent or can be framed as such to justify to the public at large spying, infiltration, disruption, and even violence against these movements. Such repression typically succeeds in eviscerating the resistance, in part because as people are arrested and tried, some will cooperate with the prosecution in return for a reduced sentence.

At the University of California, Berkeley, protestors refuse to leave the last standing tree of a grove of mostly oaks leveled to make way for construction of a campus building.

More than half of those arrested did just that during what Federal authorities dubbed "operation backfire," which led to the arrests and conviction of more than two dozen Earth Liberation Front saboteurs who had been involved in arson cases. One of the leaders, facing life in prison under post-9/11 terrorism laws, committed suicide shortly after his arrest, while several others became fugitives. The individuals convicted drew prison terms ranging from 6 to 22 years. The noncooperating activists, and those for whom terrorism enhancements had been added to the arson charges, drew the longest terms.[11]

As if this were not devastating enough to the resistance, broader radical environmental campaigns that were not using such radical tactics ebbed dramatically in the wake of these arrests. This was because movement activists who were friends and allies of those arrested rallied to provide prison support, which then took their time and resources away from their campaigns. But it was also because the resistance community was divided over whether (and if so, how) to support the defendants who, to various degrees, cooperated with investigators. Given this history, it makes little sense to base strategy and tactics on such an unlikely possibility that communities of resistance will ever be able to mount a sustained campaign to bring down industrial civilization, even if that were a desirable objective.[12]

The envisioned alternative to this objective—creating or, in the view of many activists, returning to small-scale, egalitarian, environmentally friendly lifestyles—would not be able to support the billions of people currently living on Earth, at least not at anything remotely like the levels of materialism that most people aspire to. So the most radical of the resistance prescriptions would quite naturally lead to strong and even violent counter-resistance.[13]

These ideologies, explicitly or implicitly, make unduly optimistic assumptions about our species, including about our capacity to maintain solidarity in the face of governmental suppression, as well as about the human capacity for cooperation and mutual aid. To expect such behavior to become the norm may be conceivable, and it may be exemplified by some small-scale societies, but it is not something to be expected universally, let alone during times of social stress intensified by increasing environmental scarcity.[14]

So despite the accurate assessment about the ways agricultural and industrial societies have reduced biocultural diversity, there is little reason to think that the most radical resistance tactics would be able to precipitate or hasten the collapse of such societies. Nor is there much evidence that such tactics would contribute to more-pragmatic efforts to transform modern societies. In contrast, there is significant evidence that these sorts of tactics have been and are likely to remain counterproductive.

Spiking Awareness of Biodiversity Decline

There are, nevertheless, concrete historical examples where extralegal resistance has played a significant and even decisive role in campaigns to protect natural habitats and change government policies. Examples from diverse sites of contention around the world are documented in *Ecological Resistance Movements: The Global Emergence of Radical and Popular Environmentalism*. Many other studies have documented the successes and promise of such movements, as well as the failures and often-violent resistance that they face.[15]

These dynamics were all present a few decades ago when activists aggressively, and often illegally, campaigned to halt deforestation in the forests of the Pacific Northwest and Rocky Mountains of the United States. Tree spiking, which involves putting metal or ceramic spikes in trees that are slated for logging, was among the most controversial of tactics. First used in anti-logging campaigns in Australia in the late 1970s and in Canada in 1982, radical environmentalists took up the practice with a vengeance in the United States during the 1980s and early 1990s.[16]

Tree spiking was a tactic that, it was hoped, when combined with blockades and other forms of sabotage, would bankrupt logging companies believed to be engaged in unsustainable and species-threatening logging. Failing that, the hope was that logging would slow down when some of it became unprofitable due to the additional costs of removing the spikes.

Although there have been examples of spiking leading directly to the quiet cancellation of a timber sale or to economic distress for a small logging company, the practice did not often, in a direct way, significantly reduce deforestation. It did, however, have another important impact. In a short period of time, the controversy it precipitated contributed significantly to public awareness of deforestation and related endangered species issues. As Mike Roselle, one of Earth First!'s cofounders, later claimed, before they began spiking trees nobody had even heard of the ancient forests or the threats to them. Indeed, before these campaigns the term *biodiversity* was not in the public lexicon, nor was its value advanced in public discourse. It took these campaigns to bring the very idea of biodiversity and its importance out from obscure scientific enclaves and into public view.[17]

With the occasional destruction of logging equipment, publicity stunts such as banner hangings, increasingly sophisticated blockades of logging roads, and the occupation of logging equipment or trees to prevent logging, public awareness of these issues grew. So did expressions of concern (and sometimes outrage) to public officials. In several cases, the resistance gained enough strength to orchestrate large protests that included mass arrests, as when in 1996 thousands of citizens gathered in a sparsely populated area of northern California to protest logging by the Pacific Lumber Company (PALCO) in ancient redwood groves. More than a thousand people were arrested for trespassing on land owned by the timber company.[18]

This, plus a decade of resistance to PALCO's practices, contributed to political pressures to reduce social disruption and the loss of political support, and it led to heightened scrutiny and a citation to the company for violating the law. Eventually, a deal was worked out to sell the most biologically precious old-growth groves to the state of California. Not long afterward, the company went bankrupt and was sold to another firm that promised to protect the remaining ancient groves and manage the rest of its forestland more gently.[19]

This was not the only case in which blockades of logging roads or tree occupations, which were sustained for months and even years, forced concessions from business or resource managers or provided time for attorneys to win injunctions or lawsuits against the logging. Not incidentally, one rationale for extralegal resistance is the often-accurate charge, as validated in the courts in responses to lawsuits, that industries or the government itself had broken environmental laws. Such facts allow those engaged in resistance to contend that they are actually displaying respect for laws by risking arrest and incarceration in their efforts to force companies and the government to obey existing statutes. And when governments and corporations see that they are being monitored, it contributes to improved compliance with environmental laws and regulations.[20]

Sometimes resistance movements put so much pressure on government officials that major victories are won, as when the U.S. Forest Service under President Bill Clinton issued the Roadless Area Conservation Rule in 2001, which protected some 25 million hectares (more than 58 million acres) of federal forestland. Although it took more than a decade of legal battles for opponents of this rule to exhaust their legal challenges to it, this has become the law of the land. And it is inconceivable that this rule would have been issued without more than a decade of very strong and often disruptive resistance to the Forest Service's timber program. Although the rule does not do everything that activists sought, it is a significant advance for biodiversity conservation in North America.[21]

A Time for Resistance?

People engaged in environmental causes around the world, including those who deploy resistance strategies, lose far more often than they win. But there are signs that direct action resistance is growing. Reports of desperate people resisting displacement from their lands and livelihoods for environmentally devastating projects justified under the rubrics of progress and development appear to be increasing in many regions, including in China, South America, Russia, and a variety of other sites. Increasingly, those resisting are threatening or even in a few cases resorting to violence, although such movements have generally been the object of far more violence than they have ever used against others.[22]

It is by no means certain that these movements will succeed or even survive the repression by authorities that they all too typically face. This will depend in no small measure on whether strong, international alliances are established and whether repressive acts are publicized internationally. Done in a way that minimizes or prevents reactionary counter-resistance and that does not lead to widespread public revulsion, ecological resistance has played and can continue to play a valuable and important role in environmental protection and sustainability.[23]

Indeed, direct action resistance can bring attention to issues in a way that electoral politics and lobbying cannot. It can inspire action and apply political pressure on corporate and governmental officials. Like a rowdy audience or angry coach riding a referee, it can affect the decisions that are made and even whether officials will enforce the law. More significant in the long term is that such resistance may even contribute to shifting the center of public debate more toward the positions of environmentalists.

That mainstream environmental organizations and actors are reticent to acknowledge the positive role of resistance is understandable. After all, they work within the system and by its rules, and it would seem hypocritical to work for laws, policies, and enforcement mechanisms while refusing to abide by society's existing laws. Yet there are many examples of individuals and groups honored today for obeying the overwhelming majority of existing laws while protesting highly consequential and exceptionally harmful immoral laws. Martin Luther King, Jr., for one, claimed that disobeying unjust laws and facing the consequences for doing so actually expresses the highest regard for the importance and value of the law as an institution.[24]

In August 2011, journalist and activist Bill McKibben and his group 350 .org orchestrated a protest at the White House demanding action and leadership by the United States on climate change. The action led to 143 arrests that same day and over a thousand that month. Most prominent among those arrested was James Hansen, the head of the National Aeronautics and Space Administration's Goddard Institute for Space Studies. It was not Hansen's first arrest, for he had become so alarmed about climate change and the government's anemic response that he had decided the time for resistance had come. In 2013, more such protests are being organized.[25]

But how much more powerful these protests would be if there were a march on Washington comparable to those during the civil rights era and involving thousands of arrests by individuals demanding action on climate change? And how much more powerful yet if similar marches took place in Brussels, Beijing, Brasília, London, Moscow, Cairo, Pretoria, and other centers of power around the planet? Of course, there have been some large demonstrations already, beginning most notably with the anti-globalization protests at the World Trade Organization meeting in Seattle in 1999 and continuing at other such international meetings. But the complaints and demands in these cases were diluted, ultimately unspecific, and thus easier to ignore. Climate change protest could provide a unifying focus for forcing global changes toward sustainability. Indeed, as there are many precedents where "people power" has toppled regimes, the global nature of the threat posed by climate change certainly makes it feasible that social protest and unrest could force concerted action on the part of targeted governments and businesses.

Arguably, such protests would be all the more effective if they were protracted and scrupulously nonviolent, while also disrupting business as usual. Social disruption is often a prerequisite to concessions by political elites. Yet for such a dramatic, global movement of conscience to arise and gather strength, there would need to be leadership from the most powerful environmental organizations, alliance building by them and the world's religious communities, and careful planning regarding the kind of public theater that would be the most effective. Given how high the stakes are, and how slow the global response has been, it is reasonable to ask whether the time has come for the most prominent and respected environmental organizations and individuals to add another dimension to their advocacy for environmental sanity: direct action resistance.

If there are regrets in the struggle for sustainability among those who know the facts and the stakes involved, it may well be akin to the musings of Henry David Thoreau. Toward the end of his life, after noting how out-of-step he was with the conventional wisdom of his day, he commented, "If I repent of anything, it is very likely to be my good behavior. What demon possessed me that I behaved so well?" That is a timely question for us all.[26]

CHAPTER 29

The Promises and Perils of Geoengineering

Simon Nicholson

Over the last handful of years, a set of radical ideas that have long been confined to the fringes of climate change discussions have begun to edge toward center stage. The ideas are known collectively as *geoengineering proposals*—sweeping technological schemes designed to counteract the effects of planetary warming. (See Box 29–1 for a full definition.)[1]

Many of the best-known geoengineering proposals read like science fiction. One widely circulated idea is to launch giant mirror arrays or sunshades into near-Earth orbit, in an attempt to reflect some amount of solar radiation. Other lines of research suggest that a similar effect could be achieved by depositing fine reflective particles of sulfur dioxide in the stratosphere or by deploying a host of ocean-going ships to spray cloud-whitening saltwater high into the sky. At the same time there are ongoing efforts to develop vast machines designed to suck carbon dioxide (CO_2) out of the air, to produce carbon-capturing cement, to lock carbon into soil, and to perfect the dropping of massive quantities of soluble iron into the oceans to encourage great carbon-inhaling blooms of plankton.[2]

Yet even while many geoengineering proposals sound fantastical, the field is beginning to receive sustained attention from serious people and groups. The Intergovernmental Panel on Climate Change (IPCC) has convened expert meetings to consider the topic. So too have other important scientific bodies around the world. In the United States, government agencies from the Pentagon to the Department of Energy have advocated that federal dollars be devoted to geoengineering research, and research teams in universities and the private sector in many countries are looking to move beyond theorizing about global climate control to technological development and deployment.[3]

Even as sober a scientific voice as President Obama's chief science advisor, John Holdren, who in 2007 had claimed that "belief in technological miracles is generally a mistake," seems to have come at least partly around.

Simon Nicholson is an assistant professor in the School of International Service at American University in Washington, DC.

www.sustainabilitypossible.org

A straightforward definition of geoengineeering comes from an influential report issued by the United Kingdom's Royal Society in 2009. Geoengineering, says the report, is any "deliberate large-scale manipulation of the planetary environment to counteract anthropogenic climate change."

Building on this definition, there are—as physicist David Keith has noted—two key aspects that must delineate a geoengineering enterprise: scale and intent. By these criteria, sending giant mirrors into orbit is clearly a geoengineering activity. So would be the dropping of thousands of tons of iron into the oceans or the introduction of hundreds of tons of sulfate particles into the stratosphere.

Other activities fall in a gray zone. An individual installing a reflective white roof on a house gets a check mark for "intent," but such an activity fails, by Keith's criteria, to qualify as a geoengineering effort because of limited "scale." The same can likely be said of a single coal-fired power plant that attempts to capture and sequester some portion of its emitted carbon. On the other hand, if a coordinated nationwide or international effort were made to install white roofs, or if a regulatory move required carbon sequestration from coal-fired power plants, then activity would be prompted at a large-enough scale to constitute geoengineering.

Source: See endnote 1.

Holdren suggested in 2009, when asked about the geoengineering option, that "we don't have the luxury of taking any approach off the table. . . .We might get desperate enough to want to use it."[4]

Dreams of weather and climate control are hardly new. Ancient traditions had a variety of rituals aimed at calling forth favorable weather. Since the beginning of the science age, numerous attempts have been made to create or dissipate rain, to still hurricanes, and to manage ice flows. This has not always been a venerable undertaking. Weather and climate manipulation has throughout history been a field replete with more than its share of tricksters and dreamers. Today a fresh cadre of would-be climate engineers is emerging. They have newly honed scientific understandings, increasing amounts of money, and strengthening political winds at their backs. So what, then, is to be made of geoengineering? Is it a new form of hucksterism? A dangerous folly? Or does geoengineering have some ultimately positive role to play in the transition to a sustainable future?[5]

Answering such questions is hardly straightforward. One important thing to keep in mind is that not all geoengineering proposals are alike. A catch-all category like this hides some very important distinctions. Some geoengineering ideas threaten to unleash extraordinarily high environmental or social costs or promise to concentrate political power in a troubling fashion. Other proposals, if developed in sensible and sensitive ways, hold out some real hope for a world adjusting to a changing climate. Making sense of geoengineering demands a separation of the reality from the hype—and a separation of the ideas that are altogether too risky from those that appear a good deal more benign.

A Look at the Geoengineering Landscape

In November 2007, the U.S. National Aeronautics and Space Administration (NASA) hosted a meeting of handpicked scientists at the Ames Research Center in San Francisco, California. The meeting was called to look at the innocuous-sounding enterprise of "managing solar radiation."[6]

The gathering brought together an array of geoengineering luminaries. While their main goal was development of a scientific research agenda for this developing field, a central theme over the two days of conversation was impatience and frustration with the traditional suite of measures put forward to tackle climate change. United Nations–sponsored political negotiations, carbon trading schemes, attempts to promote alternative energies—all were seen by those in attendance as doomed to fail or to be progressing far too slowly to avert disaster.[7]

In this, the tone of the Ames meeting echoed a message from a particularly influential geoengineering paper in 2006 by Nobel prize–winning chemist Paul Crutzen. There, Crutzen had labeled attempts by policymakers to bring about reductions in greenhouse gas (GHG) emissions as "grossly unsuccessful." He went on to call the hope that emissions could be brought under control rapidly enough to prevent widespread climate catastrophe a "pious wish."[8]

Such views are the entry point into the world of geoengineering. By just about any available measure, the climate situation is worsening. As Arctic ice melts, sea levels rise, wildfires increase in frequency and severity, and storms worsen, there is a growing sense in influential quarters that political and social strategies aimed at reducing GHG emissions are proving hopelessly ineffective. The stage is set for a shift in focus to dramatic, technology-based climate stabilization measures.

The technological strategies under consideration fall into two basic categories. The first are the kinds of solar radiation management (SRM) techniques that were under explicit consideration at the Ames meeting. SRM techniques are concerned with blocking or reflecting sunlight. Such a feat could, in theory, be achieved by boosting Earth's surface albedo—its reflectivity—using any of a variety of methods or by preventing some portion of solar radiation from ever reaching the earth's surface. The second category is carbon dioxide removal (CDR). Strategies under this heading are concerned with drawing CO_2 out of the atmosphere and locking it into long-term storage.

Solar Radiation Management. The central notion underlying SRM efforts is straightforward, although in its implications SRM is a recipe for audacious action. Basic atmospheric science tells us that as greenhouse gas concentrations rise, so does the atmosphere's ability to lock in heat from the sun. It is this simple fact, a brute product of chemistry and physics, that is pushing up global average temperatures. As human activity ups the planet's levels of CO_2 and other greenhouse gases, the average temperature of the planet continues to rise.[9]

The most obvious way to prevent further warming is to stop putting excessive amounts of GHGs into the atmosphere. Failing that, the warming effect of these heat-trapping gases could, in theory, be counteracted by scattering or deflecting some percentage of incoming solar radiation. Models of the climate system suggest that the heating associated with a doubling of CO_2 could be neutralized by deflecting about 1.5–2.0 percent of the sum total of the sun's energy currently striking Earth. To achieve a feat of this magnitude—to, in effect, dim the sun—would be an extraordinary undertaking. On the other hand, the enterprise is far from unimaginable.[10]

There are, in fact, some well-established options for SRM. They start at ground level, with activities focused on the world's lands, waterways, ice packs, and oceans, and extend all the way into the far reaches of space. (See Figure 29–1.)[11]

At ground level, the basic strategy is to make some portion of the planet's surface shinier. Some scientists are betting on the genetic engineering of crop varieties with more-reflective leaves. If deployed on large enough a scale, such an innovation could reflect some measurable amount of solar radiation directly back into space. Other ideas include the creation of oceanic foams or the addition of reflective bubbles to expanses of the world's seas or the placement of reflective materials in deserts, over areas of polar ice, or in the oceans. U.S. Secretary of Energy Steven Chu has called for home and business owners to whiten the roofs of their buildings. At large enough a scale, such an undertaking could have a small but discernible effect on the earth's climate.[12]

Figure 29–1. Solar Radiation Management Options

Sun Shields

Sulfate Particles

Cloud Whitening

Reflective Buildings

Reflective Leaves on Crops

Source: Graphic designed by Isabelle Rodas

Moving up to the lower atmosphere, the leading idea is to increase "oceanic cloud albedo"—that is, to make clouds whiter and more reflective. This was first proposed in the context of climate geoengineering by climatologist Jonathan Latham in 1999. It is Scottish engineer Stephen Salter, however, who has become cloud whitening's poster child. Salter has envisaged a fleet of 1,500 computer-controlled "albedo yachts." These wind-powered ocean-going vessels would draw water from the seas and deliver it in micron-sized droplets into the cloud layer. Developing precisely the right size for sprayed saltwater droplets is a big part of the engineering challenge for this scheme: too big a drop would simply rain back to earth; too small a drop would evaporate without a trace.[13]

While cloud whitening is an idea that has been receiving interest from influential financial backers, it is the upper atmosphere that has been receiving the most attention from SRM enthusiasts. Cooling the planet by introducing reflective material into the stratosphere is actually a geoengineering technique that has a direct analogy in nature. Erupting volcanoes can introduce vast quantities of material into the atmosphere, and the cooling effects of these natural events have long been noted and measured. Indeed, a real-world test of the "put sulfur in the stratosphere" idea happened relatively recently. When Mount Pinatubo in the Philippines erupted in 1991, a gaseous plume containing an estimated 20 million tons of sulfur dioxide enveloped the planet. The earth's average temperature fell by a remarkable 0.5 degrees Celsius for 18 months.[14]

The trick, for geoengineers, would be to reproduce something like the Pinatubo effect over a sustained period and in a controlled fashion. A steady supply of sulfate particles, or perhaps some other material with similar properties, could conceivably be introduced into the upper reaches of the atmosphere via ballistics—which is to say, as historian James Fleming has put it, by "declaring war on the stratosphere." Other proposals involve streaming sulfate particles through giant hoses tethered to helium-filled balloons or adding sulfates to jet fuel. The required sulfur could itself be harvested in the needed quantities from coal-fired power plants, in effect rendering two of the main contributors to climate change—jet travel and the burning of coal—central components of the fix.[15]

Paul Crutzen, in his 2006 article, suggested that the stratospheric sulfur approach to climate stabilization could be developed and implemented for $25–50 billion a year—a small fraction of the 5–20 percent of global gross domestic product that Nicholas Stern estimated, in his much-cited report for the U.K. government, climate change will cost the global economy if no remedial action is taken. One way to get more bang for these bucks would be to deploy stratospheric sulfate aerosols (or, perhaps, ground-level whitening) in a targeted fashion. Consider the Arctic. Shielding the Arctic from some percentage of solar radiation could, some suggest, rapidly reverse global warming–induced ice melt. Since melting Arctic ice sparks two very powerful and potentially dangerous feedback loops that affect the climate system—by releasing stored methane and exposing dark water that absorbs higher levels of solar radiation—arresting Arctic warming would be a logical priority for this sort of geoengineering approach.[16]

Finally, the most "way out" SRM strategy—way out in every sense—would involve launching sunshades into space. This would be by far the most technologically challenging of the options listed, but speculative accounts in support of the idea abound. A well-known proponent is astrophysicist Roger Angel. His plan is for a "cloud of many spacecraft," with each small vessel consisting of a transparent material designed to reflect solar radiation, all launched into orbit using a system of ion propulsion. Angel has suggested that such a scheme could be in place in as few as 25 years, for a cost of a few trillion dollars.[17]

Carbon Dioxide Removal. While SRM options can potentially turn down the heat, they do nothing to clear the air of CO_2 and other greenhouse gases. This means that if an SRM project were to be successfully developed, it would have to be continued indefinitely. Otherwise, the full pent-up warming effect of rising atmospheric GHG concentrations would be suddenly unleashed. SRM also does nothing to curtail ocean acidification and the other disruptions that increases in CO_2 concentrations can cause. Here is where carbon dioxide removal enters the picture. With CDR, the idea is to draw significant amounts of carbon out of the atmosphere and then store it in some benign, long-term fashion.

The United Kingdom's Royal Society, in an influential 2009 report, identified and analyzed a variety of CDR possibilities, dividing the schemes into land-based and ocean-based options. (See Figure 29–2.) One land-based idea that has captured a good deal of attention is development of a new generation of mechanical CO_2 "scrubbers." The hope for these machines is that they could pull large quantities of CO_2 directly from the air. This is quite different from most carbon capture and storage schemes currently under discussion, which aim to remove CO_2 from the flue gases that escape from fossil-fuel-driven power plants. A company calling itself Carbon Engineering, based in Alberta, Canada, and started by academic David Keith, is a leading proponent of CO_2 scrubbers that operate apart from power stations, and it has developed a functioning prototype.[18]

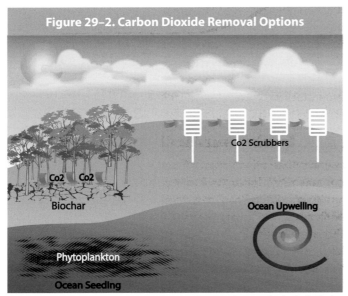

Figure 29–2. Carbon Dioxide Removal Options

Co2 Scrubbers

Co2   Co2

Biochar

Ocean Upwelling

Phytoplankton

Ocean Seeding

Source: Graphic designed by Isabelle Rodas

An alternative land-based CDR approach involves sequestering carbon in biomass. The most obvious way to do this is to plant a whole bunch of trees or, on a large-enough scale, to invest in tilling methods that encourage carbon to be taken into and stored in the soil. Finding adequate land area for such schemes is the central limiting factor. Or perhaps biomass could be grown and then converted into liquid or hydrogen fuels, with the CO_2 from combustion of those fuels then captured and stored. Another idea in which a great deal of hope has been invested is the "biochar" option, which has captured the attention of figures like James Lovelock of Gaia-hypothesis fame. Biochar involves growing biomass, combusting the living material to produce charcoal, and then burying the charcoal in the soil, which in turn serves both as carbon sink and soil enhancement.[19]

As for the oceans, the most talked-about CDR possibility is ocean seeding. Here the idea is to take advantage of the natural process whereby phytoplankton take in carbon from the atmosphere. When these plankton die, they sink to the ocean floor. Under certain conditions the carbon they contain may remain under the ocean in a benign form for many centuries. Some would-be geoengineers hope to encourage blooms of carbon-hungry phytoplankton by introducing soluble iron into areas of the oceans in which iron is in relatively short supply. Though this idea makes sense in theory, the few field trials that have been undertaken have given mixed results. In one early trial, iron dumped into the South Atlantic did indeed trigger a plankton bloom. However, most of the additional plankton was eaten by a swarm of shrimp before it reached the bottom of the ocean. Ocean seeding, as with all the geoengineering proposals just described, has all kinds of challenges associated with its successful development, including any number of problems that cannot readily be anticipated in advance of full-scale deployment.[20]

Suffice it to say, some CDR schemes on both land and water would depend on the willful augmentation and use of existing biological or chemical systems, while others would require the development of entirely new mechanical arrays. There is, ultimately, no shortage of schemes for drawing down the planet's surfeit of atmospheric carbon. The question then becomes where to put it and whether the carbon will stay where it is deposited. What was once thought to be the easy part of the "carbon capture and storage" puzzle is now turning out itself to be extraordinarily thorny.

The obvious place to put billions of tons of carbon is into the depleted oil wells from which much of it originally came or into porous rock formations deep underground. Carbon dioxide, once captured, can be transformed into a liquid and forced under pressure into such belowground formations. A handful of demonstration projects in Algeria, Canada, Norway, and the United States have shown the feasibility of this carbon storage approach.[21]

But feasibility does not mean practicality. Part of the problem is the sheer scale of the proposed undertaking. For instance, one estimate suggests that liquefying 60 percent of the CO_2 that U.S. coal-fired power plants produce annually in order for the CO_2 to be stored underground would amass about the same volume of liquid as the United States currently consumes in oil— that is, on the order of 20 million barrels a day. There is also the challenge associated with keeping the carbon in underground storage for, it is to be supposed, many thousands of years. Potential problems like groundwater contamination or the sudden release of vast quantities of CO_2 appear small but by no means negligible.[22]

The bottom line is that research into these and many other ideas has already begun. There is much hope in the geoengineering community that a real, workable techno-fix can be developed. Still, very few are pretending that the task is an easy engineering puzzle. At the Ames meeting in 2007, for instance, hope for a technological breakthrough to tackle climate change was apparently tempered with a well-honed appreciation for the extraordinary nature of the challenge. We can just hope that there was also a strong sense of irony present in the meeting room, given this anecdote related by James Fleming, who was present at the meeting: "Even as [conference participants] joked about a NASA staffer's apology for her inability to control the temperature in the meeting room, others detailed their own schemes for manipulating the earth's climate."[23]

Affixing a thermostat to the planet's climate system should be considered no small task for a species that struggles to control the temperatures in its meeting spaces.

Parsing Geoengineering's Costs

So, can human beings willfully use large-scale technologies to cool the planet? The answer is almost certainly yes. A different and altogether trickier question is, Should we? Is the geoengineering path really worth pursuing?

For some, the answer is a resounding "of course." Richard Branson, for instance, chairman of Virgin Atlantic airlines and a host of other companies, is a well-known proponent of geoengineering: "If we could come up with a geoengineering answer to this problem, then [international climate change meetings like] Copenhagen wouldn't be necessary. . . . We could carry on flying our planes and driving our cars." Branson is investing more than words in pursuit of a solution that would leave his core business—flying people around the world—intact. In 2007 he kicked off the $25 million Virgin Earth Challenge, an ongoing search for commercially viable ways to pull carbon out of the atmosphere.[24]

Others, including the vast majority of scientists involved in geoengineering research, are far more circumspect. Hugh Hunt, a professor of engineering at Cambridge University, who is part of a team working on delivery systems to introduce reflective particles into the stratosphere, has summed up the general feeling among scientists working on geoengineering in this way: "I know this [talk of geoengineering] is all unpleasant. Nobody wants it, but nobody wants to put high doses of poisonous chemicals into their bodies, either. That is what chemotherapy is, though, and for people suffering from cancer those poisons are often their only hope. Every day, tens of thousands of people take them willingly—because they are very sick or dying. This is how I prefer to look at the possibility of engineering the climate. It isn't a cure for anything. But it could very well turn out to be the least bad option we are going to have."[25]

This talk of cures suggests a critically important distinction that must be drawn, if it is not already clear. The only real way to tackle climate change is to stabilize and then work to dramatically reduce the atmospheric concentration of greenhouse gases. The surest way to achieve such a feat is to break the world's addiction to fossil fuels. (See Chapter 14.) Carbon dioxide removal schemes offer a back-end work-around—emit the carbon and then retrieve it—and so can be seen as another way to resolve the central dynamic driving climate change. In other words, it is possible to imagine that CDR really does offer a kind of "cure" to the climate malady. But with current technologies, it is hard to see a CDR scheme coming online quickly enough or being deployed at large enough a scale to make a real dent in the atmospheric carbon load.

So it is SRM, rather than CDR, strategies that are receiving the bulk of the attention in geoengineering circles. And for SRM approaches, Hunt's circumspection is absolutely warranted. Solar radiation management is not any kind of real answer to climate change. At best, SRM can reduce the planet's fever for a period, perhaps allowing time for the real roots of climate change to be tackled.

Still, such distinctions are easily lost. Talk of geoengineering is gaining traction at least in part because of Richard Branson's line of argument. That is, geoengineering has the appearance of an easy, sacrifice-free approach to tackling climate change. Finding ways to reduce the world's dependence on fossil fuels is hard and messy. In contrast, developing some kind of geoengineering techno-fix looks easy and clean. Yet it is critically important to recognize that there are sacrifices, some obvious and some harder to spot, associated with the bulk of the geoengineering schemes under serious consideration—sacrifices that can be summarized as material, political, and existential.[26]

Material Sacrifices. Perhaps the most obvious cause for concern is that geoengineering interventions could go catastrophically wrong. The great historian of technology Henry Petroski has argued in a series of books that failure is in the very nature of technological design. He once noted that while the object of engineering design is to reduce the possibility of failure, "the truly fail-proof design is chimerical." In fact, Petroski has shown in a persuasive fashion that technological development has in a very basic sense depended on failure, since the lessons learned from failed design can often teach a great deal more than successful machines and structures.[27]

Given the scope of the geoengineering endeavor, however, that calculus cannot apply. A problem with a new design for a television set or a new line of running shoes may provoke irritation. A problem with a space mirror or stratospheric sulfur deployment, on the other hand, could have truly devastating, irreparable consequences. With many of the geoengineering proposals on the table, there is scant room for error. This is a worrying notion, particularly if influential elites become hell-bent on deploying geoengineering options, since as environmental studies professor Roger Pielke, Jr., has put it, "There is no practice planet Earth on which such technologies can be implemented, evaluated, and improved."[28]

The potential for catastrophe depends, of course, on the type and scale of the planned geoengineering scheme. As a group, SRM approaches offer the biggest potential for disaster, and computer modeling is our best current tool for understanding the potential risks. Some forecasts based on computer modeling have looked anything but promising. For instance, one research team, during work for the IPCC, concluded that any large-scale attempt at SRM would likely have serious adverse climate effects, most notably a sharp decline in rainfall due to decreased evaporation at the tropics and a reduced ability of the atmosphere to transport wet tropical air to higher and lower latitudes.[29]

Along with the danger of things going wrong, there are also massive challenges associated with things going exactly as planned. Even if executed without a hitch, certain geoengineering schemes would entail extraordinarily complex trade-offs. Under an SRM scenario, rainfall—even if it were not reduced—would almost certainly be redistributed by any radical intervention in the climate system. Some regions would see more rain, some would see less. The eruption of Mount Pinatubo has been linked to disruption of the Asian monsoon. To take two other examples, shooting sulfur into the sky would cause acid rain and would promote stratospheric ozone depletion, while adding iron to the oceans would drive the overuse of important nutrients, potentially causing massive disruption of ocean ecosystems. These most promising of SRM techniques, in other words, would force those who seek to use them to choose among competing environmental disasters.[30]

Courtesy U.S.G.S.

The 1991 eruption of Mount Pinatubo in the Philippines.

With this in mind, geoengineering is, it must be said, too grand a name for the enterprise. "Geo-tinkering" is closer to the mark. The climate system is incompletely understood. Any intervention would be tentative at best, with catastrophic failures likely. And this is taking account just of the problems that are relatively easy to forecast. Complex technologies and technological systems have a habit of "biting back," as historian Edward Tenner once put it, in ways hard to predict and sometimes hard to respond to.[31]

Given the stakes and challenges, 40 years ago British meteorologist H. H. Lamb suggested that before embarking down the geoengineering path, "an essential precaution [is] to wait until a scientific system for forecasting the behavior of the natural climate . . . has been devised and operated success-fully for, perhaps, a hundred years."[32]

Political Sacrifices. Waiting 100 years for greater levels of scientific cer-tainty is sage advice, but it is unlikely to be followed. This is because the po-litical pressure to rapidly deploy geoengineering technologies may become overwhelming as the effects of climate change grow more pronounced. Mustering the political will to generate large-scale social change in response to climate change is proving, to state the obvious, difficult. However, should melting ice drive rapid sea level rise, or should climate-related food and wa-ter pressures cause great suffering in industrial countries (rather than just in developing ones, as at present), or should some other fast-moving climate calamity force the hand of rich-country elites, then swift technology-based action may suddenly be demanded.

Deploying geoengineering technologies under such circumstances would likely be met with more limited social and political resistance than might be expected, given that geoengineering fits into a broader narrative about using technologies to solve complex problems and that geoengineering ap-proaches require little buy-in or behavior change by the public.

Scientists are eager to start with small-scale geoengineering experiments rather than be forced into large-scale development. If political pressure mounts, though, starting small would be hard. If geoengineering comes to be seen as a last-gasp option, the impetus will be toward rapid, full-scale deployment. There is no guarantee in such a situation that those who end up with their metaphoric hands on the planet's thermostat would act in the global interest rather than following some other calculus. Imagine for a moment that the U.S. government could deploy stratospheric sulfur for the direct short-term benefit of the North American continent. What if that deployment threatened African rainfall patterns? Or imagine a time when the United States is having a rotten summer while Europe is experiencing a heat wave: Who gets to adjust the mirror? What, to play this scenario out, of the legal costs to societies when every bad harvest or vacation spoiled by too much rain is thought to be the fault of distant geoengineers?

Space mirrors, stratospheric sulfur schemes, and the like all require con-centration of materials and political authority. By this measure, many geo-engineering schemes have a distinctly anti-democratic flavor. Who, then, gets to call the shots in a geoengineered world? Who will receive the ben-efits? What of small countries with limited economic means and limited political voice? What of villages that happen to be situated on top of the perfect location for underground carbon storage? The questions that can be raised about such activities are endless.

The history of weather modification efforts and of technological development more generally suggests that tussles over mirror alignment might be the smallest of our problems. Militarization could be a far bigger challenge. The militaries of the world's great powers have long looked to weather modification as a potentially potent weapon of war. Given such a history, James Fleming has suggested that "it is virtually impossible to imagine governments resisting the temptation to explore military uses of any climate-altering technology."[33]

Finally, there is a very real danger that a focus on geoengineering saps the political will for other forms of action. It is, tragically, in our collective nature to hope for a miracle. It is in the natures of our politicians and business leaders to promise one. This is the case despite repeated injunctions from scientists to continue work on traditional mitigation efforts even as research on geoengineering technologies advances.

Existential Sacrifices. This leads to a third category of geoengineering sacrifice—a category that we might call "existential." The ability to control the weather was once the prerogative of a divine creator. Now it is a technique within the reach of the world's governments, large corporations, and even wealthy individuals. The transgression of previously sacred and inviolable boundaries that is the product of such a development may seem abstract in the face of climate change, but it is actually profoundly important.

This is because more technology alone does not, despite narratives to the contrary, equal progress. Progress signals movement toward some goal. The large-scale development of geoengineering technologies would render some goals realistic and others unattainable. To imagine that geoengineering is some passive, neutral enterprise, forced on humanity by a changing climate, is to ignore the other options for response that are available and to ignore the role played by the blind worship of technology in creating the current ecological mess.

Now, there is no denying that, as Stewart Brand of the Long Now Foundation has put it, "humanity is stuck with a planet stewardship role." The conversation has to be about what to do with that role. The ultimate ecological question is a deceptively straightforward one: What kind of future will we craft? Because craft it we will. Does that crafting entail a kind of global biospheric management—the geoengineering path—or something else? A different vision of the future would privilege shared sacrifice, directed toward living well and meaningfully within ecological limits. Some geoengineering options close off or render unimaginable such a pathway. Why live differently if space mirrors will come to our rescue? A few geoengineering options, though, may be compatible with a world in which sufficiency rather than domination is the guiding ethic.[34]

Political theorist Langdon Winner once coined a useful phrase that it is worth keeping in mind: *technological somnambulism.* Too often, he suggested, people tend to sleepwalk through the making of technological decisions. With geoengineering, the scope is too vast and the implications too all-encompassing for any kind of passive decisionmaking. The risks and impacts of geoengineering cannot be considered in isolation. They must be compared with the risks of doing nothing in the face of climate change, certainly, but also the risks and benefits that inhere in other forms of response.[35]

The Future of Planetary Engineering

Is geoengineering something to be avoided at all costs? Or is it, perhaps, "a bad idea whose time has come"? It is relatively easy to poke holes in the geoengineering enterprise. Humanity's track record with large-scale technological deployment hardly gives one faith in the ability of geoengineers to completely and without harm manage the entire climate system. Scientific elites have too often had a misplaced faith in their abilities to cut through complex social problems. The horrors of the early years of the nuclear age and the ongoing blight of global hunger are just two obvious examples.[36]

Still, at the same time as there is cause for real concern about the geoengineering push, doing nothing in the face of climate change is itself not an option. And the track record of recent international climate change meetings and of most efforts to wean individuals and communities from fossil fuel dependence hardly gives cause for optimism.

Perhaps the most dangerous of all future scenarios is that the climate situation becomes so bad so quickly that rogue actors try to implement some geoengineering option about which very little is understood. The specter of such a future was raised in a particularly stark way in October 2012. That month the public learned that Russ George, an American who for some time has dabbled in the world of geoengineering, had that summer taken a ship out into the Pacific Ocean and dumped something like 100 tons of iron sulfate into the water. George claimed that his actions represented "the most substantial ocean restoration project in history." Given the many risks attached to such an enterprise, a different label, proposed by writer Michael Specter, is more apt. Russ George is now, by Specter's reckoning, the world's first "geo-vigilante."[37]

He is unlikely, though, to be the last. The genie of geoengineering is not going back into any bottle any time soon. Are there ways, then, that geoengineering's development and deployment might be effectively governed? There is a difficult dance to choreograph here. Scientists need the freedom to propose and test geoengineering options without their work being used as an excuse to delay real mitigation actions. The public and the planet need to be protected from rogue geoengineering efforts and well-intentioned efforts run amok. There is a desperate need for transparency and openness in the development of geoengineering technologies, even as the deployment of those technologies is tightly managed.

With these sorts of challenges in mind, a team of scholars in the United Kingdom drafted in 2011 a short declaration that is now known as the Oxford Principles as a code of conduct for geoengineering research. (See Box 29–2.) In this way, scientists working on geoengineering are echoing the efforts of the 1975 Asilomar Conference on Recombinant DNA—trying to self-regulate by way of the establishment of clear guidelines for safe and ethical conduct. Such efforts are to be applauded and must receive further and widespread support. The straightforward and declarative nature of the Oxford proposal is as good a place as any to start the wide conversation that now must take place about the managed development of geoengineering options.[38]

Futurist Robert L. Olson has gone further to suggest a set of criteria that differentiate "soft geoengineering" technologies—those that can actually make a difference in the face of a changing climate but that have relatively few risks attached to their development—from their more dangerous cousins. (See Box 29–3.) Olson starts from the position that a sweeping dismissal of all geoengineering options may prove imprudent. Given the complexity of the climate challenge, he is almost certainly right. Far more useful than sweeping rejection is a clearheaded evaluation of the options before us.

Box 29–2. The Oxford Principles: A Code of Conduct for Geoengineering Research

• Geoengineering to be regulated as a public good.

• Public participation in geoengineering decisionmaking.

• Disclosure of geoengineering research and open publication of results.

• Independent assessment of impacts.

• Governance before deployment.

Source: See endnote 38.

Box 29–3. Criteria for "Soft Geoengineering" Technologies

• Can be applied locally.

• Scalable to larger areas.

• Low or no anticipated negative impacts on ecosystems or society.

• Rapid reversibility if problems do arise.

• Has multiple benefits beyond impacts on climate.

• Analogous to natural processes.

• Effects are large enough soon enough to be worthwhile.

• Cost-effective with mature technologies deployed at moderate scale.

Source: See endnote 39.

Are there really, as Olson believes, possibilities for geoengineering that entail "low or no significant negative impacts"? If so, then careful development of "soft geoengineering" options by credible actors should become a legitimate part of our efforts to tackle climate change.[39]

Olson's criteria focus on geoengineering's technical elements. By his reckoning, options like brightening water through the infusion of "microbubbles," blanketing vulnerable areas of ice and water in reflective fabrics, working to improve direct-air capture technologies for CO_2, and building up carbon in soil and vegetation are no-brainers, since they offer real hope for slowing the destruction of vulnerable areas while limiting potential downsides. White roofs and other such efforts to make urban spaces more reflective should also receive attention, but whether a roof whitening scheme could ever be undertaken on a scale to make any real difference is an important consideration. On the other hand, some options—like stratospheric aerosols, space mirrors, and seeding the oceans with iron—have far too many associated risks and offer far too many technical hurdles to be taken seriously, at least at present.

Another criterion can usefully be added to Olson's list: local and democratic control. Forays into geoengineering could, conceivably, be part of the move to a more just and sustainable social order—but only if the technological development that geoengineering entails is tied to the cultivation of humanity's oldest political virtues, including humility and compassion. A moratorium on geongineering is doomed to fail. At the same time, pushing ahead with the most outlandish geoengineering schemes is likely to result in catastrophic failure of a wholly different variety. The need is for a middle ground—not geoengineering as techno-fix but rather geoengineering as one small part of an effort to steer the world to a state of rightness and fitness in ecological and social terms.

CHAPTER 30

Cuba:
Lessons from a Forced Decline

Pat Murphy and Faith Morgan

The end of the Cold War in the early 1990s was not the positive turning point some thought it would be; instead it marked the start of new crises. Worldwide inequity is at record levels. Military spending is at the highest level in modern history. Fossil fuel resources have become more limited, threatening economic hardship, at the same time that their emissions are causing dangerous climate change.

Of all these challenges, climate change is arguably the most severe and daunting. Stabilizing the climate seems unlikely without a significant reduction in fossil fuel consumption. In this context, Cuba has become an important example, since in the past two decades it has reduced its carbon dioxide (CO_2) emissions by 25 percent, from 3.2 tons per person in 1990 to 2.4 tons in 2009. Cuba's focus on meeting basic human needs instead of on economic growth and consumption offers an important example to the rest of the world.[1]

The context for Cuba's current situation is set by its long history of colonization and isolation. After several hundred years of Spanish domination, control of Cuba passed to the United States in 1898, which continued interfering in Cuba's political, economic, and military affairs. Cuba achieved full independence with the overthrow of General Fulgencio Batista at the end of the Cuban Revolution (1953–59) led by Fidel Castro. Although President Dwight D. Eisenhower officially recognized the new Cuban government, the relationship cooled after Cuba began nationalizing properties owned by American-based corporations. Eisenhower authorized a CIA-managed 1961 invasion of Cuba that failed. In 1962 President John F. Kennedy imposed economic sanctions by banning all trade with Cuba except for nonsubsidized sales of food and medicine. To ward off the continuing U.S. threat and to find new trading partners, Cuba developed relations with the Soviet Union. This led to the very serious Cuban missile crisis of 1962, settled when Russia withdrew missiles from Cuba and the

Pat Murphy is research director and Faith Morgan is executive director of the Arthur Morgan Institute for Community Solutions in Yellow Springs, Ohio.

www.sustainabilitypossible.org

United States withdrew missiles from Turkey. The United States also agreed not to invade Cuba.[2]

The dissolution of the Soviet Union, beginning in 1990, had a devastating effect on Cuba. Trade with Cuba's former partners declined by more than 90 percent, cutting off 80 percent of the island's food imports. And imports of Soviet oil plummeted from 13 million tons in 1989 to 1.8 million tons in 1992. President George H. W. Bush added new economic sanctions with the 1992 Cuban Democracy Act, which prohibited foreign subsidiaries of U.S.-based companies from trading with Cuba. Travel to Cuba by U.S. citizens was banned, as were family remittances to Cuban relatives. President Bill Clinton signed the 1996 Cuban Liberty and Democratic Solidarity Act (also known as the Helms-Burton Act), further tightening the economic blockade against Cuba. This act, which is still in effect, prohibits recognition of any transitional government in Cuba that includes Fidel or Raúl Castro and provides for retaliatory action against any non-U.S.-based company that trades with Cuba.[3]

Since 1960, the United States has spent over $500 million trying to destabilize the Cuban government. This long-term U.S. effort forced Cuba to adapt to severe shortages of oil, medicine, and food after 1990. As a result of more than 20 years of such privations, Cuba now serves as an example of a country that has survived and thrived with very limited fossil fuel resources.[4]

Cuba's Special Period

Between 1989 and 1993, Cuba's gross domestic product (GDP) fell 35 percent, and in the absence of markets for its goods, exports dropped 75 percent. The decline in food imports caused severe food shortages. Electrical blackouts of 16 or more hours a day became common. In response to the crisis, Cuba announced the implementation of the *Período Especial* (Special Period) in August 1990. This was a series of contingency plans, austerity measures, and rationing schedules that had originally been developed for use during wartime.[5]

During the early years of the Special Period, daily energy intake fell from 2,899 calories to 1,863 calories per person. Fuel shortages forced people to walk or ride bicycles. The percentage of physically active adults increased from 30 percent to 67 percent. The average adult lost 9–11 pounds, or 5–6 percent of body weight.[6]

The availability of medical supplies and equipment was dramatically reduced. A report from the American Association for World Health noted that "a humanitarian catastrophe was averted only because the Cuban government has maintained a high level of budgetary support for a health care system designed to deliver primary and preventive health care to all of its citizens."[7]

Housing and transportation were similarly affected. New housing starts dropped precipitously; existing housing units deteriorated in the absence of construction materials, replacement parts, and the resources for routine maintenance. Overcrowding became common in Havana as families "doubled up" in domiciles, with adult children living much longer with their parents. Passenger transport fell by 58 percent, and Cuba's ability to move goods around the country was severely constrained.[8]

In response, major efforts were made to develop public transportation. Trucks were converted into buses, and local manufacture of buses began. Horse-drawn carriages, buggies, and carts were used extensively. Hitchhiking became a necessity, and state-owned vehicles were (and still are) required to pick up hitchhikers. An extensive taxicab service was introduced by the national government.[9]

The economic crisis also transformed Cuba's agricultural system. Prior to 1990, the country's agriculture had used a mixture of Soviet and U.S. farming techniques that were large-scale, export-oriented, heavily mechanized, and highly dependent on chemical inputs. Cuba's agriculture had used 1.3 million tons of fossil-fuel-based fertilizer annually before the economic crisis; afterward, fertilizer use dropped to 160,000 tons a year.[10]

John Morgan

An organic urban farm in Havana on land leased rent-free from the government.

Throughout the worst years, 1993 to 1995, two basic government policies kept the food crisis from becoming catastrophic. Food programs targeted the most vulnerable populations (children, the elderly, and pregnant and lactating women). And a ration-card system for distributing the greatly curtailed quantities of food supplied all Cubans with rice, beans, and other basic foods.[11]

Cuba developed agricultural techniques to deal with the lack of chemical inputs and limited fuel, electricity, and machinery. These included organic fertilizers, animal traction (oxen), mixed cropping, and biological pest control. The development of urban gardens and farms yielded a major increase in domestic fruit and vegetable production.[12]

In a 2001 report, Oxfam America stated that although inequity had increased, social unrest was minimal thanks to the government's agricultural-reform strategies of the mid-1990s. The policies encouraged private enterprise and decentralized decisionmaking by distributing large state farms (41 percent of the arable land) to thousands of smaller farmers' cooperatives and by leasing government land to private farmers in rent-free lease agreements. Given the loss of chemical inputs and fuel for machinery, smaller farms were necessary to implement the sustainable agricultural practices, such as the use of oxen and increased manual farm labor, that were vital to maintain food production.[13]

In 1994 the ad hoc urban gardening movement was recognized by the government. Laws were enacted to support, promote, and regularize the movement's practices of ecological (organic) agriculture without stifling local initiative. Distribution reform allowed private farmers markets for the first time in nearly four decades. Prices paid by the government for food were raised in order to increase production incentives. Farmers were allowed to sell high-quality produce to tourist facilities in order to reduce the country's import bill. Cuba also benefited from its stable rural sector, where small farmers' land rights had been maintained and where earlier agrarian development policies had produced a well-educated peasantry, unique in modern Latin America.[14]

Cuba's Energy Response

Fossil-fuel energy has been at the forefront of Cuban concerns since 1959, when U.S. oil companies cut off shipments to Cuba. Before then, only 56 percent of Cubans had access to electricity; by 1989, this had increased to 95 percent. This improvement was possible due to shipments of oil from the Soviet Union, which continued until 1990.[15]

In 1993 Cuba's legislature passed the National Energy Sources Development Program. Its goals include increased energy efficiency (the first priority), reduced energy imports, and maximized domestic energy sources. Cuba began a drive to save energy and use more renewable sources. Off-grid schools, health clinics, and social centers were electrified with solar panels.[16]

The Cuban Electricity Saving Program and the Energy Saving Program of the Ministry of Education were launched in 1997 to promote energy education. The goal was to reduce the consumption of electricity in all Cuban households, industries, and enterprises. Children were educated about energy and then influenced their families and the rest of the culture. In 2000, Cuba and Venezuela signed an Integral Cooperation Accord under which Venezuela sends oil to Cuba in exchange for goods and services.[17]

From 2003 to 2005, malfunctioning power plants and increased hurricane activity brought the return of massive blackouts. Historically Cuba had averaged one hurricane every other year, but in 2008 Hurricanes Gustav, Ike, and Paloma devastated the island, causing $10 billion in damage. Two million people were evacuated from the threatened areas. Many of Cuba's agricultural crops were destroyed, and that year imports had to supply 55 percent of Cubans' food, an increase from 16 percent.[18]

The effects of climate change will likely include a rise in the intensity and frequency of extreme weather events. The biggest threats to Caribbean island nations like Cuba are hurricanes, droughts, heavy rainfall, and a rise in sea level. Cuba has developed emergency preparations and evacuation plans based on the specific vulnerabilities of each of its 167 municipalities. When a hurricane approaches, power plants in its path are shut down and people are evacuated. In recent years this has affected hundreds of thousands of Cubans. After decades of energy shortages, and with the heightened danger from hurricanes, Cubans are aware of their vulnerability to both shortages and overuse of fossil fuels.[19]

In 2005 the parliament passed the Cuban Energy Revolution (CER) initiative. Its goal was to guarantee sustainable development of the economy and energy invulnerability. To meet the first of its five objectives, increased energy efficiency and conservation, Cuba distributed 9.4 million compact fluorescent light bulbs in the second half of the year to homes, businesses, and other institutions to replace nearly all of the incandescent lights used in the country. In 2006 millions of older, inefficient appliances were replaced: 1,043,709 fans; 2,404,035 old American and Russian refrigerators; 209,480 air conditioners; 216,149 televisions; and 267,568 electric motors. In addition, almost 3.5 million rice cookers and over 3 million pressure cookers were distributed for families to encourage the switch from cooking with kerosene to cooking with electricity (which also brought health and safety benefits). These were sold at a subsidized cost to about 4 million households. (The population of Cuba is 11 million.)[20]

Electricity is highly subsidized in Cuba, and prior to 2006 it was sold very cheaply to consumers. The efficiency measures reduced the government's costs by reducing electricity demand. To encourage conservation, a new electricity tariff was introduced that allows people using less than 100 kilowatt-hours (kWh) per month to continue paying the very low rate. Above that, for every increase of 50 kWh per month, the tariff goes up.[21]

The second CER priority, improving the availability and reliability of electrical service, involved changes in production, transmission, and use of electricity. In 2005 most of Cuba's 11 large oil-fired thermoelectric power plants were more than 25 years old, inefficient, and functioning only about 60 percent of the time. To improve energy security, Cuba decentralized its energy system, moving toward distributed generation. In 2006 Cuba installed 1,854 diesel and fuel-oil micro-electrical plants throughout the country and upgraded the transmission network. The new diesel generators were more efficient, using 234 grams of fuel to generate a kilowatt-hour, compared with 284 grams for the older plants. This distributed generation system provides 25 percent of Cuba's electricity. Cuba also installed over 4,000 emergency backup generators in critical areas, such as hospitals, schools, bakeries, stores, and food production facilities. These maintain power to tourist hotels and meteorological stations as well. In 2006 and 2007, Cuba saved over 960,000 tons of imported oil through these measures.[22]

Micro-hydro plant near Santiago de Cuba.

The CER's third concern, renewable energy, is a vital part of Cuba's current and future energy mix. Distributed generation is key to developing regional sources of renewables, such as wind farms, hydropower, solar photovoltaic panels, solar water heating, biogas, and biomass from reforestation and sugarcane. Renewable sources account for about 6 percent of Cuba's installed capacity. Rivers in Cuba are not long, limiting the country to micro-hydro installations. Over 8,000 independent solar-electric systems have been installed in rural areas to provide electricity where it is difficult to extend the national grid. Today, all rural areas have solar-electric systems for school lights, computers, educational television, and health centers.[23]

The fourth focus of the CER, developing Cuba's own oil and natural gas resources, has done little more than replace declining oil production. In 2010 Cuba's oil production was just over 3 million tons, compared with 2.73 million tons in 2009; natural gas output was 1 million cubic meters, compared with 1.16 million cubic meters the preceding year. The national output of oil and gas amounts to the equivalent of about half of the 150,000 barrels per day that Cuba consumes. Venezuela provides the remainder in exchange for support from Cuba in the fields of education, health care, sports, science, and technology.[24]

To meet the fifth CER goal, international cooperation, Cuba is exporting the CER to other countries. It is working with Bolivia, Honduras, Lesotho, Mali, South Africa, and Venezuela, sharing strategies for reducing energy demand. Cuba has provided and installed solar-electric panels (over 1 megawatt of total capacity) in these countries. Cuban social workers have replaced about 100 million incandescent bulbs with compact fluorescents in a dozen Latin American countries.[25]

These efforts have paid off in terms of one of the key measures of sustainability: greenhouse gas emissions. (See Table 30–1.) Cubans on average use 43 percent less energy than people in the rest of the world (1.03 tons of oil equivalent compared with 1.8 tons) and account for 44 percent less CO_2 each year (2.4 tons compared with 4.29 tons). And compared with Americans, Cubans use 85 percent less energy on average and account for 86 percent less CO_2. Cubans have far fewer material possessions than people in more industrialized countries, but due to the country's commitment to a high level of education and social services, Cubans are far richer in other resources, such as social capital and a sense of community.[26]

Human Development and Survivability

Since 1960 Cuba has been committed to maintaining a high level of social services, devoting far more of its energy and resources to human development or social capital than the former Soviet Union, which abandoned social services for privatization in the 1990s.

Medical Care. Free high-quality medical care is a key part of the Cuban revolution. Article 49 of the constitution states:

Everyone has the right to health protection and care. The state guarantees this right by providing free hospital and medical care by means of the installation of the rural medical service network, polyclinics, hospitals, preventive and specialized treatment centres, by providing free dental care and by the health publicity campaigns, health education, regular medical examinations, general vaccinations, and other measures to prevent the outbreak of disease. All the population cooperates in these activities and plans through the social and mass organizations.[27]

This commitment has placed Cuba first in the world in terms of physicians per person. In 1960 Cuba had 0.95 doctors per 1,000 people; today the ratio is 6.4 physicians per 1,000 people. The United States, in comparison, has 2.67 physicians per 1,000 people. Cuba has 5.9 hospital beds per 1,000 people while in the United States the figure is 3.1 beds. Medical expenditures in Cuba account for 11.8 percent of GDP; U.S. expenditures are 16.2 percent of GDP. Cuban doctors and other medical personnel also serve overseas, with about 37,000 Cuban doctors practicing in about 50 countries. Cuba's high ratio of doctors to patients gives family physicians more time to spend with each patient. Prevention is emphasized, with a holistic approach that seeks to integrate psychological and physical well-being. Under the U.S. blockade, acquiring needed medical supplies and equipment is very difficult, but Cuba has shown that people's health does not depend on a high-cost medical system.[28]

Cuba has also excelled in supporting the health of mothers and children. In a 2012 report from the nonprofit group Save the Children, 165

Table 30–1. Annual Energy Consumption and Carbon Dioxide Emissions per Person in Major Regions, Cuba, and the United States

Region, Country, or Economy	Population	Energy Use	Carbon Dioxide Emissions
	(millions)	(tons of oil equivalent per person)	(tons per person)
OECD countries	1,225	4.28	9.83
Middle East	195	3.03	7.76
Non-OECD Europe and Eurasia	335	3.14	7.46
China	1,338	1.70	5.14
Asia	2,208	0.66	1.43
Latin America	451	1.20	2.16
Africa	1,009	0.67	0.92
World	6,761	1.80	4.29
Cuba	11	1.03	2.40
United States	307	7.03	16.90

Source: See endnote 26.

nations were ranked according to a Mothers, Women and Children index. The Women's Index was calculated as a weighted average of health status (30 percent), educational status (30 percent), economic status (30 percent), and political status (10 percent). The Mothers' Index was calculated as a weighted average of children's well-being (30 percent) along with women's health (20 percent), education (20 percent), economic status (20 percent), and political status (10 percent). Among 80 mid-level developing countries, Cuba ranked first on the Mothers Index and second on the Women's Index.[29]

Education. Education in Cuba is free. The country ranks second in the world in the share of GDP allocated to education, at 5.5 percent. The United States, ranked first, spends 13.6 percent; the world average is 4.4 percent. Average length of time in primary, secondary, and tertiary schools is 18 years for Cuba, 15 years for the United States, and 11 years for the world. Cuba, with 2 percent of the population of Latin America, has 11 percent of the scientists. Having a well-educated population was a huge advantage in dealing with the massive social changes needed to surmount the difficulties faced during the 1990s. (See Box 30–1.) Indeed, a World Bank study notes:

> The record of Cuban education is outstanding: universal school enrollment and attendance; nearly universal adult literacy; proportional female representation at all levels, including higher education; a strong

Box 30–1. Who Was Behind Cuba's Response in the 1990s?

The crisis after the collapse of the Soviet Union was rapid and severe. Unprepared, industrial plants and factories reduced weekly work hours; some closed altogether. An additional 200 consumer goods were added to the ration list, and foods of all kinds became increasingly scarce. Cuba's mass organizations played a key role in this difficult period. The Committee for the Defense of the Revolution (CDR, founded in early 1960) had been extended to organize blood donations, vaccination campaigns, neighborhood cleanup, and recycling. There are 122,000 neighborhood CDRs in Cuba, each run by people selected from within the community.

In the crisis, CDRs took it upon themselves to find places to grow food and locate seed, quickly extending their scope to support backyard and urban gardens for cultivating produce and medicinal plants. Other mass organizations also aided during the crisis, including the Federation of Cuban Women, the Central Organization of Cuba Trade Unions, and organizations for students, writers/artists, and small farmers.

Cuba's nongovernmental organizations played a key part in overcoming the crisis. They are not anti-government but rather are smaller, more flexible groups of people working in parallel with the government to handle social, environmental, and economic programs and research. One such small research group, the Groupo de Agricultura Organica (GAO), developed integrated pest management, which was not an important part of Cuba's pre-crisis agriculture but became very valuable after the crisis. This and other GAO work on low-technological inputs was used immediately.

Coordination between government and people was critical. Television and radio were used to communicate the status of the crisis and government plans. Mass organizations played key roles in aiding people in their workplaces and neighborhoods. Out of necessity, people took spontaneous immediate action, such as hitchhiking and gardening. Later the government organized policies to support these grassroots movements.

The commonly held view of Cuba as a dictatorship slights the social solidarity of a people who have withstood invasions and colonization. This countrywide solidarity and cooperation is very much a part of the Cuban character and was important in dealing with the stresses of the Special Period.

Source: See endnote 30.

scientific training base, particularly in chemistry and medicine; consistent pedagogical quality across widely dispersed classrooms; equality of basic educational opportunity, even in impoverished areas, both rural and urban. In a recent regional study of Latin America and the Caribbean, Cuba ranked first in math and science achievement, at all grade levels, among both males and females. In many ways, Cuba's schools are the equals of schools in OECD countries, despite the fact that Cuba's economy is that of a developing country.[30]

Agriculture. Cuba has achieved high levels of success with a unique form of ecological agriculture. There are about 140,000 high-level professional and medium-level technicians, dozens of research centers, agrarian universities and their networks, government institutions such as the Ministry of Agriculture, scientific organizations supporting farmers, and farmers' organizations. Farmers and gardeners in Cuba are well educated and receive excellent remuneration.[31]

Urban farms and gardens have become a significant part of Cuba's agricultural system. There are 383,000 urban farms on 50,000 hectares of otherwise unused land. Urban farms produce 1.5 million tons of vegetables a year without using synthetic chemicals and supply 70 percent or more of the fresh vegetables consumed in Havana and other cities.[32]

In 2006, Cuban rural farmers (using 25 percent of the agricultural land) produced 65 percent of the country's food. These farmers, along with Cuban researchers, have developed a unique form of agroecology science and practice that has achieved high levels of production. Vegetable production doubled from 1994 to 1998 and doubled again in 1999. Production of tubers and plantains, staples of the Cuban diet, tripled in the same period. Bean yields increased by 60 percent and citrus by 110 percent. From 1996 to 2005 Cuba had the best food production performance in the Caribbean and Latin American region, with an annual growth rate of 4.2 percent per person. Agrochemical use from 1988 to 2007 declined 72 percent for vegetables, 55 percent for beans, and 85 percent for roots and tubers.[33]

Under the Soviet system, Cuban agriculture focused on large-scale sugar plantations. Since 1990 it has been increasing its ability to provide a wide variety of foods. Cuba's food import dependency has been dropping for decades, despite brief upturns due to natural and human-made disasters. Large amounts of cooking oil, cereals, legumes (principally rice and wheat for human consumption and corn and soybeans for livestock), and powdered milk continue to be imported.[34]

Cuba's production efforts are focused on food sovereignty, defined as the right of everyone to have access to safe, nutritious, culturally appropriate food in sufficient quantity and quality to sustain a healthy life with full human dignity. According to the U.N. Food and Agriculture Organization, Cuba's average daily per capita dietary energy supply was over 3,200 kcal in 2007—the highest of all Latin American and Caribbean nations and an increase from the 2,899 kcal before the Special Period. This has been achieved while Cuba continues to decrease its per capita CO_2 generation and without the assistance of the International Monetary Fund or the World Bank.[35]

Other Indicators. Cuba's population growth rate is negative (−0.12 percent per year). Cuban life expectancy at birth is 77.7 years, just below the U.S. figure of 78.4 years. Cuba's infant mortality rate is 4.8 deaths per 1,000 live births, significantly lower than the U.S. rate of 6.06 deaths. Cuba's obesity rate among adults is only 11.8 percent. According to the U.S. Centers for Disease Control and Prevention, 35.7 percent of U.S. adults are obese.[36]

One very important success is Cuba's provision of health care for people living with HIV/AIDS. Key to this has been the political will to act without waiting for external assistance. Cuba's HIV/AIDS program is based on its comprehensive health care system, which has facilitated control over blood transfusions and blood products. It also supports the prevention of mother-to-child transmission of HIV. Cuba has developed its own antiretroviral drugs, largely through Cuban resources due to the U.S. embargo. With its large number of scientists, the country has the skilled workforce necessary to address diverse technological and scientific areas of need, including pharmaceutical research and development. Adult HIV prevalence is 0.1 percent for Cuba, compared with 0.8 percent for the world and 0.6 percent for the United States.[37]

The Cuba Paradigm

Cuba has a very low per capita income, yet in the non-materialistic, quality-of-life domain, it excels. Thus Cuba represents a paradox. It is a materially poor country that has First World education, literacy, and health care. It is rich in human development resources and low in environmental burdens, but its standard of living, and therefore its fossil fuel use and CO_2 emissions, is very low. Cuba has maintained its human service programs—free education, old-age support, basic nutrition, and free health care—throughout its Special Period. In 2006, Cuba was the only country in the world rated as having "sustainable development" in WWF's *Living Planet Report.*[38]

Fidel Castro has said that "consumer based societies are incompatible with the saving of natural resources and energy that the development and preservation of our species require," and Cubans simply have less of all material goods than people in industrial countries. They have much smaller homes (about 150 square feet per person in Havana compared with the U.S. average of about 800 square feet). Fewer than 10 percent of Cubans have private cars. They rarely fly. The consumption of common consumer personal goods is very limited. Yet Cubans don't need to fear cancelled medical insurance. They know their children will be educated without being saddled with student loans. Cubans are not weighed down with enormous debts. They know they will not go hungry or homeless.[39]

"We need a global energy revolution," according to Mario Alberto Arrastia Avila, an energy expert with the energy information center Cubaenergia in Havana. "But in order for this to happen, we also need a revolution in consciousness." A clear revolution of consciousness would involve the acknowledgement, strongly resisted by richer nations, that CO_2 emissions are directly related to material consumption. Cuba represents an alternative, where material success as measured by energy consumption is secondary, while other quality-of-life issues are given priority. The message is clear: humanity will survive and can even thrive in a resource-constrained world if it learns from the Cuban example.[40]

CHAPTER 31

Climate Change and Displacements

Michael Renner

In late 2010, the *New York Times* reported that after four consecutive years of drought—the worst in 40 years—Syria's agricultural heartland, along with adjacent areas in Iraq, was in deep trouble: "Ancient irrigation systems have collapsed, underground water sources have run dry and hundreds of villages have been abandoned as farmlands turn to cracked desert and grazing animals die off. Sandstorms have become far more common, and vast tent cities of dispossessed farmers and their families have risen up around the larger towns and cities of Syria and Iraq."[1]

The area primarily affected by the lack of rainfall is the northeast, which accounts for 75 percent of total wheat production in Syria. The 2011 *Global Assessment Report on Disaster Risk Reduction* published by the United Nations notes that since the start of the drought, close to 75 percent of agriculture-dependent households in the northeast have experienced total crop failure. Prior to the drought, Syria's agricultural sector accounted for 40 percent of the country's workforce and 25 percent of gross domestic product. Some 2–3 million people have been pushed into extreme poverty by the lack of crop income combined with the need to sell livestock at 60–70 percent below cost. Syria's livestock herd has been decimated: it went from 21 million to an estimated 14–16 million. A number of factors have produced this calamity, including climate change, overexploitation of groundwater due to subsidies for water-thirsty crops (cotton and wheat), inefficient irrigation systems, and overgrazing.[2]

The drought has led to an exodus of hundreds of thousands of people from rural to urban areas. Syria's cities were already under economic stress, in part because of the influx of refugees from Iraq after the U.S. invasion of 2003. Growing numbers of destitute people find themselves in intense competition for scarce jobs and access to resources. Francesco Femia and Caitlin Werrell of the Center for Climate and Security write that "the role of disaffected rural communities in the Syrian opposition movement has been

Michael Renner is a senior researcher at Worldwatch Institute.

www.sustainabilitypossible.org

prominent compared to their equivalents in other 'Arab Spring' countries. Indeed, the rural farming town of Dara'a was the focal point for protests in the early stages of the opposition movement [in 2011]—a place that was especially hard hit by five years of drought and water scarcity, with little assistance from the al-Assad regime."[3]

Syria's experience suggests that environmental and resource pressures, including climate change, could become an important driver of displacement. And while deep-seated popular discontent over decades of repressive rule surely is a major driver of Syria's civil war, climate-induced pressures have added fuel to the fire. But this is the important point: the repercussions from environmental degradation do not occur in a void; they interact with a cauldron of pre-existing societal pressures and problems.

Climate Impacts

Although governments are on record as wanting to limit additional warming to 2 degrees Celsius, they have failed to pursue climate policies that can meet this goal. The U.N. Environment Programme now estimates that the "emissions gap" by 2020—the difference between greenhouse gas emissions consistent with the 2 degrees target and the levels projected by then if current reduction pledges by governments are fulfilled—will amount to 8–13 gigatons of carbon dioxide equivalent (depending on how pledges are implemented). This compares to a 6–11 gigaton gap estimated in 2011. The actual trajectory of greenhouse gas emissions therefore increases the likelihood that the earth will heat up by 4 degrees Celsius by the end of the century. A new report by the Potsdam Institute for Climate Impact Research and Climate Analytics warns that the consequences will be cataclysmic in many regions, including unprecedented heat waves, inundated coastal cities, exacerbated water scarcity, increasing risks for food production, increased intensity of tropical cyclones, and irreversible loss of biodiversity.[4]

As the world gets more of a taste of the repercussions from a destabilized climate, a key question is how physical changes will translate into social and economic changes that in turn may cause people to leave their homes, either temporarily or for good. As early as 1990 the Intergovernmental Panel on Climate Change warned that "the greatest single impact of climate change could be on human migration," with millions of people displaced by shoreline erosion, coastal flooding, and severe drought. But the precise dynamics and interactions will invariably differ from place to place, with more severe consequences in some places, greater resilience and adaptability in some areas, and diverging political responses.[5]

As this section describes, climate change looks to intensify many existing challenges. (See Figure 31–1.) More-extreme weather, water stress, and loss of land can undermine habitability, food security, and economic viability.

Figure 31–1. Climate Change, Livability, and Possible Responses

Manifestations and Impacts of Climate Change

Slow-onset disasters: drought, highly variable rainfall

Sudden-onset disasters: floods, storms, heatwaves

Sea level rise

Glacier melt

Disease burdens

Consequences for Livability

Reduced habitability

Loss of marginal land

Lower/fluctuating yields, crop loss

Rising food prices

Compromised economic viability

Adaptation and Coping Strategies

Drought-resistant crops, water-use efficiency

Economic diversification

Local relocation: higher ground, lower-impact areas

Migration: from seasonal and short-term to longer-term and permanent

Source: Author

Affected communities, regions, or countries may be able to cope with the pressures through more drought-tolerant crops, economic diversification, and other adaptation measures. But people may also feel the need to move, either as a coping strategy or out of desperation.

Extreme Weather and Habitability. The pace of disasters is likely to accelerate in a warming world, although the precise frequency and intensity of disasters is not yet known. A 2011 article in *Scientific American* observed that the frequency of natural disasters has already increased by 42 percent since the 1980s and that the share of disasters that are climate-related has risen from 50 to 82 percent.[6]

Fast-onset impacts like floods and storms affect people in different ways than more-gradual processes like drought and desertification or sea level rise. The intensity and frequency of disasters may also have different ramifications. Population movements in response to disasters may vary widely with regard to their duration, characteristic, and destination.

Extreme weather disasters are seen as typically causing short-distance, temporary displacement, with affected communities returning to rebuild once a storm or flood has subsided. But experiences like the aftermath of Hurricane Katrina in the United States suggest that displacements could well be permanent in some cases. The population of Orleans Parish dropped by more than 120,000, or 24.5 percent, between 2005 and 2010.[7]

Water Stress and Food Security. Shifting rainfall patterns, more-erratic rain, and more-severe droughts resulting from a warming climate translate into fluctuating water availability—with potentially severe impacts on agriculture. Arid and semiarid areas cover about 40 percent of Earth's land surface and are home to more than 2 billion people.[8]

Over a decade ago, scientists warned that desertification processes put an estimated 135 million people worldwide at risk of being driven from their lands. Growing water stress in parts of the world will be compounded by the effects of saltwater intrusion in coastal areas due to sea level rise, by glacier melt in regions like the Himalayas and the Andes, and by disruptions of the monsoon cycle. Water shortages could affect anywhere from 75 million to 250 million people in Africa by 2020 and more than 1 billion people in Asia by 2050.[9]

In 2012, drought devastated crops around the world, including in major producers like Argentina, Australia, Brazil, India, Russia, and the United States. The World Meteorological Organization stated in August 2012 that "climate change is projected to increase the frequency, intensity, and duration of droughts, with impacts on many sectors, in particular food, water, and energy." In a world where the average temperature has risen 4 degrees Celsius, yields for staple crops in large parts of sub-Saharan Africa are projected to drop massively, and more than one third of current cropland in eastern and southern Africa would likely become unsuitable for cultivation.[10]

Lower yields, shortened growing seasons, or outright loss of harvests undermine food security for many millions of people. They threaten household income from farming in rural areas. Oxfam notes that affected people are typically forced to change their diets, sell productive assets, incur even more debt, take their children out of school, and in some cases migrate. Price volatility is bad for planning ahead, and many small-scale farmers may not even be able to take advantage of rising prices if they lack access to credit and agricultural inputs.[11]

The repercussions of climate change will be felt by way of rising food prices—both sudden spikes and a more-gradual, longer-term rise. Already the past decade has seen a steady price rise, along with two severe spikes. (See Figure 31–2.) A recent study by the New England Complex Systems Institute argued that food prices are a key precipitating factor for social unrest. Given the reliance of many poor countries on the global food system and a limited ability of local supplies to provide a sufficient buffer, there is heightened sensitivity to global food price trends. To the extent that governments are unable to provide food security, their legitimacy suffers, and ensuing protests could become a vehicle for expressing discontent with a range of other problems. When prices first spiked in 2008, more than 60 food riots occurred in 30 different countries. Surging prices in late 2010 and early 2011 again coincided with food riots, including in the Arab Spring countries. Aside from price spikes, the underlying steady upward trend in prices observable over the past decade may be an indicator of more continuous unrest and instability to come.[12]

Rising Seas and Loss of Land. Small island states like the Maldives in the Indian Ocean and Tuvalu in the Pacific could be submerged entirely as the seas continue to rise. And more than 600 million people worldwide live in river deltas and other low-lying coastal zones. Sea level rise can lead to greater coastal erosion as well as bigger storm surges. The government of Bangladesh warns that more than 20 million of the country's inhabitants could be forced to move due to a combination of rising sea levels and a growing number of cyclones and storm surges. Modeling suggests that 40 million people in India could be displaced by a one-meter sea level rise. In Vietnam's Mekong Delta, a one-meter sea level rise could eventually displace more than 7 million residents, and a two-meter

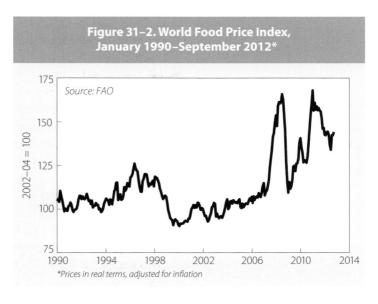

Figure 31–2. World Food Price Index, January 1990–September 2012*

Source: FAO

2002–04 = 100

*Prices in real terms, adjusted for inflation

rise would double the figure—affecting half of all delta residents.[13]

Sea level rise may have more-gradual impacts than extreme weather events, but it also has an irreversible impact. Floods recede eventually, but in a warming world the sea does not return to lower levels. Resulting displacements of people are therefore permanent ones.

To Move or Not to Move

There is still vigorous debate over whether climate change will lead to a massive increase in population movements. The International Organization for Migration rightly points out that "migration does not always occur, as the most vulnerable may lack the means to migrate." Where climate-induced population movements do take place, they can be seen as either a failure to adapt (that is, a reflection of vulnerability and inadequate resilience, and thus a more refugee-like response) or as a coping strategy (an effort to diversify sources of income and build resilience). Still, in order to move, people need financial resources, and they may need access to social networks that facilitate mobility and perhaps provide assistance at their destinations. Without such wherewithal, people may be stuck in their place of residence irrespective of the conditions. Of course, absence of movement does not equate with absence of adverse impacts.[14]

The conventional view is that even in a warming world migration will continue to be a safety valve that allows people and communities to cope. The resilience and adaptability of people should certainly not be underestimated. Still, the past is unlikely to be prologue, and for several reasons this may be an overly sanguine view.

First, the repercussions from a destabilized climate system—stronger and more-frequent disaster events—have no meaningful precedent in the human experience. Second, societies will likely not be exposed to one impact at a time but rather will experience different types of impacts—for instance, floods and droughts—simultaneously, with the possibility of cascading effects and unexpected feedback loops. Far greater numbers of people may feel the need to move than is currently the case.[15]

Third, larger populations on the move limit the maneuvering space for adaptation, as more people compete with each other and with host communities for the same opportunities, jobs, resources, and services. Fourth, in receiving areas there may also be a sharply reduced willingness to be open to an influx of people—a response that is already in evidence around the world in today's circumstances.

Fifth, migration patterns may become more permanent and less temporary. For instance, severe impacts of climate change could disrupt traditional patterns of seasonal mobility. In sub-Saharan Africa, nomadic patterns used by pastoralists to cope with droughts are already affected by rapidly changing environmental conditions. In Bangladesh, the traditional movement between different *chars* (sand and silt islands in the Padma river delta and Bay of Bengal that are home to more than 5 million people) is being disrupted by increasingly frequent and intense flash floods.[16]

Similarly, Vietnamese rice farmers who have migrated seasonally to cities during the flooding season in order to diversify their incomes have more recently been forced to settle there permanently because extreme floods have destroyed their rural livelihoods. And in Mozambique, communities along the Zambezi and Limpopo rivers have traditionally moved out of the floodplain periodically to avoid flooding. Following disastrous floods in 2000, 2001, and 2007, however, the government encouraged residents to relocate permanently. But people who have resettled lack the means to sustain themselves; heavily dependent on aid, they may need to consider moving to the new capital, Maputo, or to neighboring South Africa.[17]

New Categories and Controversies

Among the various groups of people who leave home for different reasons, some categories are well established. (See Box 31–1.) International law accords recognition to international refugees (though governments do not always live up to their responsibilities). By contrast, internally displaced persons receive far fewer protections and sometimes none at all. There have been efforts to give additional groups of displaced people—persons uprooted by natural hazards and by development projects—greater visibility, but they typically remain at the mercy of ad hoc humanitarian aid if they receive any support at all.[18]

Box 31–1. Displacement and Migration: How Many People Are Affected?

According to the 2012 edition of the *World Disasters Report* published by the Red Cross, close to 73 million people were displaced in 2011, either inside their home countries or across a border. International refugees numbered more than 16 million (see Figure below), including 10.4 million refugees under the care of the U.N. High Commissioner for Refugees (UNHCR), 5.1 million Palestinians under the care of the U.N. Relief and Works Agency for Palestine Refugees in the Near East, and close to 1 million asylum seekers. Internally displaced persons are an even larger category of displaced persons, with 26.4 million. People displaced by natural hazards were estimated at nearly 15 million, roughly the same number as those displaced by ill-considered development projects.

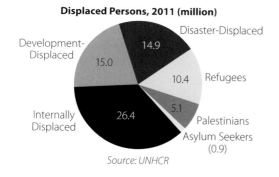

Displaced Persons, 2011 (million)

Disaster-Displaced 14.9
Development-Displaced 15.0
Refugees 10.4
Internally Displaced 26.4
Palestinians 5.1
Asylum Seekers (0.9)

Source: UNHCR

The number of people forced to flee in the face of disasters fluctuates strongly from year to year, declining from 36 million in 2008 to 17 million in 2009, surging to 42 million in 2010, but then declining again to 15 mil-

lion in 2011. The relative importance of climate-related events is also fluctuating. Among the 36 million people displaced in 2008, some 56 percent were uprooted due to climate-related events. In 2010, however, climate was regarded as the culprit for more than 90 percent of displacements.

The number of climate-displaced persons is generally expected to rise in coming years, as extreme weather events become more frequent and intense and as droughts, desertification, sea level rise, and glacier melt become more prominent. The International Organization for Migration, for example, has suggested that in a 4-degree warmer world, the commonly cited estimate of 200 million people displaced by climate change by 2050 could "easily be exceeded." However, it seems impossible to make any reliable projections about how many people may be uprooted due to climate change in coming years and decades. There are too many unknowns to be able to predict the scale of population movements to come, let alone their direction, destination, and timing.

It should be noted that at present the number of people who leave involuntarily for any reason remains considerably lower than that of people who leave more or less voluntarily. Long-term international migrants (people who live outside their home country for a year or longer) are estimated at 214 million, and internal migrants may number as many as 740 million. The ranks of both groups of migrants have grown significantly over the past half-century as economies have become more interconnected.

Source: See endnote 18.

A number of researchers have suggested for years that the world community needs to develop new categories of people on the move and that the old categories no longer adequately capture the complex reasons why and how people move. The term *environmental refugee* was proposed as early as the 1970s, but a report written by Essam El-Hinnawi for the U.N. Environment Programme in 1985 brought the term into much broader view.[19]

The emergence of this new terminology prompted a vigorous debate. Some analysts argue that the category of refugees—legally defined as people fleeing persecution without access to protection by their own country—

should not be muddied by other factors such as environmental degradation. To some extent this reflects the fact that migration studies have essentially ignored environmental factors until recently.[20]

Other analysts point out that not everyone uprooted by environmental change crosses a border—and thus does not technically become a refugee but rather an *environmentally displaced person.* Further, some people may be more aptly described as environmental migrants—moving, sometimes seasonally or temporarily, before the "push" of environmental degradation forces them to leave, motivated in part by the "pull" of an expected better life elsewhere or the prospect of remittances flowing back home to supplement local incomes made more meager or precarious by climate impacts. Climate change is likely to extend the time that seasonal migrants spend away from home, and over time "push" may outweigh "pull."[21]

Beyond the category of refugees, there is no agreed-upon—and, more important, no legally binding—definition of other groups of people on the move. The definition of internally displaced people finds some de facto recognition in guidelines adopted by the United Nations. But terms like *environmental refugees* and *environmental migrants* are wholly informal and contested. (See Table 31–1.)[22]

For now, the distinction between forced and voluntary forms of population movements remains key to international law and government policies, and the fact that there is no official recognition given to new categories of people on the move constrains the world's ability to properly deal with the situation.

There is growing recognition that it will be increasingly difficult to easily categorize the displaced by separate causes. Environmental problems are often closely intertwined with socioeconomic conditions such as poverty and inequality of land ownership, resource disputes, poorly designed development projects, and weak governance. Distinguishing in a clear-cut way between forced and voluntary movements of people is becoming harder. Instead of distinctions written in stone, it is more useful to think in terms of a continuum of causes and factors. Indeed, as the 2012 edition of the *World Disasters Report* from the Red Cross explains, the term *mixed migration* is increasingly being used. For a better understanding of the dynamics and for more-productive discussions about possible policies, it is essential that migration, refugee, and environmental experts engage with each other with an open mind.[23]

Resilience and Adaptation

Resilience is a key factor determining whether vulnerability translates into flight. The poor are typically most exposed to environmental hazards. Social marginalization often compels them to live in risky places—steep hillsides likely to be hit by landslides, low-lying areas susceptible to flooding, or

Table 31–1. Definitions of Different Types of Population Movements	
Category (Source)	**Definition**
Refugee (1951 United Nations Convention Relating to the Status of Refugees)	Someone who "owing to well-founded fear of being persecuted for reasons of race, religion, nationality, membership of a particular social group or political opinion, is outside the country of his nationality and is unable, or owing to such fear, is unwilling to avail himself of the protection of that country; or who, not having a nationality and being outside the country of his former habitual residence as a result of such events, is unable or, owing to such fear, is unwilling to return to it."
Internally displaced persons (Guiding Principles on Internal Displacement, Introduction, 1998)	"Persons or groups of persons who have been forced or obliged to flee or to leave their homes or places of habitual residence, in particular as a result of or in order to avoid the effects of armed conflict, situations of generalized violence, violations of human rights or natural or human-made disasters, and who have not crossed an internationally recognized State border."
International migrants (International Organization for Migration)	"Generally speaking, international migrants are those who cross international borders in order to settle in another country, even temporarily."
Environmental refugees (Essam El-Hinnawi, 1985)	"People who have been forced to leave their traditional habitat, temporarily or permanently, because of a marked environmental disruption (natural and/or triggered by people) that jeopardized their existence and/or seriously affected the quality of their life."
Environmental migrants (International Organization for Migration, 2007)	"Environmental migrants are persons or groups of persons who, for compelling reasons of sudden or progressive change in the environment that adversely affects their lives or living conditions, are obliged to leave their habitual homes, or choose to do so, either temporarily or permanently, and who move either within their country or abroad."

Source: See endnote 22.

coastal strips whose natural buffers (wetlands, mangroves, and coral reefs) have been stripped away. And they often have limited capacity to deal with these challenges, sometimes lacking the necessary monetary resources, family networks, or other connections needed to migrate.[24]

Adaptation measures can help reduce vulnerability: disaster and famine early warning systems, livelihood diversification, drought-tolerant crops, restoration of ecosystems, flood-defense infrastructure, crop insurance, and other measures. But even in the wake of floods or storms, well-calibrated emergency and recovery aid can make the difference between staying and leaving. Resilience is also a function of overall economic capacity, diversification to reduce dependence on one or a few economic assets, demographic pressures, governance structures and good leadership, and social and political cohesiveness.[25]

The World Bank estimates that in a 2-degrees warmer world, annual adaptation costs for developing countries will run to $70 billion by 2020 and $100 billion by 2050. Other estimates, however, make this look like a very conservative figure, and warming above that level would escalate the costs. So far, international funding for adaptation in poor countries is wholly inadequate, and commitments by richer countries seem weak and ambiguous at best. Yet timely and well-designed adaptation will be much less costly in economic and human terms than dealing with growing disasters and displacements.[26]

The U.N. High Commissioner for Refugees already struggles to provide adequate support for refugees and internally displaced persons, and the same is true for agencies providing humanitarian aid. They will be overwhelmed if large-scale, climate-related displacements come to pass. UNHCR's 2012 annual report warns of a gap in international protection when it comes to people who flee across borders to escape the impact of climate change or natural disasters, as they are not recognized as refugees under international law. High Commissioner for Refugees Antonio Guterres argues that people who are on the move to escape the reach of storms, floods, and droughts need forms of support that differ from those provided by the 1951 Refugee Convention.[27]

While it is undoubtedly important to update the world's applicable conventions and legal categories and close the yawning protection gap, it remains essential to try and ward off as much damage as possible to Earth's natural systems. Mitigation—reducing greenhouse gas emissions and scaling back other human assaults on nature—must be given much higher priority and urgency. Adaptation can only go so far, and to be effective it must be pursued now, before the worst consequences of climate instability arrive, rather than later.

Climate activists have long insisted that science should guide policymaking. Yet over the years it has become ever clearer that the biggest challenge for humanity may not be to master the intricacies of climate science but rather to answer the much more vexing questions of how political systems operate and why they are so resistant to heeding science's alarm bells. It is a deadly irony that three U.S. presidential debates took place in 2012 without the word "climate" being uttered even once, swiftly followed by nature's "last word" in the form of the devastating Superstorm Sandy that hit the eastern United States, a storm that likely was made worse by the gathering pace of climate change. If we fail to learn how to make our political systems pay attention to climate challenges, we will have to learn how to deal with massive population displacements in coming decades.

CHAPTER 32

Cultivating Resilience in a Dangerous World

Laurie Mazur

The last few years have witnessed a stunning array of calamities, both natural and human-made. A catastrophic earthquake in Haiti killed 300,000 people and left much of that nation in ruins. In Japan, an earthquake and tsunami caused 19,000 deaths and precipitated one of the most dangerous nuclear accidents in history. Drought left millions hungry in the Sahel and decimated crops in the United States. Meanwhile, the worst financial crisis since the Great Depression unspooled in every corner of the globe.[1]

Of course, disasters of all kinds are nothing new. But the current era may be one in which their frequency, scale, and impact are greater than anything our species has previously confronted. According to the Center for Research on the Epidemiology of Disasters, the number of people affected by natural disasters exploded over the last century, from just a few million in 1900 to roughly 300 million in 2011. (See Figure 32–1.) The global reinsurance firm Munich Re says 2011 was the costliest year ever for the insurance industry.[2]

Some of that increase is surely due to better reporting. And some simply reflects the growth of the human enterprise: World population more than quadrupled between 1900 and 2011, from 1.65 billion to 7 billion. Economic output grew even more rapidly, from just under $2 trillion in 1900 to nearly $51 trillion in 2008. There are more people, and collectively they have more to lose.[3]

Humanity's increased vulnerability partly reflects the changes people have made to the global environment. Climate change, species loss, and other modifications to the ecosphere have destabilized the natural world, ushering in a new and unpredictable era of storms, drought, disease, and rising seas. As climate scientist James Hansen puts it, "Ten thousand years of good weather is over."[4]

But the calamities are not all environmental. Today the planet's inhabitants are linked as never before by dense global networks of commerce and in-

Laurie Mazur is a Washington, D.C.-based writer and consultant to nonprofit organizations.

www.sustainabilitypossible.org

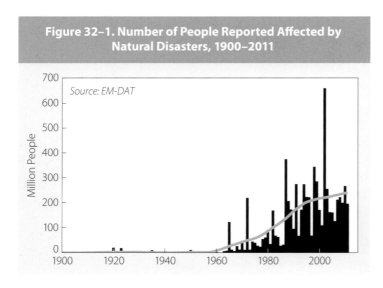

Figure 32–1. Number of People Reported Affected by Natural Disasters, 1900–2011

Source: EM-DAT

formation. Networks can amplify disturbances; the World Economic Forum has warned of "the prospect of rapid contagion through increasingly interconnected systems and the threat of disastrous impacts." The ongoing financial crisis was triggered by risky mortgage lending in the United States, for example, but in an interconnected global economy, its impacts reverberate around the world.[5]

Yet while disasters of all kinds are increasingly inevitable, it is possible to limit their impact. Some people, communities, and nations are able to weather substantial shocks and bounce back afterwards; they are, in a word, resilient. But what exactly does that mean? What characteristics confer resilience, and how can they be cultivated?

Resilience Defined

Resilience, in the simplest terms, can be defined as a system's ability to mitigate and withstand disturbances and to bounce back afterwards, while continuing to function. The question of how resilience is gained or lost has been the focus of significant research in many disciplines.

"Resilience thinking" emerged from the natural sciences with the pioneering work of ecologists C. S. Holling, Lance Gunderson, and others. Resilience thinkers explore the lifecycles of complex socio-ecological systems and the factors that make those systems robust or vulnerable. Insights gleaned from that work have been applied in a range of fields, from economics to national security. Another relevant stream of research comes from human psychology, where researchers are working to understand what enables individuals to withstand traumas of all kinds.[6]

Intriguingly, several common themes have emerged from these inquiries. While each discipline approaches the subject with a distinct perspective and terminology, there is considerable overlap between concepts of resilience in the natural and social sciences. This makes sense: human beings are inextricably entwined with nature, so the rules of the natural world may apply to us as well. And it is not surprising that the qualities that confer individual resilience would scale up to larger human systems.

Diversity. A system with diverse components will have a wide range of responses to a disturbance and is therefore unlikely to fail all at once. (See

Box 32–1.) A city with a diverse economic base, such as San Francisco, is less vulnerable to economic upheaval than one that relies on a single industry, such as Detroit. An ecosystem with healthy biodiversity can withstand more stress without "flipping" into an undesirable state, as when a coral reef is destroyed by algae.[7]

Redundancy. Similarly, a resilient system has multiple ways to perform basic functions, so that the failure of any one component does not cause the entire system to crash. This is the "belt and suspenders" approach; it is one reason why aircraft have multiple jet engines. It is also the thinking behind the design of the Internet: Originally engineered to ensure continuous communications in the event of a Soviet nuclear strike, the Internet sends data through a vast, distributed network of routers with redundant connections to other nodes in the network. If one of those routers is busy or damaged, data are simply diverted along an alternate pathway.

Modularity. Modular systems, in which individual units retain some self-sufficiency when disconnected from larger networks, will fare better in times of crisis. For example, people living in a community with a robust local food culture (nearby farms, a farmers market) will be less likely to go hungry if there is a disruption in larger supply chains. A distributed energy system, in which individual households and communities produce as well as consume power, is much less vulnerable to grid failures. Accordingly, some regions are experimenting with "microgrids" that are both diverse and modular: they rely on a variety of power sources, including renewables, and can attach to the national grid or operate independently.[8]

Reserves. Healthy resource reserves can help any system weather disturbances, just as a good supply of acorns can help squirrels survive a harsh winter. Not surprisingly, wealth matters: hence the Japanese were able to recover fairly quickly from the devastating earthquake of 2011, while Haitians are still struggling to rebuild after the much smaller quake they endured in 2010. But money isn't everything. One study of resil-

---

Box 32–1. Saving Plant Varieties to Preserve Resilience

Ecosystems that are diverse tend to be more resilient, yet one aspect of modern global agriculture is its vast expansion of monocropping that ignores or marginalizes thousands of plant varieties and thereby exposes the system to risk. The Svalbard Global Seed Vault is one effort to mitigate that risk. Located in Svalbard, Norway—an area that lies in total darkness for nearly four months a year—the vault is designed to protect thousands of seed varieties from both natural and manmade disasters. Cary Fowler, executive director of the Global Crop Diversity Trust (GCDT), explains that the seeds the vault receives are crucial to preservation of global crop diversity: "Our crop diversity is constantly under threat, from dramatic dangers such as fires, political unrest, war and tornadoes, as well as the mundane, such as failing refrigeration systems and budget cuts. But these seeds are the future of our food supply, as they carry genetic treasures such as heat resistance, drought tolerance, or disease and pest resistance."

The vault currently contains more than 700,000 samples—from wheat native to Tajikistan and old subspecies of barley from Germany to amaranth once grown by the Aztecs—and is buried deep within permafrost and thick rock that keep its interior temperature far below freezing, even without electricity. Its initial construction was funded by the Norwegian government and it is currently maintained through a partnership between the Norwegian government, Nordic Genetic Resources Center, and the GCDT.

—Danielle Nierenberg
Former Director, Nourishing the Planet Project,
Worldwatch Institute
Source: See endnote 7.

ience to climate change found that, in addition to wealth, resilience depends on "environmental capacity"—the integrity of ecosystems—as well as on "civic and human resources"—the health, education, and economic capacity of a society's citizens.[9]

Social Capital. Resilience is reinforced by social capital. For an individual, social capital is about relationships with family, friends, and colleagues. In communities, social capital can be measured by levels of trust, the strength of social networks, and the quality of leadership. At both the individual and community levels, social capital provides resilience. For example, college freshmen with large social networks have stronger immune responses than their isolated peers. And communities with abundant social capital are better able to withstand and recover from disaster.[10]

Agency. Agency—the capacity to make choices and enact them in the world—is central to individual and social resilience. Resilient people have a sense of control over their destiny; resilient communities fully engage their citizens in decisionmaking. Agency is clearly related to the capacity to adapt and thrive in the face of environmental and other changes. Fundamentally, agency is about power—personal and political. In a resilient society, power is not hoarded at the top, it is distributed broadly. The devolution of power is a moral imperative, and it has practical benefits: capable, empowered people are better able to cope with all manner of crises, from job loss to tsunamis. (See Box 32–2.)[11]

Inclusiveness. Inclusive social institutions—economic, political, and cultural—can strengthen resilience. For example, communities that practice "deliberative democracy" by involving people in problem-solving are better able to recover from disaster and rebuild for long-term sustainability. And inclusiveness is protective on a broader social level: As economist Daron Acemoglu and political scientist James Robinson have argued, societies thrive when they develop inclusive institutions that distribute power and opportunity broadly. They fail when those institutions become "extractive," serving to concentrate power and opportunity in the hands of a few.[12]

Tight Feedbacks. A resilient system has tight feedbacks, which enable it to quickly detect changes in its constituent parts and respond appropriately. Tight feedbacks mean that the consequences of someone's actions are

---

Box 32–2. Empower Women, Build Resilience

Women play a pivotal role in fostering resilience—in their roles as caregivers, resource managers, and stewards of social networks. And yet they are disproportionately affected by disasters. In some areas affected by the 2004 Indian Ocean tsunami, for example, women were four times as likely as men to die. Social roles and discrimination made women more vulnerable: they could not escape quickly because they were caring for children and the elderly, and—unlike their brothers—they had not been taught to swim. Poverty also increases vulnerability, and women account for 70 percent of the world's poor.

Where women have agency and power, they improve their own resilience and that of their families and societies. Empowering women to take leadership in their communities has been shown to improve disaster preparedness in Bangladesh, Indonesia, and Nicaragua, as well as forest management in India and Nepal and adaptation to drought in the Horn of Africa.

Source: See endnote 11.

immediately apparent. When feedbacks are loosened, mistakes are easily made: it is easier, for example, to spend assets that you do not have using a credit card rather than cash. For most of human history, tight feedbacks defined the human relationship with the natural world. If a community overexploited a fish stock it depended on for food, the stock would crash and people would go hungry. Communities learned to heed those signals and develop institutions to manage common resources sustainably. Over the centuries, however, humans have loosened critical feedback loops— a process that accelerated dramatically over the last century with the expansion of capitalist market economies to every part of the globe. The result has been a profound loss of resilience—both ecological and social. (See Box 32–3.)[13]

Box 32–3. Resilience Lost: The Coastal Mangroves of Vietnam

Mangroves are among the most productive ecosystems on Earth; they serve as hatcheries for fish and protect coastal communities from storms. Until recently, the coastal Vietnamese communities that depended on mangroves for their sustenance followed time-honored practices to preserve the integrity of the ecosystem. Because the community members benefited equally from the shared resource, they each had a stake in protecting it.

But in the 1990s the Vietnamese government embraced the market economy and privatized much of its land and marine resources. Commercial interests bought up the mangroves, converting them for agriculture or aquaculture, mostly shrimp raised for export. The tight feedback loop that once connected the ecosystem to those dependent on it was severed: now those benefiting from the ecosystem—mostly investors

and consumers in Europe and North America—are far removed from the system's signals of distress, and they bear no consequences from its destruction. And those who depended on the mangroves are no longer empowered to preserve them.

The cost of the degraded ecosystem is borne by the most vulnerable, as is usually the case. Research by Neil Adger in Quang Ninh Province showed that the poorest members of the community suffered the most from the decline in fish stocks when mangroves were destroyed. Inequality increased, and the social compact that once protected the shared resource began to unravel, undermining the resilience of the community as a whole. That loss of resilience can be measured in human lives: other researchers found that communities with depleted mangroves suffer higher mortality during cyclones.

Source: See endnote 13.

Innovation. A resilient system generates novel responses to changing conditions. In nature, this is accomplished by evolution; in human society, it requires innovation—the ability and willingness to try new things. The capacity to innovate derives from many of the qualities just described. Diversity, for example, generates more novelty than uniformity does. An inclusive society with reserves of civic and human resources is better able to engage the agency and creativity of all of its citizens. And tight feedbacks provide timely and accurate information about changing conditions, which is essential for appropriate innovation.

Systems within Systems

The growing body of research on socio-ecological systems shows that these systems grow and change, as does their vulnerability and resilience.

In both the natural and the social realm, complex adaptive systems (CAS) are made up of many components, or systems within systems. For example, the human body functions as an integral whole, but it contains subsystems dedicated to digestion, respiration, and immunity, which are connected to one another by innumerable causal links and feedbacks. At the same time, a CAS is open to, and affected by, the outside environment, as the human body is affected by, say, changes in food availability. Because of this complexity, and the capacity to be affected by systems at lesser and greater scales, complex adaptive systems exhibit nonlinear, dynamic behavior that is often difficult to predict.

But complex adaptive systems also follow cyclical patterns, and the system's resilience or vulnerability is affected by its place in that "adaptive cycle." The cycle begins with a rapid growth phase, in which the system's elements first come together and interact. Picture an open field that has recently been cleared by fire. In the growth phase, resources like soil and sunlight are up for grabs, and opportunistic "weedy" species move in and prosper. As the forest matures, more plants and animals move in, fully exploiting available resources. In this conservation phase, the forest gradually becomes more efficient but less resilient; its specialized and interconnected species are less able to weather change than the weedy generalists. Next comes release, when a disruption from outside the system (perhaps another fire, kindled by a dropped match) causes the system to collapse, dispersing its resources. Finally, in the reorganization phase, the cycle begins anew.[14]

The growth and conservation phases—in which complex systems spend most of their time—are known as the "fore loop." The release and reorganization phases, which are often brief and chaotic, constitute the "back loop."

Systems are generally more resilient in the growth phase. Children's brains, for instance, are more "plastic"—better able to reorganize patterns and connections—than mature adult brains. That means children are much less vulnerable than adults to lasting damage from brain injury. A young child can lose half of his or her brain, and the remaining half will compensate for the functions that were lost. But a child's brain is also less efficient, as anyone who has ever followed a toddler from point A to point B, or listened to a small child tell a story, can attest.[15]

As a system grows more efficient and less resilient, it may eventually reach a threshold, or tipping point, at which a disturbance can "flip" the system into a different state—the forest burns, the patient dies, the empire falls. Often, it is difficult to see when the threshold is near, because thresholds move as conditions change. A system can weather a series of disturbances with seemingly little effect, then suddenly collapse. For example, acid rain may fall on a mountain lake for decades without apparent damage. But when acidity reaches a certain level, quite suddenly the lake can no longer sustain life. When a system flips, it may regenerate itself—as when a burned forest grows anew. Or it may enter a new stable state—as when a lake acidifies or a coral reef is consumed by algae. That system will then begin a new adaptive cycle.[16]

Practicing Resilience

It is clear that a world designed to weather shocks and disturbances would look very different from today's world. The systems that supply modern societies with food, electricity, and other essentials are not diverse and modular; they are massive monocultures that grow ever more efficient and vulnerable. The natural reserves that could protect us from ecological disaster are declining. Poverty and discrimination inhibit individual agency and problem-solving capacity, while inequality weakens social cohesion. And in a thoroughly globalized economy, the feedbacks that warn of impending disaster have gone slack.

The need to withstand disaster offers a powerful reason to change. But how can resilience thinking be applied in communities, societies, and individual lives? Resilience is stubbornly contextual; there is no one-size-fits-all guide to building resilient systems. Yet a few generalizations apply.

First, do no harm. Human beings are nothing if not resilient. Over hundreds of thousands of years, humans have successfully colonized nearly every ecosystem on the planet, rebounding after plagues, famines, and other disasters. The lessons of those experiences are encoded in human immune systems and in enduring social structures. This is the good news. "Resilience does not require anything rare or extraordinary," says psychologist Ann Masten, "instead it requires that basic human adaptive systems are operating normally." Thus, any effort to foster resilience must begin with a deep understanding of existing strengths and adaptive mechanisms—and making every effort to keep them intact.[17]

Second, see the forest—and the trees. Preserving intrinsic resilience means trying to understand complex systems before tinkering with them. This requires taking a broad view: focusing myopically on one part of a system, and managing for a single outcome, will likely yield surprises from unanticipated feedbacks. For example, traditional forest management focuses on preventing forest fires. But fire is a necessary part of the forest lifecycle; it burns combustible deadwood and allows fire-resistant species to thrive. Without occasional fires, the forest accumulates dangerous amounts of fuel, so that when a fire finally does occur it is so intense that it burns deep into the soil, destroying seeds and preventing regeneration. So by focusing narrowly on suppressing fires, forest managers actually invite truly devastating conflagrations. Managing for resilience begins with a deep understanding of a system and its functions at many scales. But it also calls for a certain amount of humility, an admission of what we do not—and cannot—know.[18]

Next, embrace change. Socio-ecological systems are in constant flux. Some of those changes are fast-moving and easy to track, others are more gradual. Change can be good or bad, but it is unavoidable. In fact, trying to keep a system in the same state invariably lowers its resilience. For example, children whose parents try to protect them from disease by preventing them from playing in dirt grow up with weaker immune systems. Challenges build resilience—unless they cross critical thresholds.[19]

Finally, it is important to build both individual self-reliance and collective responsibility. A human system is only as resilient as its component parts; social resilience rests on a foundation of individual well-being and self-reliance. Yet most people are woefully ill prepared for disaster. Clearly, people need to take more ownership of building resilience in their lives—by developing contingency plans for disaster and getting to know their neighbors, for example. At the same time, no one is an island; individual resilience is of limited value if the surrounding systems are in total collapse. And in a world where poverty and social inequality are increasing, vulnerability is on the rise.[20]

Bart Everson

A house in New Orleans, two years after Katrina.

"Social vulnerability occurs when unequal exposure to risk is coupled with unequal access to resources," writes sociologist Betty Hearn Morrow. These dynamics were tragically apparent when Hurricane Katrina decimated New Orleans' poorest neighborhoods, where many residents lacked the resources to cope with disaster. Nearly one third did not own a car and alternative transportation was scarce, so 100,000 people were still in the city when Katrina made landfall. After the storm, the poorest people, families, and neighborhoods were the last to recover; many never did. When segments of a society are marginalized in this way, vulnerability increases at all levels, from individuals to communities to nations.[21]

Resilience, then, requires greater self-sufficiency but also a new commitment to social justice. A resilient society empowers all its people with access to health care, education and opportunity. It distributes power by including its citizens in governance and decisionmaking. And it shares costs and benefits equitably, fostering a sense of common responsibility and purpose.

Questions and Contradictions

Resilience isn't free; it sometimes comes at the expense of other qualities a society may value. The most glaring trade-off is between resilience and efficiency. Our industrialized market economy, which favors globalized, "just-in-time" supply chains, is efficient from a profitability perspective—but staggeringly vulnerable to disruption.

The efficiency conundrum brings us to the problem of scale. The globe-spanning monocultures that supply us with food and other essentials may be more vulnerable than diversified, decentralized systems. But there are now 7 billion people on the planet, and by mid-century our numbers will grow to anywhere from 8 billion to nearly 11 billion. Is it possible to build resilience into systems capable of sustaining 8 billion and more? That remains an open question. At the same time, the resilience imperative might argue for voluntary measures to slow population growth, especially since the most effective of those measures—educating girls, empowering women, and ensuring access to reproductive health services—could themselves promote resilience.[22]

The prescriptions for resilience cited earlier are sometimes at odds with each other. For example, open societies are good at fostering innovation, but they are also vulnerable to terrorists and other "rogue" actors. Diversity promotes innovation, but sometimes at a cost to social cohesion. And social cohesion can be protective, but it can also discourage innovation and adaptation.[23]

Again, there is no template to apply, no binary set of rules about what is and is not resilient. Instead, as futurist Andrew Zolli and journalist Ann Marie Healy conclude in *Resilience: Why Things Bounce Back*, "Goldilocks had it right all along. Resilience is often found in having just the *right* amounts of these properties—being connected, but not too connected; being diverse but not too diverse; being able to couple with other systems when it helps, but also being able to decouple from them when it hurts."[24]

A debate is raging within the environmental community about whether resilience should replace or augment sustainability as the dominant paradigm. In theory, resilience and sustainability could be mutually reinforcing. Resilience asks us to avoid passing critical thresholds that could destabilize natural systems—with disastrous results. So, using resources sustainably is resilient. And resilience is, for the most part, sustainable: many resilient systems, such as decentralized, renewable energy and local food, would also enable us to live more lightly on Earth.[25]

In practice, however, it is essential to clarify what is being sustained and what is being made resilient. The sustainability paradigm has failed, for the most part, because we have not transformed socioeconomic systems predicated on endless growth. If the resilience framework merely attempts to make those systems more robust, it, too, will fail.

Resilience thinking asks human beings to transform their relationships with the natural world and with one another. A world that fails to invest in the capabilities of its people, a world that squanders the potential contributions of women, a world of metastasizing poverty and inequality—that world will not weather the shocks and surprises of the future.

Seeds of Hope

Resilience is a concept with potentially transformative power; it could help build broad-based support for paradigm-shifting changes like distributed renewable energy, sustainable and local agriculture, greater social equity, and inclusive governance. Or, like sustainability before it, resilience could be co-opted to the point of meaninglessness. Worse, it could be deployed to strengthen social and economic arrangements that are neither sustainable nor just. The choice is ours to make.[26]

Ultimately, resilience thinking asks us to embrace change. The systems that surround and include us are forever in flux as they grow, mature, and—inevitably—collapse. This is a terrifying prospect but in some respects a hopeful one. It is often in the "back loop"—that chaotic period of release and reorganization—where meaningful change occurs. From collapse comes the release of resources, the opportunity to rebuild, and the seeds from which the world blooms anew.

CHAPTER 33

Shaping Community Responses to Catastrophe

Paula Green

Vahidin Omanovic was 15 when war came to his rural village of Hrustovo and the nearby city of Sanski Most, Bosnia, in 1993. Bosnian Serb militias dragged men and boys from their homes, shooting some immediately, capturing others for concentration camps, and deporting the elderly on buses. The homes they left behind were plundered down to the copper wiring and then dynamited. Hiding on a departing bus under his mother's ample peasant skirt, Omanovic survived in a displaced persons camp. Returning to Hrustovo after the war, he helped build a graveyard for 300 murdered villagers and reconstructed homes for surviving residents or newly arriving refugees. Twenty years later, Hrustovo is repopulated. Omanovic lives in his rebuilt home with his family, working as founder-director of a nongovernmental organization (NGO), the Center for Peacebuilding, dedicated to re-establishing relationships between Bosnian Muslims and Serbs.[1]

Although Sanski Most has lost its haunted, ghostly postwar appearance, domestic and foreign investors are deterred by an inconclusive and unstable postwar agreement, limiting recovery and opportunity. Residents rely on each other, but without economic and political capital, Bosnia's young adults are forced abroad to find work, fraying the social fabric and further depressing the region. Not many young men like Omanovic have remained to reweave the community life shattered by war.[2]

Dishani Jayaweera lives in Colombo, Sri Lanka's capital, far from the war-torn and tsunami-affected regions of that island. As a member of the dominant Sinhalese majority, she could enjoy a good life without much regard for the immense suffering of marginalized Tamil minorities in other regions. But she has chosen the difficult and somewhat dangerous path of helping fellow Sri Lankans re-establish ethnic relationships after the ravages of war and nature. Her Center for Peacebuilding seeks to heal and strengthen community by rebuilding inter-ethnic and inter-religious relations sundered by war and further damaged by unjust post-tsunami and

Paula Green is founder and senior fellow at the Karuna Center for Peacebuilding and a professor of conflict transformation at the School for International Training Graduate Institute in Vermont.

www.sustainabilitypossible.org

A fisherman and his wife tend their nets on Mathagal beach, Sri Lanka. The Mathagal fishing community has received aid to support its post-tsunami recovery from AusAID and the International Organization for Migration.

postwar resettlement decisions. Sri Lanka's transition to peace remains fragile, and Tamils experience discrimination rather than support from the government.[3]

In Rwanda, Joseph Sebarenzi survived the 1994 genocide because he was out of the country, but his parents and most of his large nuclear family were murdered, along with 1 million others in 100 days. In the 20 years since, Sebarenzi has transformed himself from someone bent on revenge into a prominent peacemaker who lectures and writes about reconciliation and forgiveness. The Rwandan government has invested heavily in rebuilding on both the physical and social levels, especially modernizing the capital, Kigali, and supplying modest housing stock nationwide. It instituted a community social healing program called *gacaca*, designed to restore broken inter-ethnic relationships between Hutus and Tutsis. A strong government keeps the peace, but fear and tension exist on all sides, and trust remains understandably low. Most Rwandans remain economically impoverished, but Rwanda's current stability has attracted foreign investment. Some educated, urban youth can imagine, if not yet attain, a better future.[4]

The victimized identity groups in Bosnia and Rwanda were unprepared for war, and Sri Lankans could not even imagine the tsunami. In our current moment, however, with climate crises already occurring, population increasing, and vital resources diminishing, social capital, resilience, and preparedness may make the difference between life and death—or between bare survival and a more ample post-disaster transition.

Denial and Resistance

Can most people even imagine life in the environmentally compromised world of the future or come to terms with our ability to destroy our nest? Why have the fires, floods, hurricanes, droughts, temperature extremes, species extinctions, toxins, cancers, and other evidence of a disordered environment not produced responses as large as the problems? What will it take to break through the collective fog of denial, passivity, ignorance, and unspoken terror that seem to underlie refusal, especially in the United States, to grapple with the reality of catastrophic change? In the past decade, approximately 200–300 million people a year were seriously affected by natural disasters or technological accidents. The U.N. International Strategy for Disaster Reduction estimates economic costs of global natural disasters will reach $300 billion by 2050 at the present rate of environmental challenges, while the Global Humanitarian Forum projects an annual cost of $600 billion from climate disasters by 2030. Clearly, a crisis is at hand—and yet denial accompanies this crisis for large segments of the population.[5]

A "cultural trance of denial" impedes the capacity to awaken to increasingly pervasive and obvious environmental destruction. Through the defense mechanism of denial, barely articulated primal fears lie below the level of consciousness, providing a false sense of reassurance that all is right with the world. Noted psychiatrist and author Robert Jay Lifton studied the paralytic effects of the prospect of nuclear annihilation on the U.S. population during the cold war period. He coined the phrase *psychic numbing* to describe how individuals and societies block awareness or minimize the dangers of issues that are too painful to comprehend. Psychic numbing, he wrote, "is a societal reaction to impending doom, chaos, and ultimately mankind's extinction."[6]

In the United States and elsewhere, the looming consequences of environmental collapse have evoked similar responses of denial and numbing, blunting awareness of climate and resource realities and creating large time gaps between evidence, acceptance of evidence, and policies based on evidence. Fear of individual and collective nonexistence can override powers of observation, constrict assimilation of critical knowledge, and protect against the "inconvenient truths" that are too painful to know. Reinforcing this denial, disinformation from those with vested interests in environmentally harmful enterprises undermines the clarity required to respond in any measure commensurate with the magnitude of the crisis. Soothed by manipulated information, confused about how to respond, and frustrated by the inadequacy of replacing lightbulbs and growing tomato plants in the face of catastrophe, feelings of helplessness, depression, or misdirected anger may arise. Relentless busyness protects against anxiety and leaves little time for sustained thought or community organizing. Diffuse worries about the future fuel demands for illusory protective boundary walls and border fortifications, which are useless against environmental deterioration.

"Humankind cannot bear very much reality," wrote T. S. Eliot. But repression and denial have severe consequences that affect our collective safety. Denial shelters unacknowledged fear and saps motivation for appropriate responses, whereas confronting reality counteracts helplessness and hopelessness, builds social capital, lessens fear, and releases energies needed for change. One hopeful initiative, known as the Degrowth Movement, advocates reducing production and consumption in order to live within local ecological, equitable, and environmentally responsible limits. Visionary alternatives like degrowth inspire others and increase demands for reasonable government-led programs dedicated to the common good. Deliberate planning for climate-altered lives involving multiple sectors of society will arouse further awareness, sideline climate deniers, and stimulate adaptive innovations.[7]

Human Behavior in Times of Crisis

Although laws and customs help guide individual responses to crisis, human behavior in times of war and natural disasters cannot be fully predicted or controlled. Within countries or across borders, in high and low social relatedness cultures, and under the pressures of environmental catastrophe or genocide, there are no guarantees of humane responses. Humans are beset with fear and greed, as well as endowed with compassion and generosity. And disaster calls forth wide-ranging reactions within individuals and communities, as witnessed in Rwanda, Sri Lanka, and Bosnia.

At a seminar for South Asian activists, Indian participants confronted Bangladeshis about their lack of preparedness for near-certain devastating floods, informing them that India would not accept massive numbers of new climate migrants, given its responsibility for over 1 billion citizens of its own, 25 percent of whom live below the poverty line. India, in fact, is slowly building a wall on its Bangladeshi border in response to an already steady flow of climate and economic refugees who are adding to instability in northeastern India. Mass climate migration in response to specific disasters in South Asia and elsewhere has not yet occurred but is widely predicted despite evidence that, for emotional and legal reasons, displaced people prefer to repatriate and rebuild rather than relocate. In Bangladesh and elsewhere, environmental conditions may make rebuilding impossible, greatly expanding climate refugee populations. [8]

After the December 2004 tsunami in Sri Lanka that resulted in over 30,000 deaths, many Tamil, Muslim, and Sinhalese residents in coastal areas rescued fellow citizens without regard for the ethnic identities that perennially enflame their communities. For a brief time, until disaster allocations favored the majority Sinhalese population, the impulse to care overshadowed the distinctions of status and affiliation. Aid discrimination provoked resentments that led to further armed conflict until in 2009 the government declared victory over the Tamil minority LTTE separatist militia, ending a 26-year insurgency. Damaged inter-ethnic relations now threaten Sri Lanka's capacity to manage expected climatic events such as sea level rise and storm surges.[9]

The destruction wrought by Hurricane Katrina in New Orleans in 2005, when approximately 80 percent of the city flooded in 18 hours, brought out the best and worst of human behavior—from rescue to robbery to racism. Neighbors did their best to help each other throughout the city, but the Lower Ninth Ward, where the population was 90 percent African-American, experienced the highest number of deaths by drowning. Evacuation by car succeeded, but no buses were provided for an estimated 200,000 to 300,000 persons who rely on public transportation, although New Orleans officials were aware of the risks facing these individuals. Despite the cohesion of the neighborhood, residents lacked connections to those in power who might have aided their evacuation.[10]

Other lessons learned after Katrina included the need for effective coordination between different levels of government, electronic protection of medical records, emergency generators, guaranteed fuel delivery for hospitals and shelters, and much more extensive community planning and drilling. When Hurricane Isaac struck in 2012, preparedness paid off: the levees held, electronic medical records were available, and generators moved throughout New Orleans on flatbed trucks.[11]

Racial and class prejudices, with their demarcations of access, visibility, and privilege, affected human behavior during and after the disasters in Sri Lanka and New Orleans. Despite pleas by residents, much of the Lower

Louisiana National Guardsmen distribute Meals Ready to Eat in preparation for Hurricane Isaac.

Ninth Ward community in New Orleans has not been rebuilt, fueling accusations about deliberate population transfer based on race and poverty. The economically impoverished half-Vietnamese Village de L'Est in New Orleans, however, was rebuilt, an accomplishment attributed to the refugee experiences of the highly networked Vietnamese community as well as the leadership role of their church.[12]

In the 1992–95 war that shattered Bosnia, previously the most ethnically diverse of Yugoslavian states, the bonds of community snapped. The same pattern emerged in 1994 in the Rwandan genocide, another nation ripped apart by political manipulation, greed, grievances, and history. Amid the strife and chaos of war or natural disasters, however, individual behaviors varied widely: not all Bosnians and Rwandans abandoned the designated enemy group. Rescuers emerged, as they do in all cultures—often un-predictably and with no previous indication of a heroic disposition. Ho-locaust rescuers, for example, who quietly risked their lives by sheltering people marked for genocide during World War II, recount their generally unplanned rescue behavior as an ordinary response to a particular request for help. Their outstanding acts stand at the pinnacle of moral behavior, in sharp contrast to the violence and betrayal committed by some and the by-stander behaviors of most, who neither contribute to nor prevent violence. Rescuers provide a counterpoint to the worst of human behavior observable in war and upheaval.[13]

In planning for climate crises, resource scarcity, and forced migration, this great variability in human behavior and resilience must be accounted for. Because catastrophes, whether natural or human-made, stimulate pri-mal survival instincts, individuals may react in ways that protect, provoke, or disrupt. Some may collapse into depression and helplessness or engage in violent behavior in response to frustration and anxiety. Others will seek to shield only those of similar affiliation. At the other extreme, there will be individuals facing calamities who will mobilize community resources for responsible and inclusive action, and some who will become rescuers. Edu-cational efforts that provide guidance for accountable and ethical behavior in times of crisis can help to control fear-based responses that might harm or destabilize the community as well as encourage mobilizers and rescuers.[14]

Cultivating Social Capital

Social capital, the sum total of resources, knowledge, and goodwill possessed by everyone in a network, provides a web of connections that communities can use to obtain relief and reconstruction aid. Participants in networked communities are best able to organize support, articulate their needs, and work together to rebuild and stabilize. In countries like Rwanda, Sri Lanka, and Bosnia, with limited geographic mobility and multigenerational attach-ments to land and ancestors, social capital tends to be high, at least in times of peace. Among the mobile populations of greatly diverse cities, social capital characteristically is built within neighborhoods, identity groups, or professional and business associations. In regions fractured by war, fear and mistrust must be overcome for communities to re-establish relationships that protect against oncoming disasters.

Political scientist Daniel Aldrich believes that social networks are the most important determinants for coping with disasters. Those who are strongly rooted in community often rely more on themselves than on their governments, whereas more individualistic, less connected populations ex-pect state services and support. In India, for example, bus passengers will disembark to repair washed-out roads that impede their journey or collectively rebuild villages destroyed by floods. Rich networks, Aldrich claims from studies of the United States, Japan, and India, energize people, improve resilience, and encourage disaster preparedness, which reduces loss and suffering in climate-related catastrophes. He cites two cities in Japan equally engulfed by post-earthquake fires in 1995. In one city, people in their neighborhoods organized rapidly to douse the flames; in the other, people did not mobilize and suffered far more harm. Similarly, after the 2004 tsunami in India's state of Tamil Nadu, one village secured disaster relief through relationships with the local government whereas another village lacked the social connections to gain access to this network of aid.[15]

Aldrich believes social capital is a more critical variable than wealth, education, or culture and that it can be cultivated through crisis preparedness events and community activities such as exist in Seattle, which offers disaster management classes and training programs for officials and civil society. Preparedness and resilience researchers Kevin Ronan and David Johnston document the advantages of working with schools, youth, and families—an existing network—to increase resilience while planning for disasters. Encouraging schools and communities to prepare and practice for a hazardous event through realistic and carefully planned scenarios builds trust, establishes mutual reliance, and increases the odds of survival far beyond what can be achieved by reliance on external hazards managers.[16]

In another cultural context, a study conducted by social scientist Ashutosh Varshney in India confirms the value of intergroup networks in areas plagued by communal violence. He found that Indian cities with positive connections between Hindus and Muslims prevented inter-ethnic riots whereas those without solid inter-religious relations could not stem the rising tide of violence. Both informal and associational robust civic links prevented riots, but the more deliberate and formal associational relations created especially sturdy bonds that helped end violence in times of threat. The Indian cities lacking sufficient Hindu-Muslim social capital capitulated under siege from violent mobs, leaving trauma and resentments to fester and reignite future violence.[17]

High social capital in besieged communities does not always prevent violence, however. Rwanda and Bosnia, for instance, experienced an onslaught of ethnic violence despite previous associational links and high rates of intermarriage. In Sri Lanka, strong ties did bind the majority Sinhalese to each other, as occurred in Bosnia and Rwanda, but that bonding deprived the less-favored Tamils of humanitarian aid when the tsunami multiplied the problems caused by their ongoing war. In postwar Bosnia, Sri Lanka, and Rwanda, Vahidin Omanovic, Dishani Jayaweera, and Joseph Sebarenzi promote social healing and reconciliation, revitalizing inter-ethnic networks that might prevent a return to armed conflict and help communities facing severe climate events. Social capital is predictive for disaster prevention or preparedness, but not necessarily sufficient. Yet it does often raise the odds of averting disaster, and it raises the quality of life before and after climate or war catastrophes.

Socially cohesive Japan is not challenged by ethnic diversity or plagued by extensive poverty. Japanese culture rewards conformity and eschews challenges to authority; its citizens expect efficient services and truthfulness in return for their obedience. The handling of Japan's 2011 massive 9.0 magnitude earthquake and tsunami, and the subsequent meltdown at its Fukushima Daiichi nuclear plant, created a disaster that may leave thousands of square miles of this overcrowded country uninhabitable for decades, producing Japan's greatest reconstruction tasks since World War II. The management of this disaster shattered people's trust in their government so severely that some Japanese citizens have purchased their own dosimeters to measure radiation penetration in the land and waters.[18]

This nuclear event, recalling the nuclear-bombed annihilation of Hiroshima and Nagasaki, spawned social networks in Japan that challenge authority and press for elimination of nuclear power. Industries relying on nuclear power are currently on a collision course with protesting citizens who no longer believe their elected officials, while the Japanese government equivocates and seeks acceptable compromises to protect the future of nuclear power. Groups of citizen activists, no longer waiting for their government to lead, are planning an alternative future based on the "four-Cs" of climate, connectivity, community, and character. Some are joining the Transition Towns movement, which has spread to 24 Japanese communities.[19]

In North America, Europe, Australia, New Zealand, and elsewhere, the rapidly spreading Transition Towns movement is a robust example of communities mobilizing in anticipation of resource and climate-based threats. In rural Putney, Vermont, bonds develop as residents acquire skills in sustainable energy, food preservation, alternate transportation, and other survival topics. Transition Town members in Totnes in the United Kingdom helped spark this global movement and initiated over 30 workshops, including bicycle maintenance, eco-construction, and local economic regeneration. "There is no cavalry coming to our rescue," remarked cofounder Rob Hopkins. "Transition says we need to come to our own rescue."[20]

Farmers markets, Transition Towns, the Occupy movement, degrowth, and many other citizen-led initiatives arise in response to an increasingly fragile planet. Combined with disaster preparedness, the mutual responsibility and resilience characteristic of well-networked communities have proved essential in times of war or climate disaster. Although bonds of friendship and community sometimes fail to survive the assaults of armed conflict or environmental catastrophe, the memory of harmony and mutual assistance helps people rebuild communities for a future that is bound to present challenges to human well-being.

Disaster Preparedness, Development Assistance, and Resilience

Crisis planning progresses beyond denial to deliberation and decisionmaking. Well-prepared communities anticipate and manage denial, helping those caught in anxiety or bewilderment. Having already deliberated as part of planning, these communities are poised for life-saving decisions and rapid action, augmented by rich social networks that organize support services. With the increasing number and severity of climate events, disaster specialists recommend drills and preparedness for cities, school systems, hospitals, and public officials.[21]

These services, however, may exist primarily for economically privileged communities and nations. In the United States, many large and some smaller cities have emergency management websites that offer, for example, Community Emergency Response Teams, first responder training, Red Cross information, guidance for businesses, and extensive publications. Minneapolis offers preparedness information in English, Hmong, Somali, and Spanish; Los Angeles provides guidance for families, children, and neighborhoods in Spanish and English. In economically developed countries, most earthquake- or hurricane-prone cities are somewhat prepared for expected natural disasters. In the global South, however, where people already live with almost constant disruption, little climate vigilance exists. And for situations of armed conflict, the best planning is mitigation of the conditions that give rise to war and violence.[22]

A four-pronged approach can create disaster-resistant communities: mitigation, preparedness, response, and recovery/reconstruction. Mitigation is concerned with the planning, building, accessibility, and maintenance of the systems and facilities in a community, such as transport, land use, and development codes. Good systems help prevent hazards from turning into disasters and make a sustained difference in the outcome. While community members build their social capital and create more environmentally responsible lives, local governments should conduct a community risk analysis, integrate planning into all decisionmaking bodies, create a local resource network, and promote public awareness. This checklist for local government can provide cities and regions a margin of safety and sustainability that will help minimize the negative impacts of future climate events.[23]

Studies show that the way communities and government officials respond in the weeks and months that follow disaster can have a strong impact on the mental health of victims. Research from Hurricanes Katrina and Rita, both in Louisiana, indicate that survivors whose lives were returned to some degree of normalcy shortly after the traumatic climate event fared much better psychologically than those for whom services and security were not restored rapidly. The longer the adversity continues, the higher the rates of mental health problems.[24]

Climate change will continue to take its most inexorable toll on the poor in both poor and rich countries—those who have the least protection from physical infrastructure failure, unreliable government institutions, faulty warning systems, and inadequate emergency health and transport facilities. Furthermore, the impoverished of the world have limited financial and material reserves for mitigation or recovery and live in vulnerable areas that are especially subject to the ravages of nature and the toxic spills and fumes from polluting industries in their neighborhoods. Both the Ninth Ward of New Orleans, which houses the poor of a rich country, and the Tamil fishing villages on Sri Lanka's coast, home to the poor of a poor country, experienced this class impact of disaster rescue and relief.[25]

Sadly, it is sometimes the overstretched but well-meaning development assistance agencies that stifle community support networks by introducing systems that feel regimented or disempowering to local residents. Aid availability can create competition for goods and services that results in antagonistic relations between communities and development workers. Aid organizations, in the country with the permission of the host government, can be limited by government priorities for aid distribution. Victims of disaster, often anxious and vulnerable, may be hypersensitive to slights and resent the erosion of cultural beliefs and local leadership.[26]

New programming in aid, known as community-driven development (CDD), shifts the equation by relegating planning, decisionmaking, and financial resources to both majority and marginalized communities, who are fully responsible for implementation. With aid agencies anticipating further demands due to increasing catastrophes, CDD approaches can empower local leaders to use and strengthen their own social capital and resilience, maximize available resources for their own welfare, and implement community recovery, reconstruction, and preparedness in accordance with local traditions.[27]

Integrated planning for risk reduction and adaptation to twenty-first century challenges must be established through cooperation by the world's governments and intergovernmental agencies, which are ultimately responsible for human safety and well-being. The U.N. Environment Programme (UNEP), for example, is tasked with minimizing environmental threats to human welfare from the environmental causes and consequences of conflicts and disasters, working within the constrained limits of U.N.

member-state mandates. For nearly 40 years it has delivered environmental expertise to U.N. members and regional partners. UNEP notes that since the start of the new millennium, more than 35 violent conflicts and perhaps 2,500 disasters have affected billions of people worldwide. UNEP works with communities on such issues as risk reduction, capacity development, transformation of resource-based conflicts, and learning to "build back better." Like CDD, it seeks to empower rather than lead, to inspire rather than direct.[28]

More than aid, planning, or even resources, what may help most of the world's people cope with catastrophe is the remarkable human capacity for resilience. Humans bear the unbearable, survive to tell the story, manage the shadows of grief, and create renewed lives. Resilience is the ability of individuals and communities to withstand shock, cope with emergencies, adapt to new realities, and heal from the experience. It exists in all human communities and is built on the strength of local relationships. Many residents of Bosnia, Rwanda, and Sri Lanka, like people in climate-battered New Orleans and Japan, have demonstrated stunning resilience in their capacity to recover from tragedies that no one should have to endure. Some have rebuilt their lives on the ashes

Hajime Nakano

A team of volunteers help with post-earthquake and tsunami recovery at Kobuchihama, Japan.

of their homes and their ancestors, planting their fields and awaiting new growth. Those with no possibility of "building back better" must establish lives in new lands, demonstrating even more resilience and coping capacity and requiring additional social services.

In *Resilient People, Resilient Planet*, the U.N. Panel on Global Sustainability noted that a resilient world requires the eradication of poverty, inequality, unsustainable consumption, and inadequate governance. Continued existence for all the world's communities demands a radical shift from resource competition to appropriate allocation of what remains; a willingness to share responsibility for climate mitigation, resource management, and vulnerable populations; and a commitment to resolve increasing sociopolitical tensions without the additional affliction of armed conflict. Civilization depends on acknowledging our capacity to destroy our common nest, focusing our collective energy on its survival, and respecting planetary limits.[29]

CHAPTER 34

Is It Too Late?

Kim Stanley Robinson

In a cartoon loved by children everywhere, there's a moment when Wile E. Coyote is chasing the Roadrunner so intently that he runs off a cliff without noticing he has left the ground. While he remains ignorant, gravity has no effect on him, although he does of course lose traction, and as his legs spin he looks out at us suspiciously. Our return gaze does not reassure him, and he gathers his courage and looks down. When he sees where he is, gravity immediately reasserts itself and he plunges faster than terminal velocity to the distant desert floor, where only a tiny puff of smoke or crack in the earth marks his impact. Oh dear; foiled again.

Being a cartoon character, Coyote is indestructible and will return to do it all over again in a different foolish way. But humans? A fall like that wouldn't end well.

So we ask this question about our present situation on Earth, Is it too late?, because we wonder: are we already in the air? Has humanity already overshot the carrying capacity of Earth so badly that we are doomed to a horrible crash after oil, or freshwater, or topsoil, or fish, or the ozone layer, or many other things—after one or all of them run out? So that no matter what we do in the meantime, it's a foregone conclusion that we're in for a fall?

No. In that sense, it is not yet too late. As demonstrated in this volume and in other analyses, including the wedge diagrams of Robert Socolow and others, if we were to do everything right, starting this year and continuing for the next several decades—did everything that has been proposed to decarbonize and to conserve, restore, protect, replace, and so on—then we could do it. It might involve so many actions that it would end up constituting the primary emphasis of civilization's efforts, but that is probably as it should be. The point is, it is physically possible. We could shift infrastructures, technological arrays, and social systems in ways that would make them so much cleaner than what exists now, especially in carbon terms, that global average temperatures would probably not rise by more than 2 degrees

Kim Stanley Robinson is a California-based science fiction writer and author of the Mars trilogy, *2312*, and other novels.

www.sustainabilitypossible.org

Celsius; extinctions would not soar, food shortages would not occur, and 7 billion or even 9 billion humans could share the planet with other living creatures in a healthy way, with all the humans living well.[1]

The measurements we can make of our physical situation in relation to the planet support this assertion, but they also make clear that we need to start most of the helpful actions pretty soon. Indeed, it would be best if we managed to do all of them as soon as possible.

Are we going to do everything right in the rest of the twenty-first century? No. Or, let's say it looks very unlikely. We just aren't that good, either as a species or as a civilization—it's hard to tell which. If we were good enough as a species, meaning smart enough as animals, then presumably we could make our civilization good enough by the sheer force of our wisdom. And maybe we will. But as we evolved, our brilliance grew with some startling holes in it, probably because we were adapting to live in small packs on a savannah. We were good at that, so good that we succeeded in spreading far beyond our original setting. Possibly as a species we have succeeded so well that we have overshot our evolved abilities. On the other hand, maybe it's just that accidents of power differentiation and accumulation have left us with a damaged ability to act in the general interest: not something in our nature, in other words, but in our history.

But either way, oh dear—at least one foot is off the cliff. Could be quite a fall. Must turn quickly in a new direction.

That's hard to do. We will do some things wrong, it is almost certain. As a result there will be human suffering, there will be suffering among the other creatures on Earth. There will be extinctions. It can hardly be denied without being unrealistic, or so it seems. We are going to do damage in the twenty-first century, possibly big damage. It might not involve a dramatic fall, it might just be our ordinary reality, doing bad things day after day.

So the question could be changed from Is it too late? to How much damage will we let happen? Then we could flip that revised question to its positive formulation: How much will we save? How much of the biosphere will we save? That's the real question.

When we ask that question, it reminds us: life is robust. Restorations can be made. Everything but extinctions can be made better. So there is reason for hope. We can think of our work as saving things that will come back stronger later. Even in the bad present, we can create inoculants and refugia for a better time.

Which does not justify any complacency. In the damage that will come first, before any better time, the poor will suffer much more than the rich, both because the rich will be better able to afford adaptations to the degrading environment and because many of the poor live in the parts of the world that will be most hammered by climate change. This human suffering is both a moral and a practical issue for the richer part of the population, which has more power to act now: morally, no one is free in an unjust system, as Abraham Lincoln pointed out; practically, there is no firewall that will protect even the rich from the kind of damage we are creating, which ranges from endocrine disruptors to food crashes to infectious diseases to political violence, meaning terrorism and war. Nor is it difficult to imagine more than one of these impacts combining.

A working question for this project of saving as much as we can: how set are we in our path dependencies?

There is infrastructural path dependency: once we build a certain transport system or power generation system, we have set a technological path in concrete and steel and are dependent on it for its working lifetime. Changes to systems this big and physical take a lot of time and effort. Shortening the normal working lifetime of such a system is not business as usual, not the expected future that is path-dependent on budgets and debt schedules, but rather a rupture in all that, a social decision.

That brings us to our social systems' path dependencies, which also exist, because deciding to change an infrastructure already built takes immense social effort. So the question becomes, how flexible are our social systems? It seems as if they could be more flexible than infrastructures, being more abstract and more sensitive to new desires on the part of people. Thus in our political lives we flex and change our laws all the time, and allocate government funds and enact laws that shape and direct private investment, and build new things; then flex again and tear those things out, and rebuild other things in their place. It happens all the time, it keeps happening. No one should object to using this normal process.

Is it "cruel optimism" to say we are flexible enough to change rapidly, or is it a realistic reading of our history and situation? Depends on what angle you take, but also on how flexible our social systems actually turn out to be, now, if we try to flex them. What if some of the most powerful elements in our decisionmaking process decide to do everything in their power to ignore new information and stick to the very infrastructure that is wrecking the biosphere? This is, of course, not a hypothetical question.

Governments are big collections of capital, among the biggest that exist, but they are still dwarfed by the totality of private capital, which is now heavily clumped into a small number of private organizations. So far, most big accumulations of private capital are saying no to the idea of rapid decarbonization, not consciously or out of malice but simply as a consequence of economic laws as they stand. If there is no financial profit to be had in decarbonization, and more generally if financial law dictates that we continue to damage the environment, this is not necessarily taken as a sign that the financial system has to change. Some people defend the current financial system no matter its effect on our biophysical situation. So far, the people doing this have not been defeated politically, nor do they seem to be changing their minds. And they hold a lot of power.

That being so, we have to hope we really are in a polyarchy and that it continues to work for us. Can we prevail over destructive private interests when the common good of humanity and the biosphere are at stake? In the polyarchy we live in, this is not at all clear. It is an open question.

Polyarchy is a useful word for our current social system, because it accurately describes the form of our governance without specifying content or intent. It's a fairly simple and general term that says that human power over human affairs is distributed across a number of different organizations of different kinds, which compete with each other to decide or influence what we do. Thus there is political power in capital as well as in government, science, religion, civil society, and military force—and in people as the embodiments of these powers, as producers and consumers and as individuals both on their own and en masse. What we do as a civilization gets decided by all these power centers in a combined effort or struggle that has many outcomes. Polyarchy as a name for our system may be more accurate than more content-specific names such as democratic capitalism, social democracy, state socialism, and the like. None of the content-based names we usually use include science, even though if we were to examine how we live now on the planet, it would seem that science ought to weigh very heavily in any accurate description of what is powerful.[2]

We never hear the current global system described as scientific capitalism. Maybe this is because the phrase sounds a bit oxymoronic, these two power centers being vaguely understood to be at cross-purposes. In fact, modern history could be understood as a struggle between these conjoined twins for primary control of humanity's affairs. One view of their fight could portray capitalism as attempting to buy science's efforts and direct them to reinforce capitalist ownership, while science could be seen as attempting to reduce human suffering, repair damage, and dismantle injustice, all by its particular method of discovering and manipulating the world. In Raymond Williams' terminology of the residual and the emergent, which says that any given historical moment consists of residual elements and emergent elements engaged in a collaboration and a struggle, we could say that capitalism is the residual of the feudal system while science is what we named the emergent next system long before we recognized it as the post-capitalism it has been from its very beginning.[3]

This is admittedly a sock puppet rendering of three or four centuries of dense action. It is only defensible as a practice because we do need basic orientation; we sometimes need to see history not as an endless number of events but as History, with big shapes that make a simple but tellable story.

There are other ways of telling modern history, but this version, science versus capitalism, clarifies a lot that is otherwise confusing. The two powers have been so intertwined through their shared existence that it is hard to see how different they are; they need destranding, so to speak. So consider science as it has been: an emerging system of health and justice, a political force of immense power and potential for good, struggling from the very start to stay independent enough to operate by its own rules, which are both utopian and highly effective in the physical world. Then think of who owns what in capitalism, and how blind that older system is to the realities of our biosphere as our ultimate life support. Given the data of history, and of our own lives now, it seems that we should be supporting science in every way we can.

Our science tells us we have to change infrastructure faster than our current polyarchy plans to. Changing infrastructure is not in itself a bad thing; with cleaner new tech we lessen our impact, and installing the new tech gives a lot of people meaningful work. It is a major investment, however, and our current economic system is telling us that it is unaffordably expensive compared with using the dirty old infrastructure; changing out would be not profitable. Economics says this, and the current laws make its analysis accurate.

Both economics and laws can change. But changing economics is not so easy, because as currently constituted it supports the present distribution of power. When we say economics we actually mean capitalist economics, because the field takes for granted and helps to justify its object of study. So it is resistant to change because its owners and clients are resistant to change.

Here is a case where science needs to emerge more fully as a form of political action, for the good of all. Science is a reiterative process that is always trying to improve its methods, and we do it better than we ever have before. Lessons have been learned from earlier mistakes, and the clarity that science can now bring, not only to the physical world but to human behaviors and desires, has made it more useful to us. This growing versatility and power is why many intellectual fields have, to one degree or another, been scientized, to their own great good: philosophy is now infused by brain science, sociology and anthropology and psychology are all collecting and analyzing big data like never before, and even history is now challenged by a newly proposed field called "cliodynamics," which tries to use historical "big data" and statistical analysis to model classes of historical events in ways that would allow predictions to be made about current similar situations.[4]

Economics should be given a similar infusion of the scientific method, which would start to turn it into a wing of ecology and of science generally, as it will include behavioral economics, biophysical economics, and so on. At that point we could formulate our economic plans within the paradigm of ecological thinking, with the biosphere regarded as the bio-infrastructure, with its estimated $33 trillion a year of unpaid services, all finally accounted for in a way that properly values and preserves it. This is a crucial project for science and society.[5]

The growing view that economics as presently practiced is a very inaccurate and damaging pseudo-science has been expressed often since the crash of 2008, most clearly from the other human sciences. Anthropology, sociology, political science, psychology, history—these disciplines are not directly powerful parts of our polyarchy, but they do help to establish a window of acceptable discourse and to provide new ideas. Economics as a field, is still so protected by power that it can ignore such critiques from the other social sciences and humanities, and it does. But in the face of evidence of the damage being done by capitalist economics, pressure is rising for change. We clearly need a more realistic working economic system to measure our efforts, evaluate them, and in essence arrange to pay ourselves for what needs doing.

If scientizing economics allowed us to analyze and subsequently direct our activities in ways that helped us to live more sustainably in the only biosphere we have, then all civilization would become in effect a project composed of a collection of experiments in improved relations with the planet. What we then looked for, what we measured, what we said was happening, and how we dealt with it—all that would change.

New terms are appearing everywhere as people try to articulate this new understanding in different ways. New names are being proposed for new systems, older ideas are being reexamined. Some of these propose small improvements, others complete transformations. Given how fixed in place the global economic system looks, how set in law and backed by force, advocating big changes can look quite unrealistic. But different time scales need to be brought into play when we talk about the future. Sometimes it helps to imagine how different our system is going to be in a thousand years, just to see what exists now from a different angle, and perhaps figure out what actions in the present can best set us in the direction of a better state of affairs. Grant that the long arc of history will keep bending, that centuries from now things will be very different, and suddenly the present too looks a little more malleable, its little changes part of that long bend.

So it helps to stop sometimes and take the long view. In one future, quite possible and we hope even likely, we will be providing ourselves with energy, power, food, water, transport, and infrastructure using an extremely clean and renewable suite of technologies. Our population will have stabilized as a result of the full extension of justice to women and to everyone alive. We will be restoring landscapes and wildlife populations while still feeding ourselves.

All these accomplishments are possible; there is no physical or technical impediment to us creating such a rich and vibrant permaculture. Having seen the possibility, humanity can make this permaculture its project.

Now, with this long-term vision or goal kept in mind, there are all kinds of things being tried or proposed to make the first little bends in the system that will curve it in that good direction. There are too many emerging projects to list here, although it would be very good to have this list, in something like a new economics *State of the World*, perhaps. In a compilation of this kind we would surely learn more about Mondragón and Kerala, Ecuador and Cuba, Bhutan and Scandinavia. We would also find out about ideas like predistributed value, microtaxing financial transactions, treating necessities as public utilities, full employment, permaculture, hedging (both kinds) for environmental repair, gross national happiness, the 2000 Watt Society, carbon taxes as truing the cost, intrinsic shareholders, land tithing, living wages, steady state economics, degrowth economics, moral hazard, systemic predatory dumping, deponzification, Leyden contentment indexes, rewilding, assisted migration, mongrel ecologies, cooperatives, open source, earth work, earth credits, the land ethic. . . .

And so on. The list will keep growing, and all these ideas will be understood as parts of a bigger thing, a global effort we have already begun to work on. We can see our present danger, and we can also see our future potential: a stable human population of some 7–9 billion, living cleanly and well on a healthy biosphere, sharing Earth with the rest of the creatures who rely on it. This is not just a dream but a responsibility, a project. And things we can do now to start on this project are all around us, waiting to be taken up and lived.

Notes

State of the World: A Year in Review

December 2011. Justin Gillis, "Carbon Emissions Show Biggest Jump Ever Recorded," *New York Times*, 4 December 2011; "Brazil Says Amazon Deforestation Down to Lowest Level," *Agence France-Presse*, 5 December 2011; Louise Gray, "Durban Climate Change Conference: Big Three of US, China and India Agree to Cut Carbon Emissions," (London) *Daily Telegraph*, 11 December 2011; "Mass Burials as Toll Hits 1000," *Agence France-Presse*, 21 December 2011; Christopher Joyce, "Turbulence as EU Court Oks Fee on Plane Emissions," *National Public Radio*, 21 December 2011.

January 2012. David Zeiler, "Oil Companies Big Winners as U.S. Becomes Net Exporter of Fuel," *Money Morning*, 4 January 2012; Juliet Eilperin, "Toxic Releases Rose 16 Percent in 2010, EPA Says," *Washington Post*, 5 January 2012; Julia Whitty, "Doomsday Clock Ticks Closer to Midnight," *Mother Jones*, 10 January 2012; "Nitrogen Pollution an Increasing Problem Globally," *Public Radio International*, 27 January 2012; Karla Zabludovsky, "Food Crisis as Drought and Cold Hit Mexico," *New York Times*, 30 January 2012.

February 2012. "Snow Blocks in Tens of Thousands as Cold Death Toll Rises," *Terra Daily*, 11 February 2012; Paul Vallely, "Special Report: The Hungry Generation," (London) *Independent*, 15 February 2012; Alan Buis, "NASA Satellite Finds Earth's Clouds are Getting Lower," *Space Daily*, 23 February 2012; David Fogarty, "World Bank Issues SOS for Oceans, Backs Alliance," *Planet Ark*, 27 February 2012.

March 2012. Rik Myslewski, "Oceans Gaining Acid Faster than Last 300 Million Years," (London) *The Register*, 2 March 2012; UNICEF, "Millennium Development Goal Drinking Water Target Met," press release (New York: 6 March 2012); "World Breakthrough on Salt-Tolerant Wheat," *Seed Daily*, 13 March 2012; Fiona Harvey, "England Faces Wildlife Tragedy as Worst Drought in 30 Years Hits Habitats," (London) *Guardian*, 18 March 2012; "International Chiefs of Environmental Compliance and Enforcement: Summit Report," INTERPOL, 27–29 March 2012.

April 2012. U.S. Environmental Protection Agency, "EPA Issues Updated, Achievable Air Pollution Standards for Oil and Natural Gas," press release (Washington, DC: 18 April 2012); Justin Gillis, "Study Indicates a Greater Threat of Extreme Weather," *New York Times*, 26 April 2012; Eleanor Bader, "A New Autism Theory," *Salon*, 26 April 2012; Isma'il Kshkush and Josh Kron, "Sudan Declares State of Emergency as Clashed Continue," *New York Times*, 29 April 2012; W. Barksdale Maynard, "An Underground Forest Offers Clues on Climate Change," *New York Times*, 30 April 2012; fossilized ferns photo credit: ©2012 University of Illinois Board of Trustees. All rights reserved. For permission information, contact the Illinois State Geological Survey. Photo courtesy of Scott Elrick.

May 2012. Eyder Peralta, "Study: Plastic Garbage in Pacific Ocean Has Increased 100-fold in 40 Years," *National Public Radio*, 9 May 2012; Doyle Rice, "Study: Many Mammals Won't be Able to Outrun Climate Change," *USA Today*, 14 May 2012; Suzanne Goldenberg, "US's Dolphin-safe Tuna Labels Banned by Court Calling Them 'Unfair' to Mexico," (London) *Guardian*, 16 May 2012; Tom Miles, "World Living Beyond its Resources, Summit Off-track: WWF," *Planet Ark*, 16 May 2012; Geoffrey Lean, "G8: Leaders Open Up Vital New Front in the Battle to Control Global Warming," (London) *Daily Telegraph*, 21 May 2012.

June 2012. Pete Spotts, "Report: Humans Near Tipping Point That Could Dramatically Change Earth," *Christian Science Monitor*, 6 June 2012; Fiona Harvey, "Fishing Discards Practice Thrown Overboard by EU," (London) *Guardian*, 13 June 2012; Richard Black, "Rio Summit: Little Progress, 20 Years On," *BBC*, 22 June 2012; David

Tuller, "BPA Linked to Brain Tumors for the First Time," *Mother Jones*, 27 June 2012; Sandy Shore, "Wheat, Corn Prices Climb as Heat Takes Toll on Crops," *Chicago Sun-Times*, 27 June 2012.

July 2012. Erik Olsen, "Growing Ship Traffic Threatens Blue Whales," *New York Times*, 2 July 2012; Kate Kelland, "Diseases from Animals Hit Over Two Billion People a Year," *Baltimore Sun*, 5 July 2012; Joseph O'Leary, "More than 2,000 Heat Records Matched or Broken," *Planet Ark*, 5 July 2012; Todd Wilkinson, "New Breed of Ranchers Shapes a Sustainable West," *Christian Science Monitor*, 29 July 2012; Brian Handwerk, "Caffeinated Seas Found off U.S. Pacific Northwest," *National Geographic*, 30 July 2012.

August 2012. Tim Newcomb, "Mutant Butterflies Found Near Fukushima," *Time*, 14 August 2012; Kelly Slivka, "Introducing the Ocean Health Index," *New York Times*, 15 August 2012; Kim Murphy, "Keystone XL Pipeline Construction Begins Amid Protests," *Los Angeles Times*, 16 August 2012; Monica Eng, "Who Determines Safety of New Food Ingredients?" *Chicago Tribune*, 25 August 2012; "Vast Reservoir of Methane Locked Beneath Antarctic Ice Sheet," (London) *Guardian*, 29 August 2012.

September 2012. Jeffrey Gettleman, "Elephants Dying in Epic Frenzy as Ivory Fuels Wars and Profits," *New York Times*, 3 September 2012; Alister Doyle, "Rising Chemicals Output a Hazard, Clean-up Needed by 2020: UN," *Planet Ark*, 6 September 2012; Jay Lindsay, "National Fishery Disaster Declared in New England by Commerce Department," *Huffington Post*, 13 September 2012; Nina Chestney, "100 Million Will Die by 2030 if World Fails to Act on Climate: Report," *Planet Ark*, 26 September 2012; Barbara Lewis, "EU Wind Capacity Hits 100 Gigawatt Mark: Industry," *Planet Ark*, 28 September 2012.

October 2012. Kenneth Weiss, "Oceans' Rising Acidity a Threat to Shellfish—and Humans," *Los Angeles Times*, 6 October 2012; Michael Lemonick, "New Study Ties Hurricane Strength to Global Warming," *Climate Central*, 15 October 2012; Alister Doyle, "Twenty-five Primates on Brink of Extinction, Study Says," *Planet Ark*, 16 October 2012; Jonathan Allen, "Pollution as Harmful as Malaria, TB in Developing World—Study," *AlertNet*, 23 October 2012; Alister Doyle, "U.N. Urges Foreign Fishing Fleets to Halt 'Ocean-Grabbing,'" *Planet Ark*, 31 October 2012.

November 2012. John Hocevar, "Looking for Hope in the Ruins as CCAMLR Talks Fizzle," *Huffington Post*, 1 November 2012; Tom Miles, "Greenhouse Gas Volumes Reached New High in 2011: Survey," *Planet Ark*, 21 November 2012; Stanglin Dough and Michael Winter, "Scattered Walmart Protests Don't Dent the Bottom Line," *USA Today*, 24 November 2012; Hilary Russ, "New York, New Jersey Put $71 Billion Price Tag on Sandy," *Chicago Tribune*, 26 November 2012; Ben Cubby, "The Top of the World Is Melting," *Brisbane Times*, 28 November 2012.

Chapter 1. Beyond Sustainababble

1. Definition (paraphrased) and etymology of sustainable from *Webster's Third New International Dictionary of the English Language Unabridged* (Springfield, MA: Merriam-Webster, 1981); World Commission on Environment and Development (WCED), *Our Common Future* (Oxford: Oxford University Press, 1987), p. 43.

2. "Sustainable Cars," *Inhabit.com*, at inhabitat.com/tag/sustainable-cars; "LOOK: PACT Sustainable Underwear," *Good Is*, at www.good.is/post/look-pact-sustainable-underwear; airline and gas-utility material collected in 2012 by the author.

3. Maria Cardona, "What Olympics Teach about Going Green," *CNN Opinion*, 28 July 2012; Figure 1–1 by Randall Munroe, at xkcd.com/1007.

4. Lester R. Brown, *Building a Sustainable Society* (New York: W. W. Norton & Company, 1981); George Perkins Marsh, *The Earth as Modified by Human Action* (London: Sampson Low, Marston, Low, and Searle, 1874).

5. Edmund Morris, *Theodore Rex* (New York: Random House, 2001), p. 76; National Environmental Policy Act, at ceq.hss.doe.gov/nepa/regs/nepa/nepaeqia.htm; Figure 1–2 from Jay N. "Ding" Darling Wildlife Society, originally published 15 September 1936.

6. Gro Harlem Brundtland, "Chairman's Foreword," in WCED, op. cit. note 1, p. xi.

7. Box 1–1 from the following: Justin Kitzes et al., *Guidebook to the National Footprint Accounts: 2008 Edition* (Oakland, CA: Global Footprint Network, 2008), pp. 9, 88; U.N. Environment Programme (UNEP), *The Emissions Gap Report 2012* (Nairobi: 2012), p. 1; U.N. Population Division, *World Population Prospects: The 2010 Revision, Volume I: Comprehensive Tables* (New York: 2011); Robert Engelman, *Profiles in Carbon: An Update on Population, Consumption and Carbon Dioxide Emissions* (Washington, DC: Population Action International, 1998); WWF et

al., *Living Planet Report 2012* (Gland, Switzerland: WWF, 2012); Robert Engelman, "Nine Strategies for Stopping Short of 9 Billion," in Worldwatch Institute, *State of the World 2012* (Washington, DC: Island Press, 2012).

8. Brundtland, op. cit. note 6.

9. UNICEF and World Health Organization, *Progress on Drinking Water and Sanitation: 2012 Update* (New York: United Nations, 2012); Homi Kharas and Andrew Rogerson, *Horizon 2025: Creative Destruction in the Aid Industry* (London: Overseas Development Institute, 2012).

10. UNEP, Ozone Secretariat, "The Montreal Protocol on Substances that Deplete the Ozone Layer," at ozone.unep .org.

11. Marc Lacey, "Across Globe, Hunger Brings Rising Anger," *New York Times*, 18 April 2008; Jim Yardley and Gardiner Harris, "India Staggered by Power Blackout; 670 Million People in Grip," *New York Times*, 1 August 2012.

12. Seth Borenstein, "World's Carbon Emissions Surpass Target," *Washington Post*, 3 December 2012; Potsdam Institute for Climate Impact Research and Climate Analytics, *Turn Down the Heat: Why a 4°C Warmer World Must Be Avoided* (Washington, DC: World Bank, 2012). Figure 1–3 based on data from BP, *BP Statistical Review of World Energy* (London: 2012).

13. Haibing Ma, "Energy Intensity Rising Slightly," *Vital Signs Online*, 20 September 2011.

14. Johan Rockström et al., "A Safe Operating Space for Humanity," *Nature*, 23 September 2009, pp. 472–75.

15. Anthony D. Barnofsky et al., "Approaching a State Shift in Earth's Biosphere," *Nature*, 7 June 2012, pp. 52–58.

16. UNEP, "World Remains on Unsustainable Track Despite Hundreds of Internationally Agreed Goals and Objectives," GEO5 press release (Rio de Janeiro, 6 June 2012).

17. Paul Epstein and Dan Ferber, *Changing Planet, Changing Health* (Berkeley: University of California Press, 2011).

18. Bill McKibben, "Global Warming's Terrifying New Math," *Rolling Stone*, 2 August 2012.

19. W. H. Auden, "September 1, 1939," in *Another Time* (New York: Random House, 1940).

20. U.N. Population Division, op. cit. note 7.

21. Potsdam Institute, op. cit. note 12, p. xviii.

22. Bill McKibben, *The End of Nature* (New York: Random House, 2006); Paul Wapner, *Living through the End of Nature* (Cambridge, MA: The MIT Press, 2010); Michiel Schaeffer et al., "Long-term Sea Level Rise Implied by 1.5°C and 2°C Warming Levels" (letter), *Nature Climate Change*, December 2012, pp. 867–70.

23. "A Wild Love for the World," Joanna Macy interview by Krista Tippett, *On Being*, American Public Media, 1 November 2012.

Chapter 2. Respecting Planetary Boundaries and Reconnecting to the Biosphere

1. Carl Folke et al., "Reconnecting to the Biosphere," *Ambio*, vol. 40, no. 7 (2011), pp. 719–38.

2. Hans Rosling, *Gapminder*, 2012, at www.gapminder.org/world/; Paul J. Crutzen, "Geology of Mankind," *Nature*, 3 January 2002, p. 23; Will Steffen, P. J. Crutzen, and J. R. McNeill, "The Anthropocene: Are Humans Now Overwhelming the Great Forces of Nature?" *Ambio*, vol. 36, no. 8 (2007), pp. 614–21.

3. Will Steffen et al., "The Anthropocene: From Global Change to Planetary Stewardship," *Ambio*, vol. 40 (2011), pp. 739–61.

4. Lisa Deutsch et al., "Feeding Aquaculture Growth through Globalization; Exploitation of Marine Ecosystems for Fishmeal," *Global Environmental Change*, May 2007, pp. 238–49; Evan D. G. Fraser and A. Rimas, "The Psychology of Food Riots," *Foreign Affairs*, 30 January 2011.

5. Folke et al., op. cit. note 1; Victor Galaz et al., "Institutional and Political Leadership Dimensions of Cascading Ecological Crises," *Public Administration*, June 2011, pp. 360–80; Brian Walker et al., "Looming Global-Scale Failures and Missing Institutions," *Science*, 11 September 2009, pp. 1,345–46.

6. F. Stuart Chapin, III et al., "Ecosystem Stewardship: Sustainability Strategies for a Rapidly Changing Planet," *Trends in Ecology and Evolution*, 24 November 2009, pp. 241–49.

7. Steffen et al., op. cit. note 3.

8. Figure 2–1 from Oran Young and W. Steffen. "The Earth System: Sustaining Planetary Life Support Systems," in F. S. Chapin III, G. P. Kofinas, and C. Folke, eds., *Principles of Ecosystem Stewardship: Resilience-Based Natural Resource Management in a Changing World* (New York: Springer-Verlag, 2009), pp. 295–315; Chapin et al., op. cit. note 6; Robert Costanza et al., "Sustainability or Collapse: What Can We Learn from Integrating History of Humans and the Rest of Nature," *Ambio*, vol. 36, no. 7 (2007), pp. 522–27.

9. Johan Rockström et al., "A Safe Operating Space for Humanity," *Nature*, 23 September 2009, pp. 472–75.

10. Table 2–1 and data in this section from Johan Rockström et al., "Planetary Boundaries: Exploring the Safe Operating Space for Humanity," *Ecology and Society*, vol. 14, no. 2 (2009).

11. Rockström et al., op. cit. note 9; Rockström et al., op. cit. note 10.

12. Stephen R. Carpenter and E. M. Bennett, "Reconsideration of the Planetary Boundary for Phosphorus," *Environmental Research Letters*, vol. 6, no. 1 (2011).

13. John M. Anderies et al., "The Topology of Non-Linear Global Carbon Dynamics: From Tipping Points to Planetary Boundaries," *Geophysical Research Letters*, forthcoming; Folke et al., op. cit. note 1; Chapin et al., op. cit. note 6; Young and Steffen, op. cit. note 8.

14. Carl Folke et al., "Resilience Thinking: Integrating Resilience, Adaptability and Transformability," *Ecology and Society*, vol. 15, no. 4 (2010).

15. Frances Westley et al., "Tipping Towards Sustainability: Emerging Pathways of Transformation," *Ambio*, vol. 40, no. 7 (2011), pp. 762–80; Melissa Leach et al., "Transforming Innovation for Sustainability," *Ecology and Society*, vol. 17, no. 2 (2012).

Chapter 3. Defining a Safe and Just Space for Humanity

1. Joseph Stiglitz, Amartya Sen, and Jean-Paul Fitoussi, *Report of the Commission on the Measurement of Economic Performance and Social Progress*, at www.stiglitz-sen-fitoussi.fr/documents/rapport_anglais.pdf.

2. Johan Rockström et al., "A Safe Operating Space for Humanity," *Nature*, 23 September 2009, pp. 472–75; Johan Rockström et al., "Planetary Boundaries: Exploring the Safe Operating Space for Humanity," *Ecology and Society*, vol. 14, no. 2 (2009), p. 32.

3. Rockström et al., "Planetary Boundaries," op. cit. note 2.

4. For the need for further conceptual clarification of planetary boundaries, see Simon L. Lewis, "We Must Set Planetary Boundaries Wisely," *Nature*, 23 May 2012, p. 417, and Ted Nordhaus, Michael Shellenberger, and Linus Blomqvist, *The Planetary Boundaries Hypothesis: A Review of the Evidence* (Oakland, CA: Breakthrough, 2012).

5. Kate Raworth, *A Safe and Just Space for Humanity: Can We Live within the Doughnut?* Oxfam Discussion Paper (Oxford: Oxfam International, 2012).

6. Figure 3–1 from Raworth, op. cit. note 5, based on Rockstrom et al., "Safe Operating Space," op. cit. note 2, and from Rockström et al., "Planetary Boundaries," op. cit. note 2.

7. See, for example, U.N. Committee on Economic, Social and Cultural Rights, General Comment No. 12: The Right to Adequate Food, 1999, and the Office of the United Nations High Commissioner for Human Rights, The Right to Water, Fact Sheet 35; on ecological economics, see Herman Daly, *Beyond Growth: The Economics of Sustainable Development* (Boston: Beacon Press, 1996), and Paul Ekins, *Economic Growth and Environmental Sustainability: The Prospects for Green Growth* (London: Routledge, 2000).

8. Table 3–1 based on the following: prevalence of undernourishment from U.N. Food and Agriculture Organization (FAO), food deficit database 2012; population living on less than $1.25 (PPP) per day from S. Chen and M. Ravallion, *The Developing World is Poorer Than We Thought But No Less Successful in the Fight against Poverty*, Policy Research Working Paper (Washington, DC: World Bank, 2008); total net primary enrollment rate from World

Bank Databank; United Nations, *The Millennium Development Goals Report 2011* (New York: 2011) for population using an improved water source, population using an improved sanitation facility, literacy rate of 15–24 year olds, gap between women's and men's share in non-agricultural wage employment, and gap between women and men holding seats in national parliaments; population without regular access to essential medicines from World Health Organization, *Equitable Access to Essential Medicines: A Framework for Collective Action* (Geneva: 2004); population lacking access to electricity and clean cooking facilities from International Energy Agency (IEA), *Energy for All: World Energy Outlook 2011* (Paris: 2011); social inequity based on national Gini coefficients exceeding 0.35, from Frederick Solt, "Standardizing the World Income Inequality Database," *Social Science Quarterly*, June 2009, pp. 231–42; SWIID Version 3.0, July 2010.

9. Figure 3–2 from Rockström et al., "Safe Operating Space," op. cit. note 2, and from Raworth, op. cit. note 5.

10. The statistic on food supply requirements is calculated for each country by multiplying the average food deficit of the undernourished population by the total undernourished population, then dividing the global total by the global food supply (per capita global food supply x global population). Data source for food-deficit and undernourished population is FAO, "Food Security Indicators," at www.fao.org/economic/ess/ess-fs/fs-data/ess-fadata /en, and the source for per capita global food supply and global population is FAO, *FAOSTAT Statistical Database*, at faostat.fao.org. Other data from FAO, *Global Food Losses and Food Waste: Extent, Causes and Prevention* (Rome: 2011), from IEA, *Energy for All: Financing Access for the Poor* (Paris: 2011), and from L. Chandy and G. Gertz, *Poverty in Numbers: The Changing State of Global Poverty from 2005 to 2015* (Washington, DC: The Brookings Institution, 2011).

11. B. Milanovic, *Global Inequality Recalculated: The Effect of New 2005 PPP Estimates on Global Inequality*, Policy Research Working Paper (Washington, DC: World Bank, 2009); S. Chakravarty et al., "Sharing Global CO_2 Emission Reductions among One Billion High Emitters," *Proceedings of the National Academy of Sciences*, 6 July 2009; S. Chakravarty, discussions with author; nitrogen use from Mark A. Sutton et al., "Too Much of a Good Thing," *Nature*, 10 April 2011, pp. 159–61.

12. H. Kharas, *The Emerging Middle Class in Developing Countries*, Working Paper (Paris: OECD Development Centre, 2010); Foresight, *The Future of Food and Farming: Challenges and Choices for Global Sustainability* (London: Government Office for Science, 2011).

13. Stiglitz, Sen, and Fitoussi, op. cit. note 1. Box 3–1 from the following: Lew Daly and Stephen Posner, *Beyond GDP: New Measures for a New Economy* (New York: Demos, 2012); Stiglitz, Sen, and Fitoussi, op. cit. note 1; "Resolution 65/309. Happiness: Towards a Holistic Approach to Development," U.N. General Assembly, 25 August 2011; WAVES Partnership, World Bank, at www.wavespartnership.org/waves; World Bank, *Moving Beyond GDP* (Washington, DC: WAVES Partnership, 2012); Office for National Statistics, *Measuring What Matters: National Statistician's Reflections on the National Debate on Measuring National Well-being* (London: 2011); Australian Bureau of Statistics, "Measures of Australia's Progress," at www.abs.gov.au/ausstats; J. Steven Landefeld et al., "GDP and Beyond: Measuring Economic Progress and Sustainability," *Survey of Current Business*, April 2010; "Maryland's Genuine Progress Indicator," at www.green.maryland.gov/mdgpi; "Vermont Establishes a Genuine Progress Indicator, Blazes a Path for Measuring What Matters," Demos, 9 May 2012.

14. Basel data from Shahra Razavi, *The Political and Social Economy of Care in a Development Context*, Gender and Development Programme Paper (Geneva: U.N. Research Institute for Social Development, 2007); U.S. household production data from Benjamin Bridgman et al., "Accounting for Household Production in the National Accounts, 1965–2010," *Survey of Current Business*, May 2012.

15. U.N. Environment Programme (UNEP)–World Conservation Monitoring Centre, *The UK National Ecosystem Assessment: Synthesis of the Key Findings* (Cambridge, U.K.: 2011).

16. United Nations University–International Human Dimensions Programme and UNEP, *Inclusive Wealth Report 2012: Measuring Progress toward Sustainability* (Cambridge, U.K.: Cambridge University Press, 2012).

17. Organisation for Economic Co-operation and Development, *Divided We Stand: Why Income Inequality Keeps Rising* (Paris: 2011).

18. Ian Gough et al., *The Distribution of Total Greenhouse Gas Emissions by Households in the UK, and Some Implications for Social Policy* (London: Centre for Analysis of Social Exclusion and the New Economics Foundation,

2011, amended 2012); Statistics Sweden, System of Environmental and Economic Accounts, *CO_2 Emission per Income Deciles 2000* (Stockholm: 2000); China from Jie Li and Yan Wang, "Income, Lifestyle and Household Carbon Footprints (Carbon-Income Relationship), a Micro-level Analysis on China's Urban and Rural Household Surveys," *Environmental Economics*, vol. 1, no. 2 (2010).

Chapter 4. Getting to One-Planet Living

1. Jared Diamond, *Collapse: How Societies Choose to Fail or Succeed* (New York: Viking Press, 2005).

2. U.N. Department of Economic and Social Affairs, *World Economic and Social Survey 2011* (New York: United Nations, 2011), p. ix.

3. Donella Meadows et al., *The Limits to Growth* (New York: Universe Books, 1972); Lance Gunderson and C. S. Holling, eds., *Panarchy: Understanding Transformations in Human and Natural Systems* (Washington, DC: Island Press, 2002); Millennium Ecosystem Assessment, *Ecosystems and Human Well-being: Synthesis* (Washington, DC: Island Press, 2005).

4. WWF et al., *Living Planet Report 2010* (Gland, Switzerland: WWF, 2010); WWF, *Living Planet Report 2012* (Gland, Switzerland: WWF, 2012); Mathis Wackernagel and William E. Rees, *Our Ecological Footprint* (Gabriola Island, Canada: New Society Publishers, 1996). Box 4–1 from Global Footprint Network, *National Footprint Accounts, 2011 Edition* (Oakland, CA: 2012), and from www.footprintnetwork.org.

5. Wackernagel and Rees, op. cit. note 4; William E. Rees, "Ecological Footprint: Concept of," in S. A. Levin, ed. in chief, *Encyclopedia of Biodiversity*, 2nd ed. (Amsterdam: Elsevier/Academic Press, forthcoming); WWF et al., *Living Planet Report 2010*, op. cit. note 4; WWF, *Living Planet Report 2012*, op. cit. note 4.

6. Anup Shah, "Poverty Facts and Stats," citing *World Development Indicators*, World Bank, 2008, at www.globalissues.org/article/26/poverty-facts-and-stats; William E. Rees, "Ecological Footprints and Biocapacity: Essential Elements in Sustainability Assessment," in J. Dewulf and H. Van Langenhove, eds., *Renewables-based Technology: Sustainability Assessment* (Chichester, U.K.: John Wiley and Sons, 2006); WWF et al., *Living Planet Report 2010*, op. cit. note 4.

7. Table 4–1 from the following: Global Footprint Network, at www.footprintnetwork.org/en/index.php/GFN/page/world_footprint; U.N. Food and Agriculture Organization, "Nutrition Country Profiles," at www.fao.org/ag/agn/nutrition/profiles_by_country_en.stm; Peter Menzel, *Material World* (San Francisco: Sierra Club Books, 1994); World Bank, "Indicators," at data.worldbank.org/indicator; International Civil Aviation Organization, "Special Report: Annual Review of Civil Aviation," *ICAO Journal*, vol. 61, no. 5 (2005); Worldmapper, at www.worldmapper.org; World Resources Institute, "EarthTrends: Environmental Information," at earthtrends.wri.org; WWF, "Footprint Interactive Graph," at wwf.panda.org/about_our_earth/all_publications/living_planet_report. Global average statistics for living space and motor vehicle travel are estimated assuming two thirds of the global population consumes at the one-planet level and one third consumes at the three-planet level.

8. Life expectancy from World Bank, op. cit. note 7.

9. Land area data from "Understanding Vancouver," at vancouver.ca/commsvcs/planning/census/index.htm; 2006 population from Statistics Canada, "Census Data: Community Profiles: Vancouver, British Columbia (Census Metropolitan Area)" (Ottawa).

10. Figure 4–1 from Jennie Moore, *Getting Serious About Sustainability: Exploring the Potential for One-planet Living in Vancouver*, submitted in partial fulfillment of requirements for PhD degree (Vancouver: School of Community and Regional Planning, University of British Columbia, forthcoming).

11. City of Vancouver, *Greenest City 2020 Action Plan* (Vancouver: 2011), pp. 48–53.

12. Figure 4–2 from Moore op. cit. note 10.

13. Moore, op. cit. note 10; British Columbia (The Province of), *Carbon Neutral BC, A First for North America*, press release (Victoria: 30 June 2011).

14. Moore, op. cit. note 10.

15. Ibid.

16. City of Vancouver, op. cit. note 11.

17. City of Vancouver, *Greenest City 2020 Action Plan (GCAP): Council Report* (Vancouver: 2011), pp. 110–11.

18. Ibid.

19. City of Vancouver op. cit. note 11.

20. Anthony Giddens, *The Politics of Climate Change* (Cambridge, U.K.: Polity Press, 2011); Norman Myers and Jennifer Kent, *Perverse Subsidies: How Tax Dollars Can Undercut the Environment and the Economy* (Washington, DC: Island Press, 2001); Ernst von Weizsäcker, Amory Lovins, and Hunter Lovins, *Factor Four* (London: Earthscan, 1997).

21. William E. Rees, "Globalization and Sustainability: Conflict or Convergence," *Bulletin of Science, Technology and Society*, August 2002, pp. 249–68; Ernst von Weizsäcker et al., *Factor 5* (London: Earthscan, 2009); U.N. Department of Economic and Social Affairs, op. cit. note 2.

22. William E. Rees, "The Way Forward: Survival 2100," *Solutions*, June 2012; William E. Rees, "What's Blocking Sustainability? Human Nature, Cognition and Denial," *Sustainability: Science, Practice, & Policy*, fall 2010; Giddens, op. cit. note 20; von Weizsäcker, Lovins, and Lovins, op. cit. note 20; World Commission on Environment and Development, *Our Common Future* (Oxford: Oxford University Press, 1987).

23. Von Weizsäcker et al., op. cit. note 21; U.N. Department of Economic and Social Affairs, op. cit. note 2.

24. World Commission on Environment and Development, op. cit. note 22, pp. 52, 89.

25. Emmanuel Saez, *Striking it Richer: The Evolution of Top Incomes in the United States* (updated with 2009 and 2010 estimates) (Berkeley: University of California, 2012); U.N. Development Programme, *Human Development Report 2010* (New York: 2010); U.N. Department of Economic and Social Affairs, op. cit. note 2; U.N. Department of Economic and Social Affairs, *World Economic and Social Survey 2006* (New York: United Nations, 2006).

26. Rees, "What's Blocking Sustainability?" op. cit. note 22; Rees, "The Way Forward," op. cit. note 22.

27. Diamond, op. cit. note 1.

Chapter 5. Sustaining Freshwater and Its Dependents

1. Figure of 250 million is approximate, per Joel E. Cohen, *How Many People Can the Earth Support?* (New York: W. W. Norton & Company, 1995), p. 77; 7 billion from U. S. Census Bureau, "U.S. & World Population Clocks," at www.census.gov/main/www/popclock.html; gross world product estimate for 2011 from U.S. Central Intelligence Agency, *The World Factbook*, at www.cia.gov/library/publications/the-world-factbook/index.html.

2. For analysis and sources, see later text.

3. Figure of 800 million from UNICEF and World Health Organization (WHO), *Progress on Drinking Water and Sanitation: 2012 Update* (New York: United Nations, 2012).

4. Igor A. Shiklomanov, *World Water Resources: A New Appraisal and Assessment for the 21st Century* (Paris: UNESCO, 1998). Box 5–1 based on National Academy of Sciences, Water Science and Technology Board, *Desalination: A National Perspective* (Washington, DC: National Academy Press, 2008); 15,000 figure from Quirin Schiermeier, "Purification with a Pinch of Salt," *Nature*, 20 March 2008, pp. 260–61.

5. Sandra L. Postel, Gretchen D. Daily, and Paul R. Ehrlich, "Human Appropriation of Renewable Fresh Water," *Science*, 9 February 1996, pp. 785–88.

6. Figures of 19 percent, 42 percent, and 15,600 cubic kilometers from ibid., adjusted for rise in water captured by dams to 10,800 cubic kilometers, from B. F. Chao, Y. H. Wu, and Y. S. Li, "Impact of Artificial Reservoir Water Impoundment on Global Sea Level," *Science*, 11 April 2008, pp. 212–14, and assumption that 64 percent of this storage capacity is actively used in the regulation of runoff, per Postel, Daily, and Ehrlich, op. cit. note 5; amount used by each sector from United Nations, *Water in a Changing World: United Nations World Water Development Report*, 3rd ed. (Paris: UNESCO, 2009).

7. Figure of 82 percent from U.N. Food and Agriculture Organization (FAO), *Aquastat Database*, at www.fao.org/NR/WATER/AQUASTAT/main/index.stm.

8.   Arjen Y. Hoekstra and Mesfin M. Mekonnen, "The Water Footprint of Humanity," *Proceedings of the National Academy of Sciences*, 28 February 2012, pp. 3,232–37.

9.   Data for Figure 5–1 from FAO, op. cit. note 7, viewed 11 September 2012, except 2010 world irrigated area of 299 million hectares from International Commission on Irrigation and Drainage (ICID), *Database*, at www.icid.org/database.html, viewed 11 September 2012, and 2010 world population of 6.9 billion from Population Reference Bureau, *2010 Population Clock*, at www.prb.org/Articles/2010/worldpopulationclock2010.aspx; most figures are for 2010, except reported irrigated areas by country are for various years.

10.   Economic Commission for Africa, *Economic Report on Africa 2007: Accelerating Africa's Development through Diversification* (Addis Ababa, Ethiopia: 2007).

11.   Intergovernmental Panel on Climate Change, *Climate Change 2007: Impacts, Adaptation and Vulnerability. Contribution of Working Group II to the Fourth Assessment Report* (Cambridge, U.K.: Cambridge University Press, 2007); P. C. D. Milly et al., "Stationarity is Dead: Whither Water Management?" *Science*, 1 February 2008, pp. 573–74.

12.   Basic sustainability definition from World Commission on Environment and Development, *Our Common Future* (Oxford: Oxford University Press, 1987), p. 43.

13.   UNICEF and WHO, op. cit. note 3; reasonable access to water is defined as the availability of 20 liters per person per day from a source within 1 kilometer of the user's dwelling; acceptable sources include household connections, public standpipes, boreholes, protected dug wells, protected spring water, and rainwater collection, from United Nations, *Water for People, Water for Life: United Nations World Water Development Report*, 1st ed. (Paris: UNESCO Publishing and Berghahn Books, 2003), p. 113; the 0.1 percent figure is author's calculation.

14.   UNICEF and WHO, op. cit. note 3. Figures reported are for 2010.

15.   Number of dams at least 15 meters high from World Commission on Dams, *Dams and Development* (London: Earthscan, 2000); 26 percent is author's calculation using estimated water impounded by dams by Chao, Wu, and Li, op. cit. note 6.

16.   Jamie Pittock et al., *Interbasin Water Transfers and Water Scarcity in a Changing World—A Solution or a Pipedream?* (Frankfurt: World Wildlife Fund Germany, 2009); Ruixiang Zhu, "China's South-North Water Transfer Project and Its Impacts on Economic and Social Development," Ministry of Water Resources, People's Republic of China, n.d.

17.   "Your Water," City of Phoenix, at phoenix.gov/waterservices/wrc/yourwater/index.html, viewed 25 July 2012.

18.   Figure of 16 percent from "Hydropower Generation," Strategic Energy Technologies Information System, European Commission, at setis.ec.europa.eu/newsroom-items-folder/hydropower-generation, viewed 25 July 2012; Energy Information Administration, U.S. Department of Energy, "International Energy Outlook 2011," at www.eia.gov/forecasts/ieo/electricity.cfm, viewed 25 July 2012.

19.   Brian D. Richter et al., "Lost in Development's Shadow: The Downstream Human Consequences of Dams," *Water Alternatives*, vol. 3, no. 2 (2010), pp. 14–42; Box 5–2 from Sandra Postel, *Liquid Assets: The Critical Need to Safeguard Freshwater Ecosystems* (Washington, DC: Worldwatch Institute, 2005).

20.   Sandra Postel and Brian Richter, *Rivers for Life: Managing Water for People and Nature* (Washington, DC: Island Press, 2003); Anthony Ricciardi and Joseph B. Rasmussen, "Extinction Rates of North American Freshwater Fauna," *Conservation Biology*, vol. 13 (1999), pp. 1,220–22; 39 percent vulnerable and 61 presumed extinct from Howard L. Jelks et al., "Conservation Status of Imperiled American Freshwater and Diadromous Fishes," *Fisheries*, August 2008, pp. 372–407.

21.   Postel, op. cit. note 19; 90 percent from Francisco Zamora-Arroyo et al., *Conservation Priorities in the Colorado River Delta: Mexico and the United States* (Sonoran Institute et al., 2005); a third of Egypt's crop from "Egypt's Fertile Nile Delta Falls Prey to Climate Change," *Agence France-Presse*, 28 January 2010.

22.   Millennium Ecosystem Assessment, *Ecosystems and Human Well-being* (Washington, DC: Island Press, 2005); wetlands drained from Rudy Rabbinge and Prem S. Bindraban, "Poverty, Agriculture, and Biodiversity," in John A. Riggs, ed., *Conserving Biodiversity* (Washington, DC: The Aspen Institute, 2005), pp. 65–77.

23. "Plan to Breach Levee in Missouri Advances," *New York Times*, 1 May 2011; wetland loss from Linda R. Wires et al., *Upper Mississippi Valley/Great Lakes Waterbird Conservation Plan*, Final Report submitted to the U.S. Fish and Wildlife Service, Fort Snelling, MN, March 2010; Donald L. Hey and Nancy S. Philippi, "Flood Reduction through Wetland Restoration: The Upper Mississippi River Basin as a Case History," *Restoration Ecology*, March 1995, pp. 4–17.

24. Matthew Rodell, Isabella Velicogna, and James S. Famiglietti, "Satellite-Based Estimates of Groundwater Depletion in India," *Nature*, 20 August 2009, pp. 999–1,002; J. S. Famiglietti et al., "Satellites Measure Recent Rates of Groundwater Depletion in California's Central Valley," *Geophysical Research Letters*, 5 February 2011.

25. Sandra Postel, *Pillar of Sand* (New York: W. W. Norton & Company, 1999); 60 percent from World Bank, *Deep Wells and Prudence: Towards Pragmatic Action for Addressing Groundwater Overexploitation in India* (Washington, DC: 2010); 15 percent from John Briscoe and R. P. S. Malik, *India's Water Economy: Bracing for a Turbulent Future* (New Delhi: Oxford University Press, 2006), p. xviii.

26. Ogallala depletion from Jennifer S. Stanton et al., *Selected Approaches to Estimate Water-Budget Components of the High Plains, 1940 through 1949 and 2000 through 2009* (Reston, VA: U. S. Geological Survey, 2011); U.S. wheat production from U.S. Department of Agriculture, Economic Research Service, "Wheat Data," at www.ers.usda.gov/data-products/wheat-data.aspx, viewed 6 September 2012.

27. Yoshihide Wada et al., "Global Depletion of Groundwater Resources," *Geophysical Research Letters*, vol. 37, L20402 (2010); other calculations are the author's, with 2000 global grain harvest from FAO, *FAOSTAT Statistical Database*, at faostat.fao.org, viewed 6 September 2012.

28. Mark Giordano, "Global Groundwater? Issues and Solutions," *Annual Review of Environmental Resources*, vol. 34 (2009), pp. 7.1–7.26, based on data from FAO, op. cit. note 7; Saudi example from FAO, *Aquastat Database*, at www.fao.org/nr/water/aquastat/countries_regions/saudi_arabia/index.stm.

29. FAO, op. cit. note 28; "Squeezing Africa Dry: Behind Every Land Grab is a Water Grab," GRAIN (Barcelona, Spain), 11 June 2012.

30. High Plains Water District, "Average Groundwater Level Decline of –2.56 Feet Recorded in 2011 Is Third Largest in High Plains Water District's 61-Year History," press release (Lubbock, TX: 2 July 2012).

31. World Bank, op. cit. note 25; other states may follow from John Grimond, "For Want of a Drink: A Special Report on Water," *The Economist*, 22 May 2010.

32. Sandra Postel, "Texas Water District Acts to Slow Depletion of the Ogallala Aquifer," *Water Currents*, National Geographic, 7 February 2012.

33. Kate Galbraith, "Texas Farmers Battle Ogallala Pumping Limits," *The Texas Tribune*, 18 March 2012; Sandra Postel, "That Sinking Feeling about Groundwater in Texas," *Water Currents*, National Geographic, 19 July 2012.

34. ICID, op. cit. note 9; Anil Jain, managing director, Jain Irrigation, discussion with author, San Antonio, TX, December 2010.

35. ICID, op. cit. note 9; Sandra Postel et al., "Drip Irrigation for Small Farmers: A New Initiative to Alleviate Hunger and Poverty," *Water International*, March 2001, pp. 3–13; 600,000 sales from "Design for the Other 90%," at other90.cooperhewitt.org/Design/drip-irrigation-system, viewed 14 July 2010.

36. Murray Darling Basin Authority (MDBA), at www.mdba.gov.au.

37. MDBA, "Plain English Summary of the Proposed Basin Plan, with Explanatory Notes," Canberra, November 2011; "The Proposed Murray-Darling Basin Plan: Scientific Statement" (32 signatories), April 2012.

38. Sandra Postel, "Lessons from the Field—Boston Conservation," National Geographic online, at environment.nationalgeographic.com/environment/freshwater/lessons-boston-conservation, March 2010.

39. Aaron Koch, Policy Advisor, Mayor's Office of Long-Term Planning and Sustainability, New York, e-mail to author, 8 June 2011.

40. Carolyn Whelan, "Liquid Asset," *Nature Conservancy*, autumn 2010, pp. 43–49.

41. Benjamin Moline, manager, Water Resources and Real Estate, Molson-Coors, Golden, Co, discussion with

author, 11 May 2012 (MillerCoors is a partnership between SABMiller and Molson Coors); Unilever, "Water Use in Agriculture," at www.unilever.com/sustainable-living/water/agriculture; tomatoes from Unilever, Sustainable Agriculture Team, *Unilever & Sustainable Agriculture—Water* (2009).

42. Water Footprint Network, "Product Gallery," at www.waterfootprint.org/?page=files/productgallery, viewed 13 September 2012; Carey W. King and Michael E. Webber, "Water Intensity of Transportation," *Environmental Science & Technology*, vol. 42, no. 21 (2008), pp. 7,866–72.

Chapter 6. Sustainable Fisheries and Seas: Preventing Ecological Collapse

1. Rachel Carson, *The Sea Around Us* (New York: Oxford University Press, 1961), p. xii.

2. Homer, *The Odyssey*, trans. Robert Fagles (New York: Viking, 1996); Tina Bishop et al., "Then and Now: The HMS Challenger Expedition and the 'Mountains in the Sea' Expedition," *NOAA Ocean Explorer Podcast RSS*, National Oceanic and Atmospheric Administration (NOAA), 16 July 2012.

3. "Ocean," *NOAA*, at www.noaa.gov/ocean.html.

4. National Ocean Economics Program, *State of the U.S. Ocean and Coastal Economies*, Center for the Blue Economy, 2009; NOAA, "The Oceans Are the Trading Routes for the Planet," *How Important Is the Ocean to Our Economy?* at oceanservice.noaa.gov/facts/oceaneconomy.html.

5. U.N. Environment Programme (UNEP)–World Conservation Monitoring Centre, *In the Front Line: Shoreline Protection and Other Ecosystem Services from Mangroves and Coral Reefs* (Cambridge, U.K.: 2006); H. Cesar et al., *Economic Valuation of Hawaiian Reefs* (Arnham, Netherlands: Cesar Environment Economics Consulting, 2002).

6. Marah Hardt and Carl Safina, "Covering Ocean Acidification: Chemistry and Considerations," Yale Forum on Climate Change & the Media, 24 June 2008.

7. Kenneth W.Bruland et al., "Iron, Macronutrients and Diatom Blooms in the Peru Upwelling Regime: Brown and Blue Waters of Peru," *Marine Chemistry*, vol. 93, no. 2 (2005), pp. 81–103; Milagros Salazar, "Peru's Vanishing Fish," Center for Public Integrity, 26 January 2012.

8. International Energy Agency, "Prospect of Limiting the Global Increase in Temperature to 2°C Is Getting Bleaker," press release (Paris: 30 May 2011); UNEP, *Keeping Track of Our Changing Environment* (Nairobi: 2011); U.S. Environmental Protection Agency, "Inventory of U.S. Greenhouse Gas Emissions and Sinks: 1990–2010 (April 2012)," *U.S. Greenhouse Gas Inventory Report* (Washington, DC: 2012).

9. Richard A. Feely, Christopher L. Sabine, and Victoria J. Fabry, "Carbon Dioxide and Our Ocean Legacy," Clear the Air and Conserve Our Ocean Legacy, April 2006; Royal Society, *Ocean Acidification Due to Increasing Atmospheric Carbon Dioxide* (London, 2005).

10. UNEP, op. cit. note 8; Bärbel Hönisch et al., "The Geological Record of Ocean Acidification," *Science*, 2 March 2012, pp. 1,058–63; NOAA, "Corals: Zooxanthallae…What's That?" *NOAA National Ocean Service Education*, 25 March 2008.

11. UNEP, op. cit. note 8; Lauretta Burke et al., *Reefs at Risk Revisited* (Washington, DC: World Resources Institute, 2011).

12. "The Biological Effects of Ocean Acidification," Princeton University, at www.princeton.edu/grandchallenges.

13. UNEP, op. cit. note 8.

14. Lothar Stramma et al., "Expansion of Oxygen Minimum Zones May Reduce Available Habitat for Tropical Pelagic Fishes" (letter), *Nature Climate Change*, January 2012, pp. 33–37.

15. Elliott L. Hazen et al., "Predicted Habitat Shifts of Pacific Top Predators in a Changing Climate" (letter), *Nature Climate Change*, online 23 September 2012.

16. Yearly cycle and lowest in 8,000 years from Institute of Environmental Physics, University of Bremen, "Arctic Sea Ice Extent Small as Never Before," 2011, at www.iup.uni-bremen.de:8084/amsr/minimum2011-en.pdf; National Snow and Ice Data Center (NSIDC), "Arctic Sea Ice Reaches Lowest Extent for the Year and the Satellite Record," press release (Boulder, CO: 19 September 2012).

17. NSIDC, "Arctic Sea Ice Extent: Area of the Ocean with at Least 15 Percent Sea Ice" (graph), at nsidc.org/data/seaice_index/images/daily_images/N_stddev_timeseries.png; Polar Science Center, "Arctic Sea Ice Volume Anomaly and Trend from PIOMAS," University of Washington, 30 July 2012; NSIDC, "Frequently Asked Questions on Arctic Sea Ice," June 2009.

18. Andrew McMinn, "Production in Sea Ice Could Fall," *Australian Antarctic Magazine*, spring 2005, p. 11.

19. U.N. Food and Agriculture Organization (FAO), *The State of World Fisheries and Aquaculture 2010* (Rome: 2010).

20. Ibid.; UNEP, op. cit. note 8.

21. B. B. Collete et al., "High Value and Long Life—Double Jeopardy for Tunas and Billfishes," *Science*, 15 July 2011, pp. 291–92; UNEP, op. cit. note 8.

22. National Academies, *Coastal Hazards* (Washington, DC: Ocean Studies Board, 2007); R. J. Nicholls et al., "Coastal Systems and Low-lying Areas," in Intergovernmental Panel on Climate Change, *Climate Change 2007: Impacts, Adaptation and Vulnerability. Contribution of Working Group II to the Fourth Assessment Report* (Cambridge, U.K.: Cambridge University Press, 2007), pp. 315–56.

23. C. M. Boerger et al., "Plastic Ingestion by Planktivorous Fishes in the North Pacific Central Gyre," *Marine Pollution Bulletin*, December 2010, pp. 2,275–78; Peter Kershaw et al., "Plastic Debris in the Ocean," in UNEP, *UNEP Yearbook 2011* (Nairobi: 2011).

24. Council on Environmental Quality, *Interim Framework for Effective Coastal and Marine Spatial Planning* (Washington, DC: White House, 2009).

25. Ibid.

26. Christopher Costello, Stephen D. Gaines, and John Lynham, "Can Catch Shares Prevent Fisheries Collapse?" *Science*, 19 September 2008, p. 1,678–81.

27. Daniel Pauly, "Major Trends in Small-Scale Marine Fisheries, with Emphasis on Developing Countries, and Some Implications for the Social Sciences," *Maritime Studies*, vol. 4, no. 2 (2006), pp. 7–22.

28. FAO, op. cit. note 19; UNEP, op. cit. note 8.

29. George Karleskint, Richard Turner, and James Small, *Introduction to Marine Biology*, 2nd ed. (Philadelphia: Saunders College Publishing, 1998).

30. UNEP, op. cit. note 8; C. Giri et al., "Mangrove Forest Distributions and Dynamics (1975–2005) of the Tsunami-Affected Region of Asia," *Journal of Biogeography*, vol. 35 (2008), pp. 519–28.

31. UNEP, op. cit. note 8; P. M. Strain and B. T. Hargrave, "Salmon Aquaculture, Nutrient Fluxes, and Ecosystem Processes in Southwestern New Brunswick," in Barry T. Hargrave, ed., *The Handbook of Environmental Chemistry: Environmental Effects of Marine Finfish Aquaculture. Volume 5: Water Pollution* (New York: Springer, 2005); NOAA and Gulf of Mexico Fishery Management Council, *Fishery Management Plan for Regulating Offshore Marine Aquaculture in the Gulf of Mexico* (St. Petersburg and Tampa, FL: 2009).

32. Sarah Simpson, ed., "10 Solutions to Save the Oceans," *Conservation Magazine*, July–September 2007.

Chapter 7. Energy as Master Resource

1. Ostwald's account is quoted in R. J. Deltete, "Wilhelm Ostwald's Energetics 1: Origins and Motivations," *Foundations of Chemistry*, January 2007, pp. 33–35.

2. Ibid., p. 33

3. Ibid., p. 34.

4. C. Hakfoort, "Science Deified: Wilhelm Ostwald's Energeticist World-view and the History of Scientism," *Annals of Science*, vol. 49, no. 6 (1992), pp. 525–44.

5. Martin J. Klein, "Thermodynamics in Einstein's Thought," *Science*, 4 August 1967, pp. 509–16; Donald Worster, *Nature's Economy*, 2nd ed. (Cambridge, U.K.: Cambridge University Press, 1994), pp. 301–06.

6. Henry Adams, "The Tendency of History," "A Letter to American Teachers of History," and "The Rule of Phase Applied to History," in Brooks Adams, ed., *The Degradation of the Democratic Dogma* (New York: Macmillan, 1919); William Frederick Cottrell, *Energy and Society* (New York: McGraw-Hill, 1955); Lewis Mumford, *The Myth of the Machine II: The Pentagon of Power* (New York: Harcourt Brace Jovanovich, 1970).

7. Daniel C. Foltz, "Does Nature Have Historical Agency? World History, Environmental History, and How Historians Can Help Save the Planet," *The History Teacher*, November 2003, pp. 9–28; Alfred Crosby, *Children of the Sun: A History of Humanity's Unappeasable Appetite for Energy* (New York: W. W. Norton & Company, 2006).

8. Lester Thurow and Rober Heilbroner, *The Economic Problem* (New York: Prentice Hall, 1981), pp. 127, 135.

9. Nicholas Georgescu-Roegen, "Energy Analysis and Economic Valuation," *Southern Economic Journal*, April 1979, p. 1,041.

10. Frederick Soddy, *Wealth, Virtual Wealth, and Debt* (London: George Allen & Unwin, 1926); Frederick Soddy, *Money Versus Man* (New York: E. P. Dutton, 1933); Frederick Soddy, *The Role of Money* (London: Routledge, 1934, 2003); see also Herman Daly, "The Economic Thought of Frederick Soddy," *History of Political Economy*, winter 1980, pp. 469–88.

11. Eric Zencey, "The Financial Crisis is the Environmental Crisis," *The Daly News* (Center for the Advancement of the Steady State Economy), 6 January 2011.

12. Frank Knight, "Money," *Saturday Review of Literature*, 16 April 1927, p. 732; Irving Fisher, *100% Money* (New York: Adelphi Company, 1935).

13. Nicholas Georgescu-Roegen, *The Entropy Law and the Economic Process* (Cambridge, MA: Harvard University Press, 1971); Herman Daly, *Steady State Economics* (New York: W. H. Freeman, 1977).

14. Georgescu-Roegen, op. cit. note 13, p. 18; Joseph Stiglitz, Amartya Sen, and Jean-Paul Fitoussi, Report of the Commission on the Measurement of Economic Performance and Social Progress, at www.stiglitz-sen-fitoussi.fr /documents/rapport_anglais.pdf.

15. James H. Keeling, *Quaero [Some Questions in Matter, Energy, Intelligence, and Evolution]* (London: Taylor and Francis, 1898), see especially pp. 7–9.

16. John Stuart Mill, "Of the Stationary State," Chapter VI of Book IV, *Principles of Political Economy* (New York: D. Appleton and Company, 1896); Herman Daly and Joshua Farley, *Ecological Economics: Principles and Applications* (Washington, DC: Island Press, 2011), pp. 6–7.

17. For the first uses of the concept and term, see C. J. Cleveland et al., "Energy and the U.S. Economy: A Biophysical Perspective," *Science*, 31 August 1984, pp. 890–97, and R. Herendeen and R. L. Plant, "Energy Analysis of Four Geothermal Technologies," *Energy*, January 1981, pp. 73–82; Georgescu-Roegen, op. cit. note 13, p. 18.

18. C. A. S. Hall and C. J. Cleveland, "Petroleum Drilling and Production in the United States: Yield per Effort and Net Energy Analysis," *Science*, 6 February 1981, pp. 576–79; Warren Davis, "A Study of the Future Productive Capacity and Probable Reserves of the U.S.," *Oil and Gas Journal*, 24 February 1958, pp. 105–19; David J. Murphy and Charles A. S. Hall, "Year in Review: EROI or Energy Return on (Energy) Invested," *Annals: Ecological Economics Review*, February 2010, pp. 102–18.

19. Murphy and Hall, op. cit. note 18.

20. Jessica Lambert et al., *EROI of Global Energy Resources: Preliminary Status and Trends* (London: U.K. Department for International Development, forthcoming), pp. 3–6.

21. Table 7–1 from Richard Heinberg, *Searching for a Miracle: 'Net Energy' Limits and the Fate of Industrial Society* (San Francisco and Santa Rosa, CA: International Forum on Globalization and the Post-Carbon Institute, 2009), p. 55.

22. Charles A. S. Hall, Stephen Balogh, and David J. R. Murphy, "What is the Minimum EROI That a Sustainable Society Must Have?" *Energies*, vol. 2, no. 1 (2009), pp. 29, 30.

23. EROI estimates from Ida Kubiszewski and Cutler Cleveland, "Energy Return on Investment (EROI) for Wind Energy," *Encyclopedia of Earth*, June 2007, revised 10 September 2011, at www.eoearth.org.

24. Tom Murphy, "The Energy Trap," *Do the Math*, at physics.ucsd.edu/do-the-math/2011/10/the-energy-trap.

25. Ibid.

26. Ibid.

27. Ibid.

28. Figure 7–1 from World Bank Data, at search.worldbank.org/data?qterm=world%20GDP%20per%20unit%20of%20energy&language=EN, viewed 10 December 2012.

29. Eric Zencey, "The New Austerity and the EROI Squeeze," *The Daly News* (Center for the Advancement of the Steady State Economy), 18 July 2011.

30. A. S. Eddington, *The Nature of the Physical World* (New York: MacMillan Company, 1928), pp. 73–75.

Chapter 8. Renewable Energy's Natural Resource Impacts

1. U.N. Framework Convention on Climate Change, Cancun Agreements, 10 December 2010, Article 1; U.S. Department of Energy (DOE), Energy Information Administration, *International Energy Statistics*, at www.eia.gov/cfapps/ipdbproject/IEDIndex3.cfm, viewed 17 July 2012; Intergovernmental Panel on Climate Change (IPCC), *Summary for Policymakers, Fourth Assessment Report* (Geneva: 2007), pp. 8, 16; A. P. Sokolov et al., *Probabilistic Forecast for 21st Century Climate Based on Uncertainties in Emissions (without Policy) and Climate Parameters* (Cambridge, MA: Massachusetts Institute of Technology, Joint Program on the Science and Policy of Global Change, 2009), p. 1; International Energy Agency (IEA), *World Energy Outlook 2011* (Paris: 2011).

2. National Research Council, *Hidden Costs of Energy: Unpriced Consequences of Energy Production and Use* (Washington, DC: National Academies Press, 2010), pp. 4–5; U.N. Environment Programme, "Mercury Control from Coal Combustion," at www.unep.org/hazardoussubstances; U.S. Environmental Protection Agency (EPA), "Mercury: Health Effects," www.epa.gov/hg/effects.htm; Union of Concerned Scientists, "The Hidden Costs of Fossil Fuels," 29 October 2002, www.ucsusa.org/clean_energy/our-energy-choices.

3. Paul R. Epstein et al., "Full Cost Accounting for the Life Cycle of Coal," in Robert Costanza, Karin Limburg, and Ida Kubiszewski, eds., "Ecological Economics Reviews," *Annals of the New York Academy of Sciences*, no. 1219 (2011), pp. 73–98; Worldwatch Institute, *Roadmap to a Sustainable Electricity System: Harnessing Haiti's Sustainable Energy Resources* (draft) (Washington, DC: forthcoming); Economic Commission for Latin America and the Caribbean, *Centroamerica: Estadisticas de Hidrocarburos* (Santiago, Chile: 2010).

4. DOE, Office of Energy Efficiency and Renewable Energy, "Wind Powering America: Installed Wind Capacity," November 2012, at www.windpoweringamerica.gov; REN21, *Renewables 2012 Global Status Report* (Paris: 2012), p. 101; BP, *BP Statistical Review of World Energy* (London: 2012), p. 108. Box 8–1 based on David L. Chandler, "What Can Make a Dent?" press release (Cambridge, MA: Massachusetts Institute of Technology, 23 October 2011), and on Janet Sawin and William Moomaw, *Renewable Revolution: Low-Carbon Energy by 2030* (Washington, DC: Worldwatch Institute, 2009), p. 18.

5. Table 8–1 from the following: global solar PV, CSP, wind, and small hydro potential and solar PV and wind land needs estimates from M. Jacobson and M. Delucchi, "Providing All Global Energy with Wind, Water, and Solar Power, Part 1: Technologies, Energy Resources, Quantities and Areas of Infrastructure, and Materials," *Energy Policy*, vol. 39 (2011), pp. 1,159–60; solar PV crystalline material requirements from A. Feltrin and A. Freundlich, "Material Considerations for Terawatt Level Development of Photovoltaics," *Renewable Energy*, February 2008, p. 182; solar thin-film PV material requirements and solar PV environmental impacts from V. M. Fthenakis, "Sustainability of Photovoltaics: The Case for Thin-Film Solar Cells," *Renewable and Sustainable Energy Reviews*, December 2009, pp. 2,749–50; CSP land and water requirements from DOE, Office of Energy Efficiency and Renewable Energy, *2010 Solar Technologies Market Report* (Washington, DC: 2011), pp. 54, 77; material requirements for wind from D. R. Wilburn, *Wind Energy in the United States and Materials Required for the Land-Based Wind Turbine Industry from 2010 Through 2030*, Scientific Investigations Report (Washington, DC: U.S. Geological Survey (USGS), 2011), p. 15; small-hydro environmental impacts from T. Abbasi and S. A. Abbasi, "Small Hydro and the Environmental Implications of Its Extensive Utilization," *Renewable and Sustainable Energy Reviews*, May 2011, pp. 2,139–40; geothermal potential estimate calculated based on data from International Geothermal Association, *Contribution of Geothermal Energy to the Sustainable Development*, Report to the U.N. Commission on Sustainable Development

(Bochum, Germany: 2001); geothermal land needs from DOE, Office of Energy Efficiency and Renewable Energy, "Geothermal Power Plants—Minimizing Land Use and Impact," at www1.eere.energy.gov/geothermal, January 2006; geothermal water needs from C. E. Clark et al., *Water Use in the Development and Operation of Geothermal Power Plants* (Lemont, IL: Argonne National Laboratory, DOE, 2010), p. 1; geothermal environmental impacts from D. Giardini, "Geothermal Quake Risks Must Be Faced," *Nature*, 17 December 2009, pp. 848–49; wave and tidal information from O. Langhamer et al., "Wave Power—Sustainable Energy or Environmentally Costly?" *Renewable and Sustainable Energy Reviews*, May 2010, pp. 1,330–33; biomass potential estimate calculated based on data from S. Ladanai and J. Vinterbäck, *Global Potential of Sustainable Biomass for Energy* (Uppsala: Swedish University of Agricultural Sciences, Department of Energy and Technology, 2009).

6. International Renewable Energy Agency, *Solar Photovoltaics, Renewable Energy Technologies: Cost Analysis Series, Vol. 1: Power Sector*, Issue 4/5, June 2012, pp. 15, 28; Fthenakis, op. cit. note 5, p. 2,747; O. Edenhofer et al., *Special Report on Renewable Energy Sources and Climate Change* (Geneva: IPCC, 2011), p. 71.

7. Jacobson and Delucchi, op. cit. note 5, p. 1,159.

8. DOE, *2010 Solar Technologies Market Report*, op. cit. note 5, p. 54.

9. Ibid., pp. 54, 77; Jacobson and Delucchi, op. cit. note 5, p. 1,160; IPCC, *Climate Change 2007: Mitigation, Contribution of Working Group III to the Fourth Assessment Report of the Intergovernmental Panel on Climate Change* (Cambridge, U.K.: Cambridge University Press, 2007), Chapter 8; U.N. Food and Agriculture Organization, *The State of the World's Land and Water Resources for Food and Agriculture* (Rome: 2011); David Beillo, "Gigalopolises: Urban Land Area May Triple by 2030," *Scientific American*, 18 September 2012; Karen C. Seto, Burak Güneralp, and Lucy R. Hutyra, "Global Forecasts of Urban Expansion to 2030 and Direct Impacts on Biodiversity and Carbon Pools," *Proceedings of the National Academy of Sciences*, 2 October 2012.

10. DOE, *2010 Solar Technologies Market Report*, op. cit. note 5, p. 77; Jacobson and Delucchi, op. cit. note 5, p. 1,160.

11. Alexander Ochs et al., *Implications of a Low-Carbon Energy Transition for U.S. National Security* (Washington, DC: Worldwatch Institute, 2010), p. 6; Feltrin and Freundlich, op. cit. note 5, pp. 182, 184.

12. Fthenakis, op. cit. note 6; Jacobson and Delucchi, op. cit. note 5, p. 1,162.

13. Fthenakis, op. cit. note 6.

14. Ochs et al., op. cit. note 11, p. 7.

15. REN21, op. cit. note 4, p. 22; BP, op. cit. note 4, p. 57.

16. "Onshore Wind Energy to Reach Parity with Fossil-fuel Electricity by 2016," *Bloomberg New Energy Finance*, 10 November 2011.

17. Jacobson and Delucchi, op. cit. note 5, pp. 1,159–61; Ochs et al., op. cit. note 11, p. 11.

18. Ochs et al., op. cit. note 11, p. 4; Jacobson and Delucchi, op. cit. note 5, p. 1,161; Wilburn, op. cit. note 5, pp. 14–15.

19. Wilburn, op. cit. note 5, p. 15; Gareth P. Hatch, "Going Green: The Growing Role of Permanent Magnets in Renewable Energy Production and Environmental Protection," presentation to the Magnetics 2008 Conference, May 2008, quoted in Ochs et al., op. cit. note 11, p. 7; Jacobson and Delucchi, op. cit. note 5, p. 1,161.

20. Jacobson and Delucchi, op. cit. note 5, p. 1,162.

21. REN21, op. cit. note 4, p. 22; BP, op. cit. note 4, p. 23; Abbasi and Abbasi, op. cit. note 5, p. 2,136; EPA, "Clean Energy: Hydroelectricity," 28 December 2007, at www.epa.gov/cleanenergy/energy-and-you/affect/hydro.html; Jacobson and Delucchi, op. cit. note 5, p. 1,159.

22. Abbasi and Abbasi, op. cit. note 5, pp. 2,139–40.

23. IEA, *Technology Roadmap: Geothermal Heat and Power* (Paris: 2011), p. 6.

24. Edenhofer et al., op. cit. note 6, p. 405; DOE, Office of Energy Efficiency and Renewable Energy, "Geothermal Technologies Program: Electricity Generation," at www1.eere.energy.gov/geothermal/powerplants.html.

25. REN21, op. cit. note 4, p. 22; BP, op. cit. note 4, p. 40; Nicaraguan Energy Institute, *Generación Bruta por Tipo de Planta, Sistema Eléctrico Nacional, Año 2012,* August 2012.

26. DOE, Office of Energy Efficiency and Renewable Energy, "What is an Enhanced Geothermal System?" September 2012, at www1.eere.energy.gov/geothermal/pdfs/egs_basics.pdf; "A Googol of Heat Beneath Our Feet," Google.org, 2011; Giardini, op. cit. note 5.

27. Clark et al., op. cit. note 5, pp. 1–2.

28. Bruce D. Green and R. Gerald Nix, *Geothermal—The Energy Under Our Feet* (Golden, CO: National Renewable Energy Laboratory, 2006).

29. Jacobson and Delucchi, op. cit. note 5, p. 1,157; Carbon Trust, "Accelerating Marine Energy," July 2011, at www.carbontrust.com/media, pp. 13–14.

30. Jacobson and Delucchi, op. cit. note 5, p. 1,159; Langhamer et al., op. cit. note 5, p. 1,330.

31. Langhamer et al., op. cit. note 5, pp. 1,331–33.

32. M. F. Demirbas et al., "Potential Contribution of Biomass to the Sustainable Energy Development," *Energy Conservation and Management,* July 2009, pp. 1,746–60; A. Evans et al., "Sustainability Considerations for Electricity Generation from Biomass," *Renewable and Sustainable Energy Reviews,* June 2010, p. 1,422.

33. V. Dornburg et al., "Bioenergy Revisited: Key Factors in Global Potentials of Bioenergy," *Energy and Environmental Science,* issue 3 (2010), pp. 258–67.

34. Table 8–2 based on the following: technology status for all technologies from IEA and International Renewable Energy Agency, "Electricity Storage Technology Brief," Paris, April 2012; lead acid battery material needs and environmental impacts from EPA, "Common Wastes & Materials: Batteries," at www.epa.gov/osw/conserve/materials; nickel-cadmium battery material needs and environmental impacts from European Parliament and Council Directive on Batteries and Accumulators and Waste Batteries and Accumulators, Brussels, 6 September 2006; lithium ion battery material needs from P. Gruber et al., "Global Lithium Availability," *Journal of Industrial Ecology,* October 2011; molten salt thermal storage and high voltage direct current transmission line material needs from A. Garcia-Olivares et al., "A Global Renewable Energy Mix with Proven Technologies and Common Materials," *Energy Policy,* February 2012, p. 566; hydrogen environmental impacts from DOE, *Energy Demands on Water Resources* (Washington DC: 2006), p. 61; high-voltage direct current transmission line and high-temperature superconducting cable environmental impacts from P. Bujis et al., "Transmission Investment Problems in Europe: Going beyond Standard Solutions," *Energy Policy,* March 2011, p. 5; high-temperature superconducting cable material needs from USGS, *Mineral Commodity Summaries 2012* (Washington, DC: 2012), p. 185.

35. IEA and International Renewable Energy Agency, op. cit. note 34, p. 4.

36. EPA, op. cit. note 34; European Parliament, op. cit. note 34.

37. IEA and International Renewable Energy Agency, op. cit. note 34, p. 15; Garcia-Olivares et al., op. cit. note 34, p. 567; USGS, *Mineral Commodity Summaries 2009* (Washington, DC: 2009), p. 95; USGS, op. cit. note 34, p. 95; Gruber et al., op. cit. note 34.

38. IEA and International Renewable Energy Agency, op. cit. note 34, p. 15.

39. Ibid.

40. David Biello, "How to Use Solar Energy at Night," *Scientific American,* 18 February 2009.

41. Garcia-Olivares et al., op. cit. note 34, p. 565.

42. DOE, Office of Energy Efficiency and Renewable Energy, "Hydrogen Storage," January 2011, at www1.eere.energy.gov/hydrogenandfuelcells/pdfs/fct_h2_storage.pdf.

43. DOE, op. cit. note 34, p. 61.

44. Bujis et al., op. cit. note 34, p. 5.

45. Garcia-Olivares et al., op. cit. note 34, p. 566.

46. Ochs et al., op. cit. note 11, p. 12; USGS, op. cit. note 34, p. 185.

47. Box 8–2 based on the following: U.S. Department of the Interior, Bureau of Land Management, "Programmatic EIS for Geothermal Resources: Frequently Asked Questions," at www.blm.gov/wo/st/en/prog/energy/geothermal /geothermal_nationwide.html; Doug Boucher et al., *The Root of the Problem: What's Driving Tropical Deforestation Today?* (Cambridge, MA: Union of Concerned Scientists, 2011), p. 51; World Commission on Dams, *Dams and Development: A New Framework for Decision-Making* (London: Earthscan, 2000), p. 102.

48. United Nations, International Decade for Action 2005–2015, "Water for Life," at www.un.org/waterforlifede cade/scarcity.shtml.

49. "Situation and Policies of China's Rare Earth Industry," *Xinhua News*, 20 June 2012.

Chapter 9. Conserving Nonrenewable Resources

1. Elisa Alonso et al., "Evaluating Rare Earth Element Availability: A Case with Revolutionary Demand from Clean Technologies," *Environmental Science and Technology*, vol. 46, no. 6 (2012), pp. 3,406–14; "China Cuts Rare-Earths Mine Permits 41% to Boost Control," *Bloomberg News*, 14 September 2012.

2. I am indebted to Christopher Clugston for this framing and for the analytical framework in his book *Scarcity*, which informed the structure of this chapter; D. Giurco et al., *Peak Minerals in Australia: A Review of Changing Impacts and Benefits* (Broadway, Australia: Institute for Sustainable Futures, 2010).

3. Elisabeth Rosenthal, "Race is On as Ice Melt Reveals Arctic Treasures," *New York Times*, 18 September 2012.

4. Share in United States from Lorie A. Wagner, Daniel E. Sullivan, and John L. Sznopek, *Economic Drivers of Mineral Supply*, Open-File Report 02-335 (Reston, VA: U.S. Geological Survey (USGS), 2002); share in China from Heming Wang et al., "Resource Use in Growing China: Past Trends, Influence Factors, and Future Demand," *Journal of Industrial Ecology*, August 2012, pp. 481–92.

5. Figure 9–1 from Thomas D. Kelly and Grecia R. Matos, *Historical Statistics for Mineral and Material Commodities in the United States*, Data Series 140 (Reston, VA: USGS, 2011). The Figure covers data for 85 metals and other nonrenewable materials.

6. Yuval Atsom et al., "Winning the $30 Trillion Decathlon: Going for Gold in Emerging Markets," *McKinsey Quarterly*, August 2012, p. 4.

7. U.N. Environment Programme (UNEP), *Recycling Rates of Metals: A Status Report* (Paris: 2011).

8. Worldwatch calculation based on data in Kelly and Matos, op. cit. note 5; Richard Dobbs, Jeremy Oppenheim, and Fraser Thompson, "Mobilizing for a Resource Revolution," *McKinsey Quarterly*, January 2012.

9. Jeremy Grantham, "Time to Wake Up: Days of Abundant Resources and Falling Prices Are Over Forever," *GMO Quarterly Letter*, April 2011.

10. Ibid.; Richard Dobbs, Jeremy Oppenheim, and Fraser Thompson, "A New Era for Commodities," *McKinsey Quarterly*, November 2011; Suwin Sandu and Arif Syed, *Trends in Energy Intensity in Australian Industry* (Canberra: Australian Bureau of Agricultural and Resource Economics, 2008).

11. Figure 9–2 from Gavin M. Mudd, e-mail to author, 11 September 2012; Canada and Russia from Gavin M. Mudd, "Global Trends and Environmental Issues in Nickel Mining: Sulfides versus Laterites," *Ore Geology Reviews*, October 2010, pp. 9–26; Gavin M. Mudd, "The Environmental Sustainability of Mining in Australia: Key Megatrends and Looming Constraints," *Resources Policy*, June 2010, pp. 106–07; Gavin M. Mudd, e-mail to author, 7 October 2012; quote is from Mudd, "Environmental Sustainability of Mining," op. cit. this note, p. 107.

12. Table 9–1 from Gavin Mudd, "Sustainability Reporting and Water Resources: A Preliminary Assessment of Embodied Water and Sustainable Mining," *Australian Journal of Mining*, August 2009.

13. Mudd, "Environmental Sustainability of Mining," op. cit. note 11, pp. 113–14.

14. Gavin M. Mudd, "Uranium," in Trevor M. Letcher and Janet L. Scott, *Materials for a Sustainable Future* (London: Royal Society of Chemistry, 2012), pp. 201–03; Mudd, "Environmental Sustainability of Mining," op. cit. note 11, p. 110.

15. Andre Dierderen, *Global Resource Depletion: Managed Austerity and the Elements of Hope* (Delft: Eburon Academic Publishers, 2010), p. 53.

16. Sandu and Syed, op. cit. note 10.

17. Gavin M. Mudd, Zhehan Weng, and Simon M. Jowitt, " A Detailed Assessment of Global Cu Resource Trends and Endowments," *Economic Geology*, forthcoming.

18. Cutler J. Cleveland, "Net Energy from Extraction of Oil and Gas in the United States," *Energy*, April 2005.

19. Carey W. King and Charles A. S. Hall, "Relating Financial and Energy Return on Investment," *Sustainability*, vol. 3, no. 10 (2011), pp. 1,810–32.

20. Dobbs, Oppenheim, and Thompson, op. cit. note 8.

21. Ernst von Weizsäcker, *Factor Five: Transforming the Global Economy Through 80% Improvements in Resource Productivity* (London: Earthscan, 2009).

22. Figure of $600 million is a 2009 estimate from Global Subsidies Initiative, at www.iisd.org/gsi/fossil-fuel-subsidies/fossil-fuels-what-cost; $775 billion to $1 trillion from Alexander Ochs, Eric Anderson, and Reese Rogers, "Fossil Fuel and Renewable Energy Subsidies on the Rise," *Vital Signs Online*, 21 August 2012; European Commission, *Roadmap to a Resource Efficient Europe* (Brussels: 2011); Kerryn Lang, *The First Year of the G-20 Commitment on Fossil-Fuel Subsidies: A Commentary on Lessons Learned and the Path Forward* (Geneva: Global Studies Initiative, International Institute for Sustainable Development, 2011).

23. Box 9–2 from the following: USGS, "Metal Stocks in Use in the United States," Fact Sheet 2050-3090 (Reston, VA: July 2005); Ben Schiller, "Trash to Cash: Mining Landfills for Energy and Profit," *Fast Company*, 7 September 2011; Group Machiels, "Enhanced Landfill Mining," at www.machiels.com, viewed 23 September 2012.

24. European Commission, op. cit. note 22; UNEP, *Green Jobs: Towards Decent Work in a Sustainable, Low-carbon World* (Nairobi: 2008).

25. Office of the Mayor, "Mayor Lee Celebrates San Francisco's Composting Achievements," press release (San Francisco: 28 March 2012); United States from U.S. Environmental Protection Agency (EPA), *Municipal Solid Waste Generation, Recycling, and Disposal in the United States: Facts and Figures for 2010* (Washington, DC: 2011); Barbara K. Reck and T. E. Graedel, "Challenges in Metal Recycling," *Science*, 10 August 2012.

26. Table 9–2 from the following: Elliot Martin, Susan A. Shaheen, and Jeffrey Lidi, "Impact of Carsharing on Household Vehicle Holdings: Results from North American Shared-Use Vehicle Survey," *Transportation Research Record*, March 2010; John A. Mathews and Hao Tan, "Progress Toward a Circular Economy in China: The Drivers (and Inhibitors) of Eco-industrial Initiative," *Journal of Industrial Ecology*, June 2011, pp. 435–57; U.S. Department of Energy and EPA, *Combined Heat and Power: A Clean Energy Solution* (Washington, DC: August 2012); Eric S. Belsky, "Planning for Inclusive and Sustainable Urban Development," in Worldwatch Institute, *State of the World 2012* (Washington, DC: Island Press, 2012), p. 45; "Neighborhood Tool Libraries in Portland Oregon," at www .neighborhoodnotes.com, viewed 23 September 2012; Osamu Kimura, *Japanese Top Runner Approach for Energy Efficiency Standards* (Tokyo: Socio-economic Research Center, Central Research Institute of Electric Power Industry, 2010).

27. Jeremy Grantham, "Welcome to Dystopia! Entering a Long-term and Politically Dangerous Food Crisis," *GMO Quarterly Letter*, July 2012.

Chapter 10. Reengineering Cultures to Create a Sustainable Civilization

1.   Erik Assadourian, "The Rise and Fall of Consumer Cultures," in Worldwatch Institute, *State of the World 2010* (New York: W. W. Norton & Company, 2010), pp. 3–20.

2.   Population Division, *World Population Prospects: The 2010 Revision* (New York: United Nations, 2011).

3.   Monika Dittrich et al., *Green Economies Around the World?* (Vienna: Sustainable Europe Research Institute, 2012); WWF et al., *Living Planet Report 2012* (Gland, Switzerland: WWF, 2012).

4.   Dittrich et al., op. cit. note 3.

5. Overweight Americans from Trust for America's Health, *F as in Fat: How Obesity Policies Are Failing in America* (Washington, DC: Robert Wood Johnson Foundation, 2008); medical and productivity costs from Society of Actuaries, "New Society of Actuaries Study Estimates $300 Billion Economic Cost Due to Overweight and Obesity," press release (Schaumburg, IL: 10 January 2011); Institute for Health Metrics and Evaluation, "Life Expectancy in Most US Counties Falls Behind World's Healthiest Nations," press release (Seattle, WA: 15 June 2011); David Brown, "Life Expectancy in the U.S. Varies Widely by Region, in Some Places Is Decreasing," *Washington Post*, 15 June 2011; S. Jay Olshansky et al., "A Potential Decline in Life Expectancy in the United States in the 21st Century," *New England Journal of Medicine*, 17 March 2005, pp. 1,138–45.

6. Global obesity from Richard Weil, "Levels of Overweight on the Rise," *Vital Signs Online*, 14 June 2011; Sarah Catherine Walpole et al., "The Weight of Nations: An Estimation of Adult Human Biomass," *BMC Public Health*, vol. 12 (2012), pp. 439–45; other ills from Erik Assadourian, "The Path to Degrowth in Overdeveloped Countries," in Worldwatch Institute, *State of the World 2012* (Washington, DC: Island Press, 2012), pp. 22–37.

7. Millennium Ecosystem Assessment, *Ecosystems and Human Well-being: Synthesis* (Washington, DC: Island Press, 2005); WWF et al., op. cit. note 3.

8. Assadourian, op. cit. note 1, pp. 11–16.

9. Ibid., p. 9.

10. Advertising from Jonathan Barnard, "ZenithOptimedia Releases September 2012 Advertising Expenditure Forecasts," press release (London: ZenithOptimedia, 1 October 2012); other expenditures from Assadourian, op. cit. note 1, pp. 13–14; new consumers from McKinsey Global Institute, *Urban World: Cities and the Rise of the Consuming Class* (McKinsey & Company, June 2012).

11. Kevin Anderson and Alice Bows, "Beyond 'Dangerous' Climate Change: Emission Scenarios for a New World," *Philosophical Transactions of the Royal Society A*, January 2011, pp. 20–44; Potsdam Institute for Climate Impact Research and Climate Analytics, *Turn Down the Heat: Why a 4°C Warmer World Must Be Avoided* (Washington, DC: World Bank, 2012); Mark G. New et al., "Four Degrees and Beyond: The Potential for a Global Temperature Increase of Four Degrees and Its Implications," *Philosophical Transactions of the Royal Society A*, January 2011, pp. 6–19 ; Joe Romm, "Royal Society Special Issue Details 'Hellish Vision' of 7°F (4°C) World—Which We May Face in the 2060s!" *Climate Progress*, 29 November 2010.

12. DARA International, *Climate Vulnerability Monitor: A Guide to the Cold Calculus of a Hot Planet*, 2nd ed. (Washington, DC: 2012). Box 10–1 from the following: Roland Stulz and Tanja Lütolf, "What Would Be the Realities of Implementing the 2,000 Watt Society in Our Communities?" presentation, Novatlantis, 23–24 November 2006; Saul Griffith, "Climate Change Recalculated," presentation at The Long Now Foundation, San Francisco, 16 January 2009; Danielle Nierenberg and Laura Reynolds, "Disease and Drought Curb Meat Production and Consumption," *Vital Signs Online*, 23 October 2012; Assadourian, op. cit. note 1; Juliet Schor, *Plenitude: The New Economics of True Wealth* (New York: Penguin Press, 2010).

13. Peter N. Stearns, *Consumerism in World History: The Global Transformation of Desire* (New York: Routledge, 2001), pp. 34–35; "Not Such a Bright Idea: Making Lighting More Efficient Could Increase Energy Use, Not Decrease It," *The Economist*, 26 August 2010; Lizabeth Cohen, *A Consumer's Republic: The Politics of Mass Consumption in Postwar America* (New York: Alfred A. Knopf, 2003).

14. Peter D. Norton, *Fighting Traffic: The Dawn of the Motor Age in the American City* (Cambridge, MA: The MIT Press, 2008); Peter Dauvergne, *The Shadows of Consumption: Consequences for the Global Environment* (Cambridge, MA: The MIT Press, 2008); automobile advertising from Stephen Williams, "Report Predicts Auto-Ad Spending Will Grow 14% This Year," *Advertising Age*, 30 April 2012; Michael Renner, "Auto Production Roars to New Records," *Vital Signs Online*, 11 September 2012.

15. McDonald's, "Company Profile," at www.aboutmcdonalds.com/mcd/investors/company_profile.html; Eric Schlosser, *Fast Food Nation* (New York: Harper Perennial Company, 2005), pp. 197–98.

16. Schlosser, op. cit. note 15; advertising from Keith O'Brien, "How McDonald's Came Back Bigger Than Ever," *New York Times*, 6 May 2012.

17. Lydia Polgreen, "Matchmaking in India: Canine Division," *New York Times*, 17 August 2009; David Lummis,

Packaged Facts Pet Analyst, *U.S. Pet Market Outlook 2009–2010: Surviving and Thriving in Challenging Times*, PowerPoint presentation; Packaged Facts, *Pet Supplies in the U.S.*, 7th ed. (Rockville, MD: August 2007), pp. 141–43.

18. Pet food from Transparency Market Research, "Global Pet Food Market is Forecasted to Reach USD 74.8 Billion by 2017," press release (Albany, NY: 10 August 2012); supplies and vet care from American Pet Products Association, "Industry Statistics & Trends," at www.americanpetproducts.org/press_industrytrends.asp, viewed 18 November 2012, and from William Grimes, "New Treatments to Save a Pet, but Questions About the Costs," *New York Times*, 5 April 2012; Packaged Facts, "Cat Litter a Nearly $2 Billion Market in the U.S.," *Pets International*, Issue 4/2010; total populations and impacts from Robert Vale and Brenda Vale, *Time to Eat the Dog: The Real Guide to Sustainable Living* (London: Thames & Hudson, 2009), pp. 235–38.

19. David W. Chen, "Shanghai Journal; A New Policy of Containment, for Baby Bottoms," *New York Times*, 5 August 2003; William Foreman, "Pork-flavored Doughnuts? A Chinese Market Beckons," *Associated Press*, 13 February 2010; calculation based on Barnard, op. cit. note 10; Assadourian, op. cit. note 1.

20. Assadourian, op. cit. note 1.

21. Adam Aston, "Patagonia Takes Fashion Week as a Time to Say: 'Buy Less, Buy Used,'" *GreenBiz*, 8 September 2011; Tim Nudd, "Ad of the Day: Patagonia," *Ad Week*, 28 November 2011.

22. Johanna Mair and Kate Ganly, "Social Entrepreneurs: Innovating Toward Sustainability," in Worldwatch Institute, op. cit. note 1, pp. 103–09.

23. Number of B Corporations from B Corps website, at www.bcorporation.net, viewed 10 November 2012; Colleen Cordes, "The Earth-Friendly Corporation: Campaigning Opportunities and Caveats for the Environmental Community," White Paper, September 2012, unpublished; $4.2 billion from Heather Carpenter, "A Scoop of Social Responsibility: Ben and Jerry's the B Corp," *Nonprofit Quarterly*, 29 October 2012.

24. George Lerner, "New York Health Board Approves Ban on Large Sodas," *CNN*, 14 September 2012; Neal Riley, "Expanded Plastic Bag Ban Takes Effect Monday," *SFGate*, 29 September 2012; Michael Maniates, "Editing Out Unsustainable Behavior," in Worldwatch Institute, op. cit. note 1, pp. 119–26; Erik Assadourian, "The Mallport and the Bibliometro" (blog), *Transforming Cultures*, 30 March 2010. Box 10–2 based on the following: Michael Grynbuam, "In Soda Fight, Industry Focuses on the Long Run," *New York Times*, 12 September 2012; Larry Gordon, "All You Can Carry: College Cafeterias Go Trayless," *Los Angeles Times*, 14 September 2009; Nate Berg, "The Math Behind Sacking Disposable Bags," *Atlantic Cities Place Matters*, 26 September 2011; "Albert Lea, MN—Blue Zones Pilot Project," Blue Zones website, at www.bluezones.com; Nancy Perry Graham, "Creating America's Healthiest Hometown," *AARP The Magazine*, September/October 2012.

25. Cormac Cullinan, "Earth Jurisprudence: From Colonization to Participation," in Worldwatch Institute, op. cit. note 1, pp. 143–48; Geoff Olson, "Bolivia's Law of Mother Earth," *Common Ground*, July 2011.

26. Ecovillages from Erik Assadourian, "Engaging Communities for a Sustainable World," in Worldwatch Institute, *State of the World 2008* (New York: W. W. Norton & Company, 2008), p. 154; transition towns from Assadourian, op. cit. note 6, p. 34.

27. Trine S. Jensen et al., *From Consumer Kids to Sustainable Childhood* (Copenhagen: Worldwatch Institute Europe, 2012), p. 53; Rome from Kevin Morgan and Roberta Sonnino, "Rethinking School Food: The Power of the Public Plate," in Worldwatch Institute, op. cit. note 1, pp. 69–74.

28. Gary Gardner, "Engaging Religions to Shape Worldviews," in Worldwatch Institute, op. cit. note 1, pp. 23–29; Gary Gardner, "Ritual and Taboo as Ecological Guardians," in Worldwatch Institute, op. cit. note 1, pp. 30–35; *shemitah* from Nina Beth Cardin, Baltimore Jewish Environmental Network, discussion with author, 16 October 2012.

29. Embalming fluid and concrete from Dave Reay, *Climate Change Begins at Home* (London: Macmillan, 2005), p. 147; steel, wood, and cost from Mark Harris, *Grave Matters: A Journey through the Modern Funeral Industry to a Natural Way of Burial* (New York: Scribner, 2007), pp. 10, 34; green burial from Joe Sehee, The Green Burial Council, presentation, 2010.

30. "Avatar," and "Memorable Quotes for Avatar," *IMDb.com*; "Crude," *IMDb.com*.

31. Bhopal from "Yes Men Hoax on BBC Reminds World of Dow Chemical's Refusal to Take Responsibility for Bhopal Disaster," *Democracy Now*, 6 December 2004; Andy Bichlbaum, "Chevron Ad Campaign Derailed" (blog), *The Yes Men*, 19 October 2010; "Chevron's $80 Million Ad Campaign Gets Flushed" (blog), *The Yes Men*, 19 October 2010.

32. Box 10–3 from Wolfgang Sachs, ed., *The Development Dictionary* (London: Zed Books, 2010).

33. Tompkins and Chouinard quotes from *180° South*, Magnolia Pictures, 2010.

Chapter 11. Building a Sustainable and Desirable Economy-in-Society-in-Nature

1. This chapter is adapted from a report commissioned by the United Nations for the 2012 Rio+20 Conference as part of the Sustainable Development in the 21st century project; see R. Costanza et al., *Building a Sustainable and Desirable Economy-in-Society-in-Nature* (New York: United Nations Division for Sustainable Development, 2012). Table 11–1 from R. Costanza et al., "The Value of the World's Ecosystem Services and Natural Capital," *Nature*, 15 May 1997, pp. 253–60

2. New research from T. Kasser, *The High Price of Materialism* (Cambridge, MA: The MIT Press, 2002).

3. R. A. Easterlin, "Explaining Happiness," *Proceedings of the National Academy of Sciences*, 16 September 2003, pp. 11,176–83; R. Layard, *Happiness: Lessons from a New Science* (New York: Penguin Press, 2005).

4. Costanza et al., "Value of the World's Ecosystem Services and Natural Capital," op. cit. note 1; R. Costanza, *Ecological Economics: The Science and Management of Sustainability* (New York: Columbia University Press, 1991); H. E. Daly and J. Farley, *Ecological Economics: Principles and Applications* (Washington, DC: Island Press, 2004).

5. Easterlin, op. cit. note 3; Layard, op. cit. note 3.

6. Figure 11–1 from R. Hernández-Murillo and C. J. Martinek, "The Dismal Science Tackles Happiness Data," *The Regional Economist*, January 2010, pp. 14–15.

7. R. Costanza et al., *Beyond GDP: The Need for New Measures of Progress* (Boston, MA: The Pardee Papers, 2009); P. A. Lawn, "A Theoretical Foundation to Support the Index of Sustainable Economic Welfare (ISEW), Genuine Progress Indicator (GPI), and Other Related Indexes," *Ecological Economics*, February 2003, pp. 105–18.

8. Figure 11–2 from J. Talberth, C. Cobb, and N. Slattery, *The Genuine Progress Indicator 2006: A Tool for Sustainable Development* (Oakland, CA: Redefining Progress, 2007).

9. K. Raworth, *A Safe and Just Space for Humanity: Can We Live within the Doughnut?* (Oxford: Oxfam International, 2012).

10. Costanza et al., *Building a Sustainable and Desirable Economy-in-Society-in-Nature*, op. cit. note 1; Great Transition initiative, at www.gtinitiative.org; The Future We Want, at www.futurewewant.org.

11. R. Costanza et al., "Principles for Sustainable Governance of the Oceans," *Science*, 10 July 1998, pp. 198–99.

12. R. Beddoe et al., "Overcoming Systemic Roadblocks to Sustainability: The Evolutionary Redesign of Worldviews, Institutions, and Technologies," *Proceedings of the National Academy of Sciences*, 24 February 2009, pp. 2,483–89.

13. R. Costanza, W. J. Mitsch, and J. W. Day, Jr., "A New Vision for New Orleans and the Mississippi Delta: Applying Ecological Economics and Ecological Engineering," *Frontiers in Ecology and the Environment*, November 2006, pp. 465–72; Intergovernmental Panel on Climate Change, *Fourth Assessment Report of the Intergovernmental Panel on Climate Change* (Cambridge, U.K.: Cambridge University Press, 2007).

14. J. B. Schor, "Sustainable Consumption and Worktime Reduction," *Journal of Industrial Ecology*, January 2005, pp. 37–50; A. Durning, *How Much Is Enough?* (New York: W. W. Norton & Company, 1992); T. Jackson, *Prosperity without Growth: Economics for a Finite Planet* (London: Earthscan/James & James, 2009).

15. D. Acemoglu and J. Robinson, "Foundations of Societal Inequality," *Science*, 30 October 2009, pp. 678–79; Jackson, op. cit. note 14.

16. H. E. Daly, "From a Failed-Growth Economy to a Steady-State Economy," *Solutions*, February 2010, pp. 37–43.

17. Ibid.; studies on giving up personal gain from I. Almås et al., "Fairness and the Development of Inequality Acceptance," *Science*, 28 May 2010, pp. 1,176–78, and from E. Fehr and A. Falk, "Psychological Foundations of Incentives," *European Economic Review*, vol. 46 (2002), pp. 687–724; Jackson, op. cit. note 14.

18. G. Hardin, "The Tragedy of the Commons," *Science*, 13 December 1968, pp. 1,243–48; E. Ostrom, *Governing the Commons: The Evolution of Institutions for Collective Action* (Cambridge, U.K.: Cambridge University Press, 1990); D. Pell, in F. Berkes, ed., *Common Property Resources: Ecology and Community-Based Sustainable Development* (London: Belhaven Press, 1989); D. Feeny et al., "The Tragedy of the Commons: Twenty-two Years Later," *Human Ecology*, vol. 18, no. 1 (1990), pp. 1–19.

19. J. Farley and R. Costanza, "Envisioning Shared Goals for Humanity: A Detailed, Shared Vision of a Sustainable and Desirable USA in 2100," *Ecological Economics*, vol. 43, no. 2–3 (2002), pp. 245–59; T. Prugh, R. Costanza, and H. E. Daly, *The Local Politics of Global Sustainability* (Washington, DC: Island Press, 2000).

20. Box 11–1 is adapted from James Gustave Speth, *America the Possible: Manifesto for a New Economy* (New Haven, CT: Yale University Press, 2012) and is based on the following: banks' control of deposits and assets from David Korten, *How to Liberate America from Wall Street Rule* (Washington, DC: New Economy Working Group, July 2011); Thomas H. Greco, Jr., *The End of Money and the Future of Civilization* (White River Junction, VT: Chelsea Green, 2009), p. 35; Mary Mellor, "Could the Money System Be the Basis of a Sufficiency Economy?" *Real World Economics Review*, no. 54 (2010), p. 79; Otto Scharmer, "Seven Acupuncture Points for Shifting Capitalism to Create a Regenerative Ecosystem Economy," Roundtable on Transforming Capitalism to Create a Regenerative Economy, MIT, Cambridge, MA, 8–9 June and 21 September 2009, p. 19; Herman E. Daly, "Moving from a Failed Growth Economy to a Steady State Economy," unpublished manuscript, forthcoming in volume from Palgrave Publishers; Daly, op. cit. note 16, p. 37.

21. Total debt from "Z.1 Statistical Release," Board of Governors of the Federal Reserve System, at www.federalreserve.gov/datadownload/Download.aspx?rel=Z1&series=654245a7abac051cc4a9060c911e1fa4&filetype=csv&label=include&layout=seriescolumn&from=01/01/1945&to=12/31/2010.

22. Daly, op. cit. note 16; H. E. Daly, *Ecological Economics and Sustainable Development, Selected Essays of Herman Daly* (Northampton, MA: Edward Elgar Publishing, 2008).

23. M. Gaffney, "The Hidden Taxable Capacity of Land: Enough and to Spare," *International Journal of Social Economics*, vol. 36, no. 4 (2009), pp. 328–411.

24. Figure 11–3 from R. G. Wilkinson and K. Pickett, *The Spirit Level: Why Greater Equality Makes Societies Stronger* (New York: Bloomsbury Press, 2009); data for Figure 11–4 from Organisation for Economic Co-operation and Development and from Wilkinson and Pickett, op. cit. this note; Paulson from M. Goldstein, "Paulson, at $4.9 Billion, Tops Hedge Fund Earner List," *Reuters*, 1 April 2011.

25. Jackson, op. cit. note 14.

26. R. Costanza et al., "Sustainability or Collapse: What Can We Learn from Integrating the History of Humans and the Rest of Nature?" *Ambio*, November 2007, pp. 522–27; J. Diamond, *Guns, Germs, and Steel: The Fates of Human Societies* (New York: W. W. Norton & Company, 2005); H. Weiss and R. S. Bradley, "What Drives Societal Collapse?" *Science*, 26 January 2001, pp. 609–10.

27. See, for example, C. Rolfsdotter-Jansson, "Malmo, Sweden," *Solutions*, January 2010, pp. 65–68, and S. M. Kristinsdottir, "Energy Solutions in Iceland," *Solutions*, May 2010, pp. 52–55.

28. D. H. Meadows et al., *The Limits to Growth* (New York: Universe Books, 1972); R. Boumans et al., "Modeling the Dynamics of the Integrated Earth System and the Value of Global Ecosystem Services Using the GUMBO Model," *Ecological Economics*, June 2002, pp. 529–60.

29. P. A. Victor and G. Rosenbluth, "Managing without Growth," *Ecological Economics*, March 2007, pp. 492–504; P. A. Victor, *Managing without Growth: Slower by Design, Not Disaster* (Northampton, MA: Edward Elgar Publishing, 2008).

30. Figure 11–5 from Victor, op. cit. note 29.

Chapter 12. Transforming the Corporation into a Driver of Sustainability

1. Johan Rockström et al., "Planetary Boundaries: Exploring the Safe Operating Space for Humanity," *Ecology and Society*, vol. 14, no. 2 (2009).

2. Private sector as percent of global gross domestic product based on 2010 global data, per IHS, "Country & Industry Forecasting: IHS Global Insight," at www.ihs.com/products/global-insight/index.aspx; Messaoud Hammouya, *Statistics on Public Sector Employment: Methodology, Structures, and Trends* (Geneva: Bureau of Statistics, International Labour Office, 1999).

3. For corporate externalities, see Principles for Responsible Investment (PRI) and U.N. Environment Programme (UNEP) Finance Initiative, *Universal Ownership: Why Environmental Externalities Matter to Institutional Investors* (London and Geneva: 2010).

4. Subsidies for fossil-fuel use include $550 billion in price subsidies and $100 billion in production subsidies, as reported in UNEP, *Towards a Green Economy* (Nairobi: 2011); see also International Energy Agency (IEA), "Analysis of the Scope of Energy Subsidies and Suggestions for the G-20 Initiative," Paris, 16 June 2010. The World Bank estimated subsidies for agriculture at around $273 billion; World Bank, *World Development Report 2008: Agriculture for Development* (Washington, DC: 2007). Subsidies for open-access pelagic fisheries are also significant.

5. World Bank, *World Development Indicators* and *Global Development Finance*, database, at data.worldbank.org; IEA, *2011 Key World Energy Statistics* (Paris: 2011).

6. "Mankind Using Earth's Resources Faster than Replenished," (London) *The Independent*, 25 November 2009.

7. Pavan Sukhdev, *Corporation 2020: Transforming Business for Tomorrow's World* (Washington, DC: Island Press, 2012), Chapter 7.

8. Romesh Sobti, chief executive office, IndusInd Bank Limited, interviewed by Pavan Sukhdev and Rafael Torres, 2011.

9. Donald DePamphilis, *Mergers, Acquisitions, and Other Restructuring Activities: An Integrated Approach to Process, Tools, Cases, and Solutions*, 5th ed. (Waltham, MA: Academic Press, 2009), Chapter 13.

10. Estimates of global advertising turnover differ among company reports. This estimate is from the Center for Media Research, *2012 Ad Spending Outlook* (New York: 2011).

11. See The Bubble Project, at www.thebubbleproject.com.

12. David Evan Harris, "São Paulo: A City without Ads," *Adbusters*, 3 August 2007; Guanaes quote from Vincent Bevins, "São Paulo Advertising Goes Underground," *Financial Times*, 6 September 2010.

13. Bob Garfield, *The Chaos Scenario* (Nashville, TN: Stielstra Publishing, 2009).

14. PRI and UNEP Finance Initiative, op. cit. note 3.

15. Matt Barney and Infosys Technologies Ltd., *Leadership @ Infosys* (New Delhi: Portfolio, 2010); see also Infosys, "Smt. Sonia Gandhi Inaugurates Infosys' Global Education Center–II in Mysore," press release (Mysore, India: 15 September 2009); human capital externality calculation done by Infosys and GIST Advisory, cited in "Human Resource Valuation," in Infosys, *Infosys Annual Report 2011–12* (2012).

16. TEEB for Business Coalition, "Natural Capital at Risk: A Study of the Top 100 Business Impacts," care of Institute of Chartered Accountants in England and Wales, June 2012.

Chapter 13. Corporate Reporting and Externalities

1. Sarah Anderson and John Cavanagh, *Top 200: The Rise of Corporate Global Power* (Washington, DC: Institute for Policy Studies, 2000), p. 1; Luca Errico and Alexander Massara, *Assessing Systemic Trade Interconnectedness—An Empirical Approach* (Washington, DC: International Monetary Fund, 2001), p. 8; The Prince's Accounting for Sustainability Project, at www.accountingforsustainability.org.

2. Kevin Wilhelm, *Return on Sustainability* (Indianapolis, IN: Dog Ear Publishing, 2009); Bob Willard, *The Sustainability Advantage: Seven Business Case Benefits of a Triple Bottom Line* (Gabriola Island, Canada: New Society Publishers, 2002); Robert G. Eccles and Michael P. Krzus, *One Report: Integrated Reporting for a Sustainable Strategy* (New York: Wiley, 2010).

3. Laurence Chandy and Geoffrey Gertz, "With Little Notice, Globalization Reduced Poverty" (blog), *Yale Global Online*, 5 July 2011; Paul Harrison and Fred Pearce, *AAAS Atlas of Population and the Environment* (Berkeley: University of California Press, for American Association for the Advancement of Science, 2011), pp. 43–46; WWF et al., *Living Planet Report 2012* (Gland, Switzerland: WWF, 2012), pp. 8–9; Millennium Ecosystem Assessment, *Living Beyond Our Means: Natural Assets and Human Well-being*, Statement from the Board (Washington, DC: Island Press, 2005), p. 3.

4. J. Hall, V. Brajer, and F. Lurmann, *The Benefits of Meeting Federal Clean Air Standards in the South Coast and San Joaquin Valley Air Basins* (Fullerton: California State University–Fullerton, Institute for Economic and Environmental Studies, 2008).

5. California Department of Toxic Substances Control, "Emerging Chemicals of Concern," at www.dtsc.ca.gov /assessingrisk/emergingcontaminants.cfm.

6. B. E. Erickson, "Bisphenol A under Scrutiny," *Chemical and Engineering News*, 2 June 2008, pp. 36–39; Public Works and Government Services Canada, *Statutory Instruments 2010, Canada Gazette Part II*, 13 October 2010, 144(21):1806–18.

7. Principles for Responsible Investment and U.N. Environment Programme Finance Initiative, *Universal Ownership: Why Environmental Externalities Matter to Institutional Investors* (London and Geneva: 2010), pp. 2–7; Theo Ferguson, "A Fossil Fuel Diet—Taking Action to Get from Here to There," *Green Money Journal*, fall 2011.

8. Securities and Exchange Commission (SEC), "Commission Statement about Management's Discussion and Analysis of Financial Condition and Results of Operations," Washington, DC, 22 January 2002.

9. E. Lynn Grayson and Patricia L. Boye-Williams, "SEC Disclosure Obligations: Increasing Scrutiny on Environmental Liabilities and Climate Change Impacts," in Lawrence P. Schnapf, ed., *Environmental Issues in Business Transactions* (Chicago: American Bar Association, 2011), pp. 447–69.

10. Global Reporting Initiative, *A New Phase: The Growth of Sustainability Reporting: GRI's Year in Review 2010/11* (2011), p. 6.

11. BP, *BP Sustainability Review 2009* (London: 2009), p. 5; BP, *BP Sustainability Review 2010* (London: 2010).

12. International Integrated Reporting Council (IIRC), at www.theiirc.org.

13. IIRC, "Draft Framework Outline," 11 July 2012, pp. 5, 7.

14. IIRC, *Capturing the Experiences of Global Businesses and Investors: The Pilot Programme 2012 Yearbook* (2012), p. 7.

15. Sustainability Accounting Standards Board, at www.sasb.org.

16. Association of Certified Chartered Accountants, Flora and Fauna International, and KPMG, *Is Natural Capital a Material Issue?* (London: 2012), p. 8.

17. "Benefits of Signing the NCD" and "The Declaration," at www.naturalcapitaldeclaration.org

18. IIRC, "Pilot Programme," at www.theiirc.org/companies-and-investors, viewed 12 December 2012.

Chapter 14. Keep Them in the Ground: Ending the Fossil Fuel Era

1. For an illustration of an idealistic energy transition, see "Shift: To A Smarter Energy Future" (documentary), Arcos Films, 2012; for a realistic, historically and empirically grounded view of energy transition, see Vaclav Smil, *Energy Transitions: History, Requirements, Prospects* (Santa Barbara, CA: Praeger, 2010), Chapter 2.

2. Percentage of world energy from Vaclav Smil, "Global Energy: The Latest Infatuations," *American Scientist*, vol. 99 (2011), pp. 212–19; national oil company portion from "Introduction and Overview," in David G. Victor, David R. Hultz, and Mark C. Thurber, eds., *Oil and Governance: State-Owned Enterprises and the World Energy Supply* (Cambridge, U.K.: Cambridge University Press, 2012), pp. 3–31; industry capitalization from Michael L. Ross, *The Oil Curse: How Petroleum Wealth Shapes the Development of Nations* (Princeton, NJ: Princeton University Press, 2012), p. 3; ExxonMobil profits from Steve Coll, *Private Empire: ExxonMobil and American Power* (New York: Penguin Press, 2012) p. 8; 2010 statistics from *Global Trends* at http://www.globaltrends.com/knowledge-center /features/shapers-and-influencers/66-corporate-clout-the-influence-of-the-worlds-largest-100-economic-

entities. Box 14–1 from the following: Smil, op. cit. this note; Smil, op. cit. note 1, p. 117; James Clark, *BP Energy Outlook 2030*, at www.bp.com; International Energy Agency, "How Will Global Energy Markets Evolve to 2035?" *World Energy Outlook 2011 Factsheet*, at www.worldenergyoutlook.org; Roni A. Neff et al., "Peak Oil, Food Systems, and Public Health," *American Journal of Public Health*, September 2011, p. 1,589; Carbon Tracker Initiative, *Unburnable Carbon: Are the World's Financial Markets Carrying a Carbon Bubble?* March 2012, at www.carbon-tracker.org; Paul R. Epstein and Jesse Selber, eds., *Oil: A Life Cycle Analysis of Its Health and Environmental Impacts* (Boston: Center for Health and the Global Environment, Harvard Medical School, 2002), p. 4.

3. Richard Dobbs et al., *Resource Revolution: Meeting the World's Energy, Materials, Food, and Water Needs* (McKinsey Global Institute and McKinsey Sustainability & Resource Productivity Practice, 2011), p. 15; German Advisory Council on Global Change, *World in Transition: A Social Contract for Sustainability: Summary for Policy-Makers* (Berlin: 2011), p. 4; petroleum royalties from *U.S. Department of the Interior News*, 20 November 2008, cited in Coll, op. cit. note 2, p. 606; Saudi percentages from Ross, op. cit. note 2, p. 45.

4. Resource curse statements from Ross, op. cit. note 2, pp. 189, 236, 253.

5. Woolsley statement from Annie Maccoby Berglof, "At Home: Jim Woosley. The Former Head of the CIA Wants to Wean the US Off Oil," *Financial Times*, 6 July 2012; John Hofmeister, *Why We Hate the Oil Companies: Straight Talk From an Energy Insider* (New York: Palgrave Macmillan, 2010), p. 48; German Advisory Council on Global Change, op. cit. note 3, p. 25.

6. Hofmeister, op. cit. note 5.

7. On moral entrepreneurs, see Ethan A. Nadelmann, "Global Prohibition Regimes: The Evolution of Norms in International Society," *International Organization*, autumn 1990, pp. 479–526.

8. Paul Hawken, *Blessed Unrest: How the Largest Movement in the World Came into Being, and Why No One Saw It Coming* (New York: Viking, 2007); Rob Nixon, *Slow Violence and the Environmentalism of the Poor* (Cambridge, MA: Harvard University Press, 2011); Robert Goodland, "Responsible Mining: The Key to Profitable Resource Development," *Sustainability*, vol. 4, no. 9 (2012), pp. 2,099–126.

9. Pamela L. Martin, *Oil in the Soil: The Politics of Paying to Preserve the Amazon* (Boulder, CO: Rowman and Littlefield Publishers, 2011).

10. Ibid.

11. "Ecological Debt and Oil Moratorium in Costa Rica," *Oilwatch International*, August 2005; "Costa Rica," in U.N. Development Programme, *Human Development Indicators*, at hdrstats.undp.org/en/countries/profiles/CRI .html; Adam Williams, "While Natural Gas Remains an Option, Chinchilla Says No to Oil Drilling in Costa Rica," *Tico Times*, 17 June 2011.

12. Alex Trembath et al., *Where the Shale Gas Revolution Came From* (Oakland, CA: Breakthrough Institute, 2012).

13. Peter Applebome, "On Drilling, Patterson Pleases Both Sides," *New York Times*, 13 December 2010.

14. Trembath et al., op. cit. note 12; New York State Department of Environmental Conservation, "Draft Supplemental Generic Environmental Impact Statement (DSGEIS) Relating to Drilling for Natural Gas in New York State Using Horizontal and Hydraulic Fracturing," at www.dec.ny.gov/energy/75370.html.

15. Dusty Horwitt, *Drilling Around the Law* (Washington, DC: Environmental Working Group, 2010); Onondaga Nation, "Traditional Native Leaders: Hydrofracking Must Be Banned," press release (Albany, NY: Onondaga Nation, 5 November 2009); Haudenosaunee Environmental Task Force, "Haudenosaunee Statement on Hydrofracking," Rooseveltown, NY, 2009; Onondaga Nation, "Onondaga Nation's Statement to NYSDEC on 'Hydro-fracking,'" press release (Syracuse, NY: 12 January 2012).

16. Appalachian coal from Laura Bozzi, Yale School of Forestry and Environmental Studies, dissertation in progress; uranium from James Goodman, "Leave It in the Ground! Eco-social Alliances for Sustainability," in Josée Johnston, Michael Gismondi, and James Goodman, eds., *Nature's Revenge: Reclaiming Sustainability in an Age of Corporate Globalization* (Peterborough, Canada: Broadview Press, 2006), pp. 155–82; gold in El Salvador from Robin Broad and John Cavanagh, "Like Water for Gold in El Salvador," *The Nation*, 1–8 August 2011; gold and diamonds from "Guyana Bans Gold, Diamond Mining In Rivers," *Jamaica-Gleaner*, 8 July 2012; oil from Berit

Kristoffersen and Stephen Young, "Geographies of Security and Statehood in Norway's 'Battle of the North,'" *Geoforum*, July 2010, pp. 577–84.

17. Donella Meadows, "Envisioning a Sustainable World," paper presented at the Third Biennial Meeting of the International Society for Ecological Economics, San Jose, Costa Rica, 24–28 October 1994, pp. 1–6.

18. Barbara Freese, *Coal: A Human History* (New York: Penguin, 2003), pp. 190–97.

19. Evidence of change in fossil fuel industries in part from informal interviews with high-level officials in the automobile and oil industries; see also George Mobus, "Industry Leaders Seem to be Showing More Openness to Energy Descent Issue" (blog), *The Oil Drum*, 4 May 2010; Hofmeister, op. cit. note 5, p. 168; evidence for end of cheap energy from Robert L. Hirsch, "The Inevitable Peaking of World Oil Production," *Atlantic Council Bulletin*, vol. XVI, no. 3, from Richard D. Kerr, "World Oil Crunch Looming?" *Science*, 21 November 2008, and from Robert Rapier, "German Military Study Warns of Potential Energy Crisis" (blog), *The Oil Drum*, 2 September 2010; Coll, op. cit. note 2; Adam Hochschild, *Bury the Chains: Prophets and Rebels in the Fight to Free an Empire's Slaves* (Boston: Houghton Mifflin, 2005); Allan M. Brandt, *The Cigarette Century: The Rise, Fall, and Deadly Persistence of the Product that Defined America* (New York: Basic Books, 2007).

Chapter 15. Beyond Fossil Fuels: Assessing Energy Alternatives

1. Early application of coal from E. A. Wrigley, *Energy and the English Industrial Revolution* (New York: Cambridge University Press, 2010), p. 45.

2. Author's estimate.

3. Annual growth at 3 percent from U.S. Energy Information Administration, *Annual Energy Review* (Washington, DC: 2011), Appendix E; current global power demand from International Energy Agency (IEA), *Key World Energy Statistics* (Paris: 2010), p. 6.

4. T. W. Murphy, Jr., "Can Economic Growth Last?" *Do the Math*, at physics.ucsd.edu/do-the-math/2011/07/can-economic-growth-last.

5. Hydroelectric power potential from Eurelectric, *Study on the Importance of Harnessing the Hydropower Resources of the World* (Brussels: April 1997).

6. Figure of 81 percent from IEA, op. cit. note 3, p. 6; automobile efficiency from R. A. Ristinen and J. J. Kraushaar, *Energy and the Environment*, 2nd ed. (New York: John Wiley and Sons, 2006), p. 71.

7. T. W. Murphy, Jr., "The Alternative Energy Matrix," *Do the Math*, at physics.ucsd.edu/do-the-math/2012/02/the-alternative-energy-matrix.

8. Photovoltaic panel production from G. Hering, "Year of the Tiger," *Photon International*, March 2011, p. 186; W. F. Pickard, "A Nation-Sized Battery," *Energy Policy*, June 2012, pp. 263–67; small number of locales from T. W. Murphy, Jr., "Pump Up the Storage," *Do the Math*, at physics.ucsd.edu/do-the-math/2011/11/pump-up-the-storage.

9. Considerable fraction of present needs from C. de Castro et al., "Global Wind Power Potential: Physical and Technological Limits," *Energy Policy*, October 2011, pp. 6,677–82.

10. Compelling possibility from E. S. Andreiadis, "Artificial Photosynthesis: From Molecular Catalysts for Light-driven Water Splitting to Photoelectrochemical Cells," *Photochemistry and Photobiology*, 8 August 2011, pp. 946–64; U.S. Department of Energy project from "Fuels from Sunlight Hub," at energy.gov/articles/fuels-sunlight-hub, 1 August 2010.

11. Judy Dempsey and Jack Ewing, "Germany, in Reversal, Will Close Nuclear Plants by 2022," *New York Times*, 30 May 2011; Risa Maeda and Aaron Sheldrick, "Japan Aims to Abandon Nuclear Power by 2030s," *Reuters*, 14 September 2012; Ayesha Rascoe, "U.S. Approves First New Nuclear Plant in a Generation," *Reuters*, 9 February 2012.

12. Fully 99.7 percent of natural uranium is U-238 while 0.7 percent is U-235; see Ristinen and Kraushaar, op. cit. note 6, p. 184.

13. Ristinen and Kraushaar, op. cit. note 6, pp. 145–50.

14. For an expanded discussion of these other sources, see Murphy, op. cit. note 7.

15. T. W. Murphy, Jr., "The Energy Trap," *Do the Math*, at physics.ucsd.edu/do-the-math/2011/10/the-energy-trap.

Chapter 16. Energy Efficiency in the Built Environment

1. Percent of energy-related emissions from U.S. Department of Energy (DOE), Energy Information Administration (EIA), *What Are Greenhouse Gas Emissions? How Much Does the US Emit?* (Washington, DC: 2012); levelized costs from DOE, EIA, *Annual Energy Outlook 2012* (Washington, DC: 2012).

2. DOE, *2009 Buildings Energy Data Book* (Washington, DC: 2009); Confederation of Indian Industry, CII-Sohrab ji Godrej Green Business Centre, "Energy Efficiency in Building Design and Construction," New Delhi, June 2005, p. 1.

3. Karim Elgendy, "The State of Energy Efficiency Policies in Middle East Buildings," at www.carboun.com; Confederation of Indian Industry, op. cit. note 2, p. 1.

4. McKinsey & Company, *Unlocking Energy Efficiency in the US Economy* (2009), p. iii; McKinsey & Company, *Impact of the Financial Crisis on Carbon Economics* (2010), p. 8.

5. Matt Krantz, "Investors Question Wisdom of 10% Rule of Thumb Rule," *USA Today*, 17 October 2011; American Council for an Energy-Efficient Economy from Institute for Building Efficiency, "Why Focus on Existing Buildings?" 2008, at www.institutebe.com/Existing-Building-Retrofits.aspx?lang=en-US; U.S. Environmental Protection Agency (EPA), *Introduction to Energy Performance Contracting* (Washington, DC: 2007), p. 11.

6. EPA, Office of Air and Radiation, *Energy Star Building Upgrade Manual* (Washington, DC: 2008), p. 2; U.S. Green Building Council, "FAQ: LEED Green Building Rating System," Washington, DC, 2012, p. 1.

7. Jonathan Miller, *Emerging Trends in Real Estate 2008* (Washington, DC: Urban Land Institute, 2007), p. 13; Charles DiRocco and Jonathan Miller, *Emerging Trends in Real Estate 2013* (Washington, DC: Urban Land Institute, 2012), p. 56.

8. Harvey M. Bernstein and Michele A Russo, *Business Case for Energy Efficient Building Retrofit and Renovation* (New York: McGraw-Hill Construction, 2011), p. 29; Piet Eichholtz, Nils Kok, and John M. Quiqley, "Doing Well by Doing Good? Green Office Buildings," *American Economic Review*, December 2010; "Green Building Market to Hit $173.5 Billion by 2015," *Environmental Leader*, 1 July 2010; Melissa Hincha-Ownby, "LEED-Certified Space Tops 2 Billion Square Feet" (blog), *Mother Nature Network*, 30 July 2012.

9. DOE, *Memorandum of Understanding Between the Appraisal Foundation and U.S. Department of Energy Office of Energy Efficiency and Renewable Energy* (Washington, DC: 2011).

10. "Commercial Building Asset Rating: A New Buzz Word or a True Driver of Transformation in Building Energy Consumption?" (blog), *Retroficiency*, 8 November 2011.

11. ICLEI Local Governments for Sustainability USA, *U.S. Local Sustainability Plans and Climate Action Plans* (Oakland, CA: 2009); DOE, "Obama Administration Announces Major Steps to Advance Energy Efficiency Efforts, Improve Access to Low-Cost Financing for States and Local Communities," press release (Washington, DC: 26 June 2012).

12. Institute for Market Transformation, "Existing Policies," at www.buildingrating.org/content/existing-policies; "China Policy Brief: Commercial and Residential Buildings," at www.buildingrating.org; New York City Building Code, "Construction Codes Update Pages," New York, 11 May 2009, p. 1.

13. Lane Burt et al., *A New Retrofit Industry* (Washington, DC: U.S. Green Building Council et al., 2011), p. 5; ICLEI USA, op. cit. note 11; "Mayors Leading the Way on Climate Protection," Climate Protection Center, at www.usmayors.org.

14. Elgendy, op. cit. note 3.

15. Charles Lockwood, "Building the Green Way," *Harvard Business Journal*, June 2006, p. 1; Ludwig Wittgenstein, *Culture and Value*, ed. G. H. von Wright, trans. P. Winch (Oxford: Blackwell, 1980), p. 74.

Chapter 17. Agriculture: Growing Food—and Solutions

1. Self Employed Women's Association (SEWA), "About Us," at www.sewa.org/About_Us.asp, viewed 4 October 2012; author's visit to SEWA Farm, Ahmedabad, India, February 2011.

2. Surajben Shankasbhai Rathwa, interview with Janeen Madan, "Women Farmers Key to End Food Insecurity" (blog), *Worldwatch: Nourishing the Planet*, 6 August 2011; SEWA Manager Ni School, at www.sewamanagerni school.org, viewed 4 October 2012; author's visit, op. cit. note 1.

3. SEWA members, Ahmedabad, India, interview with author, February 2011.

4. Figure 17–1 based on U.N. Food and Agriculture Organization (FAO), *The State of Food Insecurity in the World* (Rome: 2010) p. 8; FAO , *Obesity and Overweight*, Fact Sheet No. 311 (Rome: March 2011); World Bank, *Reduced Emissions and Enhanced Adaptation in Agricultural Landscapes*, Agricultural and Rural Development Notes (Washington, DC: World Bank, 2009), p. 1.

5. Jeffrey Delaurentis, "In Somalia Seeds of Hope and Progress Have Begun to Sprout, but They Need to be Carefully and Generously Nurtured," Security Council Meeting, United Nations, New York, 14 September 2011; FAO, "925 Million in Chronic Hunger Worldwide," press release (Rome: 14 September 2010); FAO, *The State of Food and Agriculture 2010–2011* (Rome: 2011), p. 67.

6. Figure 17–2 from FAO, Food Price Index, at www.fao.org/worldfoodsituation/wfs-home/foodpricesindex/en, updated 6 October 2011; World Bank, *Food Price Watch*, February 2011; World Bank, "Poverty Headcount Ratio at Rural Poverty Line," online database, at data.worldbank.org/indicator/SI.POV.RUHC, viewed 4 October 2012.

7. Olivier De Schutter, Eleventh Annual Edward and Nancy Dodge Lecture, Center for a Livable Future, Johns Hopkins Bloomberg School of Public Health, 27 September 2011.

8. Tristram Stuart, "Post-Harvest Losses: A Neglected Field," in Worldwatch Institute, *State of the World 2011* (New York: W. W. Norton & Company, 2011), pp. 99–108.

9. Julian Parfitt et al., *Food Waste within Food Supply Chains: Quantification and Potential for Change to 2050* (London: The Royal Society, 2010).

10. International Institute of Tropical Agriculture (IITA), "IITA, Partners Launch Initiative to Tackle Killer Aflatoxin in African Crops," press release (Ibadan, Nigeria: 4 April 2011); IITA, "Investing in Aflasafe™," 13 April 2011, at r4dreview.org/2011/04/investing-in-aflasafe%E2%84%A2.

11. Love Food, Hate Waste, an initiative of Waste and Resources Action Programme, at www.wrap.org.uk/wrap _corporate/about_wrap/resource_efficiency.

12. International Assessment of Agricultural Knowledge, Science and Technology for Development (IAASTD), *Agriculture at a Crossroads* (Washington, DC: Island Press, 2008); U.K. Government Office for Science / Foresight, *The Future of Food and Farming: Challenges and Choices for Global Sustainability* (London: 2011); Climate Change, Agriculture and Food Security, *Achieving Food Security in the Face of Climate Change* (Washington, DC: Commission on Sustainable Agriculture and Climate Change, CGIAR, 2011); Daniele Giovannucci et al., *Food and Agriculture: The Future of Sustainability* (New York: U.N. Department of Economic and Social Affairs, 2012).

13. IAASTD, *Agriculture at a Crossroads, Synthesis Report* (Washington, DC: Island Press, 2009), p. 5.

14. IAASTD, op. cit. note 12.

15. Olivier De Schutter, Office of the UN Special Rapporteur on the Right to Food, "Agroecology and the Right to Food," presented at the 16th Session of the United Nations Human Rights Council, March 2011; "Integrated Rice-duck: A New Farming System for Bangladesh," in Paul Van Mele, Ahmad Salahuddin, and Noel P. Magor, eds., *Innovations in Rural Extension: Case Studies from Bangladesh* (Cambridge, MA: CABI Publishing, 2005).

16. Eric Holt-Giménez, "Measuring Farmers' Agroecological Resistance after Hurricane Mitch in Nicaragua: A Case Study in Participatory, Sustainable Land Management Impact Monitoring," *Agriculture, Ecosystems and Environment*, December 2002, pp. 87–105.

17. Meera Shekar, "Scaling Up Nutrition: A Framework for Action," 5th Friedman School Symposium on Nutrition Security, Tufts University, Boston, 5 November 2010; Meera Shekar, *State of the World 2011* Symposium Panel

Discussion, Carnegie Endowment, Washington, DC, 12 January 2011; K. Weinberger and T. A. Lumpkin, *Horticulture for Poverty Alleviation—The Unfunded Revolution*, Working Paper No. 15 (Shanhua, Taiwan: AVRDC–The World Vegetable Center, 1995); Abdou Tenkouano, "The Nutritional and Economic Potential of Vegetables," in Worldwatch Institute, op. cit. note 8, pp. 27–35.

18. Danielle Nierenberg, "Breeding Vegetables with Farmers in Mind" (blog), *Worldwatch: Nourishing the Planet*, 3 December 2010; Monika Blössner and Mercedes de Onis, *Malnutrition: Quantifying the Health Impact at National and Local Levels*, Environmental Burden of Disease Series, No. 12 (Geneva: World Health Organization (WHO), 2005).

19. WHO, *Global Status Report on Noncommunicable Diseases 2010* (Geneva: 2011), p. 9; Rachel Nugent, *Bringing Agriculture to the Table* (Chicago: Chicago Council on Global Affairs, 2011).

20. FoodCorps, *2010–2011 Annual Report*, at www.foodcorps.org/about/files/FoodCorps-AnnualReport.pdf.

21. Hannah B. Sahud et al., "Marketing Fast Food: Impact of Fast Food Restaurants in Children's Hospitals," *Pediatrics*, 1 December 2006, pp. 2,290–97; Molly Theobold, "Innovation of the Week: Healing Hunger" (blog), *Worldwatch: Nourishing the Planet*, 13 January 2011; "Chris Hani Baragwanath Hospital, South Africa," at www.chrishanibaragwanathhospital.co.za, viewed 8 November 2011.

22. Edward Mukiibi, Project Coordinator, Developing Innovations in School Cultivation (DISC), Uganda, interview with author, November 2009; Danielle Nierenberg, "How to Keep Kids 'Down on the Farm'" (blog), *Worldwatch: Nourishing the Planet*, 9 December 2010.

23. Betty Nabukalu, student, DISC, Uganda, interview with author, November 2009; Nierenberg, op. cit. note 22; Slow Food International, "A Thousand Gardens in Africa," at www.slowfood.com/terramadreday/pagine/eng/pagina2.lasso?-id_pg=113, viewed 8 November 2011.

24. PRNewswire via COMTEX, "World Cocoa Foundation, USAID and IDH Launch the African Cocoa Initiative," press release (Washington, DC: 18 October 2011); World Cocoa Foundation, "Family Support Scholarships—Parents' Entrepreneurship for Children's Education" at www.worldcocoafoundation.org/family-support-scholarships-parents-entrepreneurship-for-childrens-education, viewed 1 November 2011; *Nurturing the Next Generation of Cocoa Farmers*, event at the Field Museum, Chicago, 4 October 2011.

25. Kristin E. Davis, "Extension in Sub-Saharan Africa: Overview and Assessment of Past and Current Models, and Future Prospects," *Journal of International Agricultural and Extension Education*, fall 2008, pp. 17–20.

26. Danielle Nierenberg, "Learning to Listen to Farmers" (blog), *Worldwatch: Nourishing the Planet*, 28 June 2011; Ernest Laryea Okorley, University of Cape Coast, School of Agriculture, Ghana, interview with author, June 2010.

27. Howard G. Buffett Foundation, *The Hungry Continent: African Agriculture and Food Insecurity* (draft) (Decatur, IL: October 2011).

28. FAO, "Increased Agricultural Investment Is Critical to Fighting Hunger," at cm.naturelabs.org/?p=242.

29. Feed the Future, at www.feedthefuture.gov, viewed 4 November 2011; Global Agriculture & Food Security Program, at www.gafspfund.org/gafsp, viewed 4 November 2011.

30. Cheryl Doss et al., *The Role of Women in Agriculture* (Rome: FAO, 2011), p. 5.

31. Rainforest Alliance, "Our Work in Sustainable Agriculture," at www.rainforestalliance.org/work/agriculture, viewed 12 October 2012.

32. Figure of $2 per day from World Bank, *World Development Indicators 2010* (Washington, DC: 2010), pp. 91–92; World Bank, "Poverty Headcount Ratio," op. cit. note 6; World Bank, *Food Price Watch*, op. cit. note 6.

33. Box 17–1 based on the following: 500 million accounts from Robert Peck Christen, Richard Rosenberg, and Veena Jayadeva, *Financial Institutions with a Double-Bottom Line: Implications for the Future of Microfinance*, Occasional Paper No. 8 (Washington, DC: Consultative Group to Assist the Poor, July 2004), p. 13; Amy Waldman, "Debts and Drought Drive India's Farmers to Despair," *New York Times*, 6 June 2004; William J. Grant and Hugh C. Allen, "CARE's Mata Masu Dubara (Women on the Move) Program in Niger: Successful Financial Intermediation in the Rural Sahel," *Journal of Microfinance*, fall 2002, pp. 189–216; Kim Wilson, Malcolm Harper, and

Matthew Griffith, eds., *Financial Promise for the Poor: How Groups Build Microsavings* (Sterling, VA: Kumerian Press, 2010); Ben Fowler and Candace Nelson, *Beyond Financial Services: Combining Savings Groups with Agricultural Marketing in Tanzania* (Aga Khan Foundation, 2011); Bob Morikawa, "Plant With Purpose Tanzania Impact Evaluation, September, 2011," unpublished, at www.plantwithpurpose.org/resources. Food prices nearly 20 percent higher in 2011 from Hazel Healy, "The Food Rush," *New Internationalist*, October 2011; Olivier De Schutter, "Food Commodities Speculation and Food Price Crisis: Regulation to Reduce the Risks of Price Volatility," *Briefing Note* (September 2010).

34. United Nations, "United Nations Launches Year-Long Celebration of Vital Role of Cooperatives in Sustainable Development," press release (New York: 31 October 2011).

35. FrontlineSMS, "About the Project," at www.frontlinesms.com/about-us/history-and-support, viewed 5 October 2012.

36. The World Food Prize, "The 2011 World Food Prize Laureates," at www.worldfoodprize.org/index.cfm?nodeID =33367&audienceID=1, viewed 12 October 2012.

Chapter 18. Protecting the Sanctity of Native Foods

1. Dennis Martinez, "The Missing Delegate at Cancún: Indigenous Peoples," *National Geographic News Watch*, 8 December 2010.

2. Dan Wildcat, *Red Alert! Saving the Planet with Indigenous Knowledge* (Golden, CO: Fulcrum Publishing, 2009).

3. Claire Cummings, "Risking Corn, Risking Culture," *World Watch Magazine*, November/December 2002; John Mohawk, "The Art of Thriving in Place," in Melissa Nelson, ed., *Original Instructions: Indigenous Teachings for a Sustainable Future* (Rochester, VT: Bear & Company, 2008), pp. 126–37; Colin Carter and Henry Miller, "Food for Food, Not Fuel," *New York Times*, 30 July 2012.

4. Hawai'i SEED, "Taro," at www.hawaiiseed.org.

5. Ibid.

6. White Earth Land Recovery Project & Native Harvest, at www.nativeharvest.com.

7. "Letter from Winona LaDuke about the Manoomin Rice Fall Harvest," Slow Food USA site, at www.slowfood usa.org/index.php/programs/presidia_product_detail/wild_rice_anishinaabeg_manoomin.

8. Claire Cummings, *Uncertain Peril: Genetic Engineering and the Future of Seeds* (Boston: Beacon Press, 2008).

9. Tirso A. Gonzales and Melissa K. Nelson, "Contemporary Native American Responses to Environmental Threats in Indian Country," in John Grim, ed., *Indigenous Traditions and Ecology—The Interbeing of Comology and Community* (Cambridge, MA: Harvard University Press, 2001), p. 504.

10. M. Kat Anderson, *Tending the Wild: Native American Knowledge and the Management of California's Natural Resources* (Berkeley: University of California Press, 2005); Dennis Martinez, "Indigenous Ecosystem-based Adaptation and Community-based Ecocultural Restoration during Rapid Climate Disruption: Lessons for Western Restorationists," presented at the 4th World Conference on Ecological Restoration, the 20th Annual Meeting of the Society for Ecological Restoration International, and the 2nd Meeting of the Ibero-American and Caribbean Ecological Restoration Network, 23 August 2011, Mérida, Yucatan, Mexico.

11. Beth Rose Middleton. *Trust in the Land: New Directions in Tribal Conservation* (Tucson: University of Arizona Press, 2011); Carlo Petrini, speech to the U.N. Permanent Forum on Indigenous Issues, New York, May 2012.

12. Cook quoted in Winona LaDuke, *All Our Relations: Native Struggles for Land and Life* (Cambridge, MA: South End Press, 1999), pp. 18–23.

13. Blue Voice, "Toxic Contamination in the Arctic," at www.bluevoice.org/news_toxicarctic.php; Lisa Charleyboy, "In the Arctic, a Hunger for Ancestral Foods," *Spirituality and Health*, November–December 2012.

14. Slow Food USA, "US Ark of Taste," at www.slowfoodusa.org/index.php/programs/details/ark_of_taste.

15. Slow Food USA, "US Presidia," at www.slowfoodusa.org/index.php/programs/details/us_presidia.

16. Slow Food USA, op. cit. note 14.

17. Lois Ellen Frank, "The Discourse and Practice of Native American Cuisine: Native American Chefs and Native American Cooks in Contemporary Southwest Kitchens" (PhD Diss., University of New Mexico, 2011), pp. 448–51.

18. Physicians' Committee for Responsible Medicine, "PCRM's Native American Diabetes Prevention Classes Focus on Ancestral Foods," August 2011, at www.pcrm.org; Physicians' Committee for Responsible Medicine, "Diabetes Success Stories," at www.pcrm.org/health/diabetes-resources.

19. Tohono O'odham Community Action, at www.tocaonline.org; Robert Bazell and Linda Carroll, "Indian Tribe Turns to Tradition to Fight Diabetes," *NBC News*, 12 December 2011; Gary Paul Nabhan, *Coming Home to Eat: The Pleasures and Politics of Local Foods* (New York: W. W. Norton & Company, 2002), pp. 289–302.

20. Author's observations.

21. Melissa Nelson, "Re-Indigenizing Our Bodies and Minds Through Native Foods," in Nelson, op. cit. note 3, pp. 180–94.

Chapter 19. Valuing Indigenous Peoples

1. First Peoples Worldwide (FPW), "Ancient Past: A Bridge over Troubled Economic Transition in Tanzania," Fredericksburg, VA, May 2010.

2. FPW, "Maasai Women Signify New Indigenous Rights in Kenya," Fredericksburg, VA, August 2010; Shadrack Kavilu, "Adoption of New Constitution in Kenya Heralds a New Dawn for Indigenous Communities," *Gáldu*, Guovdageaidnu-Kautokeino, Norway, 8 September 2010; Jerry Reynolds, "Simat Praises Indigenous Presence in New Kenya Constitution" (blog), *First Peoples Worldwide*, 30 August 2010.

3. Claudia Sobrevila, *The Role of Indigenous Peoples in Biodiversity Conservation* (Washington, DC: World Bank, 2008), pp. xii and 7; Evelyn Arce, International Funders for Indigenous People (IFIP), "The Story of IFIP" (blog); Evelyn Arce, "In Focus: A New Paradigm of Collaboration with Indigenous Peoples," *IFIP Human Rights Funding News*, 5 May 2011.

4. FPW, "Conservation's 'New Breed of Refugee' is All Too Familiar to Indian County," Fredericksburg, VA, August 2009.

5. FPW, "Conservation Evictions: First Peoples Worldwide Background Paper," Fredericksburg, VA, March 2007; Marcus Colchester, "Conservation Policy and Indigenous People," *Cultural Survival Quarterly*, spring 2004.

6. Colchester, op. cit. note 5; FPW, op. cit. note 5.

7. Table 19–1 from the following: Davinder Kumar, "Philippines' Tribes Try to Save their Forest," *Al Jazeera News*, 9 August 2011; Robyn Dixon, "Kenyan Tribe Slowly Driven Off Its Ancestral Lands," *Los Angeles Times*, 4 January 2010; Slow Food Foundation for Biodiversity, "Imraguen Women's Mullet Botargo," undated; S. Heckbert et al., "Indigenous Australians Fight Climate Change with Fire," *Solutions*, November 2011, pp. 50–56.

8. World Bank, "Food Price Hike Drives 44 Million People into Poverty," press release (Washington, DC: 15 February 2011).

9. Maurice Colchester and Maurizio Farhan Ferrari, *Making FPIC—Free, Prior and Informed Consent—Work: Challenges and Prospects for Indigenous Peoples* (Moreton-in-Marsh, U.K.: Forest Peoples Programme, June 2007).

10. United Nations, *United Nations Declaration on the Rights of Indigenous Peoples* (New York: 2008); FPW, "Corporate Engagement," at www.firstpeoples.org.

11. Shell, *Shell Sustainability Report 2011* (April 2012), pp. 9, 14.

12. Ibid.; Mark Betancourt, FPW, discussion with author Rebecca Adamson, 8 October 2012.

13. Accra Caucus on Forests and Climate Change, *Realizing Rights, Protecting Forests: An Alternative Vision for Reducing Deforestation* (June 2010).

14. FPW, "Grants: Keepers of the Earth Fund," at www.firstpeoples.org.

15. FPW, "Grants Awarded," at www.firstpeoples.org.

16. Ibid.

17. Ibid.

18. "CONGO: New Law to Protect Rights of Indigenous Peoples," *IRIN* (U.N. Office for the Coordination of Humanitarian Affairs), 7 January 2011.

19. United Nations, op. cit. note 10.

Chapter 20. Crafting a New Narrative to Support Sustainability

1. Robert Pool, *Earthrise: How Man First Saw the Earth* (New Haven, CT: Yale University Press, 2010).

2. Gary Gardner, *Inspiring Progress* (New York: W. W. Norton & Company, 2006), p. 70; "Religious Teachings on the Environment," at www.greenfaith.org; Mary Evelyn Tucker and John Grimm, "Overview of World Religions and Ecology," Forum on Religion and Ecology at Yale, at fore.research.yale.edu; "Renewal" (video), FineCut Productions, LLC, 2007.

3. E. O. Wilson, *On Human Nature*, 25th anniv. ed. (Cambridge, MA: Harvard University Press, 1979), p. 201; Brian Swimme and Thomas Berry, *The Universe Story* (New York: HarperCollins, 1992); "billions and billions" from Carl Sagan, *Cosmos*, television series, Public Broadcasting System, 1980.

4. Barry Rodrigue and Daniel Stasko, "A Big History Directory, 2009: An Introduction," *World History Connected*, October 2009; Michael Duffy, "Cosmic Education and Big History," presentation at American Montessori Society Conference, 2011.

5. Peter J. Richerson and Robert Boyd, *Not by Genes Alone* (Chicago: University of Chicago Press, 2006); Robert Boyd and Peter J. Richerson, *The Origin and Evolution of Cultures* (New York: Oxford University Press, 2005).

6. Andrew J. Revkin, "The 'Anthropocene' as Environmental Meme and/or Geological Epoch" (blog), *New York Times*, 17 September 2012; F. John Odling-Smee, Kevin N. Laland, and Marcus W. Feldman, *Niche Construction* (Princeton, NJ: Princeton University Press, 2003).

7. Jeffrey Bennett and Seth Shostak, *Life in the Universe*, 3rd ed. (Boston: Addison-Wesley, 2012).

8. David Christian, "Humanoid Histories," at www.metanexus.net/essay/humanoid-histories; see also video of the comparative humanoid histories talk, Global Futures 2045 conference, Moscow, March 2012, at www.youtube.com/watch?v=7FYfpaJ3ek0&feature=youtu.be; Peter Richerson, "Rethinking Paleoanthropology: A World Queerer Than We Supposed," in Gary Hatfield, ed, *Evolution of Mind* (Philadelphia: Penn Museum Conference Series, in press).

9. Laurie Garrett, *The Coming Plague* (New York: Farrar, Straus, and Giroux, 1994); Peter Turchin, *War and Peace and War*, reprint ed. (New York: Plume, 2007).

10. The Big History Project is a collaboration between pilot schools, teachers, and educators—supporters include Bill Gates, David Christian, and the University of Michigan, see www.bighistoryproject.com; "First Year Experience—Big History at Dominican University of California," at www.dominican.edu/academics/big-history; Ryan Wyatt et al., "Life, A Cosmic Story," California Academy of Sciences Planetarium show, 2010; Gregory C. Farrington, "Transformation of the California Academy of Sciences," in Worldwatch Institute, *State of the World 2010* (New York: W. W. Norton & Company, 2010), p. 68.

11. Dwight Collins, "The Evolutionary Account of the Universe: A Support for Behavioral Change Toward Sustainability," in Cheryl Genet et al., eds., *Science, Wisdom, and the Future* (Santa Margarita, CA: Collins Foundation Press, 2012).

12. Student quoted in David Christian, "Big History for the Era of Climate Change," *Solutions*, March 2012.

13. Matt Lappé, Director of Education, Alliance for Climate Education, discussions with authors.

14. Dwight Collins, Ron Nahser, and Art Whatley, "Sustainability as the Core Theme in Graduate Management Education: A Synopsis of Two Programs," *Management International Conference 2008: Intercultural Dialogue and Management*, Barcelona, Spain, November 2008; Ron Nahser, *Journeys to Oxford* (Global Scholarly Publications, 2008), pp.174–79, 207–11; Donella H. Meadows et al., *Limits to Growth* (New York: Universe Books, 1972); Donella

H. Meadows, *Thinking in Systems* (White River Junction, VT: Chelsea Green Publishing, 2008), pp. 25–27, 145–65.

15. Sagan, op. cit. note 3.

Chapter 21. Moving Toward a Global Moral Consensus on Environmental Action

1. Justin Gillis, "Study Finds More of Earth Is Hotter and Says Global Warming Is at Work," *New York Times*, 6 August 2012; Suzanne Goldenburg, "Greenland Ice Sheet Melted at Unprecedented Rate during July," (London) *Guardian*, 24 July 2012; "More Record Highs across Kansas Wednesday—Including Dodge City's All-Time High," *Finger on the Weather* (blog), 27 June 2012; Weather Forecast Office, "Drought Briefing Page," National Weather Service, at www.crh.noaa.gov/lsx/?n=drought; "As Wildfires Rage, the Russian Government Heads East to Battle the Crisis," *Siberian Times*, 6 August 2012; Justin Gillis, "Ending Its Summer Melt, Arctic Sea Ice Sets a New Low That Leads to Warnings," *New York Times*, 19 September 2012.

2. Severin Carrell, "NASA Scientist: Climate Change is a Moral Issue on a Par with Slavery," (London) *Guardian*, 6 April 2012; Desmond Tutu, "Foreword," in K. D. Moore and M. P. Nelson, eds., *Moral Ground: Ethical Action for a Planet in Peril* (San Antonio, TX: Trinity University Press, 2010), p. xiii; Sheila Watt-Cloutier, "The Inuit Right to Culture Based on Ice and Snow," in ibid., p. 28 (adapted from *Transcripts from Indigenous Peoples' Resistance to Economic Globalization: A Celebration of Victories, Rights and Cultures*, New York, 23 November 2006); The Dalai Lama, "A Question of Our Own Survival," in ibid., p. 19.

3. For a denial of the harms of carbon emissions, see ads by The Competitive Enterprise Institute, at www.you tube.com/watch?v=7sGKvDNdJNA.

4. United Nations, "The Universal Declaration of Human Rights," at www.un.org/en/documents/udhr/index .shtml; International Panel on Climate Change, *Fourth Assessment Report of the Intergovernmental Panel on Climate Change* (Cambridge, U.K.: Cambridge University Press, 2007); Sandra Steingraber, "Three Bets On Ecology, Economy, and Human Health," *Orion*, May/June 2009.

5. "Real Leadership for a Clean Energy Future," Remarks of Senator Barack Obama, Portsmouth, NH, 8 October 2007.

6. Aldo Leopold, *A Sand County Almanac* (New York: Oxford University Press, 1949), pp. 224–25.

7. James Gustave Speth, "The Limits of Growth," in Moore and Nelson, op. cit. note 2, p. xiii.

8. A. Leiserowitz et al., *Global Warming's Six Americas: March 2012 & Nov. 2011* (New Haven, CT: Yale Project on Climate Change Communication, 2012).

9. Quotes in Box 21–1 from Victor E. Frankl, *Man's Search for Meaning* (Boston: Beacon Press, 2006), pp. 86, 135.

10. Ezra M. Markowitz and Azim F. Shariff, "Climate Change and Moral Judgment," *Nature Climate Change*, vol. 2 (2012), pp. 243–47.

11. "The Earth Charter," Earth Charter Initiative, at www.earthcharterinaction.org/content/pages/Read-the-Charter.html; "Ecuador Rights of Nature," at www.rightsofmotherearth.com/ecuador-rights-nature; John Vidal, "Bolivia Enshrines Natural World's Rights with Equal Status for Mother Earth," (London) *Guardian*, 10 April 2011.

12. Juliette Jowit, "British Campaigner Urges UN to Accept 'Ecocide' as International Crime," (London) *Guardian*, 9 April 2010.

13. Jeremy Hance, "12,000 Surround White House to Protest Tar Sands Pipeline," *Mongabay.com*, 7 November 2011; Bill McKibben, "Global Warming's Terrifying New Math," *Rolling Stone*, 2 August 2012; Isabel Hayes, "Thousands Protest on Climate Change," *Sydney Morning Herald*, 15 August 2010.

14. Catherine Woodiwiss, "Stop the Frack Attack: Religious Leaders Kick Off First Ever Nation-wide Anti-Fracking Rally in DC" (blog), *Climate Progress*, 31 July 2012.

15. "Partner in Prayer," Evangelical Environmental Network, at prayerforcreationcare.creationcare.org; Interfaith Moral Action on Climate, "Interfaith Call to Action on Climate Change," at www.interfaithactiononclimatechange .org.

16. "The Blue River Declaration: An Ethic of the Earth," November 2011, at springcreek.oregonstate.edu/docu ments/BlueRiverDeclaraton.2012.pdf, p. 2.

17. "Certified B Corporation," at www.bcorporation.net.

18. "Joanna Macy and Her Work: The Great Turning," at www.joannamacy.net/thegreatturning.html.

Chapter 22. Pathways to Sustainability: Building Political Strategies

1. United Nations Conference on Sustainable Development, at www.uncsd2012.org; "Rio+20: At Downtown Gathering, Citizens Voice Concerns at People's Summit," UN News Centre, 20 June 2012.

2. Helpful overviews of the reformist and radical approaches include W. M. Adams, *Green Development: Environment and Sustainability in a Developing World*, 3rd ed. (London: Routledge, 2008), and A. N. H. Dobson, *Green Political Thought*, 4th ed. (London: Routledge, 2007).

3. M. Leach et al., "Transforming Innovation for Sustainability," *Ecology and Society*, vol. 17, no. 2 (2012), art. 11; for more detail of a "pathways" approach to sustainability challenges, see M. Leach, I. Scoones, and A. Stirling, *Dynamic Sustainabilities: Technology, Environment, Social Justice* (London: Earthscan, 2010).

4. World Commission on Environment and Development, *Our Common Future* (Oxford: Oxford University Press, 1987), p. 43.

5. M. Hajer and H. Wagenaar, eds., *Deliberative Policy Analysis* (Cambridge, U.K.: Cambridge University Press, 2003); F. Fischer and J. Forester, eds., *The Argumentative Turn in Policy Analysis and Planning* (Durham, NC: Duke University Press, 1993).

6. Guyana from S. Mangal and J. Forte, *Community Tradeoffs Assessment: For Culture-sensitive Planning and Evaluation*, Power Tools Series (London: International Institute for Environment and Development (IIED), 2005); India from M. P. Pimbert and T. Wakeford, *Prajateerpu: A Citizens Jury/Scenario Workshop on Food and Farming Futures for Andhra Pradesh, India* (London: IIED and Institute of Development Studies (IDS), Sussex, 2002); Box 22–1 from Sally Brooks et al., *Environmental Change and Maize Innovation in Kenya: Exploring Pathways In and Out of Maize*, STEPS Working Paper 36 (Brighton, U.K.: STEPS Centre, 2009); A. Stirling et al., *Empowering Designs: Towards More Progressive Appraisal of Sustainability*, STEPS Working Paper 3 (Brighton, U.K.: STEPS Centre, 2007).

7. International Assessment of Agricultural Knowledge, Science and Technology for Development, at www.agassessment.org.

8. "Rio+20: After Dialogues, Citizens to Make Recommendations on Rio+20 Issues," UN News Centre, 20 June 2012; Clarinha Glock, "Rio+20 Doubts over Impact of Sustainable Development Dialogues," *Inter Press Service*, 19 June 2012; Adrian Ely, "Opening up Sustainable Development Decision-making at the UN?" *The Crossing* (STEPS Centre blog), 21 June 2012.

9. For further discussion of deliberative approaches and their challenges, see F. Fischer, *Reframing Public Policy: Discursive Politics and Deliberative Practices* (Oxford: Oxford University Press, 2003), and R. Munton, "Deliberative Democracy and Environmental Decision-making," in F. Berkhout, M. Leach, and I. Scoones, eds., *Negotiating Environmental Change* (Cheltenham, U.K.: Edward Elgar, 2003).

10. Save the Narmada Movement, at www.narmada.org; see also W. F. Fisher, ed., *Toward Sustainable Development? Struggling over India's Narmada River* (Armonk, NY: M.E. Sharpe Publishers, 1995).

11. World Commission on Dams, *Dams and Development: A New Framework for Decision-Making* (London: Earthscan, 2000); L. Mehta, *The Politics and Poetics of Water: Naturalising Scarcity in Western India* (Delhi: Orient Longman, 2005); Lyla Mehta, Gert Jan Veldwisch, and Jennifer Franco, "Water Grabbing? Focus on the (Re)appropriation of Finite Water Resources," *Water Alternatives*, special issue, vol. 5, no. 2 (2012).

12. "Occupy Movement," (London) *Guardian*, at www.guardian.co.uk/world/occupy-movement.

13. For more detail on citizen mobilization and environmental social movements, see A. Jamison, *The Making of Green Knowledge: Environmental Politics and Cultural Transformation* (Cambridge, U.K.: Cambridge University Press, 2001), and M. Leach and I. Scoones, *Mobilizing Citizens: Social Movements and the Politics of Knowledge*, IDS Working Paper 276 (Brighton, U.K.: IDS, 2007).

14. La Via Campesina: International Peasant Movement, at viacampesina.org/en; World Social Forum, at

en.wikipedia.org/wiki/World_Social_Forum; "Rio+20," op. cit. note 1.

15. For discussion of the rise and operation of networked forms of governance and politics, see R. A. W. Rhodes, *Understanding Governance* (Maidenhead, U.K.: Open University Press, 1997).

16. J. Keeley and I. Scoones, *Understanding Environmental Policy Processes: Cases from Africa* (London: Earthscan, 2003); Peter Newell, "The Governance of Energy Finance: The Public, the Private and the Hybrid," *Global Policy*, September 2011, pp. 94–105; M. Betsill and H. Bulkeley, "Cities and the Multilevel Governance of Global Climate Change," *Global Governance*, April–June 2006, pp. 141–59.

17. P. Olsson et al., "Shooting the Rapids: Navigating Transitions to Adaptive Governance of Social-Ecological Systems," *Ecology and Society*, vol. 11, no. 1 (2006), art. 18.

18. Everglades management from ibid.; J. W. Kingdon, *Agendas, Alternatives, and Public Policies*, 2nd ed. (New York: Longman, 1995).

Chapter 23. Moving from Individual Change to Societal Change

1.   "Crying Indian PSA," Keep America Beautiful and The Ad Council, 1970. The one-minute ad can be seen at www.youtube.com/watch?v=j7OHG7tHrNM.

2.   Container Recycling Institute, "Keep America Beautiful: A History," Culver City, CA, undated.

3.   The Lazy Environmentalist, at www.lazyenvironmentalist.com; Recyclebank, at www.recyclebank.com.

4.   Michael Maniates, Allegheny College, email to author, 3 December 2012; The Story of Stuff Project, *The Story of Change: Why Citizens (Not Shoppers) Hold the Key to a Better Future*, 2012, available at www.storyofchange.org.

5.   Figure 23–1 from Maria Csutora, "One More Awareness Gap? The Behaviour-Impact Gap Problem," *Journal of Consumption Policy*, March 2012, p. 149.

6.   Ibid.

7.   See, for example, "The No Trash Family," *People Magazine*, 16 January 2012.

8.   Figure 23–2 from Annie Leonard, *The Story of Stuff* (New York: Free Press, 2010), based on Joel Makower, "Calculating the Gross National Trash" (blog), *Greenbiz.com*, 20 March 2009, and on Joel Makower and Cara Pike, *Strategies for a Green Economy* (New York: McGraw-Hill, 2008), p. 112.

9.   Andrew Szasz, *Shopping Our Way to Safety: How We Changed from Protecting the Environment to Protecting Ourselves* (Minneapolis: University of Minnesota Press, 2007), pp. 2–3.

10. See, for example, diverse perspectives in "Responsible Shoppers, but Bad Citizens?" Room for Debate (blog), *New York Times*, 30 July 2012

11. Lawrence Glickman, *Buying Power: A History of Consumer Activism in America* (Chicago: University of Chicago Press, 2009), p. 84.

12. Ibid.

13. See "Responsible Shoppers, but Bad Citizens?" op. cit. note 10.

14. James Gustave Speth, *America the Possible: Manifesto for a New Economy* (New Haven, CT: Yale University Press, 2012), p. 191.

15. Andy Igrejas, Safer Chemicals, Healthy Families Coalition, discussion with author, 10 November 2012.

16. Events and countries from 350.org; Bill McKibben, discussion with author, 3 December 2012; Brian Merchant, "1,252 Peaceful Protestors Arrested Opposing Tar Sands Pipeline at the White House," *TreeHugger*, 3 September 2011.

17. Monica Wilson, Global Alliance for Incinerator Alternatives, discussion with author, 4 September 2012.

18. William Martin, *Best Liberal Quotes Ever* (Naperville, IL: Sourcebooks, Inc., 2004), p. 173.

Chapter 24. Teaching for Turbulence

1. Susan Clark et al., "College and University Environmental Programs as a Policy Problem (Part 1): Integrating Knowledge, Education, and Action," and "(Part 2): Strategies for Improvement," *Environmental Management*, both online 26 February 2011.

2. Michael Soulé and Daniel Press, "What Is Environmental Studies?" *BioScience*, May 1998, pp. 397–405; Michael Maniates and John Whissel, "Environmental Studies: The Sky is Not Falling," *BioScience*, June 2000, pp. 509–17.

3. Marvin Soroos, "Adding Green to the International Studies Curriculum," *International Studies Notes*, winter 1991, pp. 37–42.

4. Shirley Vincent, *Interdisciplinary Environmental Education on the Nation's Campuses: Elements of Field Identity and Curriculum Design* (Washington, DC: National Council of Science and Environment, 2010).

5. These patterns are drawn from analysis of 41 prominent U.S. undergraduate ESS programs conducted for this chapter; these programs typically serve as models for other programs in the United States.

6. Sharon Hall, Tom Tietenberg, and Stephanie Pfirman, *Environmental Programs at Liberal Arts Colleges: Findings and Recommendations for the Andrew W. Mellon Foundation* (Washington, DC: Project Kaleidoscope, 2005).

7. Shirley Vincent, email to author, 19 October 2012.

8. Clark et al., op. cit., note 1; Richard Wallace, discussion with author, 19 October 2012; Matthew Auer, "Communication and Competition in Environmental Studies," *Policy Science*, December 2010, pp. 365–90.

9. Mark Dowie, *Losing Ground: American Environmentalism at the End of the 20th Century* (Cambridge, MA: The MIT Press, 1996).

10. Ibid.; James Gustave Speth, *The Bridge at the End of the World* (New Haven, CT: Yale University Press, 2008).

11. Samuel Rigotti, *Environmental Problem Solving: How Do We Make Change?* (Meadville, PA: Department of Environmental Science, Allegheny College, 2010).

12. Ibid.

13. Ibid.

14. G. Tyler Miller and Scott Spoolman, *Environmental Science* (Belmont, CA: Brooks Cole, 2012); Daniel Chiras, *Environmental Science* (Burlington, MA: Jones and Bartlett, 2012); Walter Rosenbaum, *Environmental Politics and Policy* (Washington, DC: CQ Press, 2010).

15. Norman Vig and Michael Kraft, *Environmental Policy: New Directions for the Twenty-First Century* (Washington, DC: CQ Press, 2012); Garrett Hardin, "The Tragedy of the Commons," *Science*, 13 December 1968, pp. 1,243–48.

16. Harris Interactive, Inc., "How Green Are We? Putting Our Money (And Our Behavior) Where Our Mouth Is," press release (New York: 13 October 2009); Harris Interactive, Inc., "One-Quarter of Americans Do Not Recycle in Their Own Homes," press release (Rochester, NY: 11 July 2007); Harris Interactive, Inc., "Fewer Americans 'Thinking Green,'" press release (New York: 18 April 2012).

17. Flexi Display Marketing, Inc., "Benefits of Using CFL Lightbulbs," *AwarenessIDEAS.com*, 8 June 2008.

18. Global Footprint Network, "Footprint Calculator," at www.footprintnetwork.org/en/index.php/GFN/page /calculators.

19. Emanuel quoted in Jeff Zelany, "Obama Weighs Quick Undoing of Bush Policy," *New York Times*, 10 November 2008; Niccolo Machiavelli, *II Principe* (1513).

20. Rebecca Solnit, *A Paradise Built in Hell: The Extraordinary Communities that Arise in Disaster* (New York: Viking Press, 2008), pp. 305–06.

21. Ibid.

22. Michael Shellenberger and Ted Nordhaus, *Break Through: From the Death of Environmentalism to the Politics*

of Possibility (New York: Houghton Mifflin, 2009), John M. Meyer, "A Democratic Politics of Sacrifice," in Michael Maniates and John M. Meyer, *The Environmental Politics of Sacrifice* (Cambridge, MA: The MIT Press, 2010), pp. 26–27.

23. Robert Reich, *Aftershock: The Next Economy and America's Future* (New York: Vintage, 2011); Ted Nordhaus and Michael Shellenberger, "The Green Bubble: Why Environmentalism Keeps Imploding," *New Republic*, 20 May 2009; Richard Hofstadter, "The Paranoid Style in American Politics," *Harper's Magazine*, November 1964; Thomas Edsall, *The Age of Austerity: How Scarcity Will Remake American Politics* (New York: Anchor Books, 2012).

24. Robert Heilbroner, *An Inquiry into the Human Prospect* (New York: W. W. Norton & Company, 1980); J. O. Hertzler, "Crises and Dictatorships," *American Sociological Review*, April 1940.

25. Box 24–1 from the following: Andrew Revkin, "The Changing (Communication) Climate" (Dot Earth blog), *New York Times*, 31 March 2011; John Meyer, Humboldt State University, email to author; Stephen Cunha, Humboldt State University, email to author; Wheaton College, "Political Science 361: Environmental Conflict Resolution," at wheatoncollege.edu/catalog/pols_361.

26. Charles Sayan and Daniel Blumstein, *The Failure of Environmental Education (And How We Can Fix It)* (Berkeley: University of California Press, 2011).

Chapter 25. Effective Crisis Governance

1. I thank Lyn Carson, Mark Diesendorf, and Steve Wright for valuable comments on earlier drafts of this chapter.

2. Michael Renner and Zoé Chafe, "Turning Disasters into Peacemaking Opportunities," in Worldwatch Institute, *State of the World 2006* (New York: W. W. Norton & Company, 2006), pp. 123–27; Eric Stover and Patrick Vinck, "Cyclone Nargis and the Politics of Relief and Reconstruction Aid in Burma (Myanmar)," *Journal of the American Medical Association*, 13 August 2008, pp. 729–31.

3. Stockholm International Peace Research Institute, *Warfare in a Fragile World: Military Impact on the Human Environment* (London: Taylor & Francis, 1980).

4. Janet Abbate, *Inventing the Internet* (Cambridge, MA: The MIT Press, 1999).

5. Jasper Becker, *Hungry Ghosts: Mao's Secret Famine* (New York: Free Press, 1996); Frank Dikötter, *Mao's Great Famine: The History of China's Most Devastating Catastrophe, 1958–62* (London: Bloomsbury, 2011); Article 19, *Starving in Silence: A Report on Famine and Censorship* (London: 1990).

6. Kenneth Bain, *Treason at Ten: Fiji at the Crossroads* (London: Hodder and Stoughton, 1989); Robert T. Robertson and Akosita Tamanisau, *Fiji—Shattered Coups* (Sydney: Pluto Press, 1988).

7. Peter Ackerman and Jack DuVall, *A Force More Powerful: A Century of Nonviolent Conflict* (New York: St. Martin's Press, 2000); Michael Randle, *People Power: The Building of a New European Home* (Stroud: Hawthorn, 1991); Kurt Schock, *Unarmed Insurrections: People Power Movements in Nondemocracies* (Minneapolis, MN: University of Minnesota Press, 2005); Stephen Zunes, "Arab Revolutions and the Power of Nonviolent Action," *National Catholic Reporter*, 25 November 2011, p. 26.

8. Erica Chenoweth and Maria J. Stephan, *Why Civil Resistance Works: The Strategic Logic of Nonviolent Conflict* (New York: Columbia University Press, 2011); Table 25–1 from ibid., p. 73.

9 Adam Roberts, "Civil Resistance to Military Coups," *Journal of Peace Research*, March 1975, pp. 19–36; D. J. Goodspeed, *The Conspirators: A Case Study in the Coup d'État* (London: Macmillan, 1962); Victoria E. Bonnell, Ann Cooper, and Gregory Freidin, eds., *Russia at the Barricades: Eyewitness Accounts of the August 1991 Coup* (Armonk, NY: M. E. Sharpe, 1994).

10. Michael Flood and Robin Grove-White, *Nuclear Prospects: A Comment on the Individual, the State and Nuclear Power* (London: Friends of the Earth, 1976); Robert Jungk, *The New Tyranny: How Nuclear Power Enslaves Us* (New York: Grosset and Dunlap, 1979).

11. David Collingridge, *The Social Control of Technology* (London: Frances Pinter, 1980).

12. C. George Benello and Dimitrios Roussopoulos, eds., *The Case for Participatory Democracy: Some Prospects for a Radical Society* (New York: Grossman, 1971); Gerry Hunnius, G. David Garson, and John Case, eds., *Workers'*

Control: A Reader on Labor and Social Change (New York: Vintage, 1973).

13. John Gastil, *Democracy in Small Groups: Participation, Decision Making, and Communication* (Philadelphia: New Society Publishers, 1993); Jane J. Mansbridge, *Beyond Adversary Democracy* (New York: Basic Books, 1980).

14. Sharon Erickson Nepstad, *Nonviolent Revolutions: Civil Resistance in the Late 20th Century* (New York: Oxford University Press, 2011).

15. Lyn Carson et al., eds., *The Australian Citizens' Parliament and the Future of Deliberative Democracy* (University Park, PA: Pennsylvania State University Press, in press); John Gastil and Peter Levine, eds., *The Deliberative Democracy Handbook: Strategies for Effective Civic Engagement in the Twenty-First Century* (San Francisco: Jossey-Bass, 2005).

16. Daniel Guérin, *Anarchism: From Theory to Practice* (New York: Monthly Review Press, 1970).

17. David E. Hoffman, *The Oligarchs: Wealth and Power in the New Russia* (New York: PublicAffairs, 2011); Vadim Volkov, *Violent Entrepreneurs: The Use of Force in the Making of Russian Capitalism* (Ithaca, NY: Cornell University Press, 2002).

18. Marina Sitrin, ed., *Horizontalism: Voices of Popular Power in Argentina* (Edinburgh: AK Press, 2006).

19. Ken Darrow and Mike Saxenian, eds., *Appropriate Technology Sourcebook: A Guide to Practical Books for Village and Small Community Technology* (Stanford, CA: Volunteers in Asia, 1986); Willem Riedijk, *Technology for Liberation: Appropriate Technology for New Employment* (Delft: Delft University Press, 1986).

20. Rob Hopkins, *The Transition Handbook: From Oil Dependency to Local Resilience* (Totnes, U.K.: Green Books, 2008).

21. Rebecca MacKinnon, *Consent of the Networked: The Worldwide Struggle for Internet Freedom* (New York: Basic Books, 2012).

22. Ziauddin Sardar and Merryl Wyn Davies, *Why Do People Hate America?* (Cambridge, U.K.: Icon, 2002); Craig R. Smith, ed., *Silencing the Opposition: How the U.S. Government Suppressed Freedom of Expression During Major Crises*, 2nd ed. (Albany, NY: State University of New York Press, 2011).

23. Daron Acemoglu and James A. Robinson, *Why Nations Fail: The Origins of Power, Prosperity, and Poverty* (New York: Crown, 2012); Shaazka Beyerle, "Civil Resistance and the Corruption-Violence Nexus," *Journal of Sociology and Social Welfare,* June 2011, pp. 53–77.

24. Bruce Stokes, *Helping Ourselves: Local Solutions to Global Problems* (New York: W. W. Norton & Company, 1981); War Resisters' International, *Handbook for Nonviolent Campaigns* (London: 2009); Karl Fogel, *Producing Open Source Software: How to Run a Successful Free Software Project* (Sebastopol, CA: O'Reilly Media, 2005).

25. Leonardo Avritzer, *Participatory Institutions in Democratic Brazil* (Baltimore, MD: Johns Hopkins University Press, 2009); Brian Wampler, *Participatory Budgeting in Brazil: Contestation, Cooperation, and Accountability* (University Park, PA: Pennsylvania State University Press, 2007).

26. Johan Galtung, *There Are Alternatives! Four Roads to Peace and Security* (Nottingham, U.K.: Spokesman, 1984), pp. 131–38.

27. Roméo Dallaire with Brent Beardsley, *Shake Hands with the Devil: The Failure of Humanity in Rwanda* (New York: Carroll & Graf, 2004).

28. Thomas Beamish, *Silent Spill: The Organization of an Industrial Crisis* (Cambridge, MA: The MIT Press, 2002).

Chapter 26. Governance in the Long Emergency

1. Svante Arrhenius, "On the Influence of Carbonic Acid in the Air upon the Temperature of the Ground," *The London, Edinburgh, and Dublin Philosophical Magazine and Journal of Science*, April 1896.

2. The phrase is from James Howard Kunstler, *The Long Emergency* (New York: Atlantic Monthly Press, 2005); Kevin A. Baumert, Timothy Herzog, and Jonathan Pershing, *Navigating the Numbers: Greenhouse Gas Data and International Climate Policy* (Washington, DC: World Resources Institute, 2005), p. 113.

3. Brian Barry, *Why Social Justice Matters* (Cambridge, U.K.: Polity Press, 2005), p. 251.

4. Thomas Homer-Dixon, *The Ingenuity Gap* (New York: Knopf, 2000); Mark Mazower, *Governing the World* (New York: Penguin, 2012), p. 424.

5. Robert Heilbroner, *An Inquiry into the Human Prospect* (New York: W. W. Norton & Company, 1980), p. 175; Robert Heilbroner, "Second Thoughts on *The Human Prospect*," *Challenge*, May-June, 1975, p. 27.

6. Anthony Giddens, *The Politics of Climate Change* (Cambridge, U.K.: Polity Press, 2009), p. 96; David Rothkopf, *Power, Inc: The Epic Rivalry Between Big Business and Government and the Reckoning that Lies Ahead* (New York: Farrar, Straus and Giroux, 2012), p. 360.

7. David W. Orr and Stuart Hill, "Leviathan, the Open Society, and the Crisis of Ecology," *Western Political Quarterly*, December 1978, pp. 457–69.

8. Amory B. Lovins et al., *Reinventing Fire* (White River Junction, VT: Chelsea Green Publishing, 2011), p. ix.

9. Value over $20 trillion from Bill McKibben, "Global Warming's Terrifying New Math," *Rolling Stone*, 2 August 2012; Robert B. Reich, *Supercapitalism* (New York: Knopf, 2007), pp. 170–01, 204.

10. Charles E. Lindblom, *Politics and Markets* (New York: Basic Books, 1977), p. 356; Charles E. Lindblom, *The Market System* (New Haven, CT: Yale University Press, 2001).

11. Karl Polanyi, *The Great Transformation* (Boston: Beacon Press, 1967), p. 73; John Dunn, *The Cunning of Unreason* (London: Harper-Collins, 2000), p. 332; David Rothkopf, *Superclass* (New York: Farrar, Straus, Giroux, 2008), p. 322; Michael Sandel, *What Money Can't Buy: The Moral Limits of Markets* (New York: Farrar, Straus and Giroux, 2012).

12. Nicholas A. Christakis and James Fowler, *Connected* (Boston: Little Brown, 2009), pp. 289–92; Steven Johnson, *Emergence* (New York: Scribners, 2001), pp. 224–26; Anne-Marie Slaughter, *A New World Order* (Princeton, NJ: Princeton University Press, 2004), p. 263.

13. Paul Hawken, *Blessed Unrest* (New York: Penguin, 2007); Steve Waddell, *Global Action Networks* (New York: Palgrave-Macmillan, 2011), p. 23.

14. Mark Mazower, *Governing the World* (New York: Penguin, 2012), pp. 420, 418; Matthew Bishop and Michael Green, *Philanthropocapitalism: How Giving Can Save the World* (New York: Bloomsbury, 2008).

15. Naomi Klein, "Capitalism vs. the Climate," *The Nation*, 21 November 2011.

16. Harold Myerson, "Foundering Fathers," *American Prospect*, October 2011, p. 16; to improve at least U.S. democracy, see Steven Hill, *10 Steps to Repair American Democracy* (Sausalito, CA: PoliPoint Press, 2006).

17. Benjamin Barber, *Strong Democracy* (Berkeley: University of California Press, 1984), pp. 117, 151; see also Thad Williamson, David Imbroscio, and Gar Alperovitz, *Making a Place for Community* (New York: Routledge, 2002); Jeffereson and Dewey from Carol Pateman, *Participation and Democratic Theory* (Cambridge, U.K.: Cambridge University Press, 1970); final quote from Barber, op. cit. this note, p. 269.

18. Amy Gutmann and Dennis Thompson, *Why Deliberative Democracy* (Princeton, NJ: Princeton University Press, 2004), pp. 7, 59; see also Susan Clark and Woden Teachout, *Slow Democracy* (White River Junction, VT: Chelsea Green, 2012). Box 26–1 from the following: Adam Liptak, "Justices, 5–4, Reject Corporate Spending Limit," *New York Times*, 22 January 2010; Robert J. Shapiro and Douglas Dowson, *Corporate Political Spending: Why the New Critics Are Wrong*, Legal Policy Report No. 15 (New York: Manhattan Institute for Policy Research, June 2012); Barber, op. cit. note 17, pp. 3, 4; Adolf G. Gundersen, *The Environmental Promise of Democratic Deliberation* (Madison: University of Wisconsin Press, 1995), pp. 9, 10, 19, and 22.

19. Bruce Ackerman and James Fishkin, *Deliberation Day* (New Haven, CT: Yale University Press, 2004), p. 171; see also James S. Fishkin, *The Voice of the People: Public Opinion and Democracy* (New Haven, CT: Yale University Press, 1995).

20. Sanford Levinson, *Framed: America's 51 Constitutions and the Crisis of Governance* (New York: Oxford University Press, 2012), p. 389; see also Derek Bok, *The Trouble with Government* (Cambridge, MA: Harvard University Press, 2001).

21. Richard J. Lazarus, *The Making of Environmental Law* (Chicago: University of Chicago Press, 2004), pp. 30, 33, 42; Richard J. Lazarus, "Super Wicked Problems and Climate Change: Restraining the Present to Liberate the Future," *Cornell Law Review*, vol. 94 (2009), pp. 1,153–234.

22. Thomas Berry, *Evening Thoughts* (San Francisco: Sierra Club Books, 2006), p. 95.

23. Ecuador from Erik Assadourian, "The Rise and Fall of Consumer Cultures," in Worldwatch Institute, *State of the World 2010* (New York: W. W. Norton & Company, 2010), p. 19; Christopher Stone, *Should Trees Have Standing: Toward Legal Rights for Natural Objects* (Los Altos, CA: William Kaufmann, 1972); Berry, op. cit. note 22, p. 44.

24. John Keane, *The Life and Death of Democracy* (New York: W. W. Norton & Company, 2009); see also Paul Woodruff, *First Democracy: The Challenge of an Ancient Idea* (New York: Oxford University Press, 2005); John Plamenatz, *Democracy and Illusion* (London: Longman, 1973), p. 9.

25. Wilson Carey McWilliams, *Redeeming Democracy in America* (Lawrence: University of Kansas Press, 2011), p. 15; Peter Burnell, *Climate Change and Democratization* (Berlin: Heinrich Böll Stiftung, 2009), p. 40.

26. See, for example, Thomas E. Mann and Norman J. Ornstein, *It's Even Worse than It Looks* (New York: Basic Books, 2012), Theda Skocpol and Vanessa Williamson, *The Tea Party and the Remaking of Republican Conservatism* (New York: Oxford University Press, 2012), and Jill Lepore, *The Whites of Their Eyes* (Princeton, NJ: Princeton University Press, 2010); Frank Bryan, *Real Democracy* (Chicago: University of Chicago Press, 2004), p. 294; see also Robert Dahl and Edward Tufte, *Size and Democracy* (Stanford, CA: Stanford University Press, 1973).

27. Richard M. Weaver, *Ideas Have Consequences* (Chicago: University of Chicago Press, 1984), p. 127; Jean M. Twenge and W. Keith Campbell, *The Narcissism Epidemic* (New York: The Free Press, 2009), p. 276.

28. Naomi Klein, *Shock Doctrine: The Rise of Disaster Capitalism* (New York: Metropolitan Books, 2007); see also Corey Robin, *Fear: The History of a Political Idea* (New York: Oxford University Press, 2004).

29. Rothkopf, op. cit. note 11; see also International Forum on Globalization, *Outing the Oligarchy: Billionaires Who Benefit from Today's Climate Crisis* (San Francisco: 2011).

30. Josh Bivens, "Inequality, Exhibit A: Walmart and the Wealth of American Families" (blog), Economic Policy Institute, 17 July 2012; Richard Wilkinson and Kate Pickett, *The Spirit Level: Why Equality is Better for Everyone* (London: Penguin Books, 2010); Jeffrey Winters, *Oligarchy* (Cambridge, U.K.: Cambridge University Press, 2011), pp. 284–85.

31. Lewis Mumford, *The Myth of the Machine: The Pentagon of Power* (New York: Harcourt, Brace, Jovanovich, 1970), pp. 413, 434.

32. Gar Alperovitz, *America Beyond Capitalism* (Takoma Park, MD: Democracy Collaborative Press, 2011); Gar Alperovitz, "Anchoring Wealth to Sustain Cities and Population Growth, *Solutions*, July 2012; James Gustave Speth, *America the Possible: Manifesto for a New Economy* (New Haven, CT: Yale University Press, 2012); Michael H. Shuman, *Going Local* (New York: Routledge, 2000); Michael H. Shuman, *Local Dollars, Local Sense* (White River Junction, VT: Chelsea Green, 2012); Greg Pahl, *Power from the People* (White River Junction, VT: Chelsea Green, 2012); Jeff Gates, *Democracy at Risk* (Cambridge, MA: Perseus, 2000).

33. William McDonough and Michael Braungart, *Cradle to Cradle* (New York: North Point Press, 2002); Janine Benyus, *Biomimicry: Innovation Inspired by Nature* (New York: William Morrow, 1996); John Lyle, *Regenerative Design for Sustainable Development* (New York: John Wiley, 1994); John R. Ehrenfeld, *Sustainability by Design* (New Haven, CT: Yale University Press, 2008); Rob Hopkins, *The Transition Handbook* (Totnes, U.K.: Greenbooks, 2008); Rob Hopkins, *The Transition Companion* (White River Junction, VT: Chelsea Green, 2011). Box 26–2 based on National Environmental Policy Act, at ceq.hss.doe.gov/laws_and_executive_orders/the_nepa_statute.html, and on David W. Orr, *The Oberlin Project: A Clinton Climate Initiative Climate Positive Project* (Oberlin, OH: undated).

34. For more on these issues, see Ron Rosenbaum, *How the End Begins: The Road to a Nuclear World War III* (New York: Simon & Schuster, 2011); Peter Barnes, *Capitalism 3.0* (San Francisco: Barrett-Koehler, 2006); Burns Weston and David Bollier, *Green Governance: Ecological Survival, Human Rights, and the Commons* (Cambridge, U.K.: Cambridge University Press, 2013); Tim Jackson, *Prosperity without Growth* (London: Earthscan, 2009); Peter Victor, *Managing without Growth: Slower by Design, Not Disaster* (Northampton, MA: Edward Elgar, 2008); Peter G. Brown, *Restoring the Public Trust* (Boston: Beacon Press, 1994), pp. 71–91; Peter G. Brown, *The Commonwealth*

of Life, 2nd ed. (Montreal: Black Rose Books, 2008); Steven Pinker, *The Better Angels of Our Nation: Why Violence Has Declined* (New York: Viking, 2011); Harald Welzer, *Climate Wars: Why People Will be Killed in the 21st Century* (Cambridge, U.K.: Polity Press, 2012).

35. Jared Diamond, *Collapse: How Societies Choose to Fail or Succeed* (New York: Viking, 2005), p. 438.

Chapter 27. Building an Enduring Environmental Movement

1. Berg quoted in Bill Devall and George Sessions, *Deep Ecology: Living as if Nature Mattered* (Layton, UT: Gibbs Smith, 1985), p. 3.

2. Michael Shellenberger and Ted Nordhaus, *The Death of Environmentalism* (Oakland, CA: Breakthrough Institute, 2004); Tom Crompton, *Weathercocks and Signposts: The Environment Movement at a Crossroads* (Godalming, U.K.: WWF-UK, 2008).

3. Shellenberger and Nordhaus, op. cit. note 2, pp. 7, 8.

4. Crompton, op. cit. note 2.

5. Michael Narberhaus, "Breaking Out of the System Trap: Civil Society Organizations," *Solutions Journal*, August 2012.

6. Jennifer Washburn, *University, Inc.* (New York: Basic Books, 2006); National Film Board of Canada, *Pink Ribbons, Inc.*, First Run Features, 2011; Christine MacDonald, *Green, Inc.* (Guilford, CT: The Lyons Press, 2008).

7. MacDonald, op. cit. note 6.

8. Ibid., pp. 25–28, 58–60; David B. Ottaway and Joe Stephens, "Nonprofit Land Bank Amasses Billions: Charity Builds Assets on Corporate Partnerships," *Washington Post*, 4 May 2003.

9. DARA International, *Climate Vulnerability Monitor: A Guide to the Cold Calculus of a Hot Planet*, 2nd ed. (Washington, DC: 2012); Fiona Harvey, "Climate Change Is Already Damaging Global Economy, Report Finds," (London) *Guardian*, 26 September 2012.

10. Anthony A. Leiserowitz and Lisa O. Fernandez, *Toward a New Consciousness: Values to Sustain Human and Natural Communities* (New Haven, CT: Yale School of Forestry & Environmental Studies, 2008).

11. Arne Naess, *The Ecology of Wisdom: Writings by Arne Naess* (Berkeley, CA: Counterpoint, 2010); Devall and Sessions, op. cit. note 1.

12. Palmer quoted in Helen Grady, "Using Religious Language to Fight Global Warming," *BBC Radio 4*, 25 January 2010.

13. Havel quoted in James Gustave Speth, "Foreword," in Leiserowitz and Fernandez, op. cit. note 10, p. 5.

14. Naess, op. cit. note 11, p. 111.

15. Aldo Leopold, *A Sand County Almanac* (New York: Oxford University Press, 1966), p. 262.

16. Naess, op. cit. note 11, p. 111.

17. Stewart J. Brown, "The Social Gospel in Britain, Germany, and the United States, 1870–1920," Ecclesiastical History Course 2D at University of Edinburgh, 1998; Roy Hattersley, *Blood and Fire: William and Catherine Booth and Their Salvation Army* (New York: Doubleday, 2000).

18. *The YMCA Blue Book* (Geneva: World Alliance of YMCAs, 2012); YMCA, "Mission," at www.ymca.int/who-we -are/mission; Salvation Army USA, *The Salvation Army 2012 Annual Report* (2012); Hattersley, op. cit. note 17; The Salvation Army International, "About Us," at www.salvationarmy.org/ihq/about.

19. Erik Assadourian, "The Living Earth Ethical Principles: Spreading Community," *World Watch Magazine*, September/October 2009, pp. 38–39; Knights of Columbus, "Knights of Columbus Tops $80 Billion of Life Insurance in Force," press release (New Haven, CT: 21 April 2011).

20. Box 27–1 based on *Ken Burns' America: The Shakers*, Public Broadcasting System, 1985.

21. Brook P. Hales, "Statistical Report, 2011," *Ensign*, May 2012; The Church of Jesus Christ of Latter-day Saints,

"One Million Missionaries, Thirteen Million Members," press release (Provo, UT: 25 June 2007).

22. Isaiah Thompson, "Idealists for Hire," *Philadelphia City Paper*, 11 August 2010; Dana R. Fisher, *Activism, Inc.* (Stanford, CA: Stanford University Press, 2006); Green Corps canvas operations, winter 2001, author's observations.

23. Uzma Anzar, "Islamic Education: A Brief History of Madrassas With Comments on Curricula and Current Pedagogical Practices," March 2003.

24. Population and area from Muchiri Karanja, "Myth Shattered: Kibera Numbers Fail to Add Up," *Daily Nation*, 3 September 2010, and from Mikel Maron, "Kibera's Census: Population, Politics, Precision," *Map Kibera* (blog), 5 September 2010; school calculation based on Map Kibera's education database at www.mapkibera.org, viewed 11 December 2012, and on Mikel Maron, Map Kibera Trust, email to author, 11 December 2012.

25. Maron, email to author, op. cit. note 24.

26. Erik Assadourian, "The Living Earth Ethical Principles: Life of Service and Prepare for a Changing World," *World Watch Magazine*, May/June 2009, pp. 34–35.

27. Erik Assadourian, "The Living Earth Ethical Principles: Right Diet and Renewing Life Rituals," *World Watch Magazine*, November/December 2008, pp. 32–33; Sarah Catherine Walpole et al., "The Weight of Nations: An Estimation of Adult Human Biomass," *BMC Public Health*, vol. 12 (2012), pp. 439–45.

28. Eduardo Porter, "Charity's Role in America, and Its Limits," *New York Times*, 13 November 2012.

29. Salvation Army USA, op. cit. note 18; Michael H. Shuman and Merrian Fuller, "Profits for Justice," *The Nation*, 24 January 2005.

30. Friends World Committee for Consultation, *Finding Quakers Around the World* (Philadelphia: 2007); A. Glenn Crothers, *Quakers Living in the Lion's Mouth* (Gainesville: University Press of Florida, 2012); see, for example, American Friends Service Committee, at afsc.org/afsc-history.

31. Box 27–2 based on the following: Gary Gardner, "Engaging Religions to Shape Worldviews," in Worldwatch Institute, *State of the World 2010* (New York: W. W. Norton & Company, 2010), pp. 23–29; Sarvodaya from Gary Gardner, *Invoking the Spirit* (Washington, DC: Worldwatch Institute, 2002), pp. 38–42.

32. Potsdam Institute for Climate Impact Research and Climate Analytics, *Turn Down the Heat: Why a 4°C Warmer World Must Be Avoided* (Washington, DC: World Bank, 2012).

33. Walter M. Miller, Jr., *A Canticle for Leibowitz* (Philadelphia: Lippincott, 1959).

34. "A Wild Love for the World," Joanna Macy interview by Krista Tippett, *On Being*, American Public Media, 1 November 2012.

Chapter 28. Resistance: Do the Ends Justify the Means?

1. "The Religion and Politics of Earth First!," *The Ecologist*, November/December 1991, pp. 258–66; "Radical Environmentalism" and "Earth First! and the Earth Liberation Front," in Bron Taylor, ed., *The Encyclopedia of Religion and Nature* (New York: Continuum, 2005), vol. 2, pp. 1,326–35, and vol. 1, pp. 518–24; Bron Raymond Taylor, ed., *Ecological Resistance Movements: The Global Emergence of Radical and Popular Environmentalism* (Albany: State University of New York Press, 1995).

2. Derrick Jensen and Lierre Keith, *Earth at Risk* (video), at PMPress/Flashpoint, 2012; Deep Green Resistance, at deepgreenresistance.org; Aric McBay, Lierre Keith, and Derrick Jensen, *Deep Green Resistance: Strategy to Save the Planet* (New York: Seven Stories Press, 2011).

3. For an example of radical prescriptions, see Alex Budd, "Time is Short: Systems Disruption and Strategic Militancy," *DGR (Dark Green Resistance) News Service*, 24 October 2012; for an influential anti-pacifism statement in 1994, see Ward Churchill, "Pacifism as Pathology," in *Pacifism as Pathology: Reflections on the Role of Armed Struggle in North America* (Oakland, CA: AK Press, 2007).

4. Bron Taylor, "Environmental Ethics," in Taylor, *Encyclopedia*, op. cit. note 1, vol. 1, pp. 597–606.

5. Bron Taylor, *Dark Green Religion: Nature Spirituality and the Planetary Future* (Berkeley: University of California Press, 2010).

6. James C. Scott, *Weapons of the Weak: Everyday Forms of Peasant Resistance* (New Haven, CT: Yale University Press, 1985); Ramachandra Guha, *The Unquiet Woods: Ecological Change and Peasant Resistance in the Himalaya* (Berkeley: University of California Press, 1989, expanded edition 2000).

7. Peter Galvin and Kieran Suckling, cofounders of the Center for Biological Diversity, discussion with author, 1 August 1992.

8. Donella H. Meadows, Jørgen Randers, and Dennis L. Meadows, *The Limits to Growth: The 30-Year Update* (White River Junction, VT: Chelsea Green Publishing Company, 2004); William Catton, *Overshoot: The Ecological Basis of Revolutionary Change* (Chicago: University of Illinois Press, 1980); Jared Diamond, *Collapse: How Societies Choose to Fail or Succeed* (New York: Viking, 2005).

9. For green anarchism, see John Zerzan, *Future Primitive* (Columbia, MO: C.A.L. Press, 1994); *Elements of Refusal* (Seattle, WA: Left Bank Books, 1988); John Zerzan, ed., *Against Civilization: Readings and Reflections*, 2nd ed. (Los Angeles, CA: Feral House, 2005); for a sense of the scale of such activism, see greenanarchy.org/earthliberation.

10. Steven Stoll, "Farm against forest," in M. L. Lewis, ed., *American Wilderness: A New History* (New York: Oxford University Press, 2007); Julie L. Lockwood and Michael L. McKinney, eds., *Biotic Homogenization* (New York: Springer, 2001); Jared Diamond, "The Worst Mistake in the History of the Human Race," *Discover*, May 1987, pp. 64–66; Jared Diamond, *Guns, Germs, and Steel* (New York: W. W. Norton & Company, 1997); Jim Mason, *An Unnatural Order: Uncovering the Roots of Our Domination of Nature and Each Other* (New York: Simon and Schuster, 1993); Paul Shepard, *Coming Home to the Pleistocene* (Washington, DC: Island Press, 1998); Clive Ponting, *A New Green History of the World: The Environment and the Collapse of Great Civilizations*, rev. ed. (New York: Penguin Books, 2007).

11. Kera Abraham, "Flames of Dissent" (five-part series), *Eugene Weekly*, 2006; Vanessa Grigoriadis, "The Rise and Fall of the Eco-Radical Underground," *Rolling Stone*, 10 August 2006, pp. 73–77, 100–07; "Operation Backfire," at en.wikipedia.org/wiki/Operation_Backfire_(FBI).

12. For rules of radical activism, see Saul David Alinsky, *Rules for Radicals* (New York: Random House, 1971), and Bron Taylor, "Experimenting with Truth," in Steven Best and Anthony J. Nocella, eds., *Igniting a Revolution: Voices in Defense of the Earth* (Oakland, CA: AK Press, 2006), pp. 1–7; interviews with affected activists in Oregon and California, September and October 2011.

13. Joel E. Cohen, *How Many People Can the Earth Support?* (New York: W. W. Norton & Company, 1995); Garrett Hardin, *Living within Limits* (New York: Oxford University Press, 1993).

14. Bron Taylor, "Deep Ecology and Its Social Philosophy: A Critique," in E. Katz, A. Light, and D. Rothenberg, eds., *Beneath the Surface: Critical Essays on Deep Ecology* (Cambridge, MA: The MIT Press, 2000), pp. 269–99.

15. Taylor, *Ecological Resistance Movements*, op. cit. note 1; David Helvarg, *The War against the Greens* (San Francisco: Sierra Club Books, 1992); Andrew Rowell, *Green Backlash: Global Subversion of the Environmental Movement* (New York: Routledge, 1996).

16. Jeni Kendell and Eddie Buivids, *Earth First: The Struggle to Save Australia's Rainforest* (Sidney, Australia: ABC Enterprises, 1987); Paul Watson, "In Defense of Tree Spiking," *Earth First!* 10.8 (1989): pp. 8–9.

17. Mike Roselle, discussion with author, 8 March 1992; Mike Roselle, *Tree Spiker: From Earth First! to Lowbagging: My Struggles in Radical Environmental Action* (New York: St. Martin's Press, 2009).

18. "Hundreds Arrested at Protest Against Redwood Logging," *Los Angeles Times*, 16 September 1996.

19. David Harris, *The Last Stand: The War between Wall Street and Main Street over California's Ancient Redwoods* (New York: Times Books/Random House, 1995); Richard Widick, *Trouble in the Forest: California's Redwood Timber Wars* (Minneapolis: University of Minnesota Press, 2009); Bill Dawson, "Redwood Protests Ease amid Reports of Deal," *Houston Chronicle*, 17 September 1996.

20. Bron Taylor, "Earth First! Fights Back," *Terra Nova* 2.2, spring 1997, pp. 29–43; Nina Witoszek, Lars Trägårdh, and Bron Taylor, eds., *Civil Society in the Age of Monitory Democracy* (New York: Berghahn Books, 2013).

21. For the October 2011 ruling in *Wyoming v. United States Department of Agriculture*, see caselaw.findlaw.com /us-10th-circuit/1583397.html; "US Supreme Court Supports Clinton's Roadless Rule," at pennfuture.blogspot

.com/2012/10/us-supreme-court-supports-clintons.html, and at wilderness.org/blog/roadless-rule-becomes-law -land.

22. Helvarg, op. cit. note 15; Will Potter, *Green Is the New Red: An Insiders Account of a Social Movement under Siege* (San Francisco: City Lights Books, 2011); Christian Parenti, *Tropic of Chaos: Climate Change and the New Geography of Violence* (New York: Nation Books, 2011).

23. On the changing political discourse, see Bron Taylor, "The Religion and Politics of Earth First!" *The Ecologist*, November/December 1991, pp. 258–66, and Taylor, *Ecological Resistance Movements*, op. cit. note 1; Lynne Davis, ed., *Alliances: Re/Envisioning Indigenous-Non-Indigenous Relationships* (Toronto, Canada: University of Toronto Press, 2010).

24. Martin Luther King, Jr., "Letter from Birmingham Jail," in S. Jonathan Bass and Martin Luther King, *Blessed Are the Peacemakers* (Baton Rouge: Louisiana State University Press, 2001).

25. Mark Drajem, "NASA's Hansen Arrested Outside White House at Pipeline Protest," *Bloomberg*, 29 August 2011; see also www.350.org.

26. Henry David Thoreau, *The Annotated Walden*, ed. Philip Van Doren Stern (New York: Barnes and Noble, 1970), p. 153.

Chapter 29. The Promises and Perils of Geoengineering

1. Box 29–1 based on Royal Society, *Geoengineering the Climate: Science, Governance and Uncertainty* (London: 2009), and on D. Keith, "Geoengineering the Climate: History and Prospects," *Annual Review of Energy and the Environment*, vol. 25 (2000), pp. 245–84.

2. For an accessible discussion of geoengineering options, see J. Goodell, *How to Cool the Planet* (New York: Mariner Books, 2010); for an authoritative statement of the current state of geoengineering research, see The Royal Society, op. cit. note 1.

3. Intergovenmental Panel on Climate Change (IPCC), Expert Meeting on Geoengineering, Lima, Peru, 20–22 June 2011; U.S. government efforts from E. Kintisch, *Hack the Planet* (Hoboken, NJ: John Wiley & Sons, 2010), p. 12.

4. Holdren quoted in A. Jha, "Obama Climate Adviser Open to Geo-engineering to Tackle Global Warming," (London) *Guardian*, 8 April 2009.

5. J. Fleming, *Fixing the Sky: The Checkered History of Weather and Climate Control* (New York: Columbia University Press, 2010).

6. L. Lane et al., eds., *Workshop Report on Managing Solar Radiation*, Ames Research Center, 18–19 November 2006 (Washington, DC: National Aeronautics and Space Administration, 2007).

7. J. Fleming, "The Climate Engineers: Playing God to Save the Planet," *Wilson Quarterly*, spring 2007, p. 46.

8. P. Crutzen, "Albedo Enhancement by Stratospheric Sulfur Injections: A Contribution to Resolve a Policy Dilemma?" (essay), *Climatic Change*, August 2006, pp. 212, 217.

9. Periodic assessments from the IPCC available at www.ipcc.ch; a useful popular primer is B. McKibben, "Global Warming's Terrifying New Math," *Rolling Stone*, 2 August 2012.

10. IPCC, *Third Assessment Report: Climate Change 2001—Working Group III: Mitigation*, section 4.7.

11. Figure 29–1 designed by Isabelle Rodas.

12. A. Ridgwell et al., "Tackling Regional Climate Change by Leaf Albedo Bio-geoengineering," *Current Biology*, vol. 19, no. 2 (2009), pp. 146–50; U.S. Department of Energy, "Secretary Chu Announces Steps to Implement Cool Roofs at DOE and Across the Federal Government," press release (Washington, DC: 19 July 2010).

13. See C. Mims, "'Albedo Yachts' and Marine Clouds: A Cure for Climate Change?" *Scientific American*, 21 October 2009.

14. Gates Foundation from O. Dorell, "Can Whiter Clouds Reduce Global Warming?" *USA Today*, 11 June 2010; for what Ken Caldeira calls the "Pinatubo option," see Kintisch, op. cit. note 3, p. 56.

15. A. Robock et al., "Benefits, Risks, and Costs of Stratospheric Geoengineering," *Geophysical Research Letters*, vol. 36, L19,703 (2009); quote from Fleming, op. cit. note 7; helium-filled balloons from Crutzen, op. cit. note 8; J. Pierce et al., "Efficient Formation of Stratospheric Aerosol for Climate Engineering by Emission of Condensible Vapor from Aircraft," *Geophysical Research Letters*, vol. 37, L18,805 (2010).

16. Crutzen, op. cit. note 8; N. Stern, *The Economics of Climate Change: The Stern Review* (Cambridge, U.K.: Cambridge University Press, 2007).

17. R. Angel, "Feasibility of Cooling the Earth with a Cloud of Small Spacecraft near the Inner Lagrange Point (L1)," *Proceedings of the National Academy of Sciences*, 14 November 2006, pp. 17,184–89.

18. Royal Society, op. cit. note 1; Carbon Engineering, at www.carbonengineering.com; Figure 29–2 designed by Isabelle Rodas.

19. Kintisch., op. cit. note 3; K. Roberts et al., "Life Cycle Assessment of Biochar Systems: Estimating the Energetic, Economic, and Climate Change Potential," *Environmental Science & Technology*, vol. 44, no. 2 (2010), pp. 827–33.

20. C. Bahric, "Hungry Shrimp Eat Climate Change Experiment," *New Scientist*, 25 March 2009.

21. IPCC, "Carbon Dioxide Capture and Storage: Summary for Policymakers," Geneva, September 2005; Global CCS Institute, *The Global Status of CCS: 2012* (Canberra: 2012).

22. See Kintisch, op. cit. note 3, p. 117; G. Shaffer, "Long-term Effectiveness and Consequences of Carbon Dioxide Sequestration" (letter), *Nature Geoscience*, July 2010, pp. 464–67.

23. Fleming, op. cit. note 7, p. 48.

24. Quoted in A. Revkin, "Branson on the Power of Biofuels and Elders" (Dot Earth blog), *New York Times*, 15 October 2009; Virgin Earth Challenge, at www.virgin.com/subsites/virginearth.

25. Quoted in M. Specter, "The Climate Fixers," *New Yorker*, 14 May 2012.

26. For earlier discussion of these categories, see S. Nicholson, "Intelligent Design? Unpacking Geoengineering's Hidden Sacrifices," in M. Maniates and J. Meyer, eds., *The Environmental Politics of Sacrifice* (Cambridge, MA: The MIT Press, 2010), pp. 271–92.

27. H. Petroski, *To Engineer Is Human: The Role of Failure in Successful Design* (New York: Vintage Books, 1985); see also H. Petroski, *Design Paradigms: Case Histories of Error and Judgment in Engineering* (Cambridge, U.K.: Cambridge University Press, 1994), and H. Petroski, *Success through Failure: The Paradox of Design* (Princeton, NJ: Princeton University Press, 2006).

28. R. Pielke, Jr., *The Climate Fix* (New York: Basic Books, 2010), p. 132.

29. H. Schmidt et al., "Solar Irradiance Reduction to Counteract Radiative Forcing from a Quadrupling of CO_2: Climate Responses Simulated by Four Earth System Models," *Earth System Dynamics*, vol. 3 (2012), pp. 63–78.

30. K. J. Anchukaitis et al., "Influence of Volcanic Eruptions on the Climate of the Asian Monsoon Region," *Geophysical Research Letters*, vol. 37, L22703 (2010).

31. E. Tenner, *Why Things Bite Back: Technology and the Revenge of Unintended Consequences* (New York: Vintage Books, 1997).

32. H. Lamb, "Climate-Engineering Schemes to Meet a Climatic Emergency," *Earth Science Reviews*, April 1971, p. 95.

33. Fleming, op. cit. 7, p. 60.

34. S. Brand, *Whole Earth Discipline: An Ecopragmatist Manifesto* (New York: Viking, 2009), p. 275; on the notion of sufficiency, see T. Princen, *The Logic of Sufficiency* (Cambridge, MA: The MIT Press, 2005).

35. L. Winner, *The Whale and the Reactor* (Chicago, IL: University of Chicago Press, 1986), p. 10; O. Edenhofer et al., eds., *IPCC Expert Meeting on Geoengineering: Meeting Report* (Potsdam, Germany: Potsdam Institute for Climate Impact Research, 2012), p. 4.

36. Quote from Kintisch, op. cit. note 3, p. 13.

37. M. Specter, "The First Geo-vigilante," *New Yorker*, 18 October 2012.

38. Box 29–2 from "'Oxford Principles' Provide a Code of Conduct for Geoengineering Research," press release (Oxford: Oxford Martin School, University of Oxford, 14 September 2011).

39. Box 29–3 from R. Olson, "Soft Geoengineering: A Gentler Approach to Addressing Climate Change," *Environment*, September-October 2012, pp. 29–39.

Chapter 30. Cuba: Lessons from a Forced Decline

1. Figure for 1990 from United Nations, *Millennium Development Goals Indicators*, at mdgs.un.org/unsd/mdg/Data.aspx?cr=192; 2009 data from International Energy Agency, *Key World Energy Statistics* (Paris: 2011).

2. "Cuba's Special Period," in Louis A. Pérez, Jr., in *Cuba: Between Reform & Revolution*, at HistoryofCuba.com; "Operation Mongoose," Spartacus Educational, at www.spartacus.schoolnet.co.uk/JFKmongoose.htm; Thomas Blanton, "Annals of Blinksmanship," *Wilson Quarterly*, summer 1997.

3. Minor Sinclair and Martha Thompson, *CUBA, Going Against the Grain: Agricultural Crisis and Transformation* (Boston: Oxfam America, 2001), p. 8; American Association for World Health, *Denial of Food and Medicine: The Impact of the U.S. Embargo on Health and Nutrition in Cuba, An Executive Summary* (Washington, DC: 1997), p. 1; Zoë Amerigian, "Radio and TV Marti Should be Prime Targets for Budget Cutters" (blog), Council on Hemispheric Affairs, 7 April 2011.

4. Amerigian, op. cit. note 3.

5. Sinclair and Thompson, op. cit. note 3, p. 8; Pan American Health Organization, "Health Situation Analysis and Trends Summary—Country Chapter Summary from *Health in the Americas, 1998*," Washington, DC.

6. M. Franco et al., "Impact of Energy Intake, Physical Activity, and Population-wide Weight Loss on Cardiovascular Disease and Diabetes Mortality in Cuba, 1980-2005," *American Journal of Epidemiology*, 15 December 2007, pp. 1,374–80; Manuel Franco et al., "Obesity Reduction and Its Possible Consequences: What Can We Learn from Cuba's Special Period?" *Canadian Medical Association Journal*, 8 April 2008, pp. 1,032–34.

7. American Association for World Health, op. cit. note 3.

8. Pérez, Jr., op. cit. note 2; Dalia Acosta, "Transport-Cuba: Nearly There," *Inter Press Service*, 17 March 2009.

9. Liliana Núñez Velis, "Taxicab Service in Cuba: A Civil Society Approach," PowerPoint presentation, May 2011.

10. Sinclair and Thompson, op. cit. note 3, p. 9.

11. Ibid., p. 10.

12. Ibid., p. 4.

13. Ibid., pp. 10, 18–19.

14. Ibid., pp. 10, 13 , 31.

15. Laurie Guevara-Stone, "La Revolucion Energetica: Cuba's Energy Revolution," *Renewable Energy World Magazine*, April 2009, p. 2.

16. Ibid.

17. Mario Alberto Arrastía Avila, "Distributed Generation in Cuba: Part of a Transition Towards a New Energy Paradigm," *Cogeneration and On-Site Power Production*, November–December 2008, pp. 61–65; Mario Alberto Arrastía Avila and Laurie Guevara-Stone, "Teaching Cuba's Energy Revolution," *Solar Today*, January/February 2009, p. 31.

18. "Hurricanes Have Added to the Woes of the Downturn," *The Economist*, 30 December 2008; Miguel A. Altieri and Fernando R. Funes-Monzote, "The Paradox of Cuban Agriculture," *Monthly Review*, January 2012.

19. Ivet González, "Abrupt Shift from Drought to Flooding in Central Cuba," Inter Press Service, 30 May 2012; "Report on 2008 Hurricane Season in Cuba," World Meteorological Organization, at www.wmo.int/pages/prog/www/tcp/Meetings/HC31/documents/Doc.4.2.8_Cuba.doc; James Hansen, Makiko Sato, and Reto Ruedy,

"Perception of Climate Change," *Proceedings of the National Academy of Sciences*, 6 August 2012.

20. Arrastía Avila, op. cit. note 17, p. 65; Mario Alberto Arrastía Avila, presentation to Global Exchange, Havana, Cuba, April 2012; Guevara-Stone, op. cit. note 15, p. 3.

21. Arrastía Avila, op. cit. note 17, p. 65.

22. Guevara-Stone, op. cit. note 15; Anita Snow, "Cuba to Restructure Electric Grid and Utilize Wind and Solar Power," *Havana Journal*, 19 January 2006; Arrastía Avila, op. cit. note 17, p. 65.

23. Guevara-Stone, op. cit. note 15, pp. 5–6.

24. Marc Frank, "Cuban 2010 Oil Output Up, Natural Gas Down," *Reuters*, 13 June 2011; "Cuba–Venezuela Relations," *Wikipedia*, viewed June 2012.

25. Arrastía Avila, op. cit. note 17, p. 65.

26. Table 30–1 from International Energy Agency, *2011 Key Energy Statistics* (Paris: 2011).

27. Peter G. Bourne, "Public Health in Cuba," PowerPoint presentation, at www.pitt.edu/~super7/9011-10001/9881.ppt.

28. "Physicians Density" and "Hospital Bed Density," in Central Intelligence Agency (CIA), *CIA World Factbook*, at www.cia.gov; "Health Statistics: Physicians > per 1,000 People (1960) by Country," NationMaster.com; Conner Gorry, Marcio Ulises, Estrada Paneque, "Global Health, Cuban Health Cooperation and Disasters," *MEDICC Review*, chart 8, at www.pitt.edu/~super4/lecture/lec32661/index.htm.

29. Save the Children, *State of the World's Mothers 2012* (Westport, CT: 2012).

30. "Education Expenditures" and "School Life Expectancy," in CIA, op. cit. note 28; World Bank quote from Lavinia Gasperini, *The Cuban Education System: Lessons and Dilemmas* (Washington, DC: World Bank, July 2000). Box 30–1 based on the following: "Cuba's Special Period," op. cit. note 2; Ministry of Foreign Affairs of Cuba, "Cuba at a Glance: Social Organizations," at www.cubaminrex.cu/english/LookCuba/Articles/Others/2005/040005.html; Isaac Saney, *Cuba—A Revolution in Motion* (Winnipeg, Canada: Fernwood Publishing, 2004), pp. 65-67; Rachel Bruhnke, email to authors, 23 October 2012; Faith Morgan, "The Power of Community: How Cuba Survived Peak Oil," documentary, Community Solutions, Yellow Springs, OH, 2006; authors' interviews with Cubans.

31. Altieri and Funes-Monzote, op. cit. note 18.

32. Ibid.

33. Ibid.; Sinclair and Thompson, op. cit. note 3, p. 33.

34. Agri-Food Trade Service, "Agri-Food: Past, Present and Future Report Cuba," Agriculture and Agri-Food Canada, March 2012; Altieri and Funes-Monzote, op. cit. note 18.

35. Altieri and Funes-Monzote, op. cit. note 18; Sinclair and Thompson, op. cit. note 3, p. 43.

36. "Total Fertility Rate," in CIA, op. cit. note 28; CIA, op. cit. note 28, pp, 168, 694; Country Templates for Cuba and United States, *CIA Factbook*, at www.cia.gov/library/publications/the-world-factbook; Melissa Healy, "Obesity in U.S. Projected to Grow, Though Pace Slows: CDC study," *Los Angeles Times*, 7 May 2012; Eric A. Finkelstein et al., "Obesity and Severe Obesity Forecasts Through 2030," *American Journal of Preventive Medicine*, June 2012, pp. 563–70.

37. Jorge Pérez et al., *Approaches to the Management of HIV/AIDS in Cuba: Case Study* (Geneva: World Health Organization, 2004); Cuba and United States Comprehensive Indicator Reports, *HIV InSite*, University of California, San Francisco, at hivinsite.ucsf.edu.

38. WWF, *Living Planet Report 2006* (Gland, Switzerland: 2006).

39. Castro quote from Mario Alberto Arrastía Ávila, "Cuba: Energy and Development," PowerPoint presentation, at www.agdf.org.au/documents/item/15; Roberto Pérez Rivero, PEACB-FANJ Director, discussion with authors, June 2012; Pat Murphy, *Plan C: Community Survival Strategies for Peak Oil and Climate Change* (Gabriola Island, Canada: New Society Publishers, 2008).

40. Arrastía Avila quote in Guevara-Stone, op. cit. note 15, p. 7.

Chapter 31. Climate Change and Displacements

1. Robert F. Worth, "Earth Is Parched Where Syrian Farms Thrived," *New York Times*, 13 October 2010.

2. Ibid.; Wadid Erian, Bassem Katlan, and Ouldbdey Babah, *Drought Vulnerability in the Arab Region: Special Case Study: Syria*, contributed to the Global Assessment Report on Disaster Risk Reduction 2011 (2010); Francesco Femia and Caitlin Werrell, "Syria: Climate Change, Drought and Social Unrest" (blog), Center for Climate and Security, 29 February 2012.

3. Femia and Werrell, op. cit. note 2.

4. Emissions gap from U.N. Environment Programme (UNEP), *The Emissions Gap Report 2012: A UNEP Synthesis Report* (Nairobi: 2012); Potsdam Institute for Climate Impact Research and Climate Analytics, *Turn Down the Heat: Why a 4°C Warmer World Must Be Avoided* (Washington, DC: World Bank, 2012).

5. Intergovernmental Panel on Climate Change, *First Assessment Report* (Cambridge, U.K.: Cambridge University Press, 1990), p. 20.

6. Alex de Sherbinin, Koko Warner, and Charles Ehrhart, "Casualties of Climate Change: Sea-level Rises Could Displace Tens of Millions," *Scientific American*, January 2011.

7. Short-distance, temporary from Frank Laczko and Christine Aghazarm, eds., *Migration, Environment and Climate Change: Assessing the Evidence* (Geneva: International Organization for Migration (IOM), 2009), p. 23; Hurricane Katrina impact from Susan L. Cutter, "CSI: The Katrina Exodus," Foresight Project, Migration and Global Environmental Change, U.K. Government, October 2011, p. 6.

8. F. Renaud et al., "Control, Adapt or Flee. How to Face Environmental Migration?" *InterSecTions No. 5* (2007), p. 24.

9. Estimate of 135 million from "The Almeria Statement on Desertification and Migration," International Symposium on Desertification and Migrations, 9–11 February 1994, Almeria, Spain; water shortage impacts from Vikram Odedra Kolmannskog, *Future Flood of Refugees: A Comment on Climate Change, Conflict and Forced Migration* (Oslo: Norwegian Refugee Council, 2008), p. 15.

10. Drought from Vikas Bajaj, "Crops in India Wilt in a Weak Monsoon Season," *New York Times*, 3 September 2012; World Meteorological Organization from "With Drought Intensifying Worldwide, UN Calls for Integrated Climate Policies," *UN News*, 21 August 2012; impacts in a 4-degrees warmer world from Actionaid et al., *Into Unknown Territory: The Limits to Adaptation and Reality of Loss and Damage from Climate Impacts* (Bonn: 2012), p. 7.

11. Impacts on household income from Laczko and Aghazarm, op. cit. note 7, pp. 3–4.

12. Figure 31–2 is based on U.N. Food and Agriculture Organization, "FAO Food Price Index," at www.fao.org/world foodsituation/wfs-home/foodpricesindex/en, viewed 19 October 2012; Marco Lagi, Karla Z. Bertrand, and Yaneer Bar-Yam, *The Food Crises and Political Instability in North Africa and the Middle East* (Cambridge, MA: New England Complex Systems Institute, 2011).

13. Coastal population from Kolmannskog, op. cit. note 9, p. 16; Bangladesh from Actionaid et al., op. cit. note 10, p. 9; India and Vietnam from de Sherbinin, Warner, and Ehrhart, op. cit. note 6.

14. Laczko and Aghazarm, op. cit. note 7, p. 24; need for resources and social networks from François Gemenne, "Climate-Induced Population Displacements in a 4 C+ World," *Philosophical Transactions of the Royal Society*, January 2011, p. 188.

15. Chris Bright, "Anticipating Environmental 'Surprise,'" in Lester R. Brown et al., *State of the World 2000* (New York: W. W. Norton & Company, 2000).

16. De Sherbinin, Warner, and Ehrhart, op. cit. note 6.

17. Ibid.

18. Box 31–1 based on the following: International Federation of Red Cross and Red Crescent Societies (IFRC), *World Disasters Report 2012* (Geneva: 2012), p. 15; United Nations Office for the Coordination of Humanitarian Affairs (OCHA) and Internal Displacement Monitoring Centre (IDMC), "42 Million Displaced by Sudden Natural

Disasters in 2010—Report," press release (Geneva and Oslo: 6 June 2011); OCHA and IDMC, *Monitoring Disaster Displacement in the Context of Climate Change* (Geneva: 2009); Actionaid et al., op. cit. note 10, p. 9; IFRC, op. cit. this note, p. 14.

19. James Morrisey, "Rethinking the 'Debate on Environmental Refugees': From 'Maximilists and Minimalists' to 'Proponents and Critics,'" *Journal of Political Ecology*, vol. 19 (2012), p. 36; Essam El-Hinnawi, *Environmental Refugees* (Nairobi: UNEP, 1985).

20. Gemenne, op. cit. note 14, p. 186.

21. Kolmannskog, op. cit. note 9, p. 9.

22. Table 31–1 from the following: refugee definition from U.N. High Commissioner for Refugees (UNHCR), "Convention Relating to the Status of Refugees," at www2.ohchr.org/english/law/refugees.htm; internally displaced persons definition from U.N. Economic and Social Council, Commission on Human Rights, "Further Promotion and Encouragement of Human Rights and Fundamental Freedoms, Including the Question of the Programme and Methods of Work of the Commission Human Rights, Mass Exoduses and Displaced Persons," 11 February 1998; definition of international migrants from IOM, "Identifying International Migrants," at www.iom.int/jahia /Jahia/about-migration/developing-migration-policy/identify-intl-migrants; proposed definition of environmental refugees from El-Hinnawi, op. cit. note 19; proposed definition of environmental migrants from Laczko and Aghazarm, op. cit. note 7, p. 19.

23. Renaud et al., op. cit. note 8; IFRC, op. cit. note 18, p. 18.

24. Kolmannskog, op. cit. note 9, p. 13; Oli Brown, *Climate Change and Forced Migrations: Observations, Projections and Implications*, Occasional Paper 2007/17 (New York: Human Development Report Office, U.N. Development Programme (UNDP), 2007), p. 15.

25. Brown, op. cit. note 24, p. 13.

26. World Bank and higher estimates from Actionaid et al., op. cit. note 10, p. 11; inadequate funding from UNDP, *Human Development Report 2007/2008* (New York: Palgrave Macmillan, 2007), p. 189.

27. UNHCR, *State of the World's Refugees 2012* (Geneva: 2012); Alister Doyle, "World Needs Refugee Re-think for Climate Victims: U.N.," *Reuters*, 6 June 2011.

Chapter 32. Cultivating Resilience in a Dangerous World

1. "Haiti Raises Quake Death Toll on Anniversary," *CBC News*, 12 January 2011; "Japan Earthquake and Tsunami of 2011," *Encyclopedia Britannica*, at www.britannica.com; U.N. Food and Agriculture Organization, "New Crisis in the Sahel Region," at www.fao.org/crisis/sahel; U.S. Department of Agriculture, Economic Research Service, "U.S. Drought 2012: Farm and Food Impacts," at www.ers.usda.gov/newsroom/us-drought-2012-farm -and-food-impacts.aspx.

2. Center for Research on the Epidemiology of Disasters, *EmDat: The International Disaster Database*, at www .emdat.be/sites/default/files/Trends/natural/world_1900_2011/affyr1.jpg; Munich Re, "Greater Uncertainty a Challenge to the Insurance Market—Munich Re Well Positioned," press release (Munich: 24 October 2011).

3. U.N. Population Division, *The World at Six Billion* (New York: 1998); U.N. Population Division, *World Population Prospects: The 2011 Revision* (New York: 2011); Angus Maddison, *Historical Statistics of the World Economy*, online database, at www.ggdc.net/maddison/Historical.../horizontal-file_02-2010.xls.

4. Intergovernmental Panel on Climate Change, *Climate Change 2007: Synthesis Report, Summary for Policymakers* (Geneva: 2007); Millennium Ecosystem Assessment, *Ecosystems and Human Well-Being: Synthesis* (Washington, DC: Island Press, 2005); Hansen quoted in "Tradition Circle of Indian Elders and Youth" (blog), Haudenosaunee Task Force, 2 August 2010.

5. World Economic Forum, *Global Risks 2011* (Geneva: 2011), p. 10.

6. For guides to this body of research, see Brian Walker and David Salt, *Resilience Thinking* (Washington, DC: Island Press, 2006), and Brian Walker and David Salt, *Resilience Practice* (Washington, DC: Island Press, 2012); Lance H. Gunderson and C. S. Holling, eds., *Panarchy: Understanding Transformations in Human and Natural*

Systems (Washington DC: Island Press, 2002); for national security applications, the Community and Regional Resilience Institute, a joint effort of the Department of Homeland Security and Oak Ridge National Laboratory, is incorporating resilience thinking into disaster preparedness; notable efforts to assess and cultivate social resilience include the Building Resilient Regions project at the University of California and the Project on Resilience and Security at Syracuse University; for research in psychological resilience, see publications of Ann S. Masten, University of Minnesota, at www.experts.scival.com/umn/expertPubs.asp?n=Ann+S+Masten&u_id=1809.

7.   Carl Folke et al., "Regime Shifts, Resilience and Biodiversity in Ecosystem Management," *Annual Review of Ecology, Evolution and Systematics*, December 2004, pp. 557–81. Box 32–1 from the following: Svalbard Global Seed Vault, at www.nordgen.org/sgsv/index.php?page=welcome; Global Crop Diversity Trust, at www.croptrust .org; Global Crop Diversity Trust, "Amaranth Grain from Ancient Aztecs, Barley Used by Modern Craft Beer Brewers, and Wheat from Pamir Mountains in Tajikistan, Among New Shipments to Arctic Seed Vault," press release (Longyearbyen, Norway: February 2012); Nordic Genetic Resource Center, at www.nordgen.org/index.php/en.

8.   Kevin Bullis, "How Power Outages in India May One Day Be Avoided," *Technology Review India*, 31 July 2012; Che Biggs, Chris Ryan, and John Wiseman, "Distributed Systems: A Design Model for Sustainable and Resilient Infrastructure," Victorian Eco-Innovation Lab, University of Melbourne, 2008.

9.   John C. Mutter, "Voices: From Haiti to Japan: A Tale of Two Disaster Recoveries," *Earth Magazine*, 9 March 2012; Richard H. Moss et al., *Vulnerability to Climate Change: A Quantitative Approach* (Washington, DC: Pacific Northwest National Laboratory, 2001).

10. Walker and Salt, *Resilience Practice*, op. cit. note 6; Sarah Pressman et al., "Loneliness, Social Network Size, and Immune Response to Influenza Vaccination in College Freshmen," *Health Psychology*, May 2005, pp. 297–306; Daniel Aldrich, *Building Resilience: Social Capital in Post-Disaster Recovery* (Chicago: University of Chicago Press, 2012).

11. Katrina Brown and Elizabeth Westaway, "Agency, Capacity, and Resilience to Environmental Change: Lessons from Human Development, Well-Being, and Disasters," *Annual Review of Environment and Resources, 2011*, pp. 321–42. Box 32–2 from the following: U.N. International Strategy for Disaster Reduction, "Women and Girls—the [in]Visible Force of Resilience," at www.unisdr.org/2012/iddr/about.html; New Course, *Women, Natural Resource Management, and Poverty* (Seattle, WA: undated); Elizabeth Frankenberg et al., "Mortality, the Family and the Indian Ocean Tsunami," University of California Los Angeles, March 2011; Oxfam International, "The Tsunami's Impact on Women," Briefing Note, Oxford, U.K., March 2005; UN Women, "Women, Poverty & Economics," at www.unifem.org/gender_issues/women_poverty_economics; World Bank, "Gender and Climate Change: 3 Things You Should Know," at go.worldbank.org/TN0KYRX8Q0.

12. Doug Millen, "Deliberative Democracy in Disaster Recovery," Centre for Citizenship and Public Policy, University of Western Sydney, 2011; Daron Acemoglu and James A. Robinson, *Why Nations Fail: The Origins of Power, Prosperity and Poverty* (New York: Random House, 2012).

13. Elinor Ostrom, *Governing the Commons* (Cambridge, U.K.: Cambridge University Press, 1990). Box 32–3 from the following: W. Neil Adger, "Social and Ecological Resilience: Are They Related?" *Progress in Human Geography*, September 2000, pp. 347–64; Saudamini Das and Jeffrey R. Vincent, "Mangroves Protected Villages and Reduced Death Toll during Indian Super Cyclone," *Proceedings of the National Academy of Sciences*, 5 May 2009, pp. 7,357–60.

14. System resilience or vulnerability from Gunderson and Holling, op. cit. note 6.

15. Joan Stiles, "Neural Plasticity and Cognitive Development," *Developmental Neuropsychology*, vol. 18, no. 2 (2002), pp. 237–72; Mike Celizic, "Meet the Girl with Half a Brain," *NBC News*, 25 March 2010.

16. John Harte, "Numbers Matter: Human Population as a Dynamic Factor in Environmental Degradation," in Laurie Mazur, ed., *A Pivotal Moment: Population, Justice and the Environmental Challenge* (Washington, DC: Island Press, 2009).

17. Ann Masten, "Ordinary Magic" (blog), *This Emotional Life*, PBS.

18. Fikret Berkes and Carl Folke, "Back to the Future: Ecosystem Dynamics and Local Knowledge," in Gunderson and Holling, op. cit. note 6.

19. Example taken from Brian Walker, "Learning How to Change in Order Not to Change: Lessons from Ecology

for an Uncertain World," lecture, University of Canberra, 20 February 2012.

20. Stephen Flynn, *The Edge of Disaster* (New York: Random House, 2007).

21. Betty Hearn Morrow, *Community Resilience: A Social Justice Perspective* (Oak Ridge, TN: Community and Regional Resilience Institute, 2008).

22. U.N. Population Division, *World Population Prospects: The 2010 Revision*, online database at esa.un.org/unpd /wpp/unpp/panel_population.htm; Malea Hoepf Young et al., "Adapting to Climate Change: The Role of Reproductive Health," in Mazur, op. cit. note 16, pp. 108–23.

23. Patricia H. Longstaff et al., *Building Resilient Communities: Tools for Assessment* (Syracuse, NY: Syracuse University, Institute for National Security and Counterterrorism, Project on Resilience and Security, 2010); chapters on the Norse settlement of Greenland in Jared Diamond, *Collapse: How Societies Choose to Fail or Succeed* (New York: Viking Penguin, 2005).

24. Andrew Zolli and Ann Marie Healy, *Resilience: Why Things Bounce Back* (New York: Free Press, 2012), p. 259.

25. See, for example, Sami Grover, "Resilience vs. Sustainability," *Treehugger*, 28 March 2011, and Jamais Cascio, "The Next Big Thing: Resilience," *Foreign Policy*, 15 April 2009.

26. Tony Juniper, "Will 2012 be the Year of the 'R' Word?" (London) *Guardian*, 14 December 2011.

Chapter 33. Shaping Community Responses to Catastrophe

1. Vahidin Omanovic, discussions with author, 1997–2002.

2. Vahidin Omanovic, discussion with author, July 2010.

3. Randall Kuhn, "Facts on the Seashore: Conflict, Population Displacement, and Coastal Vulnerability on the Eve of the Sri Lankan Tsunami," Josef Korbel School of International Studies, University of Denver, 2009.

4. Joseph Sebarenzi, discussions with author, June 2012; Joseph Sebarenzi and Laura Mullane, *God Sleeps in Rwanda* (New York: Atria Books, 2009).

5. Mark Pelling, David Manuel-Navarrete, and Michael Redclift, *Climate Change and the Crisis of Capitalism: A Chance to Reclaim Self, Society and Nature* (London: Routledge, 2011); "Disaster Preparedness and Prevention (DPP): State of Play and Strategic Orientation for EC Policy," European Commission of Humanitarian Aid Office, Brussels, 2011; John Vidal, "Global Warming Causes 300,000 Deaths a Year, says Kofi Annan Thinktank," (London) *Guardian*, 29 May 2009.

6. Richard Heinberg, *Peak Everything: Waking Up to the Century of Declines* (Gabriola Island, Canada: New Society Publishers, 2012), p. 23; Robert Jay Lifton and Richard Falk, *Indefensible Weapons* (New York: Basic Books, 1982).

7. T. S. Eliot, *The Four Quartets* (New York: Mariner Books, 1968); Research & Degrowth, at degrowth.org.

8. Index Mundi, at www.indexmundi.com; Christian Parenti, *Tropic of Chaos: Climate Change and the New Geography of Violence* (New York: Nation Books, 2011).

9. Kuhn, op. cit. note 3; Climate Change Knowledge Portal, "Sri Lanka Dashboard," World Bank, at sdwebx.world bank.org/climateportalb.

10. Victoria Transport Policy Institute, at www.vtpi.org.

11. Brendon Nafziger, "Weathering Hurricane Issac: Health Care Lessons Learned from Katrina," *DOTmed News*, 11 September 2012.

12. Juliette Landphair, "The Forgotten People of New Orleans: Community Vulnerability, and the Lower Ninth Ward," *Journal of American History*, December 2007, pp. 837–45; Karen J. Leong et al., "Resilient History and the Rebuilding of a Community: The Vietnamese American Community in New Orleans East," *Journal of American History*, December 2007, pp. 770–79.

13. Eva Fogelman, *Conscience and Courage: Rescuers of Jews During the Holocaust* (New York: Anchor Books, 1995); PROOF: Media for Social Justice, at proofmsj.com.

14. Kevin Ronan and David Johnston, *Promoting Community Resilience in Disasters: The Role for Schools, Youth, and Families* (New York: Springer Publishing, 2005).

15. Daniel P. Aldrich, *Building Resilience: Social Capital in Post-Disaster Recovery* (Chicago: University of Chicago Press, 2012); bus passengers in India from author's observations; Tamil Nadu from Daniel P. Aldrich, "How to Weather a Hurricane," *New York Times*, 28 August 2012.

16. "Office of Emergency Management," Seattle.gov, at www.seattle.gov/emergency; Ronan and Johnston, op. cit. note 14; Robert D. Putnam, *Bowling Alone: The Collapse and Revival of American Community* (New York: Simon & Schuster, 2001).

17. Ashutosh Varshney, *Ethnic Conflict and Civic Life: Hindus and Muslims in India* (New Haven, CT: Yale University Press, 2012).

18. Ken Belson, "Japanese Find Radioactivity on Their Own," *New York Times*, 31 July 2011; "Social Capital and Disaster Recovery," *Social Capital Blog*, 15 November 2011; Tina Gerhardt, "After Fukushima, Nuclear Power on Collision Course With Japanese Public," *Huff Post Green*, 26 July 2012.

19. Gerhardt, op. cit. note 18; Brendan Barret, "Japan Considers Green Future After Nuclear Disaster," *Aljazeera*, 11 February 2012.

20. Transition Putney 2.0, at www.transitionputney.net; Transition Town TOTNES, at www.transitiontowntotnes.org; quote from 3rd International Conference on Degrowth, Ecological Sustainability and Social Equity, Venice, 19–23 September 2012, at www.venezia2012.it.

21. Amanda Ripley, *The Unthinkable: Who Survives When Disaster Strikes—and Why* (New York: Three Rivers Press, 2009); Ronan and Johnston, op. cit. note 14.

22. "Minneapolis Community Emergency Response Teams (CERT)," Minneapolis, at www.minneapolismn.gov; "Community Readiness," Emergency Management Department, City of Los Angeles, at emergency.lacity.org.

23. Donald E. Geis, "Creating Sustainable and Disaster Resistant Communities," *The Aspen Global Change Institute*, 10 July 1996.

24. Pauline W. Chen, "After the Tempest Passes, Easing the Trauma It Left," *New York Times*, 6 November 2012.

25. Environmental Translation Project, at environmentaltranslation.org; DeMond Shondell Miller and Jason David Rivera, *Comparative Emergency Management: Examining Global and Regional Responses to Disasters* (Boca Raton, FL: CRC Press, 2011).

26. Anouk Ride and Diane Bretherton, *Community Resilience in Natural Disasters* (Basingstoke, U.K.: Palgrave Macmillan, 2011).

27. World Bank, "Community Driven Development," at web.worldbank.org.

28. U.N. Environment Programme, "Disasters and Conflicts," fact sheet, Geneva, undated.

29. U.N. High Level Panel on Global Sustainability, *Resilient People, Resilient Planet* (New York: United Nations, 2012).

Chapter 34. Is It Too Late?

1. Robert Socolow, "Wedges Reaffirmed," *Bulletin of the Atomic Scientists*, 27 September 2011.

2. Robert A. Dahl, *A Preface to Economic Democracy* (Berkeley: University of California Press, 1985).

3. Raymond Williams, *Marxism and Literature* (Oxford: Oxford University Press, 1977), pp. 121–27.

4. For more information on cliodynamics, see Peter Turchin, "Cliodynamics: History as Science," at cliodynamics.info.

5. Robert Costanza et al., "The Value of the World's Ecosystem Services and Natural Capital," *Nature*, 15 May 1997, pp. 253–60.

Index